CW00457435

ECONOMICS OF WORLDWIDE PETROLEUM PRODUCTION

Richard D. Seba

OGCI AND PETROSKILLS PUBLICATIONS
TULSA, OKLAHOMA USA

Printed in the United States of America.
Library of Congress Catalog Card Number 92-082865
International Standard Book Number: 978-0-930972-24-0

First Edition, February, 1993
First Edition, Second Printing, March, 1998
Second Edition, July, 2003
Third Edition, April, 2008

CONTENTS

*"That is a good book which is opened with expectation
and closed with delight and profit."*

—AMOS BRONSON ALCOTT

VIII NET INCOME, "THE BOTTOM LINE" 301-358

ILLUSTRATIONS

A picture is worth a thousand words.

—CHINESE PROVERB

TABLES

"We think in generalities, but we live in detail."

—ALFRED NORTH WHITEHEAD, BRITISH PHILOSOPHER (1861-1947)

EXAMPLES

"Example is not the main thing in life—is the only thing."

—ALBERT SCHWEITZER

MODELS

"Study the past if you would divine the future."

—CONFUCIUS

PREFACE

"General observations drawn from particulars are the jewels of knowledge, comprehending great stored in a little room."

—JOHN LOCKE

The prime objective of petroleum-producing operations is not only to supply the modern world with crude oil and natural gas but, hopefully, to make a profit while doing so. This book is a guide to the basic economics of petroleum production as practiced throughout the world.

The process of looking at an oil company from a financial and/or accounting viewpoint may involve a different perspective than the one to which most operating people in the industry may have become accustomed. The intent of this text is to demonstrate more clearly how, when, and why an oil company makes a profit, or fails to do so. This understanding should lead to an improved grasp of the various means by which any oil company's profit can be improved.

The central theme of the text focuses on adapting basic economic analysis to increase business efficiency. The format seeks to accomplish this by:

1. Emphasizing the importance of maintaining a positive cash flow.

2. Presenting how various economic evaluations are done and why they are used.

3. Presenting those aspects of economics and decision analysis, including matters of physical risk, e.g., dry hole and acts of nature; and financial risk, e.g., price and interest, that are most important to the oil and gas producing business.

4. Providing a background of how the oil and gas producing industry functions as an economic entity within the business community.

These objectives become somewhat intertwined as the text material is presented, since each objective reinforces the others toward an improved understanding of each.

Petroleum exploration, development, and production activities do not fit conventional accounting procedures. One purpose of the text is to bring together the diverse factors, including technical, financial and accounting aspects in a manner that will assist those employed in the industry to better understand the workings of an oil company as a business.

There are many factors in the categories of finding costs, oil and gas prices and volumes, and the large investments required to develop and produce petroleum that are all exceedingly important to oil industry's profit or loss. A number of these aspects are within the direct control of the firm. A significant number, however, are beyond its influence and control. The problems of fiscal and physical (geological) risk are uniquely coupled in the oil business. Chapter VI introduces the industry's methods of dealing with these.

Throughout this text, data and examples from the U.S. are used more frequently than for any other location. This is because such data is more readily available. The U.S. dollar is also used more frequently than other currency for the same reason. Since oil is priced in terms of U.S. dollars per barrel throughout the world it is also the unofficial currency of the oil industry. Even though most of the economic principles of this book are explained in terms of U.S. dollars, all of the economic principles are equally valid in any other desired currency.

The material in this book is assembled as bricks in a wall. Each chapter uses the material presented in previous chapters as the basis for information provided in subsequent chapters. By going through the book in the order it is written, sufficient background should be developed to understand each new topic.

Symbols used throughout this text conform to those established by the Society of Petroleum Engineers as published in 1986, and the update of economic symbols printed in the August 1992 issue of the Journal of Petroleum Technology. Where a standard symbol has not been established for a term used in this text, one was selected which is compatible with the standard symbols. The symbols are defined, as used, throughout the text with the appropriate units noted. A listing of all of the nomenclature used appears at the end of the book.

Dr. Fraser H. Allen, who died in 1997, created the courses for which this book was written. His over 50 years of experience in the Petroleum Industry were the source of much of the information contained in this book and its first edition, which he co-authored. Revisions of Economics of Worldwide Petroleum Production have been prepared without the benefit of his advice, review and poignant comments, which were certainly missed.

INTRODUCTION

*Business without profit is not business
any more than a pickle is candy.*

—CHARLES FREDERICK ABBOTT

Petroleum is the world's most important internationally traded commodity. Modern man takes for granted an uninterruptible supply of cheap hydrocarbon fuels for transportation, home heating, electrical power generation and industrial fuels and lubricants. The transportation industry, including automobiles, trucks, and aircraft, remain the largest consumer of petroleum products. Thus, current and future developments related to the efficiency and fuel for powering these vehicles will continue to have a major impact on the Petroleum Industry. More recently, the public which demands a cheap, unending supply of oil insists that it must have no adverse impact upon the environment. No other industry has such a complex historical, political, and economical make-up compared to that of petroleum.

The oil business differs significantly from all other types of commercial enterprise. The most noteworthy of the several factors which make the petroleum industry unique is its degree of capital intensity. This capital intensity pervades all phases of the petroleum industry from exploration through marketing.

This, in turn, requires that any study of the basic economics of this important industry must direct a great deal of attention to the massive monetary outlays for investment, and how those investments are recovered through the cash flows from operations. The three key elements in any economic evaluation are (1) income, (2) expenditures, and (3) time. These three elements are brought together in the forecast of future cash flow which will result from making an investment. Cash flow, which is the periodic recording of income and expenditures, is the most important tool for evaluating investments and choosing among alternatives. NET CASH FLOW is merely income minus expenditures, but only a positive value assures repayment of the investment and a profit. A company must generate a positive net cash flow if it is to remain solvent, pay its debts, provide a return to its investors, and to have money for new investments.

Cash flow is the basis for most management decisions. It must embrace all cost which will be incurred as a result of the operation, including both a portion of the company's overhead as well as an operation's impact on the total income tax paid by the company. Further, the effect of inflation on both income and expenditures must be properly recognized. Additionally, cash flow is important because it is the basis for buying or selling producing properties as the value of oil and gas reserves is the present value of the net cash flow realized as they are produced.

Few of the entries in this book can stand in complete isolation entirely on their own. The industry is far too complex. Yet it is possible to isolate basic examples which, when understood, cast light onto the other factors that link, or relate directly to those basics with which we deal.

In the oil and gas producing business, the cash register doesn't ring until someone puts oil in the tanks—and someone else has agreed to buy it and move it down a pipeline or into a tanker for shipment. That latter someone may be the crude oil purchasing department of an integrated oil company, or a crude oil buyer for an independent refiner. He may be a crude oil trader, or broker who makes a business of finding a "fit" for a specific number of daily barrels in some oil field. Or, the purchase may go through a series of "paper barrel" transactions on one of the world's commodity exchanges before the actual "wet barrel" is delivered to someone's refinery for processing.

Maintaining an oil and gas reserve base is essential to keeping a company in business in a natural resource industry. This requires continual additions of new reserves, either through exploration or acquisition, as reserves are produced. A number of years normally elapses between the first geological inspiration and the first commercial production if, in fact, this proves to be the successful exploration exercise. Figure 1-1 is a schematic representation of the typical exploration and production cycle, showing exploration and development expenditures as negative values and revenue and net income as positive ones. It shows the many years of expenditures which precede any income from successful ventures, and that additional development capital is required to produce the newly discovered reserves. At the end of the project, when it has reached its economic limit, there may be a significant cost of abandonment as shown in Figure 1-1 and discussed in Chapter II. Onshore abandonment costs may be nominal, but may be quite significant offshore. The long producing life of most oilfields and their depleting nature are also evident from Figure 1-1. Cumulative cash flow is included on the graph and the payout of the investments is apparent where the line crosses the zero Cumulative Net Cash Flow value. The Present Value Cumulative Net Cash Flow curve is also shown, which recognizes the time value of money over the life of a project. This is always below the undiscounted Cumulative Net Cash Flow curve and becomes almost a horizontal line for long project lives due to the servere present value discounting for long time periods. The end points on these curves represent the total undiscounted and present value profit over the entire life of the project.

The critical impact of the various methods of oil and gas pricing, projections of future producing rates, operating costs, tax considerations, and the volumes of the booked oil and gas reserves as they deal with Cash Flow, are recognized along with the means of dealing with each of them. These considerations occupy Chapters II through IV.

The most significant reason for doing economic evaluations is to make investment decisions. This process involves answering three critical questions: What will it cost? What is it worth? Will it earn enough profit? These questions cannot be answered easily or simply, but require sophisticated forecasting and calculation techniques, such as those presented in this book. The time value of money, introduced in Chapter V, must be considered when answering the last question. It involves the concept of interest, which can be regarded as rent paid to use someone else's money. Discount rates are merely a special use of interest to determine present value. Many ways of computing discount rates are discussed. For uniformity throughout the text, the mid-period convention, the one most frequently used by the petroleum producing industry, has been employed. "Economic Yardsticks" presented in Chapter V provide the tools for measuring investment opportunities.

FIGURE 1-1
E&P INVESTMENT CYCLE

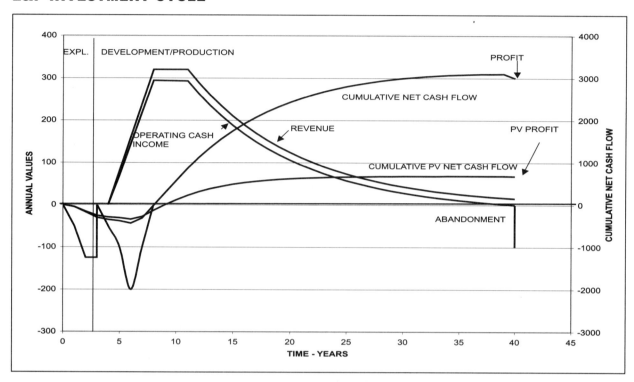

Petroleum exploration and production is often referred to as a "risk business." Chapter VI introduces the industry's methods of dealing with these important problems of fiscal and physical (geological) risk.

The industry has also acquired the reputation of being among the most "technology driven" of all commercial enterprises. However, all this new technology is not without economic constraints. Chapters VII, IX and X endeavor to introduce the reader to the basic economic analytic techniques of project evaluation and implementation; budgeting of capital, personnel, and annual expenses; and the ownership and financing of oil and gas operations which pertain to the business of crude oil and natural gas production.

The process of looking at an oil company from a financial and/or accounting viewpoint may involve a different perspective than the one to which most operating people in the industry may have become accustomed. The computations of oil company profit and sample oil company financial statements are presented in Chapter VIII.

For this purpose, the term 'profit' is employed in its more strict financial, or accounting sense. Even so, we find variation in what a private sector company in the oil business reports as profit, and what a government, or national oil company, considers to be its "bottom line." These latter organizations frequently employ their 'profit' figures in seeking public comparison as to how well they are able to compete with the private sector oil companies in achieving supposedly comparable financial objectives. However, the accounting procedures of the state and private sectors may differ widely.

Capital availability is limited in any enterprise, public or private. Budgeting these limited resources, with recognition of their costs, and the time value of their employment in a project, is important to the economic viability of the undertaking. Chapter IX seeks to draw together all of these factors in its discussion of oil industry budgeting.

The national oil companies (NOCs) around the world represent a major sector of the producing industry. Fourteen of the world's twenty largest oil companies are owned, or controlled by their home governments. There is wide variation, however, in their specific corporate objectives. These motives may include social and political goals beyond those of the private sector. For example, the NOC is by far the largest single employer in many of these countries. Nevertheless, economics must assume a fundamental role in their management. The economic analysis and planning by the NOCs is basically similar to those practiced by the private sector.

REFERENCES

Frankel, P. H., "The Structure of the Oil Industry," lecture given at the Institute of Petroleum, London July 4-6, 1988

Frankel, P. H., Common Carrier of Common Sense—A Selection of His Writings, 1989, Oxford University Press, Oxford OX2 6DP

Gallun, R. A. and Stevenson, J. W., Fundamentals of Oil&Gas Accounting, 1983, PennWell, Tulsa, OK

Huber, Peter W. & Mills, Mark P., *The Bottomless Well – The Twilight of Fuel, the Virtue of Waste, and Why We Will Never Run Out of Energy,* 2005, Basic Books, New York

Jones, Ellis P., Oil—A Practical Guide to the Economics of World Petroleum, (1988), Woodhead-Faulkner Ltd., Cambridge CB2 1QY

Parks, William R., Cost Engineering Analysis, 1973, John Wiley & Sons, New York

Pennant-Ray, R. and Emmott, B., The Pocket Economist, 1983, Martin Robertson & Company, Oxford OX4 1JF, and The Economist, London SW1A 1HG

Sampson, Anthony, *The Seven Sisters – New, Post-Gulf War Edition,* 1993, Coronet Books, Hodder and Stoughton, Ltd., London

Yergin, Daniel, The Prize—The Epic Quest for Oil, Money&Power, 1991, Simon&Schuster, New York

PRODUCTION FORECASTS AND RESERVES

"An economist's guess is liable to be as good as anybody else's."

—WILL ROGERS, AMERICAN HUMORIST (1879-1935)

It is necessary in assessing the economics of an oil producing operation to forecast the physical volumes of crude oil and natural gas by years into the future to be produced for the account of the commercial entity under study.

FORECASTING

Revenues from oil and gas producing operations result from sale of production. The most critical concerns in evaluating a producing property are its present and future producing rates, oil and gas prices to be received for that production, operating costs, taxes and its ultimate recovery. Development of forecasts of marketable crude oil production at commercial rates from individual wells, or groups of wells is an engineering responsibility within the economic confines of oil price, investment and operating costs.

Forecasting is an essential part of the preparation of any economic evaluation. It is not accomplished by consulting a crystal ball or a prophet or a mystic of some sort, but by evaluating the past, investigating current conditions and projecting these into the future based upon the best information at hand at the time the forecast is made. Since a forecast is based upon the best information available at a given time, it may be necessary to alter a forecast as additional information becomes available or as conditions change. About the only thing that can be said about a forecast is that it is unlikely to accurately predict what will actually occur. It is a rare forecast that is proved to be completely correct. Most often, it will be either optimistic or pessimistic. However, even though it is a forgone conclusion that a forecast will most likely be inaccurate to some degree, it is still necessary that the forecast be made. Without the forecast, a valid economic evaluation cannot be completed.

There are numerous forecasts that must be made during the preparation of the typical exploration and production evaluation. It is necessary to forecast the timing of oil, gas and water production. To convert hydrocarbon production to revenue, it will be necessary to prepare a forecast of the unit prices to be received as the products are sold. Finally it is necessary to estimate the magnitude and timing of capital and operating expenses which will be incurred to generate the expected revenue. These costs must be considered not only under economic conditions prevailing at the time that the estimate is made, but also under the economic conditions projected for the future. Thus, it is important to forecast changes in prices due to inflation or deflation, supply/demand and technological improvements. Some prices may escalate more rapidly than general inflation, while others more slowly, due to the individual nature of the commodity or service involved.

Petroleum, as an almost infinite series of mixtures of differing hydrocarbons in their natural state, exhibits a wide range of physical characteristics. These characteristics (Table 2-1) will have a strong influence on production rates and the price received when sold, so they must be considered when forecasting production — rates and cash flow.

TABLE 2-1
TYPICAL RESERVOIR FLUIDS CHARACTERISTICS

	Stock Tank Liquid Density API Gravity	Gas-Oil Ratio SCF/STB	Reservoir Volume Factor RB/STB	Examples
Heavy oil	to 20	nil	1.0	Kern River, U.S Athabasca, Canada
Conventional crude oil	20 - 50	50 - 2,000	1.0 - 1.8	Yates. U.S. Brent, U.K. Forties, U.K. Ghawar, Saudi Arabia
Volatile oil	45 - 55	2,000 - 5,000	3.0 - 20.0	Tirrawarra, So. Australia
Gas-condensate	45 - 55	5,000 - 50,000	3.0 - 20.0	E. Anschutz, U.S. Sharjah, UAE
Dry gas	45 - 55	50,000+	20+	Groningen, Netherlands Hugoton, U.S.

PRODUCTION RATE PROFILE DIFFERENCES AROUND THE WORLD

Worldwide exploration, and production, of crude oil and natural gas has many similarities which know no international boundaries. Petroleum geology has no limits as to nationality or region. The development of oil deposits and their commercial production on the other hand are heavily influenced by the policies and traditions of host governments. Most countries operate on the basis that all minerals (including crude oil and natural gas) belong to the government. The U.S. and parts of Canada, however, affirm that title to the minerals goes with the land, which may be privately owned. The modus operandi on these "fee" lands provides that the landowner may lease the exploration and producing rights to an oil operator in return for a signature bonus and a royalty on any production. An oil and gas lease usually provides for a limited exploration period (primary term) and an annual rental fee during that time. The primary purpose of these latter provisions is to encourage the operator to commence exploration, and thereby establish commercial production at the earliest possible date. Both the primary term and the delay rentals expire with the onset of production. The producing lease then continues in force so long as there is production of oil and/or gas in paying quantities from the leased property, unless a fixed production period is specified.

The striking impact of this fee land situation in North America vis-a-vis the rest of the oil producing world stems from the relatively small areal size of the individually owned tracts of land, and their associated minerals, in the U.S. and Canada. The size of these tracts varies, of course, over a very wide range from the small individual town lots in Kilgore producing oil from the East Texas field to Exxon's lease of the giant King Ranch sprawled over several counties in South Texas. The average, however, is probably in the order of 50 to 60 acres per tract. Commercial oilfields are much larger than that in areal extent. Crude oil in the reservoir is mobile, otherwise it could not flow to the producing wells and would have to be mined like the Athabasca tar sands. This mobility led the Supreme Court of Pennsylvania in 1889 to liken crude oil to fleeting deer, free to move from property to property, unless captured on one's own property. This "rule of capture" persists in North American land law.

The result of this capture concept directly affects the pattern, timing and cost of developing an oilfield in North America. In practice, once a successful oil or gas well is drilled by a landowner or his leaseholder, it becomes necessary that each adjoining lease be drilled promptly in order to capture its share of the oil, and thus prevent its escape to adjacent tracts.

In the early days of the petroleum industry all wells were produced at capacity. This resulted in tremendous ups and downs in supply to a growing demand. After the discoveries of the Oklahoma City field and the East Texas field in 1930, state proration was instituted in the principal producing jurisdictions, with the exception of California, to bring, and keep supply in balance with demand. State prorationing would normally curtail production throughout much of an oilfield's producing life until it could no longer sustain its allowable producing rate and normal production decline would determine its producing rate until the wells reached their "economic limit" rates of production and were abandoned. Such practices are no longer followed due to the decline in total U.S. production, unless such curtailment is necessary to conserve reservoir energy in order to maximize recovery.

Outside of North America the development pattern, producing rates, and consequent cash flow profiles are quite different. The tracts, in general are much larger, probably twenty times larger than the average North American fee tract. This means that a whole oilfield may easily lie entirely within one tract. With one landowner and a single operator, the typical North American lease drainage and migration problem normally does not exist. Determining the number, location and rates of production then becomes a study in mathematical modeling of the reservoir under various well densities and timetables of development.

The producing rate of a field is determined by the number of wells and the throughput capacity of the production facilities. These are normally at the discretion of the operator, often with the consent of the host government. If a large number of high capacity wells and adequate production equipment are available the operator might deplete a field in a relatively short time. On the other hand, if only a few low productivity wells were drilled and/or production facilities are limited, that field might produce the same ultimate volume of oil, but over a much longer period of time. By and large the ultimate recovery from fields large enough to be commercial is independent of the producing rate.

A major consideration in determining the size of facilities for platform or field is the expected productive capacity of the wells. However, the through-put capacity of the facilities will normally be something less than the expected productivity of the wells which it will handle, because the optimum size of facilities will be such that they will be fully utilized for some period of time. This recognizes the fact that larger facilities will increase the present value of production resulting from accelerating that production, but comes at a price of increased facilities cost. The optimum facilities size can be determined by studying the relationship between the capacity of the facilities and the present value of the total project, where the optimum facilities size will be the one that yields the greatest present value profit.

In practice, economic, political and market factors often combine to determine the choice of producing rate. North Sea production profiles are generally planned with a high but short-lived initial production peak, followed by a natural decline to an economic limit over 10 to 20 years. In the North Sea the high operating costs, crude oil price volatility, frequent changes in the tax structure, and technical uncertainties, such as subsidence, storms, corrosion control, bottom scouring and platform longevity, dictate maximum early-life producing rates with the shortest possible payout period. The "annual production/reserve ratio" (P/R) of 15% is common for North Sea fields followed by a fairly rapid decline in production to its economic limit in another 10 to 20 years.

The largest North Sea fields generally do not achieve P/R ratios as high as 15%. This is because the enormous initial capital requirements more than offset the NPV (net present value, as discussed in Chapter V) benefits of the higher initial capacity but higher operating costs with more wells, and shorter life. The North Sea development which took place during the 1970's had no problem marketing the production from the high P/R fields. There is no certainty, however, that this will always be the case.

Gas fields, on the other hand, generally afford only a rather low P/R because of the much lower revenues and/or contract terms. In the Gulf of Mexico, where strong water influx can reduce recovery, gas fields are sometimes produced with P/R ratios as high as 40%.

Figure 2-1 illustrates several different types of production profiles. All but the last example have a build-up of production over the first few years of production. This is the development period as facilities are constructed and wells are drilled and placed on production. In the last example production did not begin until all wells were drilled and all facilities completed. Each example has a flat top, which represents the maximum production rate that the facilities have been built to handle, pipeline capacity, or contractual constraint. The length of time that the field is expected to produce at the constant rate and the size of facilities constructed will be based upon the productivity of the reservoir and the economics of the project. In Figure 2-1 the peak rates range from an annual production of only 5% of the ultimate recovery (5% p.a.) to 20% of the ultimate recovery. Eventually the wells will no longer be capable of producing at the peak rate and will ultimately decline to the economic limit. Some fields will have long producing lives while others will be considerably shorter, depending upon the field's development and its reservoirs' characteristics.

Figure 2-1

PRODUCTION PROFILES

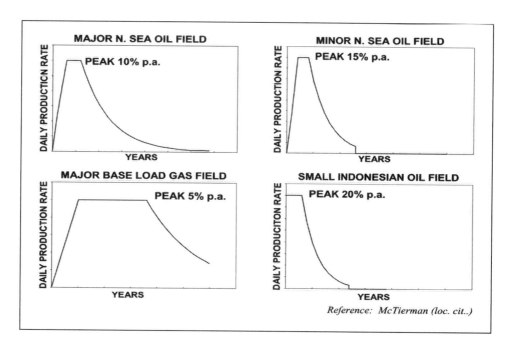

Reference: McTierman (loc. cit..)

The petroleum engineer is often called upon to furnish estimates of reserves, or estimates of ultimate recovery of oil and gas from a particular property or field. In reality this may be a somewhat less demanding assignment than the year by year projections of production, which must in total equal the reserve estimate. The ultimate recovery, (original, or initial reserve for the property), remains an estimate and the actual figures will not be known for certain until the last well is plugged and abandoned at some date in the future.

The calculation of oil and gas reserves and the prediction of the annual rates of oil recovery are the most critical and demanding aspect of any cash flow projection. The practice of reservoir engineering is almost entirely devoted to assignments of this nature. Increasingly complex methods of computation and mathematical modeling are constantly being developed. These are applied in a continuing effort to improve the accuracy of production and reserve predictions.

Early in the life of an oilfield, when there is still an abundance of primary reservoir energy, there will be little or no decline in the daily producing rate. The period of this constant production may be limited by the number of producing wells and their productive capacity, the limitations of the production equipment and pipelines, or by government proration. When reservoir productivity does begin to decline, which will normally be several years after first discovery, the matter of production forecasts becomes more complex. This is particularly true when one is evaluating the potential effects of a contemplated engineering change in a well or group of wells in, for example, a secondary recovery project or recompletions to another reservoir. The most common approach to these types of forecasts is generally by means of decline curve analysis.

PRODUCTION FORECASTS

There are many factors which affect the rate at which hydrocarbons are produced from wells and fields. Because these factors can vary widely, production rates will also vary widely. The geologic factors, i.e., type and characteristics of the rock, depth, thickness, natural resevoir energy and properties of the hydrocarbons, will have the most significant effect upon production rates. To a lesser degree, production rates will also be affected by the development of the hydrocarbon reservoir. The rate of development, well density, wellbore size, completion techniques and method of production will have an impact upon production rate as will the capacity of the facilities constructed to handle the produced fluids. Geologic factors can be altered only slightly, so the major changes in production rate over the productive life of a reservoir will be due to the number of producing wells. Therefore, a production forecast must be based upon an understanding of the geologic factors and knowledge of the plan for development of the reservoir. Other factors which should be considered are the availability of a market for the produced hydrocarbons, especially gas, and the price to be received at the point of sale. Government policy and conservation of reservoir energy may also determine the maximum rate at which a reservoir may be produced.

Forecasting gas production can be quite different than forecasting oil production. In addition to the physical factors described above for oil, which also apply to gas, gas production is very dependent upon demand for gas. Most gas goes directly from the producer to the consumer, with very little storage. Therefore, unless the customer can take the gas the wells must be cut-back or shut-in until there is a demand for the gas. For this reason gas production is cyclic during the year, being quite dependent upon the weather. If gas is produced for shipment as LNG, the annual rate may be a small fraction of the ultimate recovery so that production is typically stretched over about 25 years. This assures the purchaser and his bankers of long term constant production to amortize the large amount of capital required for such projects.

Economic Limit

It is important to know when hydrocarbon production will terminate. In most cases, production will be abandoned before it would cease due to natural causes, because production will decline to a rate at which it costs more to produce the hydrocarbons than those hydrocarbons are worth. This is the "economic limit." At the economic limit, the production costs are equal to the value of the produced hydrocarbons. To continue production beyond this point will cause economic loss.

There is a finite volume of hydrocarbons that can be produced economically from any given well, reservoir, or field. This is generally referred to as the "ultimate recovery" and is equal to the reserves before production commences. As hydrocarbons are produced, reserves will decline as cumulative production increases, but the sum of cumulative production and reserves always equals the estimated ultimate recovery at that time. This is not to say that the estimate of ultimate recovery cannot change, because this surely will occur as production performance is evaluated. But there is a single "true" ultimate recovery for every producing unit. Since there is a finite volume of hydrocarbons that can be produced from a producing unit, it follows that at some point in the productive life, the production rate will also decline, ultimately reaching the economic limit. The economic limit is the point at which the hydrocarbon production ceases to yield an operating profit. At this point the operating costs are just equal to the revenue after deducting royalty. In equation form;

$$q_{ec.lim} = \frac{DOC}{\$/BBL(1 - ROYALTY\ FRACTION)}$$

Where: $q_{ec.lim}$ = Production rate per month at economic limit
DOC = Direct operating cost per month
\$/BBL = Unit price of hydrocarbons at date of economic limit
ROYALTY FRACTION = Royalty as a fraction of gross revenue

Calculation of the economic limit should be based upon the revenue remaining after payment of all royalty at the forecast market value of oil and gas at abandonment. Operating expenses should be limited to direct operating expenses at the time of abandonment, unless a savings in indirect expenses will be realized by the abandonment. It should not be necessary to make this calculation on an after income tax basis since zero profit should not incur an income tax liability.

A reduction in direct operating costs will directly reduce the economic limit, prolonging the producing life of a well or field and add reserves, as is shown in Figure 2-2. It is common for high cost operators to sell producing properties as they approach their economic limit. The purchaser expects to realize a greater reserve potential from the property through lower operating costs. For this reason marginal producing properties have greater value to lower cost operators than to high cost operators.

FIGURE 2-2
ECONOMIC LIMIT

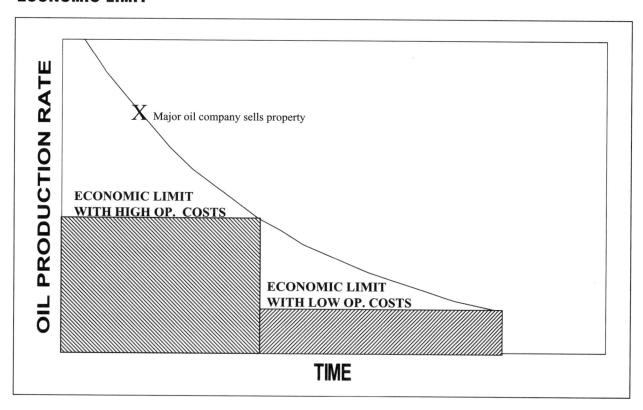

11

Field Analogs

Frequently it is necessary to make forecasts when little if any information is known upon which to base such forecasts. This is commonly the case when analyzing economics of an exploration prospect or the development of an exploration discovery. Under these conditions it is useful to identify a producing field which seems to be similar to the prospect or discovery and can be used as a model for its evaluation. The producing field may provide the basis for a production model, expected reserves, oil-in-place, recovery efficiency, natural reservoir energy, or other information necessary to develop a production forecast. Capital and operating costs can also be forecast from experience of actual producing operations. The best reservoir analogs are those of similar geologic age and setting, but do not necessarily need to be near-by. Onshore fields could be used as models for offshore prospects, as long as offshore costs are used in the economic evaluation.

For example, there is an important similarity in per-acre recoveries within the group of volumetric San Andres reservoirs of West Texas; or again, the per-acre recoveries from the Niagran Reef fields of the Michigan basin; or another analogous grouping, the per acre-foot recoveries of the Frio Sands of the Texas Gulf Coast.

It may be necessary to use more than a single analog when making a forecast. A reservoir or field analog might be used as the basis of a production forecast while the capital and operating expenses might be based upon a totally different analog or analogs.

Abandonment

Most petroleum contracts require the operator to plug and abandon all wells in accordance with recognized industry practice or in accordance with applicable laws and return the surface to its original conditions when producing operations are terminated. Abandonment of wells and facilities when oil and/or gas production ceases is a problem of increasing concern to both the producer and governments. The concerns are both environmental and financial. Onshore these costs may not be too significant. However, offshore the cost may be great, even requiring development of new technology to physically accomplish the task.

There are four options for abandoning, or as some in the industry prefer to call decommissioning, offshore structures; (1) leave in place, (2) removal of upper section to below navigation depth, (3) topple in place, or (4) complete removal. Disposal of steel structures appears to be the easiest and least costly. Frequently these can be toppled in place as part of an artificial reef program. Concrete offshore structures present a much more difficult and costly problem, as Shell's attempt to dispose of the Brent Spar structure in 1995 demonstrated. One proposal is to leave them in place, as they are expected to last for a very long time. Significant factors that must be considered in making these decisions include: (1) type of construction, (2) size, (3) distance from shore, (4) weather conditions, and (5) complexity of removal.

Abandonment costs should be forecast just as any other cost, but since they will not occur until the end of the life of the operation they will have only a slight effect on initial economics. In many cases it may be advantageous to postpone the decommissioning of platforms, especially in the North Sea, for as long as possible. This is true because of the great expense incurred in removing offshore structures. Such delays may be accomplished by redevelopment of existing reservoirs, opening additional reservoirs, using platform to service sub sea completions in nearby fields, or non-oil field uses such as wind farms, shipping beacons, radar stations, weather stations, hotels, etc. It may be possible to justify any of these activities on the basis of the reduced present value of the decommissioning cost which would result from the delayed expenditure. The risk to such a strategy would that the costs escalate faster than the company discount rate due to more strict environmental requirements; i.e. as Shell experienced in disposing of the Brent Spar. You will find an additional discussion of this topic in Chapter X. Funds for abandonment can be made available in one of two ways.

First, and probably currently the most common method, is not to worry about it until abandonment is eminent. Under this procedure funds must be provided out of then current revenue, which is sometimes supplemented by tax credits. This source may not be sufficient for smaller organizations with limited operations, so the liability may fall upon the government. Under many production sharing contracts, ownership will eventually revert to the government along with its burden of abandonment. For this reason, abandonment may be the subject of serious negotiation between oil and gas producing companies and governments when the contracts are originally signed or at the point when the producing properties revert to government ownership.

The second method which is becoming more common, as abandonment costs have become commonly recognized, is to charge a portion of the abandonment cost to each unit of production throughout a property's producing life. The advantage of this treatment is that abandonment is paid for by the oil that is produced from the operation, with all parties bearing their proportionate share of this cost. Abandonment thus becomes a part of ongoing operating costs for both accounting and tax purposes. Depending on exactly how money is provided for abandonment this may or may not be a cash flow item. These funds may just be in an identified reserve for abandonment or may be physically segregated from other funds through a trust fund or purchase of government bonds. Banks could also guarantee the availability of funds for this designated use or insurance may be purchased to guarantee abandonment. No matter which method is applied, when abandonment occurs it will be a major issue and those who have to deal with it will be very grateful if it has been provided for in advance.

DECLINE CURVE ANALYSIS

The characteristics of production decline have been studied by many and a set of generally accepted equations have been developed to describe decline performance. Decline curve analysis is an empirical statistical method of analyzing production performance. When it is applied to performance data to predict future performance from that producing unit, it is essentially using the reservoir as an analog computer. This procedure assumes that current well and reservoir conditions will remain unchanged in the future and requires a complete understanding of the field operations. Actual data points will not follow a smooth decline so it will be necessary to draw the best line through the data points, ignoring those atypical points caused by extraneous changes. It is advisable to check the validity of a decline curve by comparing the ultimate recovery calculated to a value determined by independent means, such as material balance or volumetric calculations.

Accurate records of monthly oil production and/or gas sales are usually available by individual producing properties or leases. This is largely because governments require these data for taxation and regulatory purposes. There are half a dozen ways of plotting production data depending upon the nature of the reservoir and the amount of hydrocarbon recovery to date. One basic technique is to plot the logarithm of the producing rate against time. Figure 2-3 is such a plot for an eight well fieldwide unit in the Rocky Mountains of the U.S. The declining production trend is readily discernible and can be projected into future months and years.

Decline curves terminate when producing rate and net revenue are reduced to the level that the current income from the oil and/or gas produced equals the cost of producing it. Beyond this level, further production is uneconomical. This rate of production is known as the "Economic Limit." As production approaches the economic limit, it is known as "marginal production." The experienced engineer usually resorts to individual well decline curves during these latter stages of production.

Some purists argue that the extrapolation of decline curves is an empirical, statistical method and not a true engineering approach. Others point out that, at least for the log-rate vs. cumulative production plot for a closed, or volumetric reservoir, the shape of the decline is a valid reflection of the relative permeability to oil (Kg/Ko) vs. liquid saturation relationship. In any event, the techniques of decline curve analysis are widely employed throughout the industry. They have the advantage of minimizing the time spent studying assets in the single well category which could not support an extensive, and expensive reservoir engineering project. They are also easily understood and actually comprise the only valid approach in the final stages of depletion.

The extrapolation of a trend for purposes of predicting future production from a well or group of wells assumes that the factors which caused the production changes in the past will operate in the same way in the future. The engineer has to evaluate and verify this assumption. John Campbell (loc. cit.) emphasizes the basic premise with respect to decline curve analysis that "all factors which have influenced the curve in the past will remain effective during the (remainder of) the producing life."

Graphical Methods of Production Forecasting

Extrapolations into the future can be made either graphically or mathematically. The graphical techniques will be reviewed first.

A special case in which the production history plots as a straight line on a semi-log grid of producing rate vs. time is known as "exponential decline." Exponential decline has long been a favorite of petroleum engineers because:

1. The technique is simple and easy to use

2. Exponential decline projects the most conservative estimates of ultimate recovery of the several techniques available.

3. deviation from an "exponential decline" may not occur for many years, so its impact on project economics is minimal.

FIGURE 2-3
WILLSON RANCH PRODUCTION CURVE

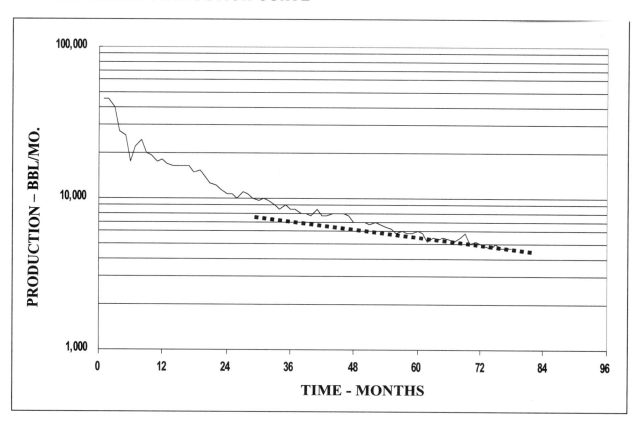

Decline curve analysis relies on either graphical or mathematical extrapolation of production decline trends. These trends are plotted against depletion of the resource, usually as a function of time on production. The longer the production history the more reliable the extrapolations will be. Decline curve extrapolations have been used extensively in the industry for more than a half a century. They have proven to be a very satisfactory method of forecasting future rates of oil and gas production.

There are a number of ways of plotting production data for interpretive purposes. The ones most frequently employed are:

1. Semi-log plot of producing rate vs. time

2. Semi-log plot of producing rate vs. cumulative production

3. Pressure divided by the gas deviation factor, (z) vs. cumulative gas production

4. Percent water in production vs. cumulative oil production

5. Cumulative gas produced vs. cumulative oil

6. Progressive rise in the elevation of the oil-water contact vs. cumulative oil production

The first two methods are used extensively in developing forecasts of oil production. The third is employed effectively for depletion-type gas reservoirs where there is little or no water encroachment.

15

The others listed are of less interest due to their quite limited, and special application. Figure 2-3 is a plot of the first type. The plot conforms to the common experience that the productivity of the wells in question decreases with time and, of course, with the total cumulative amount of oil produced. As the field is produced the energy causing the oil to flow into the wells is depleted, or becomes less effective as the hydrocarbons are removed, so that the rate of production gradually decreases. It is observed that although Figure 2-3 demonstrates a definite downward trend the function is not linear. A coordinate plot of the same data is presented in Figure 2-4, which exhibits even less linearity. A coordinate plot of the same production rate data versus cumulative production is presented in Figure 2-5.

FIGURE 2-4
COORDINATE PLOT OF WILLSON RANCH FIELD

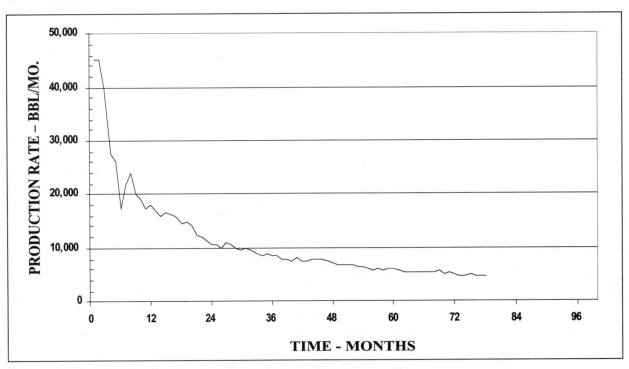

■ EXAMPLE 2-1
■ FORECASTING PRODUCTION FROM DECLINE CURVE DATA

Develop a production forecast for the Willson Ranch "J" Sand Unit for the next twelve months. (Figure 2-3) from mid-year seven.

SOLUTION:

Superimposing a straight line extrapolation on the semi-log plot of production history (Fig. 2-3), a value of 4,200 bbl/mo. is read at the start of the forecast period (month 78) and declines to 3,300 bbl/mo. twelve months later. The average of these numbers is 3,750 bbl/mo, indicating an approximate production of 45,000 barrels for the next 12 months.

Not all decline curves plot as straight lines on any of the grids mentioned. This does not preclude the usefulness of these plots. Graphical extrapolation of the decline, whether by means of a straight edge, ship's curve, or computer graphics, is an expression of the general hyperbolic rate-time equation, described in the next section. The use of personal computers has accentuated the mathematical evaluation of oil production decline relationships. Using equations 2.4(a) and 2.5, which are for production that approximates a straight line semi-log relationship, the twelve month's production is calculated to be 44,783 barrels.

FIGURE 2-5
FORECASTING PRODUCTION FROM DECLINE CURVE DATA

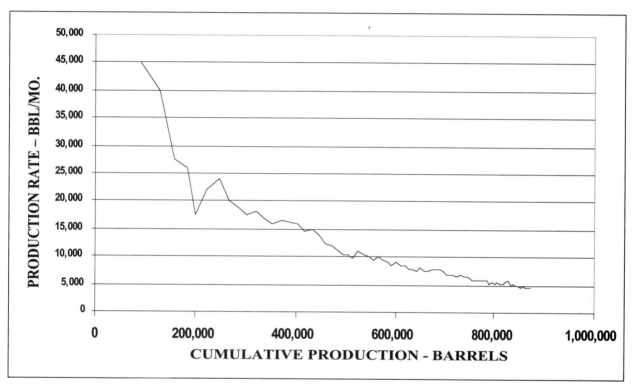

Production Forecasting by Mathematical Models

Production as a function of time generally follows a hyperbolic function in the following form:

$$q_t = q_i \left[1 + \left(a_i ht \right) \right]^{-(1/h)} \tag{2.1}$$

Where:

> q = producing rate, BOPD
> t = time from start of analysis period, years
> h = "h-factor," exponent of hyperbolic decline. Its value between
> > 0 and 1.0 characterizes a particular oil well or reservoir's production decline
> a = constant of integration, which equates to the production decline rate as a
> > fraction. This is the decline rate, fraction/year.

Subscripts:

> i = initial, at time, t = 0
> t = at end of time, t

Early reservoir engineers employed manual extrapolations by 'ship's curve', with little, or no idea of what h-factor was being introduced. Frequently the semi-log production plot of producing rate vs. time will tend to "curve up" or flatten. The hyperbolic equation 2.1 will generally model this type of production decline. In these situations the decline, instead of being a constant percentage of the previous year's production, is a complex function of time, initial decline rate and h-factor. Integration of Equation 2.1 to obtain the cumulative production (Np) from the start of the analysis period to the end of year "t" yields the following:

$$N_p = \frac{q_i}{a_i(1-h)}\left[1-\left(q_t/q_i\right)^{(1-h)}\right] \times 365 \qquad (2.2)$$

If a future time (t) is known rather than a future production rate (q t), then Equation 2.2a could be used instead of Equation 2.2.

$$N_p = \frac{q_i}{a_i(1-h)}\left[1-\left\{1+\left(a_i ht\right)\right\}^{[1-(1/h)]}\right] \times 365 \qquad (2.2a)$$

Equations 2.1 and 2.2 or 2.2a are independent equations which describe general hyperbolic decline performance. Between the two equations there are six parameters. When any four of these parameters are known, and no more than four may be specified, the two missing parameters can be calculated by using the two equations. As shown in Table 2-2, there are 15 combinations of four independent parameters which can be specified for a general hyperbolic decline. Table 2-2 also indicates that eight of the solutions will require an iterative technique because one or both of the unknown parameters does not appear to the same power in both equations.

While Equations 2.1, 2.2 and 2.2a are quite simple to compute, it may be difficult to determine values for a_i and h from actual performance data.

The large number of iterative solutions indicated in Table 2-2 also makes this type of decline less frequently used. Special cases of these general equations are frequently employed.

TABLE 2-2

DATA LIMITATIONS FOR MATHEMATICAL ANALYSIS OF HYPERBOLIC DECLINE CURVES

Case No.	Initial Rate, q_i	Final Rate, q_t	Elapsed Time t	Produced Volume N_p	Decline Rate a_i	h	Solution
1	X	X	X	X			Iterative
2	X	X	X		X		Iterative
3	X	X	X			X	Direct
4	X	X		X	X		Iterative
5	X	X		X		X	Direct
6	X	X			X	X	Direct
7	X		X	X	X		Iterative
8	X		X	X		X	Iterative
9	X		X		X	X	Direct
10	X			X	X	X	Direct
11		X	X	X	X		Iterative
12		X		X	X	X	Iterative
13		X	X		X	X	Direct
14		X	X	X		X	Iterative
15			X	X	X	X	Direct

Exponential or Constant Percentage Decline

The most commonly used special case of the hyperbolic decline equation is for a value of (h = 0) and is called "constant percentage decline." Its widespread use is principally due to two reasons. First, it is frequently observed when analyzing production data and second, it is simple to use. Commonly oil production is plotted on semi-logarithm graph paper, with production rate on the log scale and time on the coordinate scale, as was done in Figure 2-3. If the production data follows a constant percentage decline, then this plot will appear as a straight line. Future production can easily be forecast by extrapolating the straight line trend to the economic limit.

Equations for constant percentage decline are also easy to use. There are two common forms of the constant percentage decline equation, one which treats annual production as a step function and the other as a continuous exponential function. The equation for the production rate using a step function is:

$$\bar{q}_t = \bar{q}_1 \left(1-d\right)^{(t-1)} \tag{2.3}$$

Where:

$\bar{q}_1$ = average production rate or annual production for initial year

$\bar{q}_t$ = average production rate or annual production for year "t"

$$d = \left[1 - \left(\bar{q}_2 / \bar{q}_1 \right) \right]$$

= effective annual decline rate, fraction per year

Integration of Equation 2.3 to obtain the cumulative production from the start of the analysis period (N_p) to the end of year "t," where production rates are in barrels per year, yields the following:

$$N_p = \frac{\bar{q}_1 - \bar{q}_t}{d} + \bar{q}_t \qquad (2.4)$$

The exponential form of equations 2.3 and 2.4 are also quite simple to use and are presented as equations 2.3a and 2.4a. You will note that in the exponential form of the constant percentage decline equations, instantaneous production rates must be used rather than average annual values and that the decline factor is a nominal value or instantaneously applied value rather than an effective annual value.

$$q_t = q_i e^{-at} \qquad (2.3a)$$

$$N_p = \frac{q_i - q_t}{a} \qquad (2.4a)$$

$$a = \frac{Ln\left(q_i / q_t \right)}{t} \qquad (2.5)$$

where:

q_t = instantaneous production rate at end of year "t", bbl./yr.

  = $\bar{q}_t$ (a / d) (1- d)

q_i = instantaneous initial production rate, bbl./yr.

  = $\bar{q}$ (a /d)

a = nominal decline rate, fraction/year

  = Ln (q_i / q_1)

  = -Ln(1 - d)

d = 1 − (q_1 / q_i) = 1 - e⁻ᵃ

Examination of equations 2.4 and 2.4a reveals that production rate is a linear function of cumulative production for constant percentage and exponential declines. Therefore, if production rate is plotted as a function of cumulative production, using coordinate scales for both values, the resultant graph would be a straight line. The slope of that straight line is (-a), the value determined from Equation 2.5, and the zero value intercept is the initial production rate qi used in Equation 2.5. This provides another method of determining a value of "a" from actual performance data. When using both equations 2.4 and 2.4a it should be noted that since "a" and "d" are the declines per year, the production rates must also be per year. If daily production rates are used, it will be necessary to multiply the right hand side of both equations by 365 days in the year.

Equations 2.3 and 2.3a and equations 2.4 and 2.4a will yield the same results respectively if appropriate values of "a" and "d" as determined with the above relationships, are used in the proper equations. The selection and use of one or the other set of equations should be made purely on the basis of which is easier to use for the particular problem being solved. Either can be easily solved using most scientific hand held calculators. Example 2-2 shows a very simple example of a constant percentage decline using the step function and the continuous decline equations.

Using Mid-Year Values For Average Yearly Values

TABLE 2-3
ERROR USING MID-YEAR VALUES

d	Error - %
0.1	0.046
0.2	0.208
0.3	0.531
0.4	1.091
0.5	2.014
0.7	6.150
0.9	23.602

When analyzing a semi-log plot of production data, the use of mid-year values for values of $\bar{q}_t$ rather than calculating an average annual rate, introduces an error of less than one percent for decline rates of less than 40 percent per year, as shown in Table 2-3.

EXAMPLE 2-2
EFFECTIVE ANNUAL DECLINE VS. EXPONENTIAL DECLINE

The following tabular and graphically displayed data (Figure 2-6) show the correspondence between effective annual constant percentage decline (step function) for d = 0.1 and exponential decline for the equivalent a = 0.10536 over an 11 year period. Note the year to year decline is the same whether it is from average value to average value or from instantaneous to instantaneous value and the cumulative production is identical.

EFFECTIVE ANNUAL DECLINE VS. EXPONENTIAL DECLINE

Year	Effective Annual Decline d = 10.000%			Exponential Decline a = 10.536%		
t	$\bar{q}_t$	$\bar{q}_{(t+1)}/\bar{q}_t$	N_p	q_t	$q_{(t+1)}/q_t$	N_p
0				8,429		
1	8,000	90%	8,000	7,586	90%	8,000
2	7,200	90%	15,200	6,827	90%	15,200
3	6,480	90%	21,680	6,145	90%	21,680
4	5,832	90%	27,512	5,530	90%	27,512
5	5,249	90%	32,761	4,977	90%	32,761
6	4,724	90%	37,485	4,480	90%	37,485
7	4,252	90%	41,736	4,032	90%	41,736
8	3,826	90%	45,563	3,628	90%	45,563
9	3,444	90%	49,006	3,266	90%	49,006
10	3,099	90%	52,106	2,939	90%	52,106
11	2,789	90%	54,895	2,645	90%	54,895

The average production rates for each year in the above table were calculated using the relationships following equation 2.5.

FIGURE 2-6 (EXAMPLE 2-2)
COMPARISON OF PRODUCTION DECLINE RATES

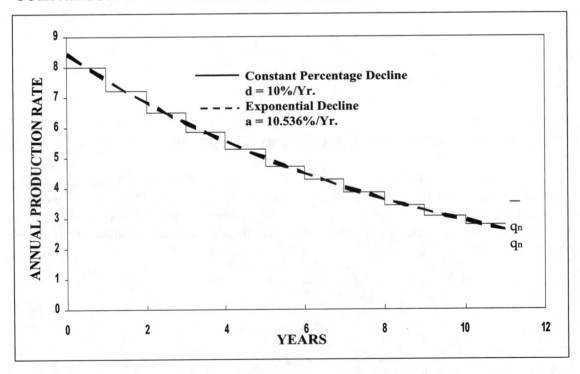

Equations 2.3 and 2.4, or 2.3a and 2.4a, are independent equations which describe constant percentage decline performance. Between the two equations in each set there are five parameters.

When any three of these parameters are known, and no more than three may be specified, the two missing parameters can be calculated by using the two equations. As shown in Table 2-4 there are ten combinations of three independent parameters which can be specified for a constant percentage decline. Table 2-4 also indicates that two solutions will require an iterative technique because one of the unknown parameters does not appear to the same power in both equations. Alternate forms of those equations are presented in Appendix II-A.

TABLE 2-4

DATA LIMITATIONS FOR MATHEMATICAL ANALYSIS OF EXPONENTIAL DECLINE CURVES

Case No.	Initial Rate, q_i	Final Rate, q_t	Elapsed Time, t	Produced Volume, N_P	Decline Rate, a	Solution
1	X	X	X			Direct
2	X	X		X		Direct
3	X	X			X	Direct
4	X		X	X		Iterative
5	X		X		X	Direct
6	X			X	X	Direct
7		X	X	X		Iterative
8		X	X		X	Direct
9		X		X	X	Direct
10			X	X	X	Direct

Note: Only three input parameters may be used. All of the solutions above are direct except for cases 4 and 7, which are iterative.

Example 2-3 shows how equations 2.3a and 2.4a can be used to calculate parameters other than the final producing rate and cumulative production. Equation 2.6 was derived by solving equations 2.3a and 2.4a simultaneously for the unknown time to reach the economic limit having produced a known quantity of oil.

■ EXAMPLE 2-3
■ ESTIMATING REMAINING PRODUCING LIFE OF A WELL

A well with an initial capacity of 85 BOPD is expected ultimately to produce 100,000 barrels continuing its exponential decline, (h = 0). Calculate the number of years that the well will produce before reaching an assumed economic limit of 5 BOPD using the equation:

$$t_a = \frac{N_p}{365(q_i - q_a)}\left[Ln\left(\frac{q_i}{q_a}\right)\right] \qquad (2.6)$$

Where:

t_a = producing time to economic limit, years

q_i = initial daily production, BOPD

q_a = economic limit, BOPD

N_p = cumulative production to econ. limit, bbls

Ln = logarithm to the base e

Thus:

$$t = \frac{100,000}{(365)(80)} \ (2.83) = 9.7 \text{ years}$$

Production Models

Earlier in this chapter, there was a discussion of various production profiles that are experienced throughout the world. Four markedly different production profiles were presented in Figure 2-1. With the equations that have just been developed for constant percentage decline, it is possible to calculate the production rate for each segment of those production curves. The convention established for "d" and "a" assume the production rate declines with time. However, the same equations can be employed when the production rate increases at a constant percentage per year by merely reversing the signs for "d" and "a". Example 2.4 shows how equations 2.3a and 2.4a can be used to calculate annual production rates and cumulative production for this type of production model. One thing you will note is that the constant percentage decline equations cannot handle a zero production rate, so it was necessary to start the production at a positive value.

■ EXAMPLE 2-4
PRODUCTION MODEL FOR AN OIL FIELD

An oil field has been discovered that has the potential of 50,000,000 barrels of oil. It is estimated that development will take about 4 years, with production beginning at the end of the second year. Facilities will be constructed to handle 15,000 barrels of oil per day. It is estimated that after production reaches the limit of the facilities, at the end of the sixth year, it will continue to produce at that rate for 5 years before beginning to decline at a constant percent per year. The economic limit has been determined to be 500 barrels of oil per day.

Determine:

A forecast of annual production rate over the entire life of the field which meets the previously outlined conditions.

Solution:

Development Period (yr. 3-6

Assume that the production at the end of the second year is 100 B/d and will be 15,000 B/d at the end of year 6.

FIGURE 2-7
PRODUCTION FORECAST MODEL

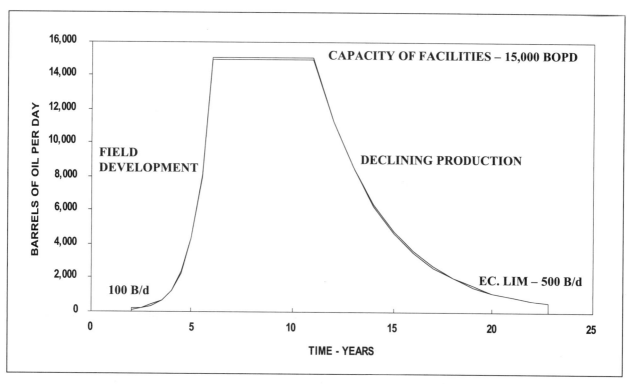

Rearranging Eq. 2.6 to calculate the total amount of oil produced during development;

$$N_{3-6} = [t(q_i - q_t)365] / [Ln(q_i/q_t)] = [4 (100-15,000)365]/[Ln(100/15,000)]$$

$$N_{3-6} = 4,341,565 \text{ barrels}$$

Using Eq. 2.5 to determine the rate of production increase during each year of development;

$$a = [Ln(q_i/q_t)] / (t) = [Ln(100/15,000)] / (4) = -1.2527/ \text{ yr.}$$

Note: The negative value of "a" indicates a growth rather than a decline with time.

Using Eq. 2.3a to calculate the year end production rate for each year during development;

$$q_t = q_i \, e^{-at} = 100 \, e^{1.2527t} \text{ (barrels per day)}$$

Using Eq. 2.4a and the production rates calculated with Eq. 2.3a to determine annual production;

$$N_{annual} = 365 \, (q_{beg. \, yr.} - q_{end \, yr.})/a \qquad \text{(barrels per year)}$$

Constant production period (yr 7-11)

$$N_{annual} = (q_{Const.})365 = 15,000*365 = 5,475,000 \text{ bbl/yr.}$$

25

Declining production period (yr. 12+)

Using Eq. 2.6 to calculate the time to reach the economic limit

$$t_{12+} = N_p [Ln (q_i/q_t)]/[(q_i-q_t)(365)]$$

$$N_p = [50,000,000 - 4,341,565 - (5 \times 5,475,000)] = 18,283,435 \text{ bbl}$$

$$t_{12+} = [18,283,435][Ln(15,000/500)]/[(15,000-500)365] = 11.75 \text{ years}$$

$$t_a = 11 + 11.75 = 22.75 \text{ years}$$

Using Eq. 2.5 to determine the annual decline rate of production;

$$a = [Ln(q_i/q_t)]/(t) = 0.2894697 \text{ (fraction per yr.)}$$

Using Eq. 2.3a to calculate production rate at end of each year;

$$q_t = (15,000) e^{-0.2894697 t} \text{ (barrels per day)}$$

Using Eq. 2.4a and the production rates calculated with Eq. 2.3a to determine annual production;

$$N_{annual} = 365 (q_{beg. yr.} - q_{end yr.}) /a \text{ (barrels per year)}$$

The following table shows the resulting production forecast for each year of the projected 22.75 year life of the oil field, producing exactly 50,000,000 barrels as the 500 BOPD economic limit is reached.

TABLE 2-5 (EXAMPLE 2-4)
PRODUCTION FORECAST

YEAR t	YEAR END DAILY OIL PRODUCTION Q, BBL/DAY	ANNUAL OIL PRODUCTION Q, BBL/YR.	CUMULATIVE OIL PRODUCTION Np, BBL.	a=
0				
1				
2	100.0			
3	350.0	72,834	72,834	-1.2527
4	1,224.7	254,894	327,728	-1.2527
5	4,286.2	892,036	1,219,764	-1.2527
6	15,000.0	3,121,801	4,341,565	-1.2527
7	15,000.0	5,475,000	9,816,565	
8	15,000.0	5,475,000	15,291,565	
9	15,000.0	5,475,000	20,766,565	
10	15,000.0	5,475,000	26,241,565	
11	15,000.0	5,475,000	31,716,565	
12	11,229.9	4,753,810	36,470,375	0.2894697
13	8,407.4	3,558,990	40,029,365	0.2894697
14	6,294.3	2,664,475	42,693,840	0.2894697
15	4,712.3	1,994,787	44,688,627	0.2894697
16	3,527.9	1,493,418	46,182,045	0.2894697
17	2,641.2	1,118,063	47,300,108	0.2894697
18	1,977.4	837,050	48,137,158	0.2894697
19	1,480.4	626,666	48,763,824	0.2894697
20	1,108.3	469,160	49,232,984	0.2894697
21	829.7	351,242	49,584,226	0.2894697
22	621.2	262,961	49,847,187	0.2894697
22.75	500.0	152,813	50,000,000	0.2894697

Hyperbolic Decline

Another special case of the Hyperbolic Decline is for a value of (h = 0.5). In this case, the equations simplify to:

$$q_t = q_i [1 + 0.5a_i t]^{-2} \tag{2.7}$$

$$N_p = [2q_i / a_i][1 - (q_t / q_i)^{0.5}] \tag{2.8}$$

$$N_p = t (q_i q_t)^{0.5} \tag{2.9}$$

These equations might prove useful when forecasting production from a well or field where an early decline is anticipated but a prolonged life is expected. Such a case cannot be adequately modeled by a constant percentage decline because of the anticipated late life flattening. Example 2-5 shows the difference that is caused by the selection of (h = 0.5) rather than an (h = 0).

EXAMPLE 2-5
CONSTANT PERCENT DECLINE VS. HYPERBOLIC DECLINE

The company's geologists and reservoir engineers have estimated that the ultimate recovery for a particular well will be 1.0 million barrels of oil. To evaluate its profitability it will be necessary to develop a forecast of oil production over the life of the well. Similar wells have produced at initial rates of 800 BOPD and it is estimated that the economic limit for such a well will be 10 BOPD. Two production forecasts are required, an optimistic one based upon a constant percent (exponential) decline and a second one, a little more conservative, using a hyperbolic decline with an "h" value of 0.5. Both declines will begin as soon as production commences.

<u>Constant percentage decline (h=0)</u>

Using Eq.2.6 to calculate the well's productive life

t = (1,000,000) [Ln(800/10)] / [(800 - 10) x 365] = 15.20 years

Rearranging Eq. 2.4a

a = $(q_i - q_t)$ / Np = [(800 - 10) x 365] / 1,000,000 = 0.2883   (2.10)

Using Eq. 2.3a to calculate year end production rates

q_t = (800) e $^{-0.2883t}$ (bbl./day)

Using Eq. 2.4a to calculate annual production

Np= $(q_i - q_t)$ / a = 365(800 - q_t)/0.2883

<u>Hyperbolic Decline (h = 0.5):</u>

Rearranging Eq. 2.9 to calculate the well's productive life:

t = $[N_p (q_i \times q_t)^{-0.5}]$ (2.11)

   = [1,000,000 (800 x 10)$^{-0.5}$] / 365 = 30.63 years

Rearranging Eq. 2.7 to calculate the initial decline rate (a_i);

a_i = $[(q_i / q_t)^{0.5} -1][2/t]$ (2.12)

   = [(800 / 10)$^{0.5}$ -1][2 /30.63] = 0.5187 / year

Or rearranging Eq. 2.8 to determine the initial decline rate (a_i);

$$a_i = [2q_i/N_p][1-(q_i / q_t)^{0.5}] \times 365 \qquad (2.13)$$

$$= [2 \times 800/1,000,000][1-(10/800)0.5] \times 365 = 0.5187$$

Using Eq. 2.7 to calculate year end production rates;

$$q_t = q_i [1 + a_i t / 2]^{-2}$$

$$= (800) [1 + (0.5187) (t / 2)]^{-2} \text{ (bbl/day)}$$

Using Eq. 2.9 to calculate annual production;

$$N_p = 365 t [(800) \times q_t]^{0.5}$$

The year by year forecast, using the equations developed above, is shown in the following table and Figure 2-8.

EXAMPLE 2-5

Year 1	Constant Percent Decline (h = 0)		Hyperbolic Decline (h = 0.5)	
	q bbl/day	Cumulative N_p bbl/year	q bbl/day	Cumultive N_p bbl/year
0	800		800	
1	600	253,672	504	231,865
2	449	443,798	347	384,538
3	337	586,298	253	492,672
4	252	693,102	193	573,276
5	189	773,151	152	635,676
6	142	833,148	122	685,414
7	106	878,115	101	725,988
8	80	911,818	85	759,717
9	60	937,079	72	788,200
10	45	956,012	62	812,571
11	34	970,202	54	833,660
12	25	980,837	47	852,090
13	19	988,808	42	868,333
14	14	994,783	37	882,757
15	11	999,261	33	895,650
15.2	10	1,000,000	33	898,030
16			30	907,245
17			27	917,728
18			25	927,252
19			23	935,942
20			21	943,904
21			19	951,225
22			18	957,980
23			16	964,232
24			15	970,035
25			14	975,436
26			13	980,475
27			12	985,187
28			12	989,603
29			11	993,751
30.63			10	1,000,000

FIGURE 2-8 (EXAMPLE 2-5)

CONSTANT PERCENTAGE VS. HYPERBOLIC DECLINE

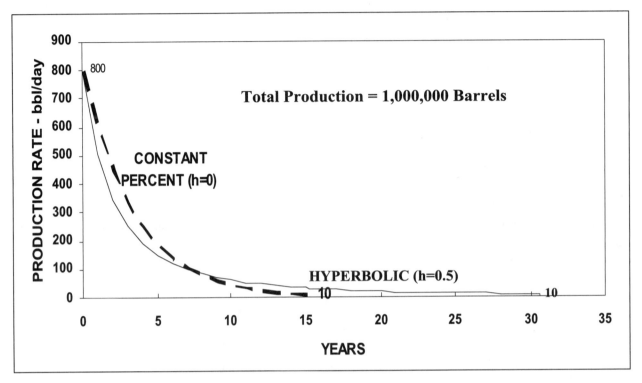

Harmonic Decline

A third special case of the hyperbolic decline is called a harmonic decline, where h = 1.0. Horizontal wells, strong water drives, and steam soak operations often follow a harmonic type of decline. The hyperbolic decline equations are again simplified yielding the following:

$$q_t = q_i \, / \, [1 + a_i t] \tag{2.14}$$

$$N_p = q_i[Ln(q_i \, / \, q_t)]/a_i \tag{2.15}$$

$$N_p = [t \, q_i q_t \, Ln(q_i \, / \, q_t)] \, / \, (q_i - q_t) \tag{2.16}$$

Summary Hyperbolic Decline

Figure 2-9 shows the effect of the different values of "h" in the hyperbolic decline equation. The same value of ai (0.25) was used for computing the values plotted for each of the five curves.

30

FIGURE 2-9
FAMILY OF HYPERBOLIC DECLINE CURVES

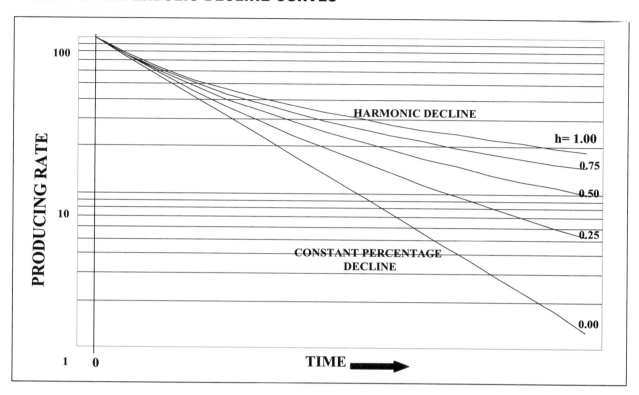

If the decline curve being analyzed does not fit one of the three special cases just discussed, it is a little more difficult to determine the appropriate value of h and ai. However, M. J. Fetkovitch (loc. cit.) in 1972 developed a type curve to simplify that determination. Figure 2-10 is a copy of the Fetkovitch type curve for hyperbolic decline curves previously discussed. By determining which of the type curves best match the production data, the correct value of h and ai can be determined. Once these values are established, then the general hyperbolic equations 2.1 and 2.2 can easily be utilized.

Use of the type curve requires that the actual production rate, in any convenient units, versus time data be plotted on log-log tracing paper of the same size cycles as the type curve to be used. The tracing paper data curve is placed over the type curve, the coordinate axes of the two curves being kept parallel, and shifted to a position which represents the best fit of the data to a type curve. The value of h on the type curve that best matches the actual data is the h value for the actual data. The value of ai is then determined by selecting a "match point" between the actual data and the typecurve. The value of ai is then the type curve dimensionless time divided by the real time at the "match point."

FIGURE 2-10
DECLINE TYPE CURVE

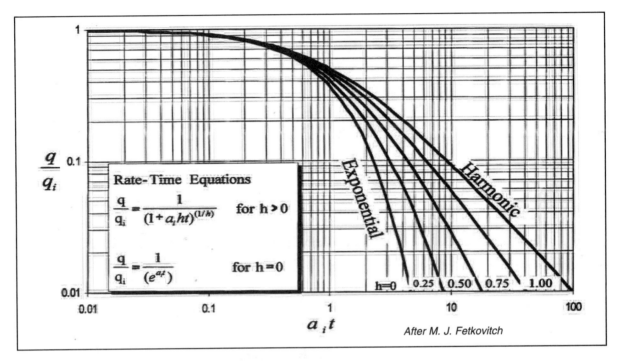

After M. J. Fetkovitch

This can also be done quite easily with a spreadsheet program that contains a graphics package, e.g. Excel or Lotus 123. Plot a log-log graph similar to that shown in Figure 2-10 using Equation 2.1 [(q/qi) vs. (ait)] for various values of h from 0 to 1.0. Create columns for real time and the real production data to be analyzed. Create a second pair of columns, one for the real time multiplied by a value of ai and the other the real production rate divided by a value of qi. Plot the column of real production rate divided by a value of qi on the same log-log graph used to plot Equation 2.1. Vary the values of qi and ai until the actual production data graph most closely matches one of the graphs plotted from Equation 2.1. The desired value of h is the value of h used to plot the theoretical graph while the desired values of qi and ai are the values which gave the closest match between the data and equation plot. Using the actual data as the analog, future production can easily be forecast for situations that are not expected to follow an exponential or constant percentage decline.

Curve fitting becomes a critical part of decline curve analysis. This is true whether it is done by straight edge, or a drafting device known as a ship's curve, or by computer. As a general observation it is found that decline trends early in the life of most reservoirs exhibit fairly strong hyperbolic tendencies (h = 0.5 to 1.0). This is to be expected as the back pressure at the sand face drops fairly rapidly with the initial radial depletion from the well bores. This hyperbolic tendency sometimes may also be due to early differential depletion of more permeable streaks or stringers within the producing formation. Most early hyperbolic declines can be expected to degenerate to exponential (h = 0) with progressive reservoir depletion.

An interesting exception to the foregoing generalization regarding degradation of the "h-factor" towards zero can occur when the reservoir is reacting to strong gravity drainage, water influx, or gas cap expansion. Any of these forces will arrest the reduction in the h-factor and cause it to hover at approximately h = 0.5. When the same decline curve data are plotted in semi-log, linear, or cumulative production format, (Figures 2-3 through 2-5), the difference between exponential and hyperbolic decline is increasingly evident during the latter stages of depletion as the wells producing from the common reservoir approach their economic limit. From the standpoint of most economic analyses, which are conducted early in the life of a property, these differences may appear to be relatively minor. Although exponential (h = 0) declines tend to be conservative, they are much simpler to handle than the hyperbolic decline in either a mathematical or graphical manner. It is generally a conservative assumption in dealing with decline curve extrapolations to proceed on the basis that recent producing trends reflect settled exponential (h = 0) decline behavior.

Production forecasts are often developed in yearly increments for inclusion in spreadsheet formats for economic analysis. (See Chapter IV.) The incremental decline steps may be approached as the repeated product of one minus the annual decline rate in the exponential case. The hyperbolic production decline case requires the solution of Equation 2.1 for each year of the study.

Estimating Remaining Productive Life of a Well or Groups of Wells

The same techniques of forecasting production decline are also useful under a number of conditions as a means of estimating the productive life of properties. This can be helpful in operations planning, and negotiation of bank loans against the property, for example. Figure 2-12 may be useful for this purpose.

Production Trends Following Major Changes in Operations

These production forecasting techniques are important in many petroleum engineering functions such as developing the economics of proposed well repairs, gas gathering system modifications, and artificial lift changes. Figure 2-11 adapted from McCray (loc. cit.) is an interesting example of a typical well workover, or repair, project. This example demonstrates the decline trends both before and after the well repair. Note that the repair has not changed the reservoir being drained by the well and has not increased the ultimate recovery. The workover can thus be looked upon as a rate acceleration type of project.

FIGURE 2-11
WELL REPAIR EXAMPLE

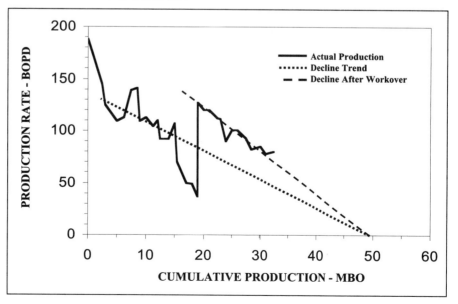

33

OIL AND GAS RESERVES

Oil and gas reserves are simply a forecast of the future volume of these products that will be produced, as of a given date. These forecasts are limited to the volume of oil and gas that can be economically produced; i.e., at a profit, under defined conditions. Normally oil and gas reserves are limited to hydrocarbon products and are subdivided into crude oil, natural gas and condensate, all of which are products that are naturally produced at the wellhead. Reserve volumes are at surface conditions of temperature and pressure, which have been standardized. Natural gas reserves include any liquids which may subsequently extracted by a gas liquids extraction plant. However, it should be recognized that there are two different kinds of oil and gas reserves that are important. One is the "technical hydrocarbon reserves" estimate made by geologists and engineers of the physical quantity of hydrocarbons that can economically be produced from a well, reservoir, or field. The other, which might be called "accounting hydrocarbon reserves", is a forecast of the future volume of hydrocarbons which will accrue to the account of some business organization or government. The sum of all future volumes of "accounting hydrocarbons" accruing to all business organizations and governments involved in a well, reservoir, or field should equal the "technical hydrocarbon reserve" estimate made by the geologists and engineers. However, disparity in hydrocarbon price forecasts between the organizations may cause some divergence in these numbers. The apportionment of "technical reserves" among all organizations involved in the production of those hydrocarbons is based on their forecast of the future volume of hydrocarbons that will accrue to the account of that organization. Thus oil and gas reserves are not only a technical estimate of future production but an accounting forecast of future production to which that organization is entitled.

As was discussed earlier in this chapter, the economic limit (see section entitled Economic Limit) determines when production ceases and is an important factor in a technical reserve estimate. The only relevant price for this calculation is that forecast for the time when the economic limit is reached. However, the apportionment of "accounting hydrocarbon reserves" among the various parties may require a forecast of prices over the entire life of those reserves, depending upon the contractual relationship of the parties involved.

Booking "accounting hydrocarbon reserves" by an organization should not be interpreted as a declaration of ownership of those volumes of hydrocarbons while they are in the subsurface reservoir. They should be considered only a forecast of hydrocarbon volumes which at some future date will accrue to the account of the organization. The issue of ownership of hydrocarbon volumes while they remain in the reservoir should be defined by the contract governing their discovery, development, and production, not whether an organization wishes to book or not book "accounting hydrocarbon reserves".

Under a typical Concession type agreement (see Chapter IV), hydrocarbon volumes in the subsurface reservoir, except for the portion to be paid to the mineral owner in the form of royalty, are considered the property of the producer, unless retained contractually by the mineral owner. Under the usual Production Sharing Contract (see Chapter IV), title to hydrocarbons in the subsurface is retained by the government, but will flow to the producer upon production to pay for cost recovery and a share of the profit from the project. Under a Service Agreement (see Chapter VII), title to hydrocarbons in the subsurface is retained by the government, but may flow to the producer upon production to pay the fee earned for services rendered to the project. That fee is usually on a per volume of hydrocarbons produced basis and may actually be paid in volumes of hydrocarbons. If non-recourse financing is used (see Chapter VII), hydrocarbon reserves are actually sold to the lending institution for the "loan" received, with the proceeds of future production to be used to repay the loan plus interest. Until the loan is paid-in-full the lending institution owns a portion of the hydrocarbon reserves representing the remaining payments on the loan.

RESERVE TO PRODUCTION RATIO (R/P)

Frequently production is characterized by the ratio of remaining reserves divided by the previous years production or simply R/P. This ratio is the inverse of P/R discussed previously when production profiles were presented. For the case of a constant percentage decline, equations 2.3 and 2.4 can be utilized to estimate the abandonment production rate, which yields an equation for R/P in terms of the decline rate and remaining life for constant percentage decline. That equation is:

$$R/P = [1 - (1-d)^t] / d \qquad\qquad (2.17)$$

Figure 2-12 is a plot of this equation which might be helpful in making quick estimates, checking data for internal consistency, or developing an insight into just what certain values of R/P imply. A very high R/P would indicate there is probably room for further increases in production rate or the current production could be sustained before a decline would be experienced. A very low R/P indicates either a small initial reserve, or that the production is very near abandonment.

FIGURE 2-12
RESERVE TO PRODUCTION RATIO (R/P)

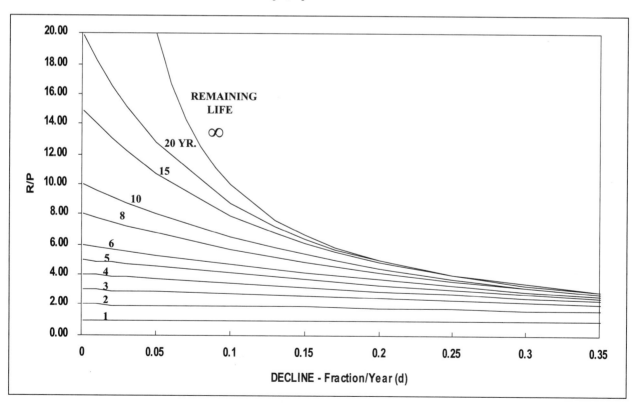

Table 2-6 is a tabulation of R/P for the major oil producing countries of the world and gives a little different prospective on the future of each.

TABLE 2-6
WORLD OIL PRODUCTION AND RESERVES

COUNTRY	ESTIMATED RESERVES** 1/01/08 (MMB)	AVERAGE PRODUCTION 2007 (MMB/d)	R/P (YEARS)	CUMULATIVE PRODUCTION 12/31/07 (MMB)	ULTIMATE RECOVERY (MMB)
United Kingdom	3,600	1.524	7.5	22,882	26,482
Norway	6,865	2.271	9.3	21,186	28,051
Mexico	11,650	3.083	11.4	36,086	47,736
United States	20,972	5.105	12.3	196,035	217,007
Argentina	2,586	0.625	12.3	5,228	7,814
China	16,000	3.739	12.7	35,084	51,084
Indonesia*	4,370	0.839	15.3	21,875	26,245
Malaysia	4,000	0.759	15.4	6,696	10,696
Angola*	9,035	1.697	15.6	6,619	15,654
Egypt	3,700	0.638	16.9	9,958	13,658
Russia	60,000	9.723	17.9	155,203	215,203
Canada***	17,392	2.621	19.2	28,662	46,054
Brazil	12,181	1.748	20.1	9,767	21,948
Oman	5,500	0.710	22.2	8,433	13,933
India	5,625	0.687	23.4	6,765	12,390
Azerbaijan	7,000	0.827	24.2	2,561	9,561
Gabon	2,000	0.230	24.8	3,326	5,326
Yemen	3,000	0.338	25.3	2,272	5,272
Algeria*	12,200	1.358	25.6	14,525	26,725
Ecuador****	4,517	0.499	25.8	4,190	8,707
Sudan	5,000	0.473	30.0	380	5,380
Nigeria*	36,220	2.166	46.8	26,438	62,658
Qatar*	15,207	0.799	53.1	8,167	23,374
Libya*	41,464	1.708	67.5	25,842	67,306
Kazakhstan	30,000	1.088	76.5	6,217	36,217
Saudi Arabia*	266,751	8.625	85.7	113,825	380,576
Iran*	138,400	3.932	97.4	61,421	199,821
Venezuela*	87,035	2.398	100.4	58,307	145,342
UAE*	97,800	2.532	106.8	26,681	124,481
Kuwait*	104,000	2.444	117.6	38,440	142,440
Iraq*	115,000	2.093	151.5	30,718	145,718
Other	21,428	4.994	12.8	60,837	82,265
WORLD TOTAL	**1,170,498**	**72.273**	**45.4**	**1,054,624**	**2,225,122**
TOTAL OPEC*	**927,482**	**30.591**	**84.1**	**432,857**	**1,360,339**
	79.2%	**42.3%**		**41.0%**	**61.1%**

Source: O&GJ December 24, 2007 & March 10, 2008

* OPEC countries

** proved reserves, with present technology and prices, except Russian figures which are "explored reserves" including proved plus some probable reserves.

*** Canada - plus 161.2 billion bbl of bitumen in tar sands

**** Ecuador rejoined OPEC 11/01/07 (not included in OPEC totals)

UAE = Abu Dhabi + Dubai + Ras al Khaimah + Sarjah

Kuwait & Saudi Arabia equally share Neutral Zone

Since 1990 the total world proved crude oil reserves have remained almost constant at about one trillion barrels with OPEC'S reserves constant at about 800 billion barrels, as can be seen in Figure 2-13. This indicates that during that period on a worldwide basis reserve additions have just kept up with production. How long this will continue is the subject of much speculation.

FIGURE 2-13
WORLD CRUDE OIL RESERVES

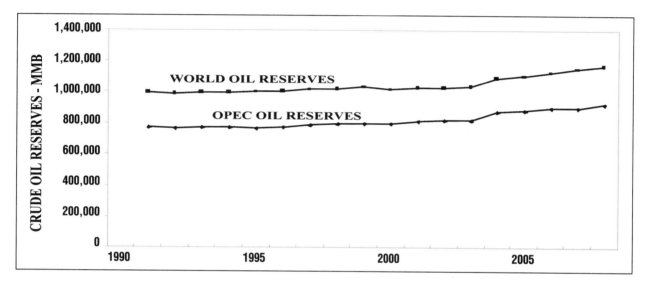

Figure 2-14 is a plot of year-end proved reserves versus production for a number of the large international oil companies. Lines representing constant values of R/P of 6.0, 10 and 14 are also shown. It is interesting to note that most of these companies have an R/P near the average value of 11.1.

FIGURE 2-14
MAJOR COMPANY RANKINGS

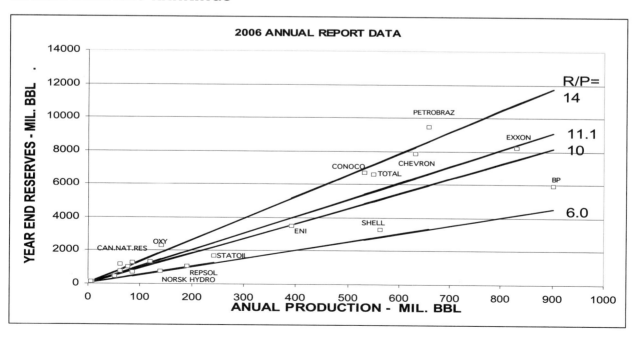

OIL AND GAS RESERVE DEFINITIONS

There are generally accepted rules and procedures for determining oil and gas reserves. The two most commonly used guides are the ones proposed by the Society of Petroleum Engineers (SPE), in association with several other organizations, and that required by the U. S. Securities and Exchange Commission (SEC) of all companies under its jurisdiction. Both of these reserve definitions will be discussed in following sections. The definition of oil and gas reserves as adopted by the Society of Petroleum Engineers and the World Petroleum Conference in 1997, which is presented in Appendix II-A at the end of this chapter, has been generally accepted by the financial community, the petroleum industry and many governments.

Since an oil and gas reserve is just a forecast of future production, the act of booking a reserve does not establish title to those volumes of hydrocarbons as they reside in the ground. Title to those volumes of hydrocarbons is established by the contractual relationship between the owner of the minerals and the producer of those minerals. Reserves may be thought of in two ways. First as a technical estimate, based upon geological and engineering data, it is the volume of oil and gas that will be produced from a reservoir or field over some future period of time. Such a number is necessary to properly design the facilities required for the production of those hydrocarbons and for economic evaluations to justify the expenditure needed to develop those reserves. Second a reserve is an accounting number. It is used directly by the accountants to calculate depreciation, depletion, and amortization, as is discussed in Chapter VIII, and thus has a direct impact on a company's annual profit. It also provides the financial community a measure of the future value of a company. As an accounting number it must recognize the contractual relationship between the owner of the minerals and the producer as well as the technical aspects of the hydrocarbon reservoir. Thus the accounting reserve number will recognize only that portion of the technical reserves which will accrue to the benefit of the producer over the life of the contractual relationship. This benefit may be in the form of actual volumes of hydrocarbons earned by the producer or the equivalent of money paid to the producer. This too will be in accordance with the contractual relationship of the parties.

Procedures for determining hydrocarbon reserves deal with the pricing of those hydrocarbons to establish the point at which production becomes uneconomical as well as the technical information necessary to establish that those volumes can reasonably be expected to be produced. The two most commonly used definitions, SPE and SEC, differ significantly on both of these items, as will be seen in the following brief discussion of these two reserve definitions. Studies have shown that 'proved reserves' determined following the SEC definition are generally lower than that determined using the SPE definition, because of the outdated and restrictive position of the SEC.

Society of Petroleum Engineers (SPE) Reserve Definition

Attempts to standardize reserves terminology began in the early 1040's when the American Petroleum Institute considered classifications for petroleum and definitions of various reserve categories. In 1965 the SPE formally adopted the Definitions for Proved Reserves. Definitions for crude oil and natural gas reserves were codified and approved by the SPE Board of Directors in February 1987. During the World Petroleum Congress (WPC) in June 1994 it was recognized that the effort to establish a worldwide nomenclature should be increased. In March 1997 the SPE and the WPC jointly approved a set of petroleum reserves definitions, which represented a major step forward in their mutual desire to improve the level of consistency in reserves estimation and reporting on a worldwide basis. In April 2007 the SPE Board approved the "Petroleum Resources Management System" (PRMS), which built on, coordinated, and replaced previous definitions. Nothing in the resource definitions should be construed as modifying the existing definitions for petroleum reserves as approved by the SPE/WPC in March 1987, a complete copy of which is presented in Appendix II-B at the end of this chapter. The SPE was joined in this effort by the American Association of Petroleum Geologists (AAPG), the Society of Petroleum Evaluation Engineers (SPEE), and the World Petroleum Council (WPC). The Financial Accounting Standards Board (FASB), International Accounting Standards Board (ASB) and the United Nations Economic Commission for Europe (UNECE) participated in this undertaking.

PRMS is a project-based system, where classification and categorization are separated, it is based upon the evaluated forecast of future conditions and also applies to unconventional resources. PRMS provides the following;

• Affords users greater granularity, the ability to move more-detailed levels of information or to alternative classification and identifying systems for internal analyses

• Allows the use of either deterministic or probabilistic calculation methods

• Formalizes and defines more fully the categories of proven, probable, and possible reserves (termed 1P, 2P and 3P reserves respectively)

• Provides a new definition of contingent resources

The objective of the revised PRMS is to better align the classification guidelines with the actual commercial evaluation processes. There is general agreement that companies make investment decisions based on their internal forecast of those conditions that are projected to apply during the development schedule. Thus, the new system focuses on a base economics case that uses forecast conditions defined by evaluators.

It is also intended that the new system will be appropriate for all types of petroleum accumulations regardless of their in-place characteristics, the extraction method applied or the degree of processing required. Accordingly, the system also specifically recognizes and accommodates unconventional sources.

FIGURE 2-15
PETROLEUM RESOURCE MANAGEMENT SYSTEM (PRMS)

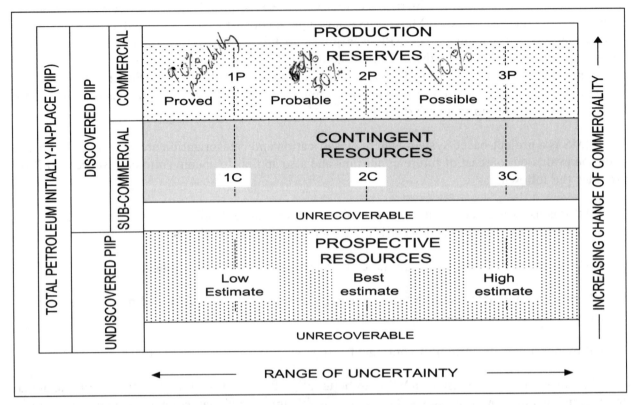

Figure 2-15 is a graphical representation of the SPE/WPC/AAPG/SPEE resources classification framework.

The following definitions apply to Figure 2-15.

TOTAL PETROLEUM-INITIALLY-IN-PLACE is that quantity of petroleum that is estimated to exist originally in naturally occurring accumulations. It includes that quantity of petroleum that is estimated as of a given date, to be contained in known accumulations prior to production plus those estimated quantities in accumulations yet to be discovered (equivalent to "total resource base")

DISCOVERED PETROLEUM-INITIALLY-IN-PLACE is that quantity of petroleum that is estimated, as of a given date, to be contained in know accumulations prior to production.

PRODUCTION is the cumulative quantity of petroleum that has been recovered over a defined time period. While all recoverable resource estimates and production are measured in terms of the sales product specifications, raw production quantities are also measured to support engineering analyses based on reservoir voidage.

RESERVES are those quantities of petroleum anticipated to be commercially recoverable from know accumulations from a given date forward under defined conditions. Reserves must satisfy four criteria; they must be discovered, recoverable, commercial, and remaining based on the development project(s) applied. Reserves are further subdivided in accordance with the level of certainty associated with the estimates and their development and production status.

Proved Reserves = 90% probability
Probable - 50% probability
Possible - 10% probability

CONTINGENT RESOURCES are those quantities of petroleum estimated, as of a given date, to be potentially recoverable from known accumulations, but which are not currently considered commercially recoverable. Contingent resources may include, for example, projects for which there are no current viable markets, or where commercial recovery is dependent on technology under development, or where evaluation of the accumulation is still at an early stage. Contingent Resources are further subdivided in accordance with the level of certainty associated with the estimates and may be sub-classified by the status of the applied development project(s).

UNRECOVERABLE includes those quantities of petroleum which are estimated, as of a given date, not to be recoverable from naturally occurring accumulations. A portion of these remaining in-place quantities may become recoverable in the future as commercial circumstances change and/or technological advances are made.

UNDISCOVERED PETROLEUM-INITIALLY-IN-PLACE is that quantity of petroleum estimated, on a given date, to be contained within accumulations yet to b e discovered.

PROSPECTIVE RESOURCES are those quantities of petroleum estimated, as of a given date, to be potentially recoverable from undiscovered accumulations by future development projects. Prospective Resources have both an associated chance of discovery and a chance of development. Prospective Resources are further subdivided in accordance with the level of certainty associated with recoverable estimates assuming their discovery and development.

ESTIMATED ULTIMATE RECOVERY (EUR) is not a resources category as such, but a term that may be applied to an individual accumulation of any status/maturity (discovered or undiscovered) to define those quantities of petroleum estimated, as of a given date, to be potentially recoverable from the accumulation under defined conditions, plus those quantities already produced.

The full text of Petroleum Reserves and Resources Classification, Definitions and Guidelines is available at www.spe.org

U.S. Securities and Exchange Commission (SEC) Reserve Definition

The SEC Reserve Definition has not been revised since 1978 and thus does not recognize the technological advances that have occurred since then, which have improved the geological and engineering evaluation of petroleum reservoirs. These technologies can help address some of the uncertainties that were inherent in 1978, thereby providing the investor with a much more reliable disclosure of recoverable volumes of hydrocarbons. However, most large non-state owned international oil companies and many smaller ones must use the SEC definition. Use of this definition by all oil companies that come under the jurisdiction of the SEC, which includes any company whose stock is traded on any of the U.S. stock exchanges, is mandated by the Sarbanes-Oxley Act of 2002 for the required public disclosure of oil and gas reserves.

The SEC Reserve Definition recognizes only a single category, proved, but does subdivide it into developed and undeveloped. This definition does not associate a probability with proved reserves, but explicitly presents the criteria which must be met for hydrocarbons to be classified as proved reserves. Therefore, most publicly disclosed reserves are limited to "proved reserves".

In 1978 natural gas was sold through long-term contracts with defined pricing, and crude oil was marketed through refinery postings. The current volatility in oil prices is the result of the development of a spot market, unknown in 1978, and often requires wide swings of proved reserve estimates due to the application of specified year-end pricing. This volatility in quantities is confusing to the investor.

The following is excerpted from the SEC Reserve Definition;

(a) Proved oil and gas reserves are the estimated quantities of crude oil, natural gas, and natural gas liquids which geological and engineering data demonstrate with reasonable certainty to be recoverable in future years from known reservoirs under existing economic and operating conditions, i.e., prices and costs as of the date the estimate is made. Prices include consideration of changes in existing prices provided by contractual arrangements, but not an escalation based upon future conditions.

The determination of reasonable certainty is generated by supporting geological and engineering data. There must be data available which indicate that assumptions such as decline rates, recovery factors, reservoir limits, recovery mechanism and volumetric estimates, gas-oil ratios or liquid yield are valid.

If oil and gas prices are so low that production is actually shut-in because of uneconomic conditions, the reserves attributed to the shut-in properties can no longer be classified as proved and must be subtracted from the proved reserve data base as a negative revision.

(b) Reservoirs are considered proved if economic producibility is supported by either actual production or conclusive formation test. The area of a reservoir considered proved includes that portion delineated by drilling and defined by gas-oil and/or oil-water contacts, if any, and the immediately adjoining portions not yet drilled, but which can be reasonable judged as economically productive on the basis of available geological and engineering data. In the absence of information on fluid contacts, the lowest known structural occurrence of hydrocarbons controls the lower proved limits of the reservoir.

In order to attribute proved reserves to legal locations adjacent to such a well (i.e. offsets), there must be conclusive, unambiguous technical data which supports reasonable certainty of production of such volumes and sufficient legal acreage to economically justify the development without going below the shallower of the fluid contact or the LKH (lowest known hydrocarbons). In the absence of a fluid contact, no offsetting reservoir volume below the LKH from a well penetration shall be classified as proved.

(c) Reserves which can be produced economically through applications of improved recovery techniques (such as fluid injection) are included in the "proved" classification when successful testing by a pilot project, or the operation of an installed program in the reservoir, provides support for the engineering analysis on which the project or program was based.

(d) Proved developed oil and gas reserves are reserves that can be expected to be recovered through existing wells with existing equipment and operating methods. Currently producing wells and wells awaiting minor sales connection expenditures, recompletion, additional perforations or borehole stimulation treatment would be examples of properties with proved developed reserves, since the majority of the expenditures to develop the reserves has already been spent.

(e) Proved undeveloped oil and gas reserves are reserves that are expected to be recovered from new wells on undrilled acreage, or from existing wells where a relatively major expenditure is required for recompletion. Reserves on undrilled acreage shall be limited to those drilling units offsetting productive units that are reasonably certain of production when drilled.

Reserves cannot be classified as proved undeveloped reserves based on improved recovery techniques until such time that they have been proved effective in that reservoir or an analogous reservoir in the same geologic formation in the immediate area.

The full text of the SEC reserve definition and interpretations can be obtained at; www.sec.gov/divisions/corpfin/guidance/cfactfaq.htm

REFERENCES

Amyx, J. W., Bass, D. M., Jr., and Whiting, R. L., Petroleum Reservoir Engineering, 1960, McGraw-Hill, New York.

Arps, J. J., "Decline Curve Analysis," Trans., AIME, 1945, p. 228.

Campbell, John M., Oil Property Evaluation, 1959, Prentice-Hall, Englewood Cliffs, NJ.

Campbell, John M., "Petroleum Evaluation for Financial Disclosures," 1982, Campbell Petroleum Series, Norman, OK 73069.

Craft, B. C., and Hawkins, M. F., Applied Petroleum Reservoir Engineering, 1959, Prentice-Hall Englewood Cliffs, NJ

Cutler, W. W., Jr., "Estimation of Underground Oil Reserves by Oil-Well Production Curves," Dept of Interior, Bull. 228, 1924

Fetkovitch, M. J., "Decline Curve Analysis Using Type Curves," 1972, AIME, New York

McKechnie, G., Energy Finance, 1983, Euromoney Publications, London EC4

McCray, Arthur W., Petroleum Evaluations and Economic Decisions, 1975, Prentice Hall, Englewood Cliffs, NJ

McGill, R. E., "Business Side of Geology Series," AAPG Explorer, April, 1989, p. 46

Slider, H. C., Worldwide Practical Petroleum Reservoir Engineering Methods, 1983, PennWell, Tulsa, OK 74101

Thompson, R. S., and Wright, J. D., Oil Property Evaluation, 1983, Thompson-Wright Assoc., Golden, CO 80402

APPENDIX II-A
MATHEMATICS OF PRODUCTION DECLINE AND RECOVERIES

The following equations were derived from the basic decline and cumulative production relationships.

Constant Percentage Decline (Step Function)

$$\overline{q}_t = \overline{q}_1(1-d)^{(t-1)} \tag{2.3}$$

$$N_p = \frac{\overline{q}_1 - \overline{q}_t}{d} + \overline{q}_t \tag{2.4}$$

Relationships derived from Eqn. 2.3 and 2.4

$$d = 1 - \left(\overline{q}_t / \overline{q}_1\right)^{1/(t-1)}$$

$$d = \left(\overline{q}_1 - \overline{q}_t\right) / \left(N_p - \overline{q}_t\right)$$

$$d = \overline{q}_1\left[1 - (1-d)^t\right] / N_p \quad \text{(Iterate solve for d)}$$

$$t = \left[\text{Ln}\left(\overline{q}_t / \overline{q}_1\right) / \text{Ln}(1-d)\right] + 1$$

$$t = \left\{\text{Ln}\left[1 - \left(dN_p / \overline{q}_1\right)\right]\right\} / \left[\text{Ln}(1-d)\right]$$

$$t = \text{Ln}\left(\overline{q}_t / \overline{q}_1\right) / \text{Ln}\left\{1 - \left[\left(\overline{q}_1 - \overline{q}_t\right) / \left(N_p - \overline{q}_t\right)\right]\right\} + 1$$

$$\overline{q}_1 = \overline{q}_t(1-d)^{(1-t)}$$

$$\overline{q}_1 = \left(N_p - \overline{q}_t\right)d + \overline{q}_t$$

$$\overline{q}_1 = dN_p / \left[1 - (1-d)^t\right]$$

$$\overline{q}_t = dN_p / \left[(1-d)^{(1-t)} - (1-d)\right]$$

$$\overline{q}_t = \left(\overline{q}_1 - dN_p\right) / (1-d)$$

$$N_p = \overline{q}_1\left[1 - (1-d)^t\right] / d$$

$$N_p = \overline{q}_t(1-d)\left[(1-d)^{-t} - 1\right] / d$$

$$N_p = \left\{\left(\overline{q}_1 - \overline{q}_t\right) / \left[1 - \left(\overline{q}_t / \overline{q}_1\right)^{1/(t-1)}\right]\right\} + \overline{q}_t$$

Exponential Decline

$$q_t = q_i(e)^{-at} \qquad (2.3a)$$

$$N_p = (q_i - q_t) / a \qquad (2.4a)$$

Relationships derived from Eqns. 2.3a and 2.4a

$$a = \text{Ln}\,(q_i / q_t) / t$$

$$a = (q_i - q_t) / N_p$$

$$a = (q_i / N_p)(1 - e^{-at}) \quad \text{(Iterate to solve for a)}$$

$$t = \text{Ln}(q_t / q_i) / (-a)$$

$$t = \{\text{Ln}[1 - (aN_p / q_i)]\} / (-a)$$

$$t = N_p[\text{Ln}(q_i / q_t)] / (q_i - q_t)$$

$$q_i = q_t\,(e)^{at}$$

$$q_i = a\,N_p + q_t$$

$$q_i = aN_p / [1 - (e^{-at})]$$

$$q_t = aN_p / [(e^{at}) - 1]$$

$$q_t = (q_i - aN_p)$$

$$N_p = q_i(1 - e^{-at}) / a$$

$$N_p = q_t(e^{at} - 1) / a$$

$$N_p = t(q_i - q_t) / [\text{Ln}(q_i / q_t)]$$

APPENDIX II-B
PETROLEUM RESERVES DEFINITIONS

SOCIETY OF PETROLEUM ENGINEERS (SPE) AND WORLD PETROLEUM CONGRESSES (WPC)

PREAMBLE

Petroleum is the world's major source of energy and is a key factor in the continued development of world economies. It is essential for future planning that governments and industry have a clear assessment of the quantities of petroleum available for production and quantities which are anticipated to become available within a practical time frame through additional field development, technological advances, or exploration. To achieve such an assessment, it is imperative that the industry adopt a consistent nomenclature for assessing the current and future quantities of petroleum expected to be recovered from naturally occurring underground accumulations. Such quantities are defined as reserves, and their assessment is of considerable importance to governments, international agencies, economists, bankers, and the international energy industry.

The terminology used in classifying petroleum substances and the various categories of reserves have been the subject of much study and discussion for many years. Attempts to standardize reserves terminology began in the mid 1930's when the American Petroleum Institute considered classification for petroleum and definitions of various reserves categories. Since then, the evolution of technology has yielded more precise engineering methods to determine reserves and has intensified the need for an improved nomenclature to achieve consistency among professionals working with reserves terminology. Working entirely separately, the Society of Petroleum Engineers (SPE) and the World Petroleum Congresses (WPC) produced strikingly similar sets of petroleum reserve definitions for known accumulations which were introduced in early 1987. These have become the preferred standards for reserves classification across the industry. Soon after, it became apparent to both organizations that these could be combined into a single set of definitions which could be used by the industry worldwide. Contacts between representatives of the two organizations started in 1987, shortly after the publication of the initial sets of definitions. During the World Petroleum Congress in June 1994, it was recognized that while any revisions to the current definitions would require the approval of the respective Boards of Directors, the effort to establish a worldwide nomenclature should be increased. A common nomenclature would present an enhanced opportunity for acceptance and would signify a common and unique stance on an essential technical and professional issue facing the international petroleum industry.

As a first step in the process, the organizations issued a joint statement which presented a broad set of principles on which reserves estimations and definitions should be based. A task force was established by the Boards of SPE and WPC to develop a common set of definitions based on this statement of principles. The following joint statement of principles was published in the January 1996 issue of the SPE Journal of Petroleum Technology and in the June 1996 issue of the WPC Newsletter:

There is a growing awareness worldwide of the need for a consistent set of reserves definitions for use by governments and industry in the classification of petroleum reserves. Since their introduction in 1987, the Society of Petroleum Engineers and the World Petroleum Congresses reserves definitions have been standards for reserves classification and evaluation worldwide.

SPE and WPC have begun efforts toward achieving consistency in the classification of reserves. As a first step in this process, SPE and WPC issue the following joint statement of principles.

The SPE and the WPC recognize that both organizations have developed a widely accepted and simple nomenclature of petroleum reserves.

The SPE and the WPC emphasize that the definitions are intended as standard general guidelines for petroleum reserves classification which should allow for the proper comparison of quantities on a worldwide basis.

The SPE and the WPC emphasize that although the definition of petroleum reserves should not in any manner be construed to be compulsory or obligatory, countries and organizations should be encouraged to use the core definitions as defined in these principles and also to expand on these definitions according to special local conditions and circumstances.

The SPE and the WPC recognize that suitable mathematical techniques can be used as required and that it is left to the country to fix the exact criteria for reasonable certainty of existence of petroleum reserves. No methods of calculation are excluded, however, if probabilistic methods are used, the chosen percentages should be unequivocally stated.

The SPE and the WPC agree that the petroleum nomenclature as proposed applies only to known discovered hydrocarbon accumulations and their associated potential deposits.

The SPE and the WPC stress that petroleum proved reserves should be based on current economic conditions, including all factors affecting the viability of the projects. The SPE and the WPC recognize that the term is general and not restricted to costs and price only. Probable and possible reserves could be based on anticipated developments and/or the extrapolation of current economic conditions.

The SPE and the WPC accept that petroleum reserves definitions are not static and will evolve.

A conscious effort was made to keep the recommended terminology as close to current common usage as possible in order to minimize the impact of previously reported quantities and changes required to bring about wide acceptance. The proposed terminology is not intended as a precise system of definitions and evaluation procedures to satisfy all situations. Due to the many forms of occurrence of petroleum, the wide range of characteristics, the uncertainty associated with the geological environment, and the constant evolution of evaluation technologies, a precise classification system is not practical. Furthermore, the complexity required for a precise system would detract from its understanding by those involved in petroleum matters. As a result, the recommended definitions do not represent a major change from the current SPE and WPC definitions which have become the standards across the industry. It is hoped that the recommended terminology will integrate the two sets of definitions and achieve better consistency in reserves data across the international industry.

Reserves derived under these definitions rely on the integrity, skill, and judgment of the evaluator and are affected by the geological complexity, stage of development, degree of depletion of the reservoirs, and amount of available data. Use of these definitions should sharpen the distinction between the various classifications and provide more consistent reserves reporting.

DEFINITIONS

Reserves are those quantities of petroleum which are anticipated to be commercially recovered from known accumulations from a given date forward. All reserve estimates involve some degree of uncertainty. The uncertainty depends chiefly on the amount of reliable geologic and engineering data available at the time of the estimate and the interpretation of these data. The relative degree of uncertainty may be conveyed by placing reserves into one of two Principal classifications, either proved or unproved. Unproved reserves are less certain to be recovered than proved reserves and may be further sub-classified as probable and possible reserves to denote progressively increasing uncertainty in their recoverability.

The intent of the SPE and WPC in approving additional classifications beyond proved reserves is to facilitate consistency among professionals using such terms. In presenting these definitions, neither organization is recommending public disclosure of reserves classified as unproved. Public disclosure of the quantities classified as unproved reserves is left to the discretion of the countries or companies involved.

Estimation of reserves is done under conditions of uncertainty. The method of estimation is called deterministic if a single best estimate of reserves is made based on known geological, engineering, and economic data. The method of estimation is called probabilistic when the known geological, engineering, and economic data are used to generate a range of estimates and their associated probabilities. Identifying reserves as proved, probable, and possible has been the most frequent classification method and gives an indication of the probability of recovery. Because of potential differences in uncertainty, caution should be exercised when aggregating reserves of different classifications.

Reserves estimates will generally be revised as additional geologic or engineering data becomes available or as economic conditions change. Reserves do not include quantities of petroleum being held in inventory, and may be reduced for usage or processing losses if required for financial reporting.

Reserves may be attributed to either natural energy or improved recovery methods. Improved recovery methods include all methods for supplementing natural energy or altering natural forces in the reservoir to increase ultimate recovery. Examples of such methods are pressure maintenance, cycling, waterflooding, thermal methods, chemical flooding, and the use of miscible and immiscible displacement fluids. Other improved recovery methods may be developed in the future as petroleum technology continues to evolve.

PROVED RESERVES

Proved reserves are those quantities of petroleum which, by analysis of geological and engineering data, can be estimated with reasonable certainty to be commercially recoverable, from a given date forward, from known reservoirs and under current economic conditions, operating methods, and government regulations. Proved reserves can be categorized as developed or undeveloped.

If deterministic methods are used, the term reasonable certainty is intended to express a high degree of confidence that the quantities will be recovered. If probabilistic methods are used, there should be at least a 90% probability that the quantities actually recovered will equal or exceed the estimate.

Establishment of current economic conditions should include relevant historical petroleum prices and associated costs and may involve an averaging period that is consistent with the purpose of the reserve estimate, appropriate contract obligations, corporate procedures, and government regulations involved in reporting these reserves.

In general, reserves are considered proved if the commercial producibility of the reservoir is supported by actual production or formation tests. In this context, the term proved refers to the actual quantities of petroleum reserves and not just the productivity of the well or reservoir. In certain cases, proved reserves may be assigned on the basis of well logs and/or core analysis that indicate the analogous to reservoirs in the same area that are producing or have demonstrated the ability to produce on formation tests.

The area of the reservoir considered as proved includes (1) the area delineated by drilling and defined by fluid contacts, if any, and (2) the undrilled portions of the reservoir that can reasonably be judged as commercially productive on the basis of available geological and engineering data. In the absence of data on fluid contacts, the lowest known occurrence of hydrocarbons controls the proved limit unless otherwise indicated by definitive geological, engineering or performance data.

Reserves may be classified as proved if facilities to process and transport those reserves to market are operational at the time of the estimate or there is a reasonable expectation that such facilities will be installed. Reserves in undeveloped locations may be classified as proved undeveloped provided (1) the locations are direct offsets to wells that have indicated commercial production in the objective formation, (2) it is reasonably certain such locations are within the known proved productive limits of the objective formation, (3) the locations conform to existing well spacing regulations where applicable, and (4) it is reasonably certain the locations will be developed. Reserves from other locations are categorized as proved undeveloped only where interpretations of geological and engineering data from wells indicate with reasonable certainty that the objective formation is laterally continuous and contains commercially recoverable petroleum at locations beyond direct offsets.

Reserves which are to be produced through the application of established improved recovery methods are included in the proved classification when (1) successful testing by a pilot project or favorable response of an installed program in the same or an analogous reservoir with similar rock and fluid properties provides support for the analysis on which the project was based, and, (2) it is reasonably certain that the project will proceed. Reserves to be recovered by improved recovery methods that have yet to be established through commercially successful applications are included in the proved classification only (1) after a favorable production response from the subject reservoir from either (a) a representative pilot or (b) an installed program where the response provides support for the analysis on which the project is based and (2) it is reasonably certain the project will proceed.

UNPROVED RESERVES

Unproved reserves are based on geologic and/or engineering data similar to that used in estimates of proved reserves; but technical, contractual, economic, or regulatory uncertainties preclude such reserves being classified as proved. Unproved reserves may be further classified as probable reserves and possible reserves.

Unproved reserves may be estimated assuming future economic conditions different from those prevailing at the time of the estimate. The effect of possible future improvements in economic conditions and technological developments can be expressed by allocating appropriate quantities of reserves to the probable and possible classifications.

PROBABLE RESERVES

Probable reserves are those unproved reserves which analysis of geological and engineering data suggests are more likely than not to be recoverable. In this context, when probabilistic methods are used, there should be at least a 50% probability that the quantities actually recovered will equal or exceed the sum of estimated proved plus probable reserves.

In general, probable reserves may include (1) reserves anticipated to be proved by normal step-out drilling where sub-surface control is inadequate to classify these reserves as proved, (2) reserves in formations that appear to be productive based on well log characteristics but lack core data or definitive tests and which are not analogous to producing or proved reservoirs in the area, (3) incremental reserves attributable to infill drilling that could have been classified as proved if closer statutory spacing had been approved at the time of the estimate, (4) reserves attributable to improved recovery methods that have been established by repeated commercially successful applications when (a) a project or pilot is planned but not in operation and (b) rock, fluid, and reservoir characteristics appear favorable for commercial application, (5) reserves in an area of the formation that appears to be separated from the proved area by faulting and the geologic interpretation indicates the subject area is structurally higher than the proved area, (6) reserves attributable to a future workover, treatment, re-treatment, change of equipment, or other mechanical procedures, where such procedure has not been proved successful in wells which exhibit similar behavior in analogous reservoirs, and (7) incremental reserves in proved reservoirs where an alternative interpretation of performance or volumetric data indicates more reserves than can be classified as proved.

POSSIBLE RESERVES

Possible reserves are those unproved reserves which analysis of geological and engineering data suggests are less likely to be recoverable than probable reserves. In this context, when probabilistic methods are used, there should be at least a 10% Probability that the quantities actually recovered will equal or exceed the sum of estimated proved plus probable plus possible reserves.

In general, possible reserves may include (1) reserves which, based on geological interpretations, could possibly exist beyond areas classified as probable, (2) reserves in formations that appear to be petroleum bearing based on log and core analysis but may not be productive at commercial rates, (3) incremental reserves attributed to infill drilling that are subject to technical uncertainty, (4) reserves attributed to improved recovery methods when (a) a project or pilot is planned but not in operation and (b) rock, fluid, and reservoir characteristics are such that a reasonable doubt exists that the project will be commercial,and (5) reserves in an area of the formation that appears to be separated from the proved area by faulting and geological interpretation indicates the subject area is structurally lower than the proved area.

RESERVE STATUS CATEGORIES

Reserve status categories define the development and producing status of wells and reservoirs.

DEVELOPED: Developed reserves are expected to be recovered from existing wells. including reserves behind pipe. Improved recovery reserves are considered developed only after the necessary equipment has been installed, or when the costs to do so are relatively minor. Developed reserves may be subcategorized as producing or non-producing.

PRODUCING: Reserves subcategorized as producing are expected to be recovered from completion intervals which are open and producing at the time of the estimate. Improved recovery reserves are considered producing only after the improved recovery project is in operation.

NON-PRODUCING: Reserves subcategorized as non-producing include shut-in and behind-pipe reserves. Shut-in reserves are expected to be recovered from (1) completion intervals which are open at the time of the estimate but which have not started producing, (2) wells which were shut-in for market conditions or pipeline connections, or (3) wells not capable of production for mechanical reasons. Behind-pipe reserves are expected to be recovered from zones in existing wells, which will require additional completion work or future recompletion prior to the start of production.

UNDEVELOPED RESERVES: Undeveloped reserves are expected to be recovered: (1) from new wells on undrilled acreage, (2) from deepening existing wells to a different reservoir, or (3) where a relatively large expenditure is required to (a) recomplete an existing well or (b) install production or transportation facilities for primary or improved recovery projects.

Approved by the Board of Directors, Society of Petroleum Engineers (SPE), Inc. March 7, 1997

OIL AND GAS PRICES

Politics determines oil price in the short term; economics determines oil price in the long term.

There are five factors that determine the price of crude oil. They are in the order of importance:

1. Market (supply/demand)
2. Reliability (production rate)
3. Location (transportation)
4. Quality (refining cost and yield)
5. Availability (reserves)

The first four items in the above list currently have the greatest effect on crude oil price. Reserves will only exert an influence on price when they are insufficient to support the desired worldwide production rate. Supply/demand must include both crude oil and petroleum products made from it. Crude oil quality reflects the products that can be refined from a particular crude oil and the cost to the refiner to do so. Location will determine the transportation cost to move crude oil and/or petroleum products from the point of production/refining to the customer. Reliability is controlled by production rate and productive capacity, while availability refers to reserves. In addition to the maxim in the title block of this chapter, it can also be said that; productive capacity influences prices in the short term while reserves influence prices in the long term.

THE SUPPLY / PRICE / DEMAND RELATIONSHIP

One of the old sayings in the oilfield is that "crude oil in the field tanks is like a fat steer on the range—it needs to be taken somewhere and made into something useful.* After foodstuffs, crude oil is probably the world's most vital raw material both in terms of volume and value. Crude oil buying, selling and transport is a large and complex business. The worldwide movements of crude oil have to meld the disproportionate local demands of some of the most heavily populated parts of the earth to the oil productive capacities located in other far removed parts of the world.

It should be realized that supply/price/demand are tied together as a package. It is impossible to change any one of these three without affecting one or both of the other two. The relationship

* Ball, Max: "This Fascinating Oil Business" 1949, Bobbs-Merrill Co., New York

between supply and demand effects the price paid for oil and gas. Conversely a significant change in price will affect both supply and demand. At times it is all quite cyclical. For example, a major price increase commonly leads to an increase in supply as the result of additional drilling. The same price increase frequently reduces demand as various conservation measures are taken. The result is an imbalance of supply and demand, which may led to a lower price. The oil industry has experienced several of these cycles over the past 35 years. There are also marked seasonal cycles of consumption – heavy gasoline demand for summer driving and heavy home heating loads in winter – which must be anticipated in planning refinery runs months ahead of time. The gas industry experiences a semi-annual cycle with peak loads in summer and winter and low demand in spring and fall.

The unprecedented price volatility which has occurred since the "price shocks" of the 1970's, and more particularly since crude oil began to trade as a commodity in the 1980's, has had a dramatic impact on the industry. The results have severely affected oil company profits, the revenues of oil-exporting countries, and the availability of investment funds. This, in turn, has led to great ingenuity by the industry and the financial community in their efforts to manage price risk.

Timing of available crude oil supply can be an important pricing factor. Imported oil may take more than a month enroute on the high seas, even more if the owners of the cargo choose to "slow steam" to conserve ship's fuel, and perhaps to wait for an improved price at the receiving end of the voyage. The inverse can also occur in a declining market. This has led to extensive use of options and futures for price hedging for the longer haul crudes. These, and other price risk control techniques, are discussed in detail later in this Chapter.

Longer range timing of available supply is also a principal factor in regard to the industry's exploration program. High prices make funds available for new exploration. When prices plummet exploration is the first activity to be curtailed in order to preserve oil company profits under the western world's system of accounting. The long lead times, five to ten years, between exploration and discovery and actual production imposes another almost irreversible cyclic factor into the equation. High prices beget increased exploration, and supply, which can easily exceed overall demand.

There is no single benchmark pricing source for crude oil. The tremendous volume of trading in crude oils has evolved several major price references. These are Saudi Arabian Arab Light, West Texas Intermediate (WTI), Forties and Brent from the U.K.waters of the North Sea, Fateh from Dubai, and more recently the Urals-Mediterranean for the Russian production entering the western markets. Singapore quotations are also increasingly employed as a reference for crude oil pricing in the Far East.

OIL AS A COMMODITY, OR IS IT?

Two schools persist on the question of whether crude oil is a commodity, like wheat (Theory A), or a unique product whose price can be controlled, perhaps politically, (Theory B). The difference involves the amount of surplus crude oil supply at the particular point in time.

During the 1970's when OPEC dominated the supply situation, with Saudi Arabia as the "swing producer" Theory B predominated. After the price fall of 1986 Theory A took over and the crude oil futures markets of the world have since succeeded in handling crude oil as a commodity. This situation should persist so long as crude oil remains in surplus capacity.

Several earlier efforts at futures trading in crude oil failed. Crude oil was not yet a commodity in the commercial sense. Two Chicago exchanges tried without success. Other efforts in New York, including NYMEX in 1973, were unsuccessful. In London the International Petroleum Exchange has succeeded in trading futures in gas oil and Brent crude oil. More recently, successful commodities trading has been conducted in Singapore using Dubai and Brent crude as its marker, and in Rotterdam based on Brent but permitting a wide variety of substitutions of other crudes in settlement.

The Value of Crude Oil as a Raw Material

A barrel of crude oil of itself is relatively worthless even though it will burn with a low smokey yellow flame. As a raw material, crude oil's mixture of hydrocarbons, of greatly varying molecular weights, has great commercial value. The refining of crude oil into the hundreds of industrial and consumer products is hardly a cottage industry. Petroleum refining has to be done in sufficiently large industrial plants (called refineries) so that even the smallest product volumes can have sufficient economy of scale to permit the whole refining process to be economically feasible.

In the larger oil companies the internal value of each individual type of crude oil as a raw material is subject to intense study and computation by a group of experts. On the basis of laboratory distillations and statistical calculations these refinery engineers determine the optimum volumes and values of all the fractions of a specific crude oil, including, for example, such attributes as the octane number of its gasoline fraction and the pour point of its diesel fuel component.

The location of the production, its transportation requirements, and its eventual market also affect the value of a particular crude oil. The lighter, fairly sweet crudes of North America and the North Sea find favor in the American market where gasoline motor fuels have large demand. The heavier Mid-East crudes fit well into the Japanese product market mix.

Crude oil is considered a transportation fuel, because as much of 90% of it is used to power vehicles; namely gasoline, diesel oil and jet fuel. Modern refineries can make these products from almost any quality of oil, but the cost of doing so is greater for low gravity high sulfur crudes. Thus, there is a definite linkage between the price of crude oil, the cost of refining, and the price of transportation fuels.

CRUDE OIL CHARACTERISTICS

Although the geologist, geophysicist and petroleum engineer may tend to think of all crude oils as alike, the refining engineer knows differently. No two crude oils are physically identical. Table 2-1 in the previous chapter summarized some of these differences. Consequently, the products that can be recovered and manufactured also differ significantly from one crude oil to another. Lighter (i.e., higher API gravity) such as Norwegian North Sea Ekofisk tends to have more gasoline by volume than heavy crudes such as Persian Gulf Dubai Fateh which has proportionately more gas-oil (diesel) and residue cracking stock.

Crude oils from different sources are categorized according to the API gravity (a rule of thumb index of the proportion of straight run gasoline to be expected), and the weight percent of sulfur incorporated in the crude. The sulfur imparts undesirable odor to the refined products and also increases their corrosiveness. A crude oil is classified as a "sweet" crude if it contains less than 0.5% by weight of sulfur and as a "sour" crude if it contains a greater amount. The price for a sour crude may be significantly less than that for a sweet crude of similar API gravity.

With these simple parameters at hand the refinery process engineer first subjects a sample of the crude oil to a laboratory distillation procedure recovering a series of "cuts." These cuts are portions of the original crude oil, each of which has a limited boiling point range. The laboratory distillation "cuts" roughly approximate the crude oil fractions that are drawn off from the various sections of a typical refinery primary distillation column.

The refinery distillation cuts are then individually subjected to a number of physical tests such as vapor pressure, octane number (for the lighter cuts), and sulfur content. This information permits the refinery engineer to begin to obtain a feel for what he can make from the crude oil and the respective volumes of the products to be obtained. At this stage he can also begin to equate the total value of those products against the landed cost of the crude oil. This leads to a determination as to whether the particular crude oil represents a good purchase for the refinery.

In addition to product yield of the crude, the presence of certain impurities will also reduce its value. The most common impurities encountered in oils are sulfur, nickel, and vanadium. The heavy metals (nickel and vanadium) will reduce the effectiveness of catalysts used in many refinery processes and adds significantly to the cost of refining. Sulfur must be removed from refinery products to meet air quality standards and to eliminate the corrosion which it causes. The amount of entrained water is an additional basis for discounting the price. In the international trade saltwater contamination is generally expressed in "pounds of salt per barrel" rather than "BS&W" (basic sediment and water) content as used in North America. All of these factors impact on the actual price that the refiner is willing to pay for each barrel of crude oil regardless of where in the world it is produced.

Some of the other tests which may be conducted on the individual laboratory cuts include the Ni/Va (nickel/vanadium) content in parts per million, the P/N/A (paraffin/nephthene/aromatic) ratio, and the pour point of the heavier cuts. Nickel and vanadium are particularly detrimental to the catalyst life in certain of the refining processes. P/N/A is an index of crude oil type in the fraction and can be used as a guide to its response to specific refining processes. Pour point is obviously a matter of concern with respect to diesel fuels in the wintertime or aviation fuels at high altitudes.

Each of the laboratory cuts and, in fact, the entire crude oil specimen, is made up of a great mixture of hydrocarbons.

Typical crude oil incremental distillation assays for three quite different crude oils are shown in Tables 3A-1, 3A-2 and 3A-3 of the Appendix to this chapter. The two samples which are included in this text as examples represent a distinct contrast in types of crude oils. Ekofisk from the Norwegian sector of the North Sea is a relatively light, 43.4 API gravity sweet, i.e., sulfur-free, crude. The Dubai

FIGURE 3-1
ETANKER ROUTES TO U.S. GULF COAST

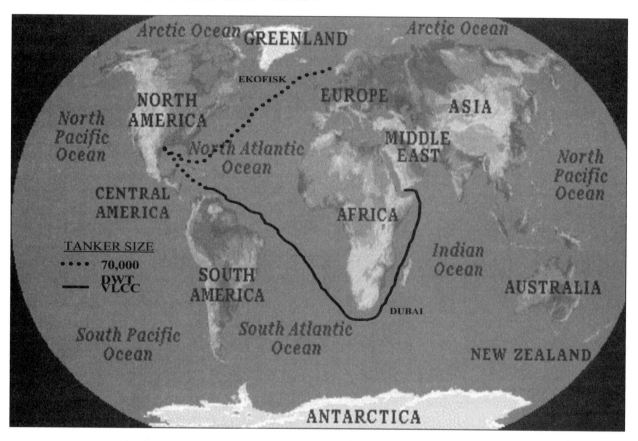

crude from the giant Fateh field in Dubai, U.A.E. is also offshore production. It is a heavy, 31.05 API sour, (i.e., high, in this case, 2.0 percent sulfur) oil, loaded into tankers from an SBM (single buoy mooring) in the Persian Gulf. Both crude oils are produced from offshore fields with correspondingly high costs of operation. The postings of the two crudes are on a more or less world wide pricing basis at their respective field locations. For this example, these are $67.69 per barrel for Dubai, and $70.76 for Ekofisk.

In order to permit our economic comparison of the two crude oils, we have assumed that they are both transported to a typical refinery on the Gulf Coast of Texas for processing. The transportation arrangements to get the two crudes to the refinery differs, and is recognized in the economic analysis.

Shipping of the Ekofisk crude is relatively straight forward. The oil is loaded at a charge of $0.17 per barrel from the terminal at Teeside on the English Coast into 65,000 DWT (dead weight tons) tankers for the voyage directly to the jetty of the refinery in the U.S. for a total transportation cost, including the loading, of $2.73/bbl. Worldscale 70 was assumed.

TABLE 3-1
OIL IMPORT FRIGHT COSTS (DECEMBER 2007)

Source	Discharge	Cargo	Cargo Size 1,000 bbl.	Freight (spot rate) Worldscale	$/bbl.
Caribbean	New York	Distillate	200	202	1.70
Caribbean	Houston	Residual	380	281	2.67
Caribbean	Houston	Residual	500	279	2.65
N. Europe	New York	Distillate	200	330	4.51
N. Europe	Houston	Crude	400	242	4.89
W. Africa	Houston	Crude	910	261	5.79
Arabian Gulf	Houston	Crude	1,900	171	7.05
W. Africa	N. Europe	Crude	910	246	4.05
Arabian Gulf	N. Europe	Crude	1,900	176	5.26
Arabian Gulf	Japan	Crude	1,750	76	1.85
Source: Drewry Shipping Consultants, London, England (O&GJ 1/21/08)					

The last column of Table 3-1 gives an idea of current tanker rates over some of the most well traveled routes in the world. There is obviously an economy-of-scale advantage for the large tankers loaded in the Persian Gulf. Table 3-1 also shows the fraction of Worldscale Freight Rates used to determine each of these costs. These rates are subject to change, reflecting the daily supply/demand situation for tankers.

Worldscale rates represent a uniform base from which freight charges are negotiated for the worldwide shipment of crude oils. Rates are quoted as percent of Worldscale (WS). Worldscale equates the daily earning capability of a tanker independent of any specified route. Thus, if a tanker owner has a tanker in the Persian Gulf and has two offers—one to transport a cargo to Japan and the other to the Caribbean for transshipment, both at Worldscale 100—in theory, he would be indifferent because both offers should create the same net cash flow on a daily basis. A number of other factors may enter the negotiation, not the least of which is the cost of positioning the vessel for its next commercial run. Worldscale 100 rates are adjusted every six months for changes in the costs of operation including bunkering (ship's fuel) and port charges.

Worldscale "points" represent the daily gauge of market fluctuation around the standard, or 100% Worldscale rate. Thus, if the flat, or 100 WS rate per long ton, Persian Gulf to Rotterdam is $27.79, it would convert to $3.71 per barrel of Arabian Light (7.49 bbl per ton of 34°API crude). The spot transport cost at WS 22 for a 200,000 DWT VLCC (very large crude carriers) tanker would be (0.22 * $3.71), or $0.82 per barrel.

The movement of the Dubai crude is a bit more complicated. The oil is loaded from floating storage in the Persian Gulf into VLCC's at Worldscale 25 and later transshipped off Bonaire in the Caribbean for $0.17 per barrel into 70,000 DWT LR-1's at WS 75 for the remaining voyage to the refinery at a total transportation cost of $1.94/bbl.

EXAMPLE 3-1
CRUDE OIL EVALUATIONS

Develop an economic comparison of the two crude oils from the Ekofisk complex offshore Norway and U.A.E., offshore Dubai. The results of simple laboratory distillations of both crudes are presented in tables 3A-1 and 3A-2.

1. ASSUMPTIONS:

Both crude oils are assumed to be processed separately in a typical refinery with the following configuration:

Atmospheric Crude Unit - which separates the various hydrocarbons in the crude oil into fractions having similar boiling ranges

Vacuum Unit - which completes the separation in the heaviest end of the barrel

Reformer - which employs heat and catalysts to effect the rearrangement of certain hydrocarbon molecules without altering their total composition appreciably. This changes the low octane gasoline from the first distillation into a higher octane product.

Catalytic Cracking Unit - which converts medium heavy hydrocarbons into lighter ones using a catalyst (usually platinum) along with heat to produce high octane gasoline. Alkylation - which is also a catalytic process that combines lighter hydrocarbons such as butanes into hydrocarbons of the gasoline range.

Hydrotreating - which chemically adds hydrogen and removes sulfur from the stream being treated. Most of the sulfur is then emitted as hydrogen sulfide into the refinery's fuel gas stream. Hydrotreating is important in treating the naphtha feedstock to the catalytic reformer. Sulfur, nitrogenous compounds and metals poison, i.e., deactivate, the reformer catalyst.

Along with the crude, the refiner must purchase other charging stocks which are readily available from the producing side of the oil business such as normal and iso-butane and natural gasoline in order to meet product specifications in an economical manner.

These added purchases are important in adjusting for seasonal variation in gasoline vapor pressure requirements and as supplemental feed to the alkylation unit. This blending and processing supplemental charge stocks in addition to the crude oil is an integral part of oil refinery economics

The product specifications assumed in the analysis are those of a U.S. Gulf Coast, or Mid-Continent refiner. The product prices are wholesale, or "rack prices." Similar calculations can, of course, be performed for a European, or other refinery location with different transportation costs and product demand.

Most often production refinery runs consist of blends of different crudes, so that the product values are somewhat more complicated than the simple analysis of a single crude oil.

TABLE 3-2

COMPOSITION OF EKOFISK AND DUBAI CRUDE OILS (from Appendix Tables 3A-1 & 3A-2):

	43.4°API Ekofisk		31.05°API Dubai	
	Boiling Range, °F	Volume %	Boiling Range, °F	Volume %
Propane/Butane		4.4		0.5
Straight run gasoline	ibp 176	11.9		
Naphtha	176-302	20.2	59-293	16.4
Kerosene	302-400	10.6	293-446	15.0
Light Gas Oil (Diesel)	400-662	23.4	446-662	36.6
Gas oil (catalytic cracker feed)	302-707	37.3	347-752	39.9
Vacuum residue	707+	25.2	752+	45.0
Fuel gas, BFOE*		0.6		0.1

*Barrels of fuel oil equivalent. This is a BTU conversion of fuel gas to barrels of No. 6 fuel oil on the basis of 6.4 million BTU's per barrel. Since Ekofisk yields more fuel gas than Dubai, the liquid yield from Ekofisk is slightly lower. **This is a blend of feed stock materials from other cuts and is not additive to the total volumes.

Two factors favor the economics of Ekofisk:

1. Ekofisk is a lighter crude oil, 43.4°API. It contains more valuable lighter cuts, and less of the "bottoms" of the heavier distillate fractions of Dubai.

2. It is a sweet crude, meaning that products made from Ekofisk meet sulfur specifications without hydrotreating (sulfur removal). The much heavier Dubai crude, 31.05°API requires hydrotreating which involves higher processing costs.

Conversely, other factors favor the economics of Dubai:

1. Dubai crude oil contains more gasoil, which, when cracked in the FCC (fluid catalytic cracking) unit, produces a substantial yield of light material for the motor fuel mix. Typically, liquid product yield from the catalytic cracking unit equals or exceed 100 volume percent.

Refining engineers generally prefer to utilize graphical plots of these data as shown in Figure 3-2 to compare crude oil distillation characteristics. This plot shows that it takes a higher temperature (more heat) to progressively vaporize the hydrocarbon mix of the heavier Dubai crude oil.

2. This particular crude oil also contains less low octane naphthas which require reforming in the catalytic reformer in order to increase octane number. The reforming process typically yields 85 to 92 percent high octane gasoline depending upon the reforming severity. There is, consequently, an overall liquid yield loss from the reforming operation.

FIGURE 3-2
CRUDE OIL DISTILLATION CURVES

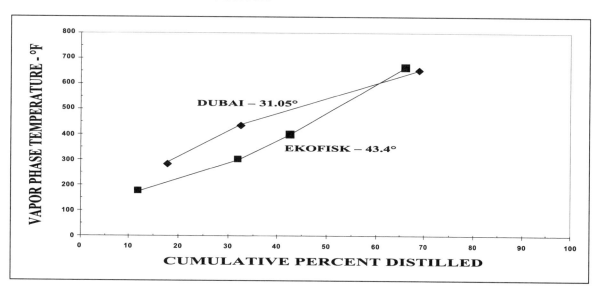

A simple look at the crude oil compositions does not permit an immediate conclusion regarding the relative profitability of the crudes and thus a complete economic evaluation is necessary.

Comments:

Dubai permits the blending of larger volumes of much cheaper iso-butane and n-butane than is the case with Ekofisk. Thus, more total barrels of crude oil can be processed. This reduces slightly the fixed expenses per barrel both of crude oil and total charge.

Dubai yields 78.0 volume percent of light products vs. 84.1 percent for Ekofisk. However, due to the high fraction of cat cracker charge in the Dubai crude, a larger proportion of high octane catalytically cracked gasoline is available for blending into unleaded motor fuel.

ECONOMIC ANALYSIS
(EXAMPLE 3-1)

CRUDE COSTS PER BARREL:

	EKOFISK	DUBAI
CRUDE OIL PRICE FOB	$70.76	$67.69
OCEAN FREIGHT	$2.73	$1.94
IMPORT TARIFF	$0.10	$0.10
	$73.59	$69.73

EKOFISK Input:

	%	$/Bbl.	$
CRUDE OIL	98.6	73.59	72.56
ISO-BUTANE	1.4	60.73	0.85
N-BUTANE	0.0	40.23	0.00
	100.00%		73.41

Products:

	%	¢/gal	$/Bbl.	$
LPG*	2.2	161.53		1.49
REGULAR GASOLINE	15.6	208.09		13.63
PREMIUM GASOLINE	31.0	236.34		30.77
#1 FUEL OIL (ATMOSPHERIC KEROSENE)	9.0	230.24		8.70
#2 FUEL OIL (DIESEL)	26.3	220.75		24.38
#6 FUEL OIL (1%S)	2.9		65.25	1.89
#6 FUEL OIL (3%S)	0.0		57.76	0.00
ASPHALT	9.5		43.36	4.12
SALEABLE YIELD	96.5			85.00
REFINERY FUEL AND PROCESS LOSS:	3.5			
	100.0			
GROSS CRACK SPREAD/BBL:				11.59

Refinery Expenses:

	$
FIXED ($/BBL OF CHARGE)	1.48
VARIABLE OPERATING COST	0.66
PURCHASED NATURAL GAS FOR FUEL	
(1298 MM BTU/DAY)	0.34
TOTAL REFINERARY COST PER BARREL OF CHARGE...	2.48

MARGIN:

PER BARREL OF CHARGE $	9.11
PER BARREL OF CRUDE OIL $	9.24

DUBAI FATEH

Refinery Input:

	%	$/Bbl.	$
CRUDE OIL	96.1	69.73	67.01
ISO-BUTANE	3.3	60.73	2.00
N-BUTANE	0.6	40.23	0.24
	100.0		69.26

Products:

	%	¢/gal	$/Bbl.	$
LPG	1.7	161.53		1.15
REGULAR GASOLINE	10.0	208.09		8.74
PREMIUM GASOLINE	29.8	236.34		29.58
#1 FUEL OIL (ATMOSPHERIC KEROSENE)	3.8	230.24		3.67
#2 FUEL OIL (DIESEL)	32.70	220.75		30.32
#6 FUEL OIL (1%S)	0.00		65.25	0
#6 FUEL OIL (3%S)	10.5		57.76	6.06
ASPHALT	8.8		43.36	3.82
SALEABLE YIELD	97.3			83.35
REFINERY FUEL AND PROCESS LOSS:	2.7			
	100.0			
GROSS CRACK SPREAD/BBL				14.09

Refinery Expenses:

	$
FIXED ($/BBL OF CHARGE)	1.48
VARIABLE OPERATING COST	0.74
PURCHASED NATURAL GAS FOR FUEL	0.88
(2293 MM BTU/DAY)	
TOTAL REFINERARY COST PER BARREL OF CHARGE	3.10

MARGIN:

PER BARREL OF CHARGE	10.99
PER BARREL OF CRUDE OIL	11.44

FIGURE 3-3
PROCESS DIAGRAM-DUBAI CRUDE

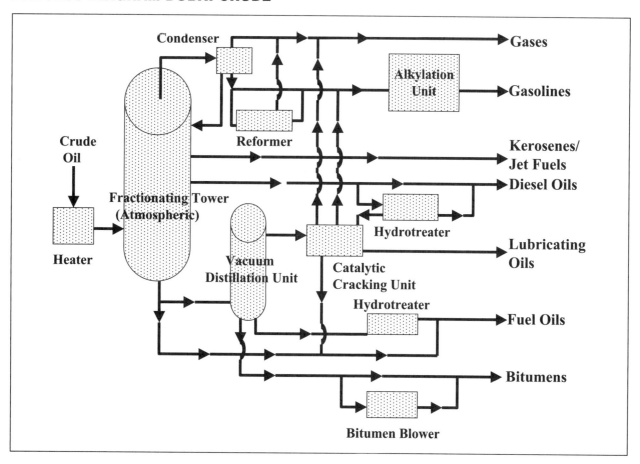

The hydrogen produced in the reformer can be used for fuel gas in the case of the Ekofisk crude which requires no hydrotreating. This is in contrast to the hydrogen requirements for hydrotreating Dubai. The saleable yield of Ekofisk is accordingly lower. This is offset by the lower requirement for outside purchased refinery fuel (natural gas) to process the Dubai crude. The variable refinery operating expenses are also lower for Ekofisk since the hydrotreating unit is not operated.

Most often refinery charges actually consist of mixtures of several different crudes so that the product yields are somewhat more complicated than the foregoing analysis of two individual crude oils.

The foregoing discussion compares two quite different crude oils from different parts of the world in terms of their product yields from a single refinery. It should also be recognized that neither are any two refineries identical as to their physical equipment and product yields. Refining installations may vary. The simplest plants are called hydroskimming refineries, which consist merely of atmospheric distillation and perhaps hydrotreating and reforming units. Skimming plants produce about one third of the crude oil barrel as gasoline.

Much more expensive to build and operate are the so called complex refineries with catalytic cracking and related facilities. These yield 50 to 60 percent gasoline motor fuel. Topping the list in terms of refining capability are the high conversion refineries which squeeze about 72 percent of the crude oil barrel into high priced gasoline. The relative economics of these several levels of refining capability are very sensitive to the prices realized from the various product streams over the life of the plant. Product demand and therefore prices vary seasonally with greater gasoline usage in the warm weather months and heavier heating oil demand in the winter. Refinery operations are switched within a limited range to adjust to these seasonal variations. According to the president of the National Petrochemical & Refiners Association (WSJ 10/16/06) the estimated cost of building a new refinery in the U.S. would be $21,000 per barrel of daily capacity, while the expansion of an existing refinery would cost between $9,000 and $12,000 per barrel of daily capacity. These high costs are at least in part due to strict state and federal environmental regulations.

WORLD OIL PRICING

Prices for crude oil in international trade are universally quoted in U.S. dollars per API barrel of 42 U.S. gallons at 60°F. The derivation of the 42 gallon measure goes back to the earliest days of the industry in Pennsylvania when oil was transported by wagon in used 50 gallon wine barrels. There was a good deal of spillage and the pattern of the trade was to pay for only 42 gallons at the destination without resort to further measurement. Producers soon learned to ship that way as well.

Some international statistics regarding production are quoted in metric tons. This method of measurement derives from Europe, where most crude oil has been received by oceangoing tanker, and weight (displacement) was an easier gauge. Conversion of one measure to the other requires knowledge of the density, or API gravity, of the specific crude oil. Actual monetary settlements are generally made in currencies other than U.S. dollars.

More and more often in the present state of the world economy, and most particularly in the developing world, settlements for crude oil imports may involve barter arrangements for exportable goods from the purchasing country.

The term "world price" occurs frequently in economic discussions of the petroleum business. This is a quite general term. In the eastern hemisphere it is often taken to mean the per barrel price paid for spot, or unscheduled tanker load purchases of crude oil in Rotterdam harbor. Elsewhere it may mean the OPEC posted price for "reference" Saudi Arabian light crude. In a number of oil-importing industrialized countries "world price" may refer to the actual, or historical cost of the average imported barrel of crude oil into that country. For the first seventy five years of the industry after the Drake well in Pennsylvania the prices of domestic North American and overseas crude oils each demonstrated a high degree of stability and similarity.

FIGURE 3-4

AVERAGE WORLD CRUDE OIL PRICES

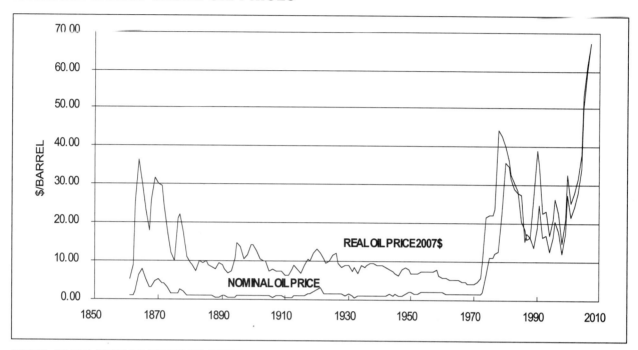

MODEL BASED UPON HOUSTON OIL PRICE AND DISCUSSION

Crude Oil Supply and Demand

In the early days of the industry the price was seen to dip precipitously with each major new oilfield. The most dramatic of these price drops come on the heels of the Spindletop discovery in 1901 and the East Texas Field in 1930. In both of these instances the price plummeted from well over one dollar a barrel to ten cents.

The world's petroleum industry has endured a cyclic supply/price/demand situation since its inception in the last century. Individual cycles of Supply, Demand, and Price operate simultaneously but not in unison. Often outside political and cooperative influences have obscured parts of the cycles.

By the 1930's, when demand for motor fuel had established a much larger and more stable domestic U.S. industry, the price of crude oil held to a remarkably constant one dollar plus per barrel figure.

For nearly forty years, until the early 1970's, the State of Texas through its Railroad Commission and the international major oil companies, dubbed the "Seven Sisters," set the pace of the world oil business. Together they controlled over 80 percent of all the oil produced outside of North America. The U.S., and Texas in particular, was the world's dominant producer of crude oil, and the U.S. the world's largest market. The Railroad Commission of Texas is an elected three man board which in those days set the state's allowable producing rate to conform to the U.S. market demand based on its monthly statewide hearings. Although its legal basis was through the conservation statutes "to prevent surface and subsurface waste," its activity was the principal force stabilizing the price of crude oil, both domestically in the U.S. as well as overseas. Oil was still relatively easy to find and the industry maintained an exploration rate more or less designed to maintain a twenty-year domestic supply.

In 1928 chief executives of the Seven Sisters meeting ostensibly to shoot grouse at Achnacarry Castle near Inverness in the highlands of Scotland drew up the secret (not fully revealed until 1952) "As-if" agreement whereby the world's principal producers outside of the U.S. agreed to cooperate in limiting production to balance demand; to freeze the market shares of each important product in each market; and incidentally, of course, their prices. Under this "Uniform C.I.F. Pricing System" prices throughout the world were quoted as if emanating from the Gulf Coast of the U.S. even though they might be produced elsewhere much closer to their eventual market. This provided a nice "cushion" under company posted prices for the development of Persian Gulf supplies.

Despite this extensive and involved arrangement which the Seven Sisters had created it could not withstand the entry during the 1950's of the independent companies into the world markets with their newly found crude oil supplies particularly in Libya. In fact, Leeman (loc. cit.) observes that "there is little evidence to support the view that the "As-if" agreements continued after the Second World War."

The stabilizing influence of the Texas RRC and the "Seven Sisters" was markedly demonstrated during the closing of the Suez Canal in 1956. Although there was serious disruption of supply, particularly between the Persian Gulf and Europe, there was only the most minor price fluctuation or interruption of supply to the consumer, worldwide.

World War II, and the industrial boom which followed, created rapid growth in demand for crude oil which temporarily obscured the traditional supply patterns. Oil prices in the U.S. were tempered by increased imports of lower priced crudes, primarily from the Middle East and ene uela. Wartime demand for petroleum had encouraged exploration in those areas.

By the latter part of the 1960's Texas production pretty well reached the state's capacity to produce and so the Railroad Commission's ability to set production, and indirectly to stabilize oil prices, diminished. Up until 1973 the U.S. had been a net exporter of petroleum and its products.

All of this relative stability ended suddenly in October of 1973 when the OPEC countries forced their first substantial price increase to take advantage of peak energy demands in the western world coincident with open hostilities in the Middle East. This was a 70 percent increase to $5.11 per barrel. In December of that year OPEC had increased the posted price for Arab Light to $11.65. The resulting history of world oil prices is depicted in Figure 3-4.

OPEC's domination of world pricing lasted for a decade. By 1979, however, the worldwide recession and the results of stringent conservation programs throughout the western world produced a decline in demand each year until 1984 when a slight upward reversal was observed. OPEC's pricing also gave rise to an unparalleled push in the non-OPEC countries to find and develop their own indigenous supplies of crude oil. OPEC had produced half of the free world's oil in the 1970's. The growing non-OPEC production and OPEC's rather desperate effort to hold its price line by curtailing production created an unstable situation which by the mid-1980's saw oil prices sliding and then tumbling to $10 at the end of 1985.

This tremendous buildup of production capacity in the non-OPEC countries and the price drop that followed can be seen in figures 3-4 and 3-8. OPEC was able to raise prices during the 1973 and 1979 crisis, or "oil shocks," as the economists like to call them. During those periods demand . This chart also demonstrates why OPEC was able to raise prices during the 1973 and 1979 crises, or "oil shocks," as the economists like to refer to them. During those periods demand was approximately 80 percent of capacity and OPEC controlled enough of the production to force the price rises. By 1986 the results of the new exploration and development stimulated by the first two "oil shocks" had deprived OPEC of their ability to set world oil prices.

OPEC quickly realized that they no longer had the ability to unilaterally set the price of crude oil and could only influence it by controlling the amount of oil their members produced. This led to the establishment of the OPEC quota system whereby a maximum daily crude oil production rate, excluding condensate, which is approved by its membership is assigned to each member country, based solely upon that country's reserves. Since the mid-1980's most OPEC countries have produced at rates below what they would have in the absence of the quota system, while most non-OPEC countries produced as much as they could. However recently, as the result of political and operational problems, a number of OPEC members have been unable to meet their quotas. For these reasons OPEC 'spare' capacity has declined measurably in the past few years as can be seen in Figure 3-6. 'Spare' capacity refers to production capacity that can be brought online within 30 days and sustained for 90 days. The cost of 'spare' capacity has risen significantly in the past few years, which has further discouraged such development. As of mid-2007, OPEC 'spare' capacity is estimated to be about 3 million barrels of crude per day and may decline to less than 2 million barrels per day by 2012.

An easy way to visualize the relationship between supply, demand and the price of crude oil is to think of it as a beam balance with supply and demand on each end and price as the balance point, as is shown in Figure 3-5. If either supply or demand change without a corresponding change in the other, then the only way that it will balance is with a change in the position of the balance point; i.e., price. Thus it is obvious that supply, demand and price are interlinked such that a change in any one of these items will cause a change in one or both of the other items. As OPEC changes its production quota it will affect both supply and the price of crude oil.

A number of fundamental characteristics of crude oil's unique supply-demand relationship are outlined by Petro Finance in their World Bank report (loc.cit.).

- "Crude oil capacity, either from a single field or from a producing region consisting of numerous fields, is high in the early years of production and then rapidly falls. The discovery of major new fields or producing regions creates waves of new capacity which are not sustainable over an extended period of time. These waves of capacity first operate to crowd out market shares of existing producers, thereby depressing prices, only subsequently to crest and then recede, in the absence of high and sustained rates of reserve replacement.

FIGURE 3-5
SUPPLY / DEMAND / PRICE RELATIONSHIP

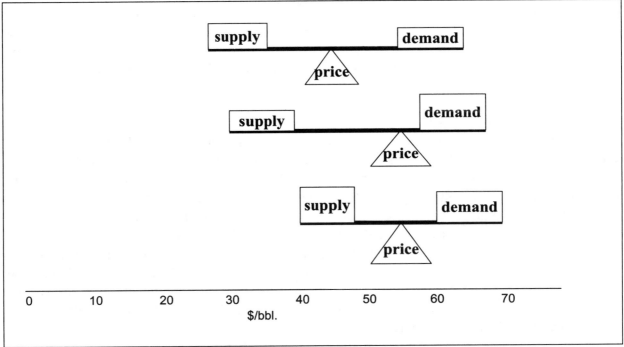

- Crude oil reserves are concentrated in producing regions and large fields within producing regions. The discovery of a new producing region generally triggers rapid exploration and development of the region. The resulting wave of new source production capacity is not necessarily related to the state of excess demand. In short, the crude supply function is lumpy or discontinuous. Rather than increasing smoothly with rising prices, (crude oil) supply capacity increases by fits and starts.

- The petroleum markets are intensely competitive and involve a large number of current and potential future producers. As prices rise, industry activity follows and arrests the decline of production from existing sources, at the same time adding new reserves. Competition within the market ensures that no internal (or market) mechanism capable of smoothing supply exists. This process will continue so long as the price of oil remains at or above current replacement costs.

- The flow of capital to the petroleum industry follows the industry cycle. The cyclical upturn creates a ready pool of new capital available for investment in exploration and production. This flood of new capital derives from both the rising cash flow associated with production from existing capacity and from external capital markets. This capital leads to an increase in real investment but, also, rapid escalation of acreage acquisition and oil finding costs, as the demand for new acreage increases relative to available supply and as the demand for oilfield equipment and services also escalates in comparison to supply.

● Petroleum supply responds to changes in oil prices only after potentially long lags attributable to at least two factors: the discovery and development of new supply sources is subject to long lead times, and industry price expectations are adaptive, slow to respond to changing conditions, and lag those conditions. Industry activity follows the cycle, creating a sustained overhang of capacity in the period immediately following a cyclical peak, and a sustained capacity shortage following a trough as activity lags behind the changing market.

FIGURE 3-6
OPEC 'SPARE' CAPACITY

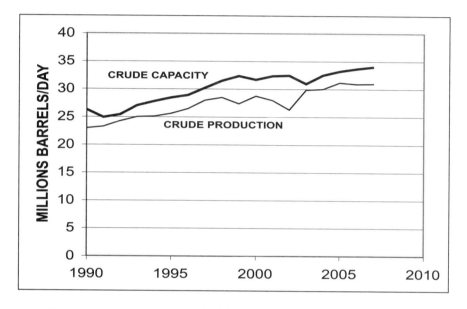

These supply factors cause capacity to overreact consistently during both industry upturns and downturns."

The OPEC oil price shocks of the 1970's had an immediate and very significant impact upon worldwide demand for crude oil, as can be seen in Figures 3-7 and 3-8. Before the OPEC oil price shocks of the 1970's demand was increasing at 7.4% per year, but it dropped abruptly to less than 0.5% per year shortly after the first price increase and has continued on that trend ever since. A growth rate of 7.4% per year was unsustainable, so maybe the oil price shocks just precipitated a change in demand that was inevitable. In any event, the oil industry was abruptly and permanently changed by those events.

The Cyclic Nature of Crude Oil Prices

One school of petroleum economists concentrates its price forecasting approach on the cyclic ups and downs of crude oil prices over the industry's 140 year history. The crude oil price cycles are basically a supply-demand relationship which, in turn, results from the extent of exploration and development investment which concentrates in periods of high oil prices.

Expenditures for exploration have more immediate impact on oil company profit than almost any other segment of its operation. Thus exploration is the first thing to curtail when oil prices and income decline. New exploration then awaits management's perception that improved prices will persist. This affects the supply situation five to ten years hence at which time the demand cycle may be in an entirely different phase.

Oil companies have traditionally increased their exploration and development expenditures during price peaks. This has the effect of increasing supply with consequent price reduction. OPEC endeavored to prolong the price peak in 1980 by establishing quotas in an effort to maintain the high prices.Unfortunately the high prices had also encouraged unprecedented exploration and development in non-OPEC countries at the same time as shown in Figure 3-7. Other political events such as the October War and the Arab Embargo in 1973, the Iranian Revolution in 1979, and the Middle East War in 1991 have each served to accelerate the run up to the price peaks.

FIGURE 3-7
TOTAL WORLD CRUDE OIL PRODUCTION

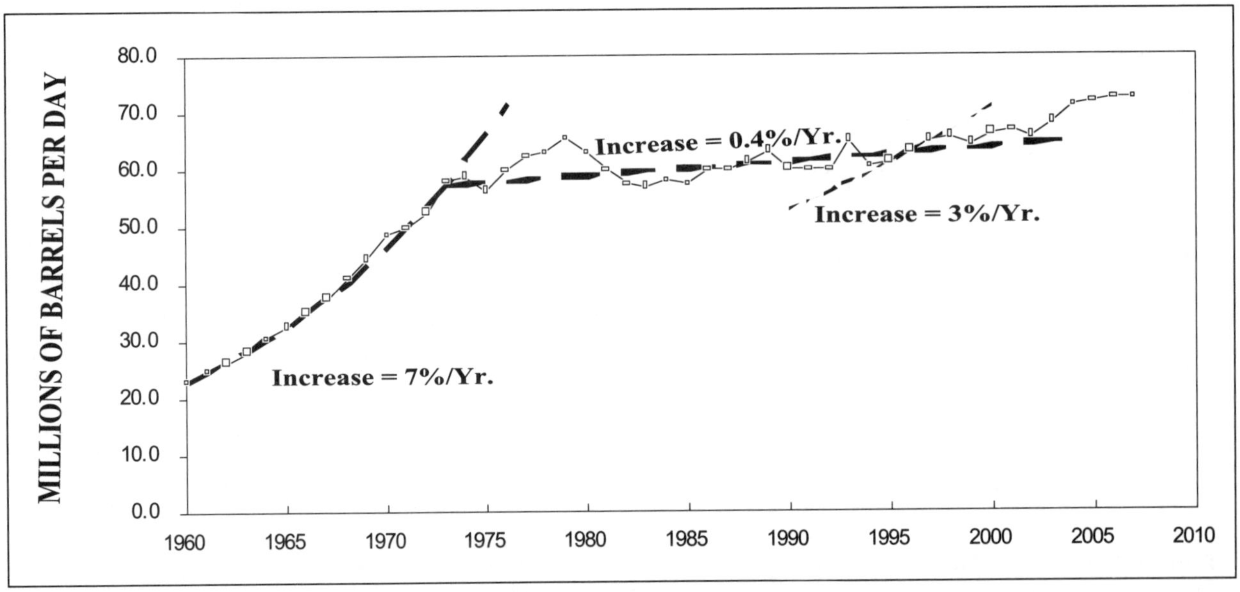

Figures do not include NGL's and oil from non-conventional sources

The industry's cyclical changes are now being felt in ways which are more typical of normal commodity markets with prices more in keeping with short term supply and demand. As Morse (loc.cit.) points out, "OPEC's unwillingness to relinquish the goals and roles that it established in the 1970's has placed its members back in the place they were in before 1973. Once again it is they who have borne an undue share of the cyclical adjustment." OPEC, and particularly Saudi Arabia as its largest producer, has consistently played the role of the industry's marginal, or "swing" producer.

The Middle East Economic Survey for December 6, 1982, in a report sympathetic to OPEC, suggested that the following forms of competition for market share either have, or might be tried, within OPEC:

- Straight price discounts;
- Extended credit terms;
- Netback processing deals by which the exporter gets the income realized from final product sales, sales, after refining and freight costs are deducted;
- Package deals, whereby crude oil is sold at official prices but is tied to sales of refined products or NGLs (natural gas liquids) at discount prices;

- Sales made on a c.i.f. basis (i.e., the seller absorbs the freight costs);
- Improved fiscal terms for those companies still with an equity stake in production.

FIGURE 3-8
WORLD CRUDE OIL PRODUCTION

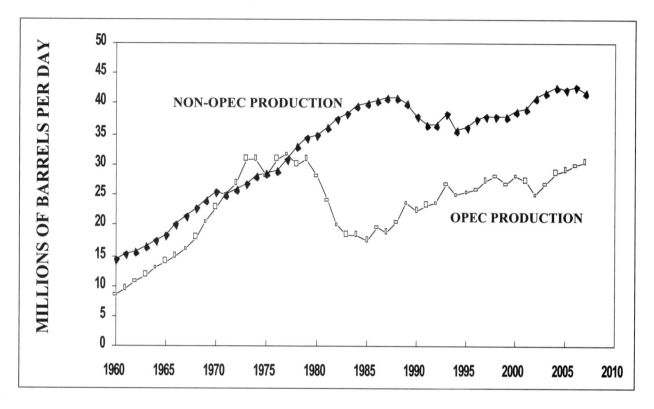

During this period of reduced oil prices OPEC has also had considerable difficulty within its organization in handling the price differentials for crude quality. This is accentuated by the fact that its members produce a variety of types of crude oil. Traditionally high API gravity low sulfur crudes brought premium prices. With the improvements in refining technology and dislocation of supplies most refineries have now invested the money to convert their processes to handle the supposedly cheaper lower gravity sour crudes. This reduces the overall demand for the high gravity sweet crudes which its North African OPEC producers still feel should command a premium price.

Inventory patterns also have a significant effect on the short term supply/demand relationship. Inventories are fundamentally cyclic on a seasonal basis due to the variations in product demand. The large demand for motor fuels occurs during the summer months during which much of the necessary fuel oil production goes to storage. The fuel oil goes to market during the cold weather months making room for gasoline buildups for the following driving season. Superimposed on this seasonal cycle are industry crude oil stocks which increase when it is perceived that prices may rise. The increase and subsequent decrease in inventory levels, which can be of tremendous volume, tends to disrupt the actual production and demand relationship. Statistics on crude oil and products in storage around the world at any point in time are problematic at best.

In the high oil price period of, 1980-85, there was a great deal of effort toward reducing fuel consumption in automobiles and industrial usage in general. Large amounts were spent to permit substitution of alternate fuels. Now a majority of the largest industrial users in North America and Europe can switch on very short notice between fuel oil, coal or natural gas depending merely upon price. Here again statistics on instant switching among alternate fuels are not the best. Crude oil futures prices on the commodities exchanges are directly affected, however, by such statistics as are released periodically regarding both petroleum stocks and fuel substitution.

Transporting Crude Oil

Since crude oil production and local consumer demand are generally far removed from one another, transportation is also a factor in setting the demand for certain crudes. Crude oil has been moved from the oilfield to refineries by wagon, truck, rail car, pipeline, barge, and tankship. Transportation costs vary with the means of transport and volumes. Political jurisdictions, and sometimes the physical characteristics of the crude oil, can also have an effect on its transportation. If there is sufficient daily volume over land pipelines are the cheapest method. Crude oil that can be moved in sufficient volume by pipeline will out-compete those crudes which have to be moved overland by other methods.

Pipelines require large investments, and once installed must be kept full, in order to realize their low throughput costs. In North America pipelines are probably subject to more governmental regulation than are the other forms of crude oil transportation due in large part to their having to cross so many tracts of individually owned land. In other parts of the world operating pipelines have sometimes been permanently shut down due to political problems.

Intercontinental movements of crude oil to market are handled by oceangoing tanker. Economy of scale has been a factor here. The size of crew to man a 500,000 DWT ULCC is not significantly larger than that required to handle a 60,000DWT"handy size" tanker. Nor does the investment vary directly with the size of the vessel. Several factors do effectively limit the practical size of tank ships. First is the depth of the Suez Canal which limits tankers to 250,000 DWT loaded and about 400,000 DWT in light ballast on their eastward backhaul. These limits are expected to rise to 300,000 DWT and 500,000 DWT respectively by 2012. Secondly, is the fact that there are very few sufficiently deepwater ports at the receiving locations to accommodate the largest tankers forcing them to offload to smaller ships, or use offshore mono-mooring facilities.

TABLE 3-3
TYPICAL TANKER TRANSIT TIME IN DAYS

OPEC Sources Middle East	Europe		North America		Japan
	North West	South	Gulf Coast	West Coast	
Via Cape	46	45	52	47	27
Via Suez	26	19	40	—	—
Red Sea	14	7	28	—	35
North Africa	11	4	25	—	38
West Africa	18	16	26	—	45
Caribbean	18	18	8	—	35
Indonesia	—	—	54	32	13
Non-OPEC					
North Sea	2	6	17	—	—
Mexico	18	18	3	—	—

Source: PIW; McKenzie; & Int. Pet. Encyclopedia; based on typical tanker speed of 10 knots, or about 2/3 of design speed.

Tankers do enjoy freedom of the seas and under flags of convenience have wide latitude with respect to governmental regulation. There is also a fair degree of flexibility in operating cost per ton mile depending on the tanker's speed while underway which is controlled by the owner and his captain.

The question is often asked, "Who pays the transportation cost for oil?" The obvious answer is the ultimate consumer of petroleum products as they pay all costs. But the more important question is "Who bears the burden of oil transportation costs?" The answer to that question is the oil producer. Because the price of oil into a refinery is determined by the quality of the crude and supply/demand conditions, not where it is produced. The wellhead price for oil is the refinery gate price minus the cost to transport it there. Houston, Texas is still generally considered the central location for worldwide pricing of crude oil, as indicated by the Worldwide Oil Pricing Model of Figure 3-9.

FIGURE 3-9
WORLDWIDE OIL PRICING MODEL

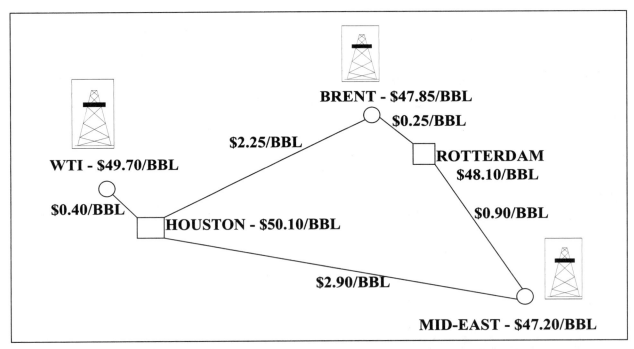

Figure 3-9 shows the general scheme for worldwide pricing of crude. It is based upon the premise that all crude is transported to Houston, even if it is not. The difference in price for equal quality crude oils is then the cost of transportation from that location to Houston, Texas. Over long periods of time this model works fairly well, but there may be local supply/demand situations that upset this scheme from time to time. This model was not established by law or treaty, but by tradition this is the way the oil markets work. As can be seen in Figure 3-12, for three prominent marker crudes, they all move together with the difference in price being roughly the cost of transportation to Houston, as described in Figure 3-9.

MARKETING CRUDE OIL

The buying, selling and trading of crude oil has undergone a number of changes during the industry's approximately 140 year history. The most profound changes have occurred during the past decade, however.

Crude Oil Postings

Since the days following Col. Drake's well in Titusville, which was the start of the U.S. oil producing industry, it has been the practice of the purchasers within the industry, normally the refiners, to publish "Crude Oil Postings." This is the price that the buyer making the posting is willing to pay for each barrel of oil that he takes, or "lifts," provided it meets the specifications for API gravity, sulfur content and BS&W.

These postings are the oil purchaser's way of paying everyone on a uniform basis. This both simplifies the buyer's bookkeeping and keeps him free of possible legal problems from a discrimination standpoint. In most jurisdictions it is necessary to notify everyone affected every time there is a price change. This becomes quite a chore in North America where there are so many operators, joint-interest owners, royalty owners, taxing authorities, etc.

The situation has been somewhat different outside of North America. For many years the producing companies, who were also the purchasers, made similar postings for the purchase of their production. In more recent times host governments have insisted that the crude oil prices for their country's production have to be the subject of formal agreement between the government and the producer. In the OPEC countries the price is established solely between governments. Thus, the seller makes the postings in these countries, rather than the purchaser, which is still the case in North America.

North American agreements for buying crude oil from an individual lease or property according to a specified purchaser's current posting tend to be for thirty days at a time. There is sufficient competition among oil purchasers that there are remarkably few terminations of producer-purchaser agreements. Once the title to a certain number of daily barrels is in the hands of the first purchaser there can be a fair amount of trading and exchange until the barrel is finally delivered to a specific refinery.

Crude Oil Pricing Model

As has been previously discussed and shown in Example 3-1, two major factors in pricing crude oil are its API gravity and the amount of sulfur it contains. The general pricing scheme for crude oil is displayed in Figure 3-10. This shows that generally the highest price paid for crude oil is in the range of 40° to 45° API with reduced prices paid for gravities above and below that range. The slopes of the curve on both sides of the plateau are a function of the construction of the refinery and can change from one refinery to another and from one area of the world to another, depending upon the type crude that the refineries are built to process. Figure 3-10 could be considered the distribution of crude oil price as a function of API gravity for a sweet oil, with a series of parallel curves below it to account for the presence of sulfur compounds in the oil. These lines will move up and down as the spot price of oil changes, but the slopes of the lines will remain relatively constant for long periods of time.

These factors are simply and easily incorporated in the price of crude oil through a commonly used model or formula. This approach to pricing crude oil is sometimes referred to as the "Quality Bank System". The formula, shown below as Equation 3.1, makes a linear adjustment for the oil's API gravity and the percent sulfur it contains.

FIGURE 3-10
NORMAL CRUDE OIL PRICING SCHEME

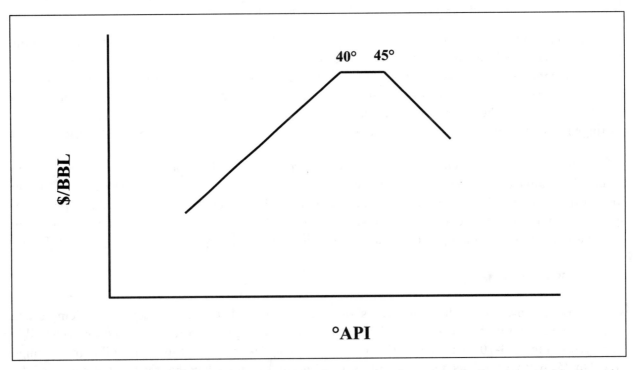

The effect of API gravity and sulfur contamination are simply and easily incorporated in the price of crude oil through a commonly used model or formula. This approach to pricing crude oil is sometimes referred to as the "Quality Bank System". The formula, shown below as Equation 3.1, makes a linear adjustment for the oils API gravity and the percent sulfur that it contains.

eg H₂S.

$$\text{Oil Price / bbl} = \text{Base Price / bbl} + A(°API) - B(\%S) \tag{3.1}$$

Where:

Base price/bbl = Current price for 0° API sweet oil

A = Scale factor for API gravity of the oil - $/°API

B = Markdown factor for presence of Sulfur - $/% Sulfur

FIGURE 3-11
QUALITY BANK EXAMPLE

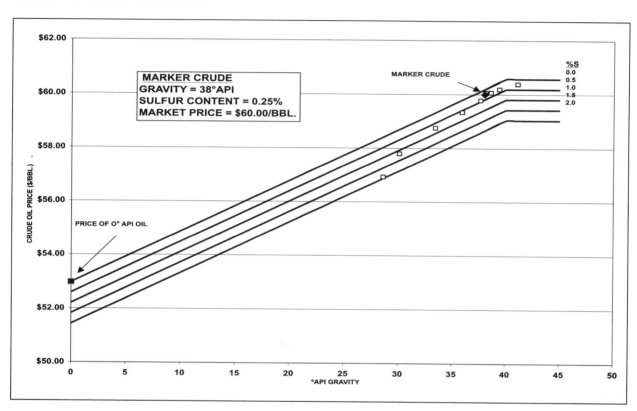

Figure 3-11 is a plot of Equation 3.1 for values of API Gravity below 40° and sulfur contents between 0% and 2%. As is normal for the oil industry, the price of crude is constant between 40° and 45°. The open square symbols represent actual crude oils and as is usually the case, lower API oils have higher sulfur content. The marker crude which was the basis of this graph is identified with a diamond symbol and the theoretical 0° API oil is indicated on the left side of the graph.

Factors "A" and "B" are determined by the refinery or pipeline system buying crude oil, based upon various tests made to determine the value of crude oil to them. The values determined reflect the configuration of the particular refinery(s) and commercial value of the refined products. The base price used will reflect the spot market value of crude (for an oil of 0° API) at any particular time and will change quite frequently, whereas "A" and "B" will remain unchanged for a period of time and may actually be fixed by agreements between the parties.

Long Term Commitments

Firm commitments to buy oil for periods in excess of one year are rather unusual in the industry. A number of older concession agreements overseas did have such provisions. These date from the days when oil was hard to sell and the host governments wanted to make sure that their oil would be produced and sold so that they would realize the income. A variation of this arrangement is the "Service Contract" type of agreement, popular in countries in which internal consumption of crude oil exceeds the country's domestic production. In these arrangements, such as has been popular in effect in Argentina, the operating company is paid a fee on the basis of each barrel found and produced. The company never actually takes title to the oil, and all of the production goes to the state oil company.

This is sometimes categorized as a long term commitment for the sale of production. The fact that the operating company never legally has title to the oil has definite nationalistic political appeal. Early in the 1990s, Argentina dramatically reversed its policy on foreign ownership of its petroleum resources and now leads the region in privatization.

In the domestic North American pattern of the industry the royalty owner has the life-of-the-lease right to take his share of the production in kind. Taking production in kind is not common, but there is ample precedence for doing so. Another long term commitment situation may be involved in the case of carved-out production payments, in which a large volume of oil is in effect sold in place as a means of financing before it is actually produced. A long term commitment ensues requiring the operator to produce the oil and market it, or sometimes to deliver it in kind to that account until the obligation is satisfied. Overrides and carried interests, in which the owner does not share in the costs of the operation, also fall marginally into this long term commitment category.

Calls on Production

As oil became a bit less plentiful on the world market in the late 1970's, another type of contractual arrangement became popular, known as "Call on Production." These arrangements are basically 'First Rights of Refusal' for the purchase of all, or a specific portion of the production from a well, or an operation. These agreements have frequently been incorporated in host government takeovers of operating company interests overseas as a means of buffering the company from a sudden and complete loss of a sizable stream of its production.

Calls on Production can be important to domestic crude oil traders as a hedge, or protection against being caught with a commitment to deliver, with nothing to put into the pipeline. These calls are generally rather vague in respect to the price to be paid for the production, in the event that such action is actually taken. The price may be worked out as a separate negotiation generally related to published spot or commodity future prices for similar production.

SPOT AND CASH MARKETS

There are two basic types of markets in crude oil: the "wet" barrel or cash market where oil is bought and sold in individual deals between buyer and seller, and the futures market where trades are made through a formal commodities exchange for some specified future delivery date. The term "Spot Market" is generally understood to denote a one time short-term transaction. Dealing is done by telephone or telex. The agreed prices for each tanker load of crude traded generally are known only to the buyer and seller. Even the identity of the buyer and seller may only pass by rumor.

Between November 1985 and early April 1986 the price of West Texas Intermediate (WTI) crude oil fell from $32 to $10 per barrel. Fixed term contracts for the sale and purchase of crude became untenable. As a result, the spot market's role in pricing crude oil sales became an increasingly important segment of the petroleum industry. During the past several years spot and spot-related transactions for the sale and purchase of crude oil in the international market have grown from some fifteen percent of all trades in 1980 to over 85 percent.

The history of the development of most commodities in international trade has been to start from individual, or spot market, sales and then to evolve gradually to longer term contracts as the markets become established. The world market in crude oil has had just the opposite history with long term purchase contracts giving way to spot sales as old concessionary agreements have been superseded by national oil company takeovers.

Spot trading takes advantage of the day to day changes in market conditions. The principal disadvantage of spot trading lies in the fact that neither the seller nor the purchaser can predict crude oil prices far enough ahead of time to do much effective planning. As a result, the industry trend now undertaken by most companies is to use a combination of contract and spot trading. The exact mix between contract and spot varies from company to company and from time to time.

Typically a spot sale in the international market involves an entire cargo of crude oil. The trades normally originate at the source, i.e., FOB country of origin. Such a cargo of crude oil may easily be worth $10 million. These trades can hardly be considered as small business undertakings. Several of the larger trading companies buy or sell 200 Brent cargoes, worth in excess of $2 billion, in a month. Response time has to be fast, however. Traders who a few years ago might ponder a deal while phoning around to test the resale market now find that the shelf-life of a transaction may be only a few seconds on a video monitor. Prices in these international trades are invariably quoted in U.S. dollars.

The majority of sales and purchase of crude oil in North America are now spot sales with no enduring contractual undertakings. Thousands of such sales take place domestically in North America every day. These are a normal outgrowth of the traditional simultaneous exchanges of crude oil which have been going on since the early days of the industry in Pennsylvania. Long before any consideration of crude oil as a commodity, the industry recognized that the physical delivery of a producer's oil to his own refinery, which might require moving it right past a competitor's refinery was poor economy. To solve this problem a series of physical crude oil exchanges have evolved to minimize the overall transportation expense. Frequent monetary or volume adjustments were necessary, and expected, in order to permit this traditional North American exchange, or trade system to work. These crude oil exchanges were generally tied to a contract term varying from several months to a year, or more, with continuing "evergreen" provisions.

Spot trading in North America has since evolved merely by dropping the contractual ties and letting the cash transactions take over. Spot sales are generally arranged by telephone in volumes of ten thousand to perhaps 100 thousand barrels per day until the total agreed upon amount is delivered. Although a very large number of contractual arrangements for crude oil sales and deliveries still exist in North America, most contain some provision for frequent price redetermination on the basis of current market reports.

MARKET PARTICIPANTS

The players in the spot marketing of crude oil include:
- The Major international oil companies
- Traders
- Brokers
- Independent oil companies

Major Oil Companies

Through the years the majors looked with some disdain upon any spot marketing of crude oil in the international market. Almost the entire supply of overseas crude oil emanated from large concession areas controlled by the "Seven Sisters." This changed during the Arab Embargo in 1973 when traditional supplies were curtailed and the majors led a mad, but temporary, scramble for any available spot market supplies.

In more recent years the majors have come to rely on spot trading for a significant but constantly varying portion of their crude oil "slate." There is always a mismatch between the locations of a company's production and its refinery requirements. Spot trading serves to alleviate this problem. The general role of the majors in the spot market is to augment, or balance their crude oil supplies.

Traders

Prior to the 1970's there was little opportunity and demand for professional traders in the international crude oil markets. During the early years of that decade, with the growing diversity and complexity of the market, a number of trading companies moved in to fill the gap. Following the 1973-74 crisis there were as many as 300 traders operating in Rotterdam alone. The number declined substantially until the next crisis following the Iranian Revolution in 1979 brought renewed and more sustained activity to the spot market. In the current crude oil marketing environment a new type of trader has emerged. Investment houses now trade and arbitrage oil just as they deal in the more traditional commodities.

Traders assume the risk of taking title to the cargoes, or barrels they are trading. This is referred to as "taking up positions." Once a trader has taken title to a shipment of crude oil he must either resell it, (he hopes for a profit), exchange it, or arrange for its delivery and storage once it reaches its point of delivery. If the trader anticipates a price rise he may hold title to the shipment as long as he feels he can afford to do so. This is known as taking a "long position." Conversely, when prices are falling, the trader might choose to sell a cargo, or a pipeline batch shipment, without having arranged a resale or exchange, in the expectation of subsequently buying the necessary volumes at a lower price, and thus pocketing the difference. This is referred to as "selling short." In another type of activity some traders will arrange with a refiner to process his oil for a fee after which the trader will sell off the products rather than the crude oil raw material.

The industry — major oil companies, NOC's, and others — often use traders as intermediaries for their supply and/or crude oil marketing requirements. The principal trading companies, particularly those with close operational contacts in key producing and refining areas, coupled with crude oil shipping expertise, provide an important service function to the international petroleum industry.

Brokers

Brokerage firms differ from traders in the fundamental distinction that brokers do not take title to the crude oil, but rather bring buyer and seller together for which they receive a commission.

All trading requires current, and if possible, future price information. Otherwise striking a bargain with any confidence is hardly possible. One fundamental problem with spot prices, since they do not emanate from a trading floor in a regulated exchange setting, is that spot price data are necessarily incomplete. Brokers are generally a good source of current crude oil prices in contrast to the traders. The trader is understandably somewhat reluctant to disclose his purchase price to a prospective purchaser. The broker functions more or less from position of neutrality and doesn't have the same problem. It should be recognized, however, that neither the trader nor the broker has any responsibility to report, or publicize his dealings. Such price information as they do release is voluntary on their part.

Independents

The independent oil companies have always played a leading role in the international crude oil spot market. The significant beginning of the Rotterdam spot market coincided with the first production of Libyan crude oil by U.S. independents in the early 1960's. U.S. import restrictions at the time forced this production into overseas markets with little or no help from the "Seven Sisters." Rotterdam became a natural focal point for the spot sales of this Libyan production. As the embargo of 1973 ended, these same independents operating in Libya, with no where else to go, continued to sell more and more of their crude into the Rotterdam spot market.

The Rotterdam Market

Rotterdam has excellent deep water harbor facilities capable of handling large oil tankers. It also has extensive crude oil storage and refining facilities. There has always been an extensive petroleum transshipping and barge trade linking Rotterdam harbor and the Rhine delta to the Netherlands, Belgium, Switzerland, France and Germany. These factors all combined to support Rotterdam's rapid growth as the location of the world's number one spot market in crude oil.

The Rotterdam market is actually just a loose network of traders, brokers and oil companies officing in and around Rotterdam. There is no formal membership or organized trading floor. Deals are made privately between the participants by telephone or telex.

During the 1973-74 embargo the majors quickly turned to the Rotterdam market as buyers in efforts to fulfill their overseas crude oil requirements. This established Rotterdam and its spot market activity as a permanent part of the international petroleum industry. From 1978 onwards, the rapidly increasing oil prices, and the phasing out of the Seven Sister's long term purchase and lifting commitments, further strengthened Rotterdam's position as the world's principal center of spot trading in crude oil. The market also took on a recognition as the only true "free market barometer" of crude oil supply and demand, as reflected in its transaction prices.

Other Spot Markets

Singapore is now developing into the world's second most important petroleum trading center. This Southeast Asian port city and refining center also has excellent location adjacent to the world's most heavily travelled tanker route with good deep water berthing facilities and is well established as a refining, storage and transshipping point. A third region of world scale spot trading, although not so specifically defined as to location, is the Texas Gulf Coast of the U.S. This area encompasses a number of large refineries originally built to process Texas crude for ocean shipment to Europe and the East Coast of the U.S. Most of these plants now refine imported crude. As a consequence, a substantial amount of spot trading takes place in the area both in crude oil and refined products.

MARKER CRUDES

In the 1970's Saudi, or Arab Light, was looked upon as the "marker crude" against which crude oils from other locations with lower and higher quality were universally compared in the world market. This steady stream of production was typical of many crude oils in the Middle East and represented 6.5 million barrels per day in 1977. Currently, OPEC's major reference crude is the "OPEC basket", which is an average of contract prices for all OPEC countries. It is this price which OPEC is trying to keep within a desired range by increasing and decreasing their quota.

In recent years West Texas Intermediate (WTI) in the U.S., Dubai's Fateh crude from the Persian Gulf, and the highly important Brent from the U.K. have become the principal marker, or reference, crudes whose prices are considered to be the most indicative of the world's market conditions. The high visibility of prices being paid for WTI due to its trading on the NYMEX has resulted in its becoming the industry's principal reference crude. This is in spite of WTI being landlocked to the midcontinent and southwest pipeline systems of the U.S. and its being non-exportable. The central delivery point for WTI is Cushing, Oklahoma. Several major pipelines, with a total capacity of over a million barrels a day, converge or cross at Cushing. This permits extensive physical exchange of crudes gathered throughout West Texas and Oklahoma and their ongoing shipment to refineries in the Chicago and mid-western areas of the country or to the refining centers on the Texas Gulf Coast.

Dubai's Fateh is gaining distinction as the principal marker crude from the Persian Gulf area. This crude which is available in substantial volumes, 350,000 BOPD, possesses similar physical qualities to Arab Light.

Brent crude oil blossomed as an important marker crude with the British Government's decision to abolish BNOC as the state trading company. This North Sea field with its over 400,000 daily barrels of production has been involved in the speculative trading of very large volumes of forward production. Spot prices of Brent Blend, primarily a mix of the crudes from the Brent and Ninian pipelines, are monitored closely by the producers in West Africa and the Mediterranean.

Trading in Brent crude began in terms of actual physical, or "wet" barrels. Trading soon evolved to sales and purchases of production forward of the transaction date in 500,000 barrel cargo lots. Crude oil from Brent more than 15 days forward has no physical reality and, so, is known as "paper" barrels.

The great majority of spot crude oil transactions are now concluded on floating prices adjusted for quality from one of the several marker crudes. For example, North African crudes are usually quoted against Brent, although a U.S. refiner might insist on pricing relative to WTI. Equation 3.1 and Figure 3-11 present a method for adjusting crude prices to reflect quality differences.

Figure 3-12 shows the recent price performance of three commonly reported marker crudes; i.e., Brent (North Sea), West Texas Intermediate (@Cushing, Oklahoma), and Arabian Light (Saudi Arabia). It can be seen that they move relative to each other, but supply/demand relationships for each crude varies and for this reason there is not a fixed correlation between these prices.

FIGURE 3-12
MARKER CRUDE PRICES

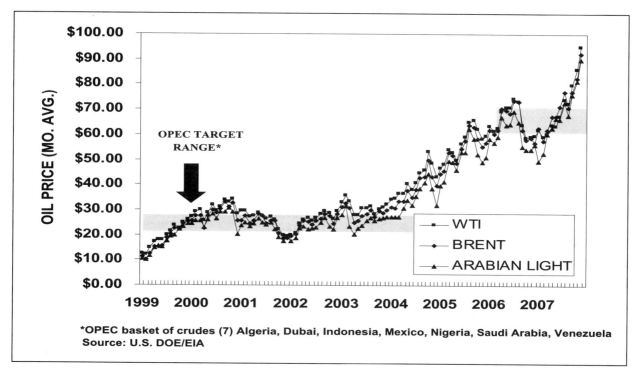

*OPEC basket of crudes (7) Algeria, Dubai, Indonesia, Mexico, Nigeria, Saudi Arabia, Venezuela
Source: U.S. DOE/EIA

A marker crude is an oil from a specific field or region which is traded in spot markets and is considered a standard. Marker crudes are not set by law, any conventions, or even by agreements among producers or consumers, but are generally recognized by the oil trade. They are strategically located throughout the world to facilitate pricing of crude produced from other oil fields in the general area of the marker crude. Marker crudes have the following general characteristics:

1. PERCEIVED TO REPRESENT "FAIR VALUE"
2. IT IS TRADED IN LIQUID AND TRANSPARENT MARKETS
3. THERE ARE A WIDE RANGE OF BOTH BUYERS AND SELLERS
4. ITS SUPPLY IS FREELY TRADABLE
5. THERE ARE ADEQUATE PHYSICAL RESERVES
6. ITS PRODUCTION IS STRATEGICALLY SITUATED
7. IT IS POLITICALLY ACCEPTABLE TO PRODUCERS AND END USERS
8. THE SPOT PRICE IS WIDELY REPORTED
9. IT IS REASONABLY IMMUNE TO MANIPULATION.

Spot markets in marker crudes are much like the stock exchanges of the world. They are essentially self regulated, conducted in full view of anyone who wishes to observe, and the price of all transactions are reported quickly and widely throughout the world. Spot prices respond quickly and efficiently to actual or perceived supply/demand relationships, thus allowing crude oil to be bought and sold as a true commodity.

Price Differentials Against Marker Crudes

The current practice in pricing crude oils in the international market is by means of fairly constant differentials. For instance, April Brent may be quoted as $0.95-$1.00 below WTI. This translates a buyer is willing to pay $1.00 per barrel below the April WTI price to purchase a cargo of April Brent. Alternatively, he offers to sell the same cargo for $0.95/bbl below the April WTI futures price quoted on the NYMEX.

Similarly, Dubai has traditionally been quoted as $2.10 under Brent. Since the Brent market is mainly determined by the WTI prices on the NYMEX, the prices of Dubai crudes are also indirectly determined from the daily closing prices for WTI. Usually an average of closing prices for an agreed period of several days close to the time of delivery is used for settlement.

The price differentials between Brent and WTI, and Brent and Dubai do not remain fixed. In the short term a $1 move in WTI may translate into a corresponding one dollar move in Dubai and Brent, as well as most of the other crudes moving in the spot market. In the longer term, however, these differentials will change in accord with the market's perception of the supply and demand of the individual crude oils. (See Figure 3-12)

FORWARD CONTRACTS

Sales and purchase contracts involving deliveries at some specified time in the future are referred to as "forward trading." This type of arrangement actually dates back to medieval times when firm contracts were sometimes made in the cities of Europe with the trading caravans from the East for goods to be purchased and delivered as much as a year or two hence.

Petroleum transactions in the "forward market" are arranged as freely-negotiated contracts between parties forming a sort of "club" to which the members are co-opted. Members are crude oil producing companies, petroleum refiners, commodity trading companies and investment banks. The transactions are carried out either directly between the two contracting parties, or through brokers over the telephone.

Contracts for the purchase of crude oil need not require its immediate delivery. Many spot trades call for delivery several months hence. They normally specify the point of delivery since this may feature in the settlement price.

Forward trades fall under the general category of spot trades as distinct from the much more formalized and regulated "futures markets." Most forward trades deal in the principal marker crudes: WTI, Brent, and Dubai.

Forward, or paper, transactions afford opportunities to both producers and refiners to insure against future price fluctuations. The producer may sell forward to protect himself against a drop in price. Refiners may buy forward to guard themselves from a possible price rise. The Brent market has become characterized by a number of substantial speculators who interpose themselves between producers and refiners. Although market speculators have a bad name, they might do better to call themselves underwriters since they do serve a very useful and worthwhile purpose in providing stability to the forward market. A producer who wishes to insure himself against a falling market may sell an agreed amount of his future production to a speculator/underwriter. This entrepreneur takes on the price risk, expecting in return, a discount on the current price.

Typically, the speculator/underwriter then proceeds to lay off some or all of the purchase either by sale in Brent forwards at some slightly earlier date, or by sales in other markets, possibly WTI in the NYMEX, or in products or other commodities. Reinsurance deals, as they might be called, of this type account for the large segments of the market. Total transactions have been estimated at 10 to 15 times the volume of actual wet barrel deliveries. Those with actual physical interest (producers and refiners) also participate in holding some degree of self-insurance forward positions. As there is no formal organization of the traders in Brent crude, anyone with the resources to buy an $18 million cargo of crude oil, and a personal acquaintance with one or two participating traders with whom to deal, can join in.

DAISY CHAINS

Forward trading in Brent crude may begin three or four months ahead of the stipulated delivery date. It could start with a trader who assumes a forward commitment to supply crude which he intends to cover at some later date, hopefully at a better price. Soon other participants associate, perhaps to hedge positions in other markets or crudes. Brent positions are in tanker cargo lots of approximately 600,000 barrels.

The great majority of crude oil trades are never actually delivered anywhere. Once the cargo has been designated, if the original purchaser feels that he can sell it, even if it is enroute, and turn a profit, he will normally do so. As a result of so many spot trades being undertaken on a daily basis, a single barrel of crude may change hands many times between the well and refinery. The large number of transactions make the changing ownership of each individual barrel difficult to trace. Because of this sale and resale of so many individual cargoes, or batches, the volume of reported trades always exceeds the volumes of crude oil actually delivered. Still other trades may go unreported. In the U.K., however, between the years 1984-86 "daisy chains" were important to Inland Revenue in determining the tax reference price. The frequency of spot trades depends upon prevailing market conditions and its volatility.

A remarkable example, at a peak of market volatility, is the "daisy chain" of a cargo of Brent crude from the North Sea which was traced by Petroleum Intelligence Weekly in 1984. These trades are diagrammed in Figure 3-13. The buying and selling by 24 trading entities in the 36 transactions in the daisy chain occurred over three months it took to transport the oil to Marcus Hook, Pennsylvania on the East Coast of the U.S.

The twenty four companies involved in the trading included a mix of international oil companies, national oil companies, refiners, and traders. Shell appeared three times (as Shell UK, Shell International and Pecten). BP showed twice, along with Texaco and Chevron once each. The NOC's were Finland's Neste and BNOC (now BP). The U.S based companies included Sohio (now B P America), Oxy, and Sun, who was the eventual buyer. The trader, Phibro, showed up six times along with Charter, once for its own account and four times through its trading affiliate, Acron. The list included two Japanese firms, Idemitsu and C. Itoh.

A typical cargo is seldom subjected to this number of players. The list is indicative, however, of the nature of the chain of ownerships which can develop. Many of the repeaters in these chains probably do not know that they are actually trading for the same cargo. The default early in 1986 of several key players in the London-based Brent market due to price gyrations, combined with unexpected spot shortages of the crude, has tended to reduce the length of these "daisy chains."

FIGURE 3-13
DAISY CHAIN CHART OF CRUDE OIL SPOT SALES

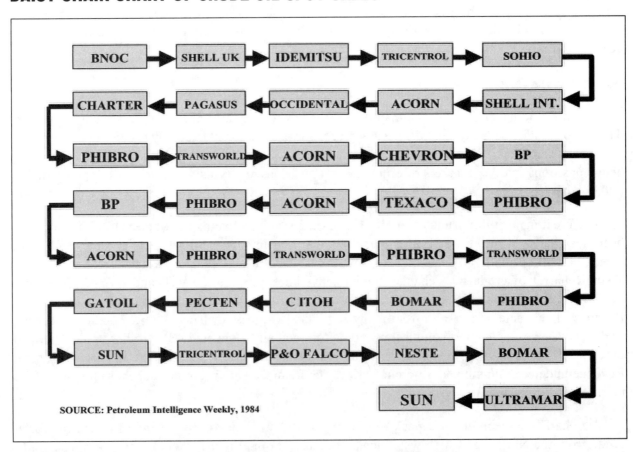

SOURCE: Petroleum Intelligence Weekly, 1984

Although it is an appealing thought, the number of exchanges of ownership that a barrel of crude oil goes through in a given voyage does not necessarily mean that it costs more at the end of a series of trades. Recognizing the volatility of both the futures and spot markets, the price can easily go up and down several times in even a few hours, dragging traders with it in both directions. Several years ago Brent trades might have remained open for a day, or so. Today, Brent trades may have a video screen-life of only seconds.

DATED CRUDE

Spot trading in Brent crude has been an essentially a free market without formal organization or restriction. A standardized contract has emerged, however. Most of its provisions stem from the physical requirements of the terminal at Sollum Voe in the Shetland Islands where the Brent production arrives by offshore pipeline and is transshipped to ocean going tankers. The terminal operator, being confronted with the desires of many buyers and sellers, allocates lifting dates to producers in the middle of the month for the following month with a day or two of flexibility for operational reasons. One of the few provisions of the standard contract that substantially affects the spot trading in Brent crude oil is the requirement that the buyer is entitled to 15 days notice of the three-day window in which he will be required to lift his purchase.

Daisy chains do not terminate with the 15-day notice. Since most traders are not really interested in taking physical delivery, the last days of trading before lifting must eventually establish its ownership for loading. A new daisy chain may then take over while the cargo is in transit.

OTHER SPOT PRICE RELATED DEALS

Tenders

In addition to the now conventional spot sales of cargoes or specific pipeline batches of crude oil there is a growing number of tender sales. A tender sale involves a purchaser, frequently a government entity, who solicits bids for certain volumes of crude oil. These volumes are scheduled for delivery over a period of time at an agreed upon price. The price may be fixed for the term, or may be varied according to a pre-defined source of published price quotes.

Countertrade and Barter

Countertrade can take on a number of varying forms. There have been a series of straight barter deals between governments exchanging crude oil for Boeing 747 airliners, French Mirage jet fighters, lamb from New Zealand, vehicles from Japan, and so on. Another related type of transaction might better be called counter-financing. In these arrangements crude oil is the medium of exchange in return for which the seller agrees to take payment in exportable products of the purchasing country. In other instances crude oil again has served as the medium for paying off past debts as in the case of Iraq in its past obligations to Italy, France, and India. Libya has also made payments in oil to the USSR, Italy and India.

The volatility of crude oil prices has added another dimension of risk to countertrades. British Aerospace reportedly signed a $15 billion deal with Saudi Arabia for 132 aircraft partly in exchange for crude oil. This was in 1986 and the price collapse necessitated renegotiation of the contract with the Saudis agreeing to supply 400,000 bpd instead of the original 300,000.

NETBACK AND FORMULA PRICING CONTRACTS

The restructuring of crude oil marketing also led to the so called "netback" arrangements. In a netback transaction, crude is sold on the basis of the price the buyer expects to receive for his final products, rather than at a price set by the producer at the time of the crude sale.

The most important calculation by a refiner, or trader who is processing crude, is the "netback" to be realized from the sale of the refined products. A crude oil's netback value is the price a refiner is willing to pay for a barrel of crude recognizing refinery configuration and efficiency, anticipated product prices and profit objectives. Netback pricing transfers the downstream market risk to the producer.

Algeria instigated the application of these crude oil pricing formulas during the 1970's. Oil companies operating in Libya also experimented with netback arrangements on a limited basis in the early 1980's in efforts to move host government's equity crude into an unstable market. It was in the summer of 1985, when Saudi Arabia undertook its most intensive netback sales effort, that this type of sales arrangement reached significant proportions in the international petroleum market. This move was occasioned by the Saudi's announced decision to forego their traditional role as OPEC's swing producer and increase their sales to 3.5 million barrels per day. A change in marketing technique of this nature was needed to win back the Saudi's market which had been lost to other OPEC and non-OPEC producers.

Basically, the netback contract sets the purchase price of the crude as the residual of the realization from the sale of all of the products less a series of agreed costs. Most netback deals can be classed as "deemed netbacks"in which the computations are made on an assumed, or deemed, refinery yield. Some netbacks are negotiated on the basis of actual "after the fact" refinery yields and product sales. Another variation is the Realization type of netback agreement in which the seller of the crude oil agrees to take a percentage split of the actual profit from the refining operation and the sale of the products. All of these types of netback deals have five basic constituents:

1. REFINERY YIELDS

2. PRODUCT PRICES

3. TIMING

4. TRANSPORTATION

5. PROFIT MARGIN, AND OTHER FEES

Refinery yield denotes the apportionment of each individually saleable refined product, which when added to the other products, represents the total barrel of refined crude oil.

An agreed published product price reference source provides the value of each refined product yield which is totaled for the whole barrel. This value is known as the crude's GPW (gross product worth). The GPW of a refined barrel is not strictly comparable to the crude oil's spot price at the receiving port. The costs of transportation, refining and some other agreed fees have to be deducted from the GPW to arrive at the crude's netback price.

The timing component of the netback incorporates the agreed time lapse after loading at which the netback computation will be figured. This is typically 10 to 60 days from the time the cargo is lifted. Timing factors also include the number of days over which the product price quotes will be averaged in determining the product values.

Transportation cost allowances can be a significant factor. These are based on "Worldscale" benchmarks which are published rates for transporting crude oil between specific loading and receiving ports. The crude transport factor is normally the cost of a spot market charter of an appropriate sized tanker for a single voyage. This cost will be on the basis of a negotiated fraction of the current Worldscale rate.

Refining costs are generally considered to be the "out-of-pocket" operating expenses. They do not include any refinery capitalization or fixed costs. PIW figures the "out-of-pocket" refining costs approximating $0.40 to $0.65 per barrel in the U.S., depending on the complexity of the particular plant, $0.30 for a major upgraded European refinery, and $0.20 elsewhere.

The marketing costs associated with sales to the ultimate consumer of the products is not a factor in netbacks since it is assumed that the finished products will be moved at wholesale, or "rack" prices from the refinery gate. The rack prices for the lighter finished products are easily matched with published spot and futures market quotes for regular, and premium gasolines. The specifications for diesel and jet fuels are sufficiently stringent that they are priced currently with little difficulty. The value of the fuel oil fraction of the barrel is highly dependent on its sulfur content. This may require an adjustment in realized price for the purposes of netback computation if the fuel oil sulfur content for the particular crude oil exceeds that of the fuel oil sold on the spot market whose quotes are being used as reference.

Saudi Arabia was very effective in using the netback approach in regaining market share for its crude in 1985-86. Other OPEC members took great exception to the approach and Saudi Arabia agreed in late 1986 to drop the technique, at least temporarily. Other members, however, have tried similar methods during periods of tight market conditions. Netback contracts may again become an important marketing tool in the future.

The netback technique has a number of shortcomings, however. It is not truly a price per se, but rather an estimate of a particular crude oil's value to refiners with unique costs and processing capabilities. Netbacks are fundamentally based on calculations of incremental barrels of refinery runs and thus do not directly relate either to the refinery's average costs or profits. Also, the netback calculation relates to a specific spot market quote for a particular marker crude and in many cases does not take into account the necessary location and quality adjustments.

Netback arrangements also have an inherent problem in adapting to the variable summer and winter product mixes that refiners typically produce. The netback technique has been further criticized as providing little incentive for the refiner to improve his processing efficiency. The netback sales made to date have not recognized a capital cost for the downstream refining and marketing segment of the business. Eventually, this aspect will have to be recognized.

FUTURES MARKETS IN CRUDE OIL AND PRODUCTS

These are organized commodity markets as distinct from the forward market already described.

The parties act through brokers who carry out the transaction on the trading floor. Each transaction is recorded. The clearing house, or exchange, defines the rules governing transactions and assures that they are observed by all parties. The clearing house also assumes the role of counterpart responsible to all the players guaranteeing their solvency.

The three principal commodity exchanges for petroleum are:

- The NYMEX (New York Mercantile Exchange) is located in the World Financial Center in lower Manhattan. Its Web site is NYMEX.COM. It trades basic crude oil contract for West Texas Intermediate Oil delivered to Cushing, Oklahoma. Contracts for heating oil and unleaded gasoline are traded for delivery in New York harbor. The volume of WTI traded in a typical day may exceed 150 million barrels. This is more than twice world consumption, and fifty times the daily production of WTI. Brent futures are also traded on the NYMEX.

- The IPE (International Petroleum Exchange) is located adjacent to the Tower of London on the Thames. The basic crude oil contract is for Brent Blend delivered to the marine terminal at Sullom Voe in the Shetland Islands of the North Sea. Heating oils are also traded for delivery at various North Sea ports. The volumes traded on the IPE approximate 5 million barrels per day, considerably less than on the NYMEX.

- SIMEX (Singapore Mercantile Exchange) which trades in Persian Gulf Dubai crude, North Sea Brent crude and heavy fuel oils for delivery in the far east. Neither the IPE nor the Simex provide for physical deliveries in the manner of the NYMEX. Their transactions being entirely in "paper barrels".

A futures contract is a means of transferring price risk. In its simplest terms, a futures market provides the means for a buyer or seller to agree in advance to buy or sell quantities of a commodity at a specified future date at a price established by free open bargaining. Futures trading, as distinct from forward trades, takes place on a physical trading floor with deals struck by open outcry. Trading is restricted to specific times between the official opening and closing of the day's trading. Open outcry means that those permitted to trade on the floor of the exchange face one another somewhat in a ring and bids and offers are shouted aloud until a match is reached. Everyone present is thus aware of the prices and volumes being traded as well as future time of delivery. This information is fully available outside the market to the financial press and internationally within seconds over the Reuters and the Telerate electronic services.

The trading of crude oil futures contracts gets pretty active and spirited with the many hand gestures and required "open outcry." Figure 3-14 seeks to depict some of this, which is obviously somewhat difficult to describe in words.

The contracts are referred to as "cleared," which means that they are entered into with the exchange rather than with a specific wet barrel seller or purchaser. It must therefore conform to definite quantities, quality, delivery point and dates specified in the standard contract. Actual delivery is more a possibility than a probability. Most futures contracts are closed out as paper barrels rather than delivered as wet barrels. The great majority of crude oil trades on either the commodity futures, or the cash spot markets is not delivered anywhere. Rather the trade is closed, or liquidated, by taking an equal and opposite position in the market. What remains after each trade is trading profit—or loss.

FIGURE 3-14
CRUDE OIL FUTURES TRADING IN THE NYMEX

When selling a delivery contract on the futures market the party undertakes to deliver crude oil in accordance with the contractual provisions. Repurchase of this contract enables the selling party to free itself of this obligation. The transaction then reverts merely to an exchange of paper. The futures markets concentrate their attention on providing price assurance rather than security of delivery.

Futures are contracts to take or make delivery, at a future date, of a specified quantity and quality of goods (e.g., crude oil) at a predetermined delivery point, and at a firm price, even though the material being sold may not yet have been produced. The concept is reported to have flourished in Holland during the "tulip mania" about 1636 when bulbs were sold and exchanged with high degrees of speculation in open bidding before the plants had even matured. The rapid growth of trading and exchange of contracts for future deliveries of crude oil by open bidding reflects a fundamental change in the structure of the world oil market.

While spot trading is fundamentally a true free market phenomenon, being uninhibited by regulation, it does suffer from lack of complete and timely publication of price data. Typically, the buyer and seller know each other, their likes and dislikes, negotiating strategies, etc. Futures trading, on the other hand, removes most of the impediments to the access of outside speculators in crude oil trading. The organized futures markets provide standardization, price transparency and impersonality. Participants do not have to know very much, if anything, about the technical and physical nature of petroleum. The clearing house of the futures exchange assumes the role of buyer to all sellers, and the role of seller to all buyers, of its futures contracts. Historically, futures markets in commodities have evolved from spot trading when the commercial need has developed sufficiently to support the more costly and extensive arrangements required for open trading in futures contracts.

The major oil companies have been reluctant to enter into futures trading but now many are recognized traders on the NYMEX. Jack Montgomery, Amoco's former Vice President for Crude Oil Supply, now retired, is quoted in late 1986 (loc. cit.):

"Let's say we need to buy a million barrels of crude oil for refining a month from now. If we think the crude price might go down, we hate to have to buy crude oil today, but we may have to buy it now to allow for delivery time to our refinery. Suppose we would have to buy the crude at a current price of $72, but think the price could drop to $62 by the time the crude gets used.

"What we do is go ahead and buy the million barrels now on the open market at $72 per barrel. Then, on the futures exchange, we sell an equal amount of crude for the current futures price of, say $68. When we run the crude in a refinery a month from now, we buy a million barrels on the futures market, which by then sells for $58 a barrel. If, as we feared, the real per-barrel price of crude has gone down to $62 we lose $10 a barrel on the oil we bought. But when we cancel our position on the exchange we find that we have made $10 per barrel offsetting the loss on the crude.

"If we need to buy some West Texas Intermediate, for instance, all we have to do is to buy what we need on the futures market and take delivery. It's a different kind of market, but it's a market. We use it like any other supply source."

Crude oil with all of its variation in types and quality is now viewed as a basic commodity. As has been the history with other commodities, crude oil futures are now traded on the New York NYMEX, the IPE in London, and the SIMEX in Singapore. This means that crude oil is actually traded around the clock, 24 hours a day, as a world commodity. Some market analysts view this as the return of oil pricing to a true public market for the first time since Spindletop, on the Texas Gulf Coast where Capt. Lucas' gusher, and the ensuing drilling in 1901, caused a severe crash in the market for crude oil. Probably the most important aspect of the futures market in crude oil has been what analysts call "price discovery," i.e., a daily public listing of what the market is willing to pay for a particular commodity. The world's futures markets deal in several "marker crudes" in their exchanges from paper to wet barrels. These market exchanges, or swaps, also provide the industry with important information on the value of whole classes of close-substitute crude oil sources.

Contracts on the NYMEX are for 1,000 barrels of 40°API West Texas Intermediate crude oil with maximum sulfur content of 0.4% to be delivered at Cushing, Oklahoma at a specified date and price. Alternate delivery to the Gulf Coast, or New York harbor are accepted. Delivery times are generally three to six months. They can be listed for a maximum of 36 months. It will be interesting to see whether insurance for payout of production payments and crude oil futures may eventually be brought to cover equal periods of time and/or barrels of production. If that were to come to pass it would be a short step to guaranteeing the payout of a new drilling venture.

Crude oil futures trading on the NYMEX reached 65 million barrels per calendar day in 1990— significantly more than the total world demand for crude oil. Almost all futures contracts are closed out in cash.

Futures prices have now become the most visible aspect of the entire international petroleum marketing operation. The market's short term figures become the reference price in essentially all crude oil trading. In fact, very few spot trades are undertaken until the NYMEXopens for morning trading at 10:30AMEastern Time. Contract and price postings now are almost universally related to the NYMEX.

The current NYMEX has its limitations as a worldwide crude oil price reference, however. The exchange deals with only a specific U.S. crude delivered by pipeline to a specific U.S. location. The "wet" barrel trade volumes for this West Texas Intermediate (WTI) crude grew rapidly reaching an average of 1.1 million barrels per day in 1987. This compares to typical individual international cargo trades in the half- to one million barrel range. Trading parties dealing with other crude oils are thus forced to develop means of converting other crudes, locations and volumes to NYMEX prices. It should also be recognized that as futures contracts approach expiration the cash and futures prices converge and the futures contract basically becomes a spot market contract.

Whereas many of the functions of the spot market can and are being taken over by the futures markets, one fundamental deficiency insofar as the majors are concerned stems from the fact that the present futures markets cannot serve as an effective channel of crude oil supply. The futures markets are not a convenient way of trading five thousand daily barrels of Williston basin crude for an agreed equivalent of California steamflood oil to be delivered to a coastal refinery under an "evergreen" arrangement that may persist for months, or years.

Product price futures are also important to crude oil trading whether as key figures in netback dealings or merely as general reference material. The NYMEX handles trades in unleaded regular gasoline and heating oil. There is no reference price for heavy, or residual, fuel oil ("resid") which is also an important concern in the overall picture. This is particularly so because of price competition due to its physical interchangeability with natural gas and coal as basic industrial fuels. For this reason there is no simple correlation between heavy fuel oil prices and the NYMEX quotes for the rest of the barrel.

Exchange of Futures For Physicals

When all the trades and exchanges of "paper" barrels have run their course and delivery time approaches, "paper" barrels have to be converted to "wet" barrels. The procedures and legalities of this exercise are known as "exchange of futures for physicals" (EFP). It should be emphasized again that the trading of futures contracts which has been going on for a number of sessions has been in terms of "contracts for later delivery" rather than actual barrels of oil.

EFP invokes futures contract provisions regarding where and when the crude oil is to be delivered and adjustments for quality and any other variations from the standard contract specifications including point of delivery. The EFP exchange is from one futures market participant to another and assumes equal and opposite futures positions by the two participants.

At any time during the life of a futures contract, and up until 12:00 noon on the first trading day after the term of the futures contract period, the actual buyers and sellers may agree on an EFP. This allows the futures seller, who is long physicals (paper barrels) to exchange positions with the futures buyer, who is short physicals and desires "wet" barrels. Thus, both physical and futures positions are offset at the same time.

The futures transactions are closed out at a price agreed to between the buyer and seller to reflect the minor adjustments mentioned above for delivery location, quality, etc. The NYMEX then has to be notified of the EFP, with a report including:

1. A statement that the exchange has resulted in an actual change of ownership

2. The date of the exchange

3. The quality and quantity of the physical product

4. The designated delivery month of the contracts remains. Each passing day erodes an option's time value. This works in favor of the sellers and against the buyers. Actually most options, both puts and calls, expire worthless.

5. The number of contracts, of 1,000 barrels each

6. The price at which the futures transaction is to be cleared from the exchange

7. The names of the clearing members

The buyer then, of course, has to arrange to take the newly purchased "wet" barrels from the delivery point, e. g., a pipeline junction at Cushing, Oklahoma or the Houston Ship Channel, and arrange for their transport to a previously arranged storage facility.

Only the NYMEX provides for this rather involved EFP. With the IPE and the SIMEX it is merely expected that the final owner of the contract will arrange to take delivery of the wet barrels.

Options

NYMEX has a pit for trading options on crude oil futures. It is physically smaller than the exchanges trading pits for other commodities but its activity is growing rapidly. An option is a contractual right, but not the obligation, to buy (call), or sell (put) a futures contract at a fixed price—the strike price—for a specified period into the future.

The buyer of an option pays a premium to the seller for this arrangement to buy or sell a futures contract. The option buyer's risk is limited to the premium he pays. The buyer is never assessed additional funds beyond the purchase price. If the market moves against the holder of an option (up for a "put" or down for a "call") the holder just lets the option expire as worthless.

The option seller, who funds the option (initial margin through his broker), assumes the market risk and can earn the amount of the premium. On the other side of the equation, however, the losses can be considerable if the market goes against the option sellers. Option 100 sellers caught in such a situation normally buy back open option contracts in order to nullify their exposure, taking a moderate loss rather than letting it grow.

The basic price of an option depends upon the initial spread, and that which is anticipated throughout the option period, between the futures price and the spot price. In a particularly volatile market purchasers of options are willing to pay higher premiums for this form of price protection and sellers require higher premiums due to their increased risk. Basically, the greater the volatility of the market, and the longer the time until expiration, the higher the premium. Options provide additional means at much lower cost to "go long" or "go short" the market. Option price quotes also serve as a guide to price movements during coming months.

The day to day price quotes for a particular option reflects the current volatility of the crude oil market. The price is determined in the trading pit. Each call and each put represents a thousand barrel futures contract. Crude oil options expire earlier than the underlying futures contract. If, in fact, the option is to be exercised there has to be time for the holder of the underlying futures contract either to liquidate it or arrange for wet barrel delivery. The Expiration Day for an option is the second Friday of the month prior to the delivery month of the underlying futures contract.

Oil and gas options are a form of forward contract which convey to the holder (purchaser) the right, but not the obligation, to buy or sell a certain quantity of oil or gas at a specified price (strike price) on a scheduled date (settlement date). The value of an option depends on four elements: time, prices, interest rates and volatility. Crude oil options are traded on theNYMEXand reported regularly in the financial press. Options give producing oil companies a means of protecting themselves against downward price movements from a fixed price without foregoing the chance of increased profits if crude oil prices rise. These transactions deal in what are known as "paper barrels."

Any perceived value that a put or call option develops above its numerical, or intrinsic, strike price is known as time value. As the expiration date of the option approaches, time value elapses and only the intrinsic value remains. Each passing day erodes the option's time value. This works in favor of the sellers and against the buyers. Actually, most options, both calls and puts, finally expire worthless.

Example 3-2 shows how an oil producer might use a "put" option to insure against the risk of an oil price drop;

EXAMPLE 3-2
THE PURCHASE OF A CRUDE OIL PUT OPTION

The following table shows the settlement price for a few selected put and call options on the New York Mercantile Exchange (NYMEX) on January 18, 2008. The price indicated as the spot price for West Texas Intermediate (WTI) was reported in the Wall Street Journal on Saturday January 19, 2008.

There is an inferred decimal between the second and third digit of the strike price. The line for a strike price of 9050 ($90.50) is shaded as the nearest options to the spot oil price on January 18, 2008.

Date: Friday January 18, 2008

Crude Oil (NYMEX)
1,000 bbl. Contracts (WTI) - $ per bbl.

Strike Price	PUTS – SETTLE			CALLS – SETTLE		
	APRIL 2008	JULY 2008	OCT. 2008	APRIL 2008	JULY 2008	OCT. 2008
8000	0.90	2.32	3.37	10.34	10.73	10.86
8600	2.42	4.44	5.87	5.89	6.94	7.48
8900	3.74	5.93	7.44	4.23	5.47	6.12
9050	4.54	6.77	8.32	3.53	4.83	5.53
9200	5.42	7.67	9.29	2.92	4.25	5.03
9500	7.48	9.73	11.36	2.00	3.35	4.17

Source: NYMEX.com Light Sweet Crude Oil

Spot Market – West Texas Intermediate Crude @ Cushing, Oklahoma - $90.58/bbl.

A producer wants to be sure that he will receive no less than the current (1/18/08) price for oil to be produced in September 2008. To accomplish this, he decides to purchase an October 2008 "put" option (options settle in the middle of the month prior to the option date) on January 18, 2008, with a strike price of $90.50. This would guarantee the January 18, 2008 price. According to the table above the cost of an October 2008 "put" option on January 18, 2008 is $8.32 per barrel. If the price drops to $70.00 per barrel on the settlement date in September 2008 the holder of the option would sell the oil for the market price of $70.00 per barrel and exercise the "put" option, realizing a net value for the oil of $82.18 per barrel (i.e., $70.00 + $20.50 - $8.32.). He would be $12.18/bbl (i.e., $82.18 - $70.00) better off than if he hadn't bought the "put" option. This would be a paper transaction and the unhappy seller of the "put" option would pay the commodity exchange's clearing house $20.50 times the number of barrels in the contract (1,000). The clearing house, in turn, would pay this amount to the owner (purchaser) of the "put" option. The seller of the "put" option would thus loose $20.50 minus the $8.32 he was paid for the "put" option or $12.18/barrel. However, if the price is above $90.50 per barrel on the settlement date in September 2008, the option would expire unexercised and owner (purchaser) of the option would sell the oil in the open market at the then current price. The price of oil would have to exceed $98.82/barrel on the settlement date for the producer to realize the $90.50 (January 18, 2008 price), because of the $8.32 paid for the insurance ("put"). All of these numbers ignore any brokerage commissions that must be paid.

Hedging

Daily price fluctuations involving millions of barrels of crude oil have meant that end users and traders have had to develop means of protection against this potentially catastrophic impact on their profitability. Airlines and other large users of fuel, as well as the independent petroleum refiners, are increasingly relying on financial arrangements designed to minimize the pricing risk.

The advent of a working futures market in crude oil opened the door to hedging operations in a manner similar to the trading in other commodities. Hedging is a futures market technique by which investors insure against the risk of market changes in price. There are two prime reasons for hedging. These are to protect the value of inventory; or to fix ahead of time the cost of purchases. A crude oil trader, who may wish to avoid the risk of significant decline in the price of crude oil over say the next ninety days, can contract to deliver a specified number of future barrels at a fixed agreed price. If the price does drop, he is protected because he will receive the previously determined price. If the price goes up during the period he has lost the opportunity to take advantage of the change since he will receive only the previously agreed price. In the terminology of the financial community he has hedged the specified volume of future crude oil sales against a price decline. The price the trader gets for his future sale is not necessarily the current spot market price, but rather the price in today's commodities market at which some other trader is willing to sell the matching volume on that future date.

Hedging, by definition, is the simultaneous initiation of equal and opposite positions in the cash and futures market. Hedging is effective because the futures and the spot, or cash, markets move together. The technique is a strategy employed as a protection against adverse price movements in the cash, or spot market. A "short hedge" involves a futures contract to deliver crude oil in a future month at a specified price. A "long hedge" involves buying a futures contract for delivery in a designated month. The opportunities for hedging only occur, however, if there are enough speculators in the market to take an "uncovered" risk, i.e., not to match every spot purchase with a forward sale.

An independent producer with available oil reserves is said to be "long" in the cash, or physical, market. By selling futures, i.e., the opposite of his long position in the cash market, he can fix the price he will receive at the time of delivery. Perhaps the most notorious successful application of a major hedging operation in crude oil prices occurred late in 1985 when prices hovered around $20 per barrel. Mesa Petroleum anticipating a price decline decided to hedge its entire 1986 production of approximately 4 million barrels and locked in a price of $26.50 for each of those barrels while the market tumbled to a $10 per barrel low over the next seven months.

For the majority of operating companies in the petroleum industry, hedging is the most important function of the futures market. By fixing prices in advance the hedger limits his potential loss in a volatile crude oil market. At the same time, however, he foregoes the chance of gains from any upward price swing at delivery time.

Crude oil traders in the international market frequently hedge the price of cargoes enroute to the U.S. which are to be sold on arrival on the spot market by selling WTI futures as protection against possible price declines while the cargo is in transit. If the spot price should increase while the cargo is on the high seas the trader will realize that gain but must also absorb the loss from his WTI futures contract.

Spreads

Approximately 80 percent of the ultimate value of a barrel of crude oil to the refiner is derived from its gasoline and heating oil fractions. The NYMEX has installed a number of simultaneous buy and sell "spreads" designed to lock in, or fix, a profit spread at a future date between the then cost of the crude oil raw material and the sale of the principal finished products. The most frequently quoted of these is the "3-2-1 Crack Spread." In this spread a short (i.e., buy) futures position is taken in three volumes of crude oil, and long (i.e., sell) positions of two volumes of unleaded gasoline and one of heating oil. Traditionally all three futures positions expire the same month. Recognizing that there is a time lapse between the time when a crude oil purchase settlement must be paid out and the income from the product sales is actually realized, a further refinement of the crack spread involves a month or two time lapse between the long and short positions. This "intermonth" variant can also be applied as a hedge against price declines for products in off season inventory.

Open Interest

Open interest is a measure of market liquidity and also of the size of the futures market at that point in time. It consists of the total number of futures contracts which, at the end of each trading session, remain open—i.e., have not been liquidated by an offsetting futures contract trade or by actual delivery (see bottom of table in Example 3-2).

Consider the following example (Source NYMEX, "Energy in the News," 3rd Quarter, 1986),, which demonstrates how open interest is computed:

> When trading in a futures contract first begins, open interest is zero. Open interest is created as trading proceeds. For example, if Trader A buys one October crude oil futures contract from Trader B, one new October futures has been created, and open interest is one. If Trader C buys one contract from a new Trader D, who is selling short, then open interest rises to two.

> What happens when existing contract owners sell out, either to new buyers or to other buyers who are covering short positions? If Trader A decided to liquidate his long position by making an offsetting sale to Trader E, a new buyer, there is no change in open interest. Trader E has simply replaced A as owner of that long contract.

> On the other hand, if Trader D had accepted the offer of the contract for sale by A, then open interest would have declined to one. This is because one long position and one short position would have been liquidated by offset, resulting in the elimination of one contract. Open interest also would decline if Trader A, who is short futures, made delivery of crude oil to Trader D, who is long futures.

Crude Oil Swaps Agreements

A wide variety of financial arrangements to manage and reduce crude oil price risk have evolved. Basically these techniques permit transfer of the price risk to others. The most firmly established of these are the forward and futures trades which have already been described.

Crude oil price swaps, which permit users to set a fixed price for the produced oil over prolonged periods in the future, are preeminent among this new generation of risk management techniques. These financial arrangements adapt the procedures which the international banking houses have developed over the years to deal with the long term currency exchange risks. Oil swaps allow a widely diverse collection of buyers and sellers to exchange their varying exposures to long term price risk over a specified, and generally prolonged, period of time. Most oil swaps are in the one to two million barrel range and for periods of three years, or more. Longer periods are certainly possible if each party is satisfied with the credit worthiness of the other.

There are two elements incorporated in the traditional crude oil price swap as diagrammed in Figure 3-15. These are a producer's sales hedge; and a purchase hedge for the end user of oil. A financial intermediary assumes the role of providing long term fixed prices to both parties. This is usually an investment banking house which employs the risk assumed from each party to offset the other. Legally the two deals are distinct. Each side of the arrangement is a separate contract. The pairing of two offsetting market positions may not be a simple matter, and may not happen simultaneously. It often happens that the firm, or institution, providing the swaps may simply hold the unbalanced risk of underwriting for its own account.

The volatility of crude oil prices in recent years and the very large amounts of money involved have led the financial institutions to adapt some of their international currency exchange risk reduction techniques to the crude oil market. The swaps arrangement is one of these.

In a typical swap arrangement a buyer is guaranteed a purchase price but agrees to pay the swaps intermediary any subsequent difference between the crude oil acquisition cost and a lower market price. On the other side of the transaction, the oil seller would agree to pay the swaps provider, i.e. financial intermediary, any difference between a guaranteed fixed price and any possible higher market level.

FIGURE 3-15
CRUDE OIL
PRICE SWAP

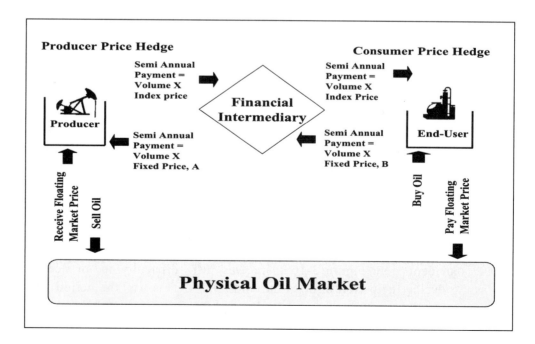

The two partially offsetting financial deals between the intermediary and the buyer or seller of the crude oil is the heart of the swap. Basically, both the buyer and seller of the oil have exchanged, or swapped, a floating market price with its inherent risk for a guaranteed fixed price. The guaranteed fixed price may not be identical for both buyer and seller. The difference can provide the required financial incentive to the intermediary. It is also possible that the price difference may be the best that the intermediary is able to negotiate, and he takes over that portion of the market pricing risk.

Payments between the intermediary and the buyer and the seller, in opposite directions, to compensate for the divergence between the contract and market prices are generally on a semi-annual basis. The compensation arrangement to the swaps provider varies from transaction to transaction.

Airlines, railroads and shipping firms, for whom fuel costs are the largest element of their operating cost, have been the biggest users of swaps from the start. Most swaps transactions are for large amounts of oil, in the millions of barrels, and for periods of three years, or more.

The "Synthetic Oilfield"

The NYMEX places a 36-month limit on crude oil price futures. This does not satisfy the much longer term requirements of the modern petroleum business, which concerns itself with long life production of twenty years, or more. The consequence has been the development of an oil-indexed security trust arrangement that, on paper, imitates the pay-out of an oilfield. These trusts are founded on the future delivery of forward crude oil "volumes" in regular amounts, much like an actual oilfield. Units of the trust can be purchased and traded at prices which fluctuate with then current oil prices. Depending on the terms of the actual agreement the trust may include a price floor for part of the period to ensure the investor a positive level of income.

These trust agreements, or "synthetic oilfields," as they are dubbed by the "Wall Street refiners," may serve as the offset to a crude oil swap agreement if a physical wet barrel producer isn't available. They also have appeal to investors who anticipate a long term growth in the prices of crude oil but do not wish to undertake the daily hassles and pitfalls of futures trading in the commodity markets.

FUTURE CRUDE OIL PRICES

Forecasting crude oil prices over the productive life of an oil field which is anticipated to produce for another twenty, or even fifty years has never been a particularly easy task. Economic theory tells us that in the long run prices will seek a level whereby the efficient firms will continue to replace their depletable reserves and earn a reasonable return on their shareholder's equity commensurate with the geological and financial risks involved. This may hold true ten, or more properly twenty, years into the future. Unfortunately, in the critical near term, the forecasting of world crude oil prices in order to arrive at predictions of cash flow, has become a little more than a guessing game with inordinately high stakes.

As is evident from Figure 3-16 there was little difficulty in, or need for, long term price forecasting during the period prior to the mid-1970's. This was the period when the actions of the "Seven Sister" oil companies, internationally, and the Texas Railroad Commission in the U.S. served to maintain worldwide stability in crude oil prices.

The petroleum industry's experience in predicting future crude oil prices since the first oil price shock of 1973 has been one of continual and costly surprises. There is a general agreement among the leading forecasters that crude oil will be more valuable in the future than it is at present.

In recent years consensus among the industry's forecasters has been the norm,—and regrettably, in every case has proven seriously incorrect as depicted in Figure 3-16. The costs of erroneously predicting in 1979-81 that crude oil prices would increase continuously for many years into the future has staggered the world's economy. The investments, loans, and much of the industry's accelerated activity, all of which were based on these price forecasts, are characterized as one of the most expensive business blunders in history.

FIGURE 3-16
CRUDE OIL PRICE FORECASTS

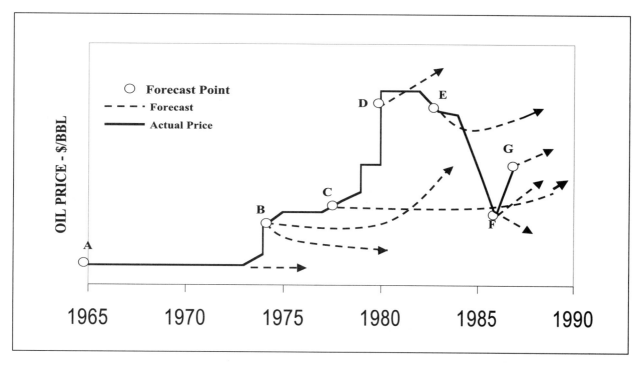

Basically, these faulty forecasts were the result of attempts at mathematically modeling crude oil price (a dependent variable) against time as the single independent (or determining) variable using linear regression. With such a model, and with oil price data from the years 1974-1980, oil prices would appear to be headed for constant increase. It must be realized, however, that oil prices are influenced by many other independent factors. These include such items as conservation incentives, the state of the world economy (recession vs. growth), successful exploration and production in non-OPEC countries, wars between OPEC members, the substitution of alternate fuels for industrial and space heating, and so on.

As the forecasters predicted that the price trend of the late 1970's would continue, they overlooked the effect of the rapid price increases on the consumer. The oil price shocks of the 1970's were the catalyst to improve energy efficiency throughout the world. Cars were built to use less gasoline, plants were designed to conserve energy, homes were insulated to loose less heat in the winter and stay cooler in the summer. All of these things significantly reduced the demand for oil and gas, which could only be corrected by a severe price decline. Figure 3-17 shows how energy efficiency improved in the U.S., and it is apparent that the big looser was oil. Not only was less oil required per dollar of "Gross Domestic Product," but oil and gas also lost out to other forms of energy. The forecasters had failed to recognize the very high degree of interchangeability between gas, coal and residual fuel oil from petroleum. Figure 3-17 shows that this trend is not as steep as it was during the 1970's, but that technology continues to improve the efficiency of energy use and that oil and gas are retaining a constant share of the energy market.

FIGURE 3-17

U.S. ENERGY CONSUMPTION AND EFFICIENCY PER GDP DOLLAR

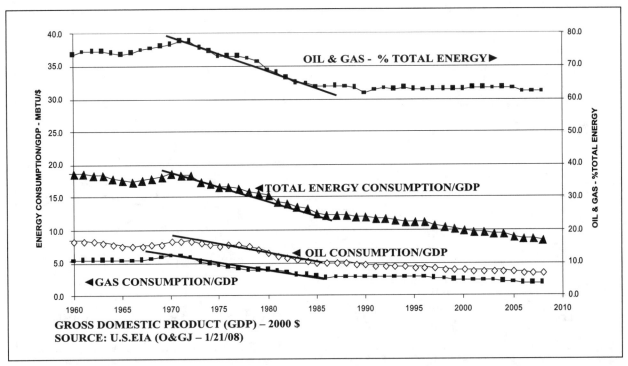

Another difficulty has been trying to project when non-OPEC production will decline to the point that OPEC's production would again dominate the world's productive capacity. This would give the cartel renewed power to set oil prices. The time that this would occur has been delayed by investors willing to continue successful exploration for oil and gas in non-OPEC countries. These investments continue to be attractive because new technology has lowered exploration and development costs and retarded productivity declines in existing fields.

The inevitable frustration has led to a great deal of consternation among the practitioners of price forecasting, and a new consensus, which is more defensive than real. It has become far easier in the present circumstance to justify one's price forecast on the basis that it is midway between Shell and Exxon than to try to explain that it is based on any new theory and insight into the world's political and economic forces on crude oil supply and demand.

Nevertheless some price projection is obligatory for any economic analysis. One acceptable approach may be to assume as a base case that the price will remain constant for the next year or so, and then to apply a series of deviations from that base assumption in a sensitivity analysis of the matter. It is evident that at present the uncertainties of crude oil price forecasting are so great that it is imprudent to base one's cash flow projections on a single price scenario.

Breakeven Crude Oil Price

Another type of sensitivity analysis that is often performed to supplement an analysis which includes a forecast of future prices is to determine the "breakeven crude oil price." This is the minimum price that would have to be obtained for all crude oil sold from a project so that the project would just meet the minimum corporate investment criteria. The specific criteria which would be considered and the method of evaluation will be discussed in detail in Chapter V. However, at this point it should be noted that the "breakeven price" is constant over the entire project life, not escalated for inflation and in terms of the value of money of the day for each year of the project. The "breakeven price" should be compared to the expected price of oil over the life of the project. If it is significantly less than the expectation of future oil prices it is a desirable project. If the breakeven price is greater than the expected future price the project is not a viable one. If the breakeven price is near the price forecast, this type of sensitivity analysis will be of little help in making investment decisions.

Another use of breakeven price is in screening of investment opportunities or in long range forecasting to determine when certain types of projects might become economic ventures. Such as what price of crude oil would be required for various enhanced oil recovery techniques, marginal discoveries, synthetic oil development, or even certain ultra deep water development.

Effects of Inflation on Crude Oil Pricing

Crude oil as a commodity has suffered from price inflation through the years. Throughout its history the price of oil has practically never kept up with inflation—a great boon to the consumer, but of great concern to the exporting countries. As can be seen in Figure 3-4, the general trend of real oil prices declined until the first oil price shock of 1973.

The real price of oil fell sharply during the 1980's, but had regained part of that loss by early 1990, when oil was selling for $22 per barrel. During the 1990/1991 Middle East War the actual (nominal) price of oil, in fact, exceeded the 1980 nominal price for a short period of time, but still fell short of the real 1980 price.

It is reasonable to expect that these inflation effects will continue, superimposed on all of the other political and supply-demand factors which have already been discussed. The general subject of inflation as it affects the petroleum production business is discussed in Chapter IV. If one chooses to confine one's efforts to predicting merely the inflationary effects on the price of crude oil, it is possible to apply an adaptation of Equation 2.3 from the last chapter.

EXAMPLE 3-3
FORECAST OF FUTURE OIL PRICES DUE TO INFLATION

Oil is currently being sold for $60 per barrel from an oilfield for which an economic analysis is being prepared. It is anticipated that the price will increase at a rate which is 1.0 percent per year greater than the rate of general inflation. Inflation is forecast to be at the rate of 2.0 percent per year for the next five years and then increase to an annual rate of 3.0 percent thereafter.

Required:

Develop a forecast of oil prices from this field for the next ten years in accordance with the foregoing premises.

Solution:

If: P_{io} = Value of crude oil at the base year ($/bbl)

P_{to} = Value of crude oil at the end of year, "t" ($/bbl)

I_I = General inflation rate (%/yr)

I_E = Escalation rate above inflation (%/yr)

Then:

$$P_{to} (1\text{-}5) = P_{io} (1+ I_I)^t (1+ I_E)^t = \$60.0[(1.02)(1.01)]^t$$

$$P_{to} (6\text{-}10) = \$60.0[(1.02)(1.01)]^5[(1.03)(1.01)]^{(t-5)}$$

(3.2)

Years t	Inflation I_I (%/yr.)	Escalation I_E (%/yr.)	Oil Price p_o ($/bbl)
0	-	-	60.00
1	2	1	61.81
2	2	1	63.68
3	2	1	65.62
4	2	1	67.58
5	2	1	69.62
6	3	1	72.25
7	3	1	75.34
8	3	1	78.39
9	3	1	81.54
10	3	1	84.83

A Future Scenario

Future pricing of oil will be influenced by several very strong forces working in opposite directions. Increasing costs will continue to push the price upward, while competition and technology will act to limit the rise in price. So the average price of oil will probably remain in a very narrow range.

Almost all cost associated with the upstream oil industry are rising. To find new reserves it is necessary to explore in deeper waters of the oceans. New reserves are also being discovered in reservoirs at ever increasing depths. These factors increase both the cost of finding and developing new reserves. As new conventional sources of petroleum become more difficult to find it will be necessary to rely upon other sources, such as enhanced oil recovery, oil shales, tar sands, bitumen deposits, coal bed methane, gas hydrates, and gas-to-liquid conversion. All of these sources will cost more than primary recovery which dominates current production activity. The cost of continued operation of currently producing aging fields will continue to rise as production rates decline, artificial lift becomes necessary, equipment wears out, more water is produced and chemicals are required to combat corrosion, emulsion and paraffin problems. And as exploration and development activities are conducted it is necessary to spend increasing amounts of money to protect the environment. On top of all of these rising costs, most countries of the world continue to experience or encourage at least a low rate of inflation, so even if industry costs did not rise for the myriad of reasons cited these costs would increase with time.

On the other hand the strong force of competition will tend to limit the increase in oil prices. Higher real prices for oil will stimulate exploration and development of gas and natural-gas hydrates in addition to conventional and non-conventional sources of petroleum as previously discussed. M.J. Economides, et al have made a very strong case for increasing demand for natural gas. In a sense, the petroleum industry, which produces both oil and gas is one of its own biggest competitors. Higher liquid petroleum prices will accelerate development of gas-to-liquid technology. Every gallon of gasoline, diesel, and jet fuel made from natural gas is one less gallon made from crude oil, which in turn reduces the demand for oil providing downward pressure on prices. If the auto industry does eventually turn to fuel cells and use hydrogen to power their cars, this will reduce the demand for gasoline and other liquid hydrocarbon fuels. The good news out of this development is that the best source of hydrogen is natural gas and oil. So the petroleum industry would still be the major supplier of automotive energy. Higher prices for petroleum products will also encourage development of renewable sources of energy such as solar, wind, tidal and biomas sources as well as hydro-electric, and nuclear.

New technology will have a positive impact upon the oil industry, but will also affect future prices. New technology reduces costs, lowers risk, and improves the probability of success. This will result in improved exploration, enabling development and production of marginal fields and increased reserves through lower economic limits. All of these factors will tend to reduce or limit future increases in the price of oil by allowing the industry to work smarter, more effectively and meet the environmental restrictions placed upon it.

However, as can be seen by examining the trend of "Real Oil Price 2007$ of Figure 3-4 between 1920 and 1970, where oil producers lost 2/3 of the value of the oil sold due to inflation, a constant price for oil is not desirable. Therefore, in the future OPEC might find themselves caught between trying to maintain the "Real" value of oil for its members through increased oil prices without encouraging alternate sources of hydrocarbons, which may become economically viable at the higher price.

NATURAL GAS SALES

Natural gas is one of the world's principal fuels. It has long been looked upon as a by-product of the production of crude oil. The industrialized countries of Europe, North America and the Far East make extensive use of natural gas. They utilize indigenous supplies fed to the local markets through vast networks of natural gas pipelines. Japan is the exception, having to rely on imported liquified natural gas (LNG). Most other countries around the world have found gas in connection with their efforts to discover crude oil for domestic use and export. Although a potential market for natural gas exists, starting with electric power generation, the supply is frequently remote from the market. The tremendous capital requirement for transportation of gas have precluded the proper development of this resource. Some tremendous shut-in reserves of natural gas have been found in such places as Algeria, Australia, Indonesia, Iran, Peru, Qatar, Russia, and Trinidad.

In North America, natural gas traditionally was marketed in the field to pipeline purchasers. These natural gas pipeline companies generally did not operate as common carriers. Now with deregulation of natural gas, producers can sell directly to a distant consumer and use the pipelines as a common carrier to transport their gas. In the U.S., gas pipelines are under the Federal Energy Regulatory Commission's (FERC) jurisdiction with regard to public convenience and necessity and regarded and regulated as public utilities. In Canada the National Energy Board (NEB) performs much the same functions.

In recent years natural gas producers have succeeded in developing direct sales spot markets among major gas users while employing the natural gas pipelines merely for transport. Essentially all of the long natural gas pipelines in North America have relied upon the "right of eminent domain" for permission to cross certain private lands. This right is afforded only to common carrier transport systems, e.g., railroads, public highways, and the like. Thus the long line natural gas pipelines can be required to transport any party's gas from point to point on their system for a reasonable fee.

Spot sales of natural gas also need a major interconnected pipeline grid crisscrossing the country so that exchanges can be effected. The industry is in the process of building pipeline "hubs" at key intersection points. These facilitate drawing gas, not just from producing wells, but also from storage facilities. The storage fields act just like gas wells and provide additional sources of gas supply. More than 100 hub projects have been announced in North America in recent years. Many of these are closer to where gas is consumed, which allows the commodity to move faster when demand peaks. Hubs are already well developed in Texas, Louisiana and Alberta. Projects are under way both in supply centers such as the Rocky Mountains and in consuming regions like the U.S. Northeast and Europe. Gas buyers can rent space in these storage fields, inject them with gas in the off-season when prices are cheaper, and withdraw it during the winter.

In concept a gas hub, it is similar to Cushing, Oklahoma, or Scotland's Sullom Voe terminal — the delivery points from which West Texas Intermediate (WTI) and Brent crudes, respectively, are traded in highly liquid spot and forward markets. As hubs spring up around the U.S., Canada, and Western Europe, gas markets are developing a strong resemblance to the freewheeling crude oil trade. And as governments in other parts of the world move to privatize and break up the monopolies that typically buy and aggregate incoming gas supplies and move them to consuming centers, it is anticipated that "hubs" will spread to those areas, providing commodity-type trading, as well as profit opportunities for hub owners.

The Chevron operated Henry Hub on the Louisiana Gulf Coast, which has the distinction of being the delivery point of the New York Mercantile Exchange's natural gas futures contract, has become a key pricing point for the whole U.S. gas market. Besides its NYMEX function, Chevron has made Henry Hub and its adjoining Gulf Coast Star Center the focal point for gathering the company's own substantial U.S. gas production. In addition, it provides storage services to other offshore Gulf and Gulf Coast producers and buyers. "Hedging services," allow consumers to lock in a fixed or bracketed price for future deliveries. Other companies are providing similar full-service facilities elsewhere. The important aggregation and storage services which these hubs provide used to be the sole purview of the interstate pipeline companies which are no longer allowed to mix their transport and marketing functions.

The typical long term contract in North America defines the specific area, i.e., lease or field, from which the gas is to be purchased. If the contract covers a larger area it will contain a reserve limitation. The agreement will also clarify that the operator has first call on the amount of gas required to conduct the field operations, and that this gas is not part of the sales volume.

The daily quantities, or rates of purchase, are also specified. These quantities may be in terms of a minimum volume per year which may be important in the amortization of pipelines to a remote area such as the Pointed Mountain Field in the Yukon. More typically the minimum sales rate is in terms of a stated daily volume for each 7.2 billion cubic feet of dedicated gas reserve (20 years at 1.0 MMCF/d). Provisions for reserve determinations, and redetermination, and arbitration comprise important parts of these gas sales contracts.

The producer usually insists on maximum and minimum rates being written into the contract to protect him from having to drill extra wells merely to take care of the purchaser's sudden peak demands. These maximums and minimums are generally stated in terms of "deliverability" of the dedicated reserve. Typical limits would be a maximum of 120% and a minimum of 80% of deliverability, which must then be defined. It is normal practice to determine this value from a thirty day test at some mutually agreed time when all wells producing from the dedicated reserve can be produced simultaneously at capacity. The contracts provide for some type of overage/underage balancing arrangement. In many cases there are also "Take or Pay for" provisions which are designed to assure an income flow to the owners of interest.

Processing rights for the recovery of natural gas liquids can be economically important. Specific provisions for these rights are carefully spelled out in the natural gas sales contract.

Differences in Production of Natural Gas and Crude Oil

Natural gas is generally much less expensive to drill and develop than crude oil once a pipeline connection is established. It does not involve artificial lift or secondary recovery such as waterflooding and consequently is less capital and labor intensive than crude oil production.

The preferred method of transporting natural gas both from an economic and operational point of view is by pipeline. However, gas pipelines are usually limited to distances of only a few thousand kilometers, primarily over land, although it is technically feasible to lay underwater pipelines, with some having been installed in great water depths.

The more economic method of transporting natural gas great distances is in the form of LNG, which will be discussed more fully later in this chapter. Capital requirements for LNG projects are much greater than for pipelines and may require significantly more time to develop a gas discovery where this mode of transportation will be required. A prime example of this is the North Field offshore of Qatar in the Persian Gulf. It was discovered in 1972, but the first gas sale from this field did not occur until early 1997. The first tanker of LNG was loaded in late 1996 for delivery to Japan in January, 1997, under a 25 year contract. The delay in development was not only due to the long construction time for such a major project, but primarily to a secure long term market for the gas.

Natural gas is measured, by an orifice flow meter or electronic probe to determine its velocity, as it leaves the lease, or unit. The measurements are corrected to the pressure and temperature base noted in the contract. Provisions are included for periodic recalibration of the measuring equipment. Due to the number and nature of impurities that natural gases may contain, it is customary to include quite stringent specifications in the contract. These deal with the minimum BTU/mcf, the maximum hydrogen sulfide and carbon dioxide contents, the maximum moisture; and the penalty for deviations.

In contrast to crude oil, which can come to rest in lease stock tanks after it is produced, natural gas flows continuously from the well to the pipeline. The flow rates are controlled at the wellhead in response to the requests of the pipeline dispatcher which are telecommunicated to the operator's pumper in the field. Flow rates are also affected by the line pressure that the pipe line chooses to maintain at any point in time. Natural gas pipeline systems have a higher degree of flexibility in their operation than do oil lines. This is mainly due to the compressibility of the gas which permits the pipeline operator to "pack" the line with extra gas in short term anticipation of drops in ambient temperature downstream and consequent increases in heating demand.

Settlement for gas purchases is made within the stated period, ninety days to six months. Part of the delay in making settlement for natural gas in contrast to the 30 day payment tradition for crude oil purchases is due to the time required for collecting and integrating the circular charts from the orifice meters located in the field. Payments may be made directly to individual interest owners, sometimes including state taxing authorities, employing a division order arrangement similar to the procedures for crude oil settlements. In some cases it may be left to the operator to handle the division order payments to the other owners of interest.

There is sometimes confusion regarding nomenclature of various products associated with the natural gas industry. Each product is a function of its chemical composition which determines its physical characteristics. Figure 3-18 endeavors to clarify this matter and also includes some of the physical data and conversion factors commonly used in the industry.

FIGURE 3-18
HYDROCARBON TERMINOLOGY

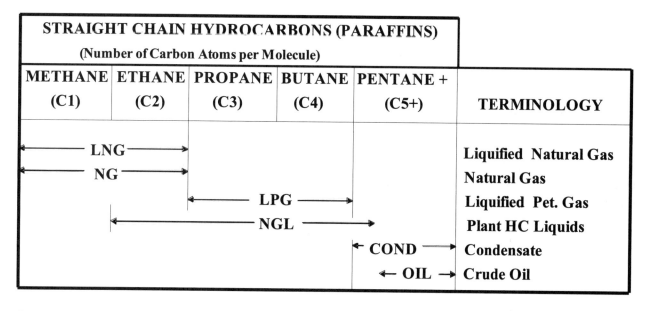

One MSCF (thousand) Methane Gas = 45.41 pounds of LNG

One MMSCF (million) Methane Gas = 20.59 Metric Tonnes LNG

One Cubic Meter Nat. Gas = 35.31 cu.ft. = 0.727 Kg LNG

One Cubic Meter LNG = 21,200 SCF Natural Gas

One Metric Tonne of LNG = 1,375 cu.meters Methane Gas = 48.558 MSCF

One Metric Tonne = 2,205 Pounds (avdp)

One Equivalent Barrel of oil = 5.8 MSCF Natural Gas (heat content)

Heat of Combustion of Methane = 1,011 BTU / cubic foot

Boiling point of Methane = -258.5 °F or -161.4 °C

Critical conditions for Methane:

    Temperature = -116°F / -82.5°C

    Pressure = 673 psia / 45.8 atm

Natural Gas Liquids

Natural Gas Liquids (NGL) is a generic term for all the easily condensable liquid hydrocarbons associated with natural gas production. Liquified Petroleum Gases (LPG), (consisting of ethane, propane, normal and iso-butane) which are separated into individual components in the gas processing plant, are a subset of the NGL stream. The other NGL components are natural gasoline, and condensate. NGL hydrocarbons are marketed under purchase contracts or on the spot market to major users or brokers as petro-chemical feedstock, refinery blending stock, bottled gas for domestic use and crop drying. The sales contracts are medium term or annual "evergreen." The prices are quoted on a U.S. gallon basis for each product. There is appreciable volatility in the price of NGL in response to the several markets listed. While there is a significant NGL pipeline network from the Gulf Coast into the mid-continent, most LPG moves by rail or truck transport.

In the U.S., government regulated prices of natural gas liquids were imposed along with those for crude oil during the 1970's. The seasonal variation in natural gas production does not match the seasonal uses of the products. This seasonal mismatch results in a major LPG storage requirement, usually in underground artificial storage caverns. There is still substantial price fluctuation for these products from day to day throughout the year. The NYMEX conducts futures trading in propane. The contract specifies delivery of 42,000 gallons (1,000 barrels) at Mount Belvue, Texas.

Value of Natural Gas

The value of natural gas, as with any commodity, is the price that the ultimate consumer is willing to pay for it in an open market. If it is the base material used in the manufacture of chemicals it will have to compete with other materials which can be used for those purposes. If it is to be used for power generation or space heating the price paid will reflect the value of heat of combustion of the gas which is obtained by burning it. The same energy can be obtained from other sources, i.e., fuel oil, coal, solar, hydro, nuclear, fermentation, etc., so the price of gas must be competitive with other sources of energy. The fact that gas is easy to handle and expensive to store, compared with other sources of energy, affects its value. Natural gas also has an environmental value premium over fuel oil and coal because it is clean burning. It produces less carbon dioxide and essentially no nitrous oxides, hydrogen sulfide, carbon monoxide, or particulates as may be the case with fuel oil and coal. However, to date no significant environmental premium has been added to the value of natural gas. Increasing environmental concerns worldwide may encourage inclusion of this additional value in future pricing of natural gas. Commonly, if gas is sold under a long term contract, the value will be tied directly to some marker crude price and allowed to move with movements in the price of oil.

The value to the ultimate consumer of Liquified Natural Gas (LNG), after regasification at the port of delivery, should be the same as described above for natural gas. However, there may be an added value for LNG at the point of regasification that is not currently recognized by the industry. A great deal of energy is utilized to compress and cool natural gas to reach a liquid phase. It should be possible to recover at least some of this energy, principally in the form of refrigeration, during regasification. Such a process has not yet been developed, but some possibilities are: freezing food, cryogenic separation of atmospheric gasses, or cooling electrical equipment to eliminate resistance. If an economical process is developed this could add additional value to LNG.

The price of natural gas at the wellhead must reflect the value to the end user minus all costs incurred to process and move the gas from the wellhead to the consumer, including a profit for performing those functions. These costs will include treating of the gas to remove undesirable contaminants, compression, transportation by pipeline or ship and in the case of LNG liquefaction and regassification.

Figure 3-19 shows the history of U.S. wellhead oil and gas prices since 1950 and recent coal prices. It can be seen that oil, gas, and coal prices moved in a similar manner over this period, but not identical as they each have different supply/demand forces. The price of these sources of energy are closely linked, as was previously discussed, but differ reflecting the difference in energy production per measurement unit. For a number of years it has been predicted that the price of oil and gas would equilibrate on the basis of their heating values, which would mean that it would take 5,800 cubic feet of natural gas to be the equivalent of one barrel of crude oil (BEQ). The fourth line on the graph of

Figure 3-19 is the ratio of oil price per barrel to gas price per thousand cubic feet. The equilibrium ratio would be 5.8 as shown by the horizontal line. It is only recently that the price ratio has approached the equilibrium value, indicating that gas is now being priced as an energy source and not as an undesirable by-product of oil production.

FIGURE 3-19
CRUDE OIL, NATURAL GAS AND COAL PRICES

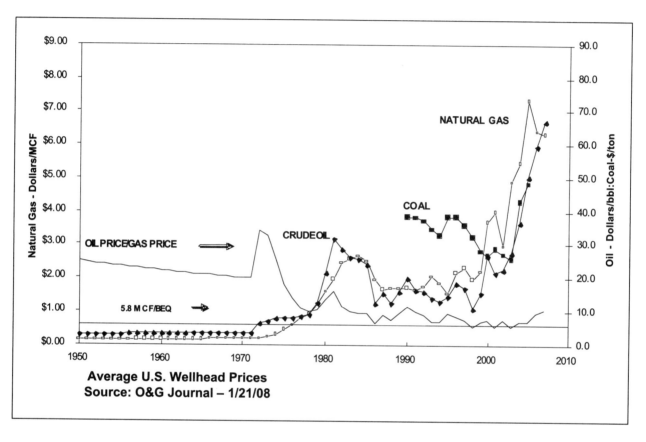

Average U.S. Wellhead Prices
Source: O&G Journal – 1/21/08

Natural gas has long been considered a byproduct of crude oil, either as casinghead often flared, or found by accident when looking for oil. Nevertheless, natural gas is becoming an increasingly important fuel throughout the world. Its principal competitor is fuel oil.

Forecasting Natural Gas Prices

Whereas the price of crude oil in the U. S. was government controlled from 1973 until 1981; the price of natural gas moving in interstate commerce was under government control from 1954 to the late 1980's. This would seem to make it easier for the analyst to predict future prices of natural gas. Deregulation and the growing trend of producers to sell directly to major natural gas purchasers rather than to the gas pipeline companies makes past history of gas prices a poor basis for future price projections. Opinions differ widely as to the future pricing situation for natural gas, and one may be pretty much on his own in establishing the price estimates which he will apply in his net cash projections.

One important factor which should be considered in price projections is the basic geological realization that as the industry drills deeper the hydrocarbons tend to be of lighter molecular weight and therefore more gaseous. Thus, the industry finds more and more gas relative to oil as time goes on. This may affect the supply and demand price relationship with the growing market for this product. A further stabilizing factor has to be the tremendous reserves of natural gas shut-in for lack of market in overseas locations such as the Persian Gulf, Russia, Australia, Algeria and Indonesia. The political consequences which would result from any rapid increase in price to domestic North American consumers, who have grown accustomed to distressed by-product prices for their home heating fuel, must also have to be kept in mind.

Outside of North America the marketing of gas is not much easier. There is same increasing frequency of discovering natural gas as the industry probes deeper in its exploratory effort. The greatest difficulty, however, involves the very large investment for a pipeline installation and the fact that markets can't really be developed until the line is installed and in operation. North America, Europe, Argentina and more recently Egypt have overcome this fundamental hurdle but many other countries have not. China has developed plans to move recently discovered offshore gas from the South China Seas by pipeline across Hainan Island to the mainland and on to the industrial city of Guangzhou (Canton). Higher priorities for the hard currency required have forced postponement of the project, however. The World Bank has become aware of the basic problem and is moving to help out in a number of situations.

Casinghead Gas Marketing Practice

Gas separated at low pressure from solution in produced crude oil has historically been valuable for its LPG, or natural gas liquids content. Most jurisdictions require that this gas can not be flared on the lease, but must either be sold or reinjected into the producing formation.

Sales contracts for casinghead gas differ in several important respects from the agreements for the purchase of high pressure natural gas. The term of the contract is normally for the life of the production. It is usually made clear that the purchaser will not prorate casinghead gas supplies to his overall gas demand but must take the full volumes offered. This is necessary so as not to curtail oil production in order to avoid flaring of gas. Casinghead gas is by nature very rich in NGL (propane, normal and iso-butanes, and pentanes-plus). The prices paid for casinghead gas in the U.S. depend in large measure upon the split, if any, in the revenue from the natural gasoline plant recoveries and disposition of the natural gas liquids which may be associated with the methane portion of the gas stream.

Liquified Natural Gas (LNG)

The increasing ratio of finding gas instead of oil from new exploration has already been mentioned. On the worldwide scene this results in more and more gas being found in the less developed countries overseas. The fundamental concept is that gas does not store handily and for practical purposes, once produced from the wells has to be kept moving, even when it is being measured. Natural gas pipelines are unfortunately rare in the developing world. The only way around this dilemma of how to monetize natural gas from remote areas may be by liquifying it and moving it by especially designed and constructed tankers to the industrial ports of countries with little or no indigenous supply of natural gas. The technology is available to accomplish this. Compression for liquefaction is one of the major costs of this system, consuming approximately 25% of the gas produced.

The pioneering work in LNG was spearheaded by the British Gas Council in a joint venture with Conoco and the Chicago Stock Yards. The project finally became operational by moving LNG from Arzu in Algeria to Canvey Island in the mouth of the Thames in England, where it was regasified and put into the mains of the Gas Council for distribution in southern England. All of this occurred just shortly before Amoco and Shell-Esso brought in the Leman and Inde Gas Fields in the North Sea in the mid-1960's. As soon as production from those fields was piped ashore to Bacton on the English east coast, the LNG project was shelved as uneconomical and the storage and regasification facilities put in "mothballs."

The following table, compiled from Oil and Gas Journal data, indicates the origin of LNG and the major markets for this product.

COUNTRY OF ORIGIN	2006 LNG EXPORTS (BILL. CUBIC METERS	MARKET
U.S.A.	1.72	SE Asia
Trinidad	16.25	U.S.A., Europe, Caribbean
Algeria	24.68	U.S.A.,SE Asia, Europe
Egypt	14.97	U.S.A., SE Asia, Europe
Libya	0.72	Spain
Nigeria	17.58	U.S.A., SE Asia, Europe
Abu Dhabi	7.08	SE Asia
Oman	11.54	SE Asia
Qatar	31.09	SE Asia, Europe, Mexico
Australia	18.03	SE Asia
Brunei	9.81	SE Asia
Indonesia	29.57	SE Asia
Malaysia	28.04	SE Asia
Total	211.08	

As can be seen in Figure 3-20, global trade in LNG has increased at an average of 7.5% per year since 1980, with over 70% going to Japan, South Korea, and Taiwan. This is used mainly for the generation of electricity. Based upon new plants under construction and planned, this growth rate should continue. The cost of LNG has declined significantly, but most of this decline is attributed to "economy of scale" as larger plants have been constructed. The current cost of delivering LNG from new capacity to the Eastern U.S. Ranges from $1.95 to $3.10 per MCF, depending upon the source.

Demand for LNG has recently increased in Europe to supplement pipeline gas from the North Sea, Russia, and North Africa. There are gas pipelines crossing the Mediterranean Sea from Africa to Italy and another across the Straits of Gibraltar to Spain. Historically only a small amount of LNG has been imported into the U.S., but as U.S. Gas reserves decline LNG imports will increase. There are several LNG terminals on the East Coast, Gulf Coast, and West Coast that will see increased use in the future to supplement gas being imported into the lower 48 States by pipeline from Canada and eventually Alaska.

FIGURE 3-20
WORLD LNG IMPORTS

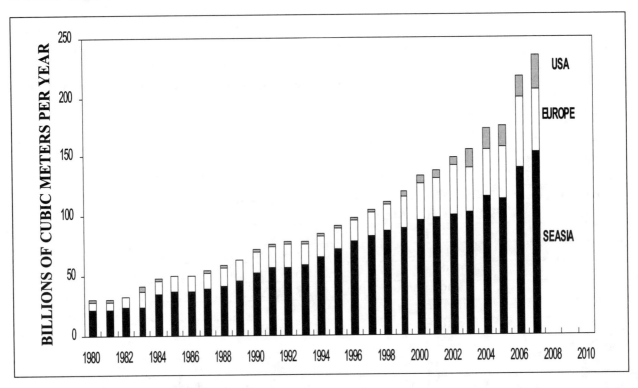

The world's LNG market is split into three geographic areas. The principal demand area, led by Japan and the other Asian Pacific Rim countries, has essentially no indigenous gas supply coupled with heavy industrial needs. The other two areas are Western Europe and the U.S., where the opportunity to deliver natural gas in major quantities directly to the local industrial market by pipeline at competitive prices offers a growing LNG marketing opportunity.

The world's LNG supply capability is increasing rapidly with essentially every major exporter planning capacity expansion over the next few years. Several major new discoveries of natural gas in areas of restricted local marketing opportunity have also added a number of new LNG projects to those currently on the drawing boards. Nigeria, Norway, the USSR, and Trinidad, are examples.

The production, transport and marketing of LNG comprises a chain of events, each of which involves a new level of high tech development in order to handle the very low temperature requirement to keep methane liquid at near atmospheric pressure.

Unfortunately, this temperature of -258°F embrittles most materials normally used to transport hydrocarbons. Technology was developed to overcome this problem, enabling the construction of modern LNG tankers that can transport 3-5 bcf (at Standard conditions) of liquified natural gas. These tankers are normally fueled by the small amount of gas that vaporizes during the trip. The physical chain of events includes:

A.   – Gas Production

      – Liquefaction and Tanker Loading

B.   Shipping

C.   – Unloading and Regasification

It is interesting to recognize that LNG is almost universally an international business, and that A, B and C occur in different countries or under differing governmental jurisdictions. This could make an interesting heading under the caption of Political Risk discussed later in Chapter VI.

Gas-to-Liquids (GTL)

Many of the storage and transportation problems associated with natural gas could be solved more easily and at a lower cost if gas could be transformed directly to a product which is a liquid at normal temperatures and pressures, rather than at the extremely low temperatures required by the LNG process. Natural gas can be converted to alcohol, but there is only a limited demand for this product. Natural gas can also be converted to liquid petroleum through the gas-to-liquid (GTL) Fischer-Tropsch process, developed by Franz Fischer and Hans Tropsch at the Kaiser Wilhelm Institut für Kohlenforschung in Mühlheim-Ruhr, Germany, in 1923. However, this process is currently more costly and less efficient than the LNG process. Currently it takes about 10,000 standard cubic feet of natural gas to make one barrel of liquid hydrocarbon product.

Recent efforts to improve the GTL process have focused primarily on improving the efficiency of catalysts used in this process. Significant investments by several giant international oil companies indicate that it may not be too long before the GTL process can economically provide liquid petroleum products from natural gas. Even though GTL is less energy efficient than LNG this may be more than offset by the value of creating a product that can be shipped more easily, requires no long-term take-or-pay contracts and can be produced by facilities requiring shorter financing and construction schedules. Additionally, GTL products may have a premium value because of their environmentally friendly properties, i.e., the absence of aromatics, sulfur, or metal contaminants. An economic GTL process would also make it possible to develop remote gas discoveries and associated gas whose reserves are less than the minimum essential to warrant an LNG facility. Although existing LNG infrastructure may not be threatened by GTL technologies, it may be a viable alternative to future LNG and pipeline projects.

WORLD GAS PRODUCTION AND RESERVES

Table 3-4 is a summary of gas production and reserves for the major producing countries of the world. The countries have been listed according to their R/P ratio, as was done for oil in Table 2-6 in the previous chapter. The percentages of worldwide reserves and production is shown for each of the countries.

TABLE 3-4
LNG EXPORTING COUNTRIES - 2006

COUNTRY	1/1/08 ESTIMATED RESERVES** (tcf)	2007 PRODUCTION (tcf)	WORLDWIDE SHARE RESERVES	PRODUCTION	(R/P)
United Kingdom	14.6	2.696	0.24%	2.67%	6.4
Mexico	13.9	2.211	0.22%	2.19%	7.3
Canada	58.2	5.935	0.94%	5.88%	10.8
Argentina	15.8	1.518	0.25%	1.50%	11.4
U.S.	211.1	19.951	3.41%	19.77%	11.6
Trinidad & Tobago	18.8	1.410	0.30%	1.40%	14.3
Netherlands	50.0	2.705	0.81%	2.68%	19.5
Pakistan	28.0	1.414	0.45%	1.40%	20.8
Australia	30.0	1.408	0.48%	1.39%	22.3
Norway	79.1	3.162	1.28%	3.13%	26.0
China	80.0	2.432	1.29%	2.41%	33.9
Indonesia*	93.9	2.350	1.52%	2.33%	41.0
India	38.0	0.935	0.61%	0.93%	41.6
Algeria*	159.0	3.310	2.57%	3.28%	49.0
Malaysia	83.0	1.713	1.34%	1.70%	49.5
Bolivia	26.5	0.500	0.43%	0.50%	54.0
Russia	1,680.0	22.550	27.16%	22.34%	75.5
Azerbaijan	30.0	0.330	0.48%	0.33%	91.9
Kazakhstan	100.0	0.940	1.62%	0.93%	107.4
Saudi Arabia*	253.1	1.955	4.09%	1.94%	130.5
U.A.E.*	214.4	1.577	3.47%	1.56%	137.0
Kuwait*	56.0	0.371	0.91%	0.37%	152.0
Libya*	50.1	0.266	0.81%	0.26%	189.4
Venezuela*	166.3	0.867	2.69%	0.86%	192.8
Nigeria*	184.0	0.860	2.97%	0.85%	214.9
Angola*	9.5	0.041	0.15%	0.04%	234.0
Iran*	948.2	2.970	15.33%	2.94%	320.3
Qatar*	905.3	1.825	14.64%	1.81%	497.1
Iraq*	111.9	0.059	1.81%	0.06%	1,914.5
Others	477.2	12.678	7.71%	12.56%	38.6
WORLD TOTAL	**6,185.7**	**100.938**	**100.00%**	**100.00%**	**62.3**
TOTAL OPEC*	**3,151.7**	**16.450**	**50.95%**	**16.30%**	**192.6**

SOURCE: O&GJ, DECEMBER 24, 2007 & MARCH 10, 2008

* OPEC COUNTRIES
** PROVED RESERVES, WITH PRESENT TECHNOLOGY AND PRICES, EXCEPT RUSSIAN FIGURES WHICH ARE "EXPLORED RESERVES" INCLUDING PROVED PLUS SOME PROBABLE RESERVES.

UAE = ABU DHABI + DUBAI + RAS AL KHAIMAH + SARJAH

KUWAIT & SAUDI ARABIA EQUALLY SHARE NEUTRAL ZONE

REFERENCES

Ball, Max, This Fascinating Oil Business, (1940), Bobbs-Merrill Co., New York

Breton, T. R., "World Oil Market," Petroleum Economist, Dec. 1985

Economides, M.J., and Oligney, R.E., and Demarchos, A.S., "Natural Gas: The Revolution is Coming", JPT, May 2001, Society of Petroleum Engineers, Richardson, TX 75083-3836

Ellis Jones, P., Oil A Practical Guide to the Economics of World Petroleum, 1988, Woodhead Faulkner Ltd., Cambridge CB2 1QY

Enron Capital & Trade Resources, Lou Pai, Managing director, Managing Energy Price Risk, 1995, Risk Publications, London W1M 5FU

Fesharaki, F. and Razavi, H., Spot Oil, Netbacks and Petroleum Futures, 1986, The Economist Press, London W1A 1DW

Frankel, P. H., "The Structure of the Oil Industry," lecture given at the Institute of Petroleum, London July 4-6, 1988

Hampton, Michael, "Cycling Towards Low Prices," Petroleum Economist, Apr. 1991

Hough, G. V.,"World Survey - Liquified Natural Gas," Petroleum Economist, Dec. 1988

Leeman, W. A., The Price of Middle East Oil, 1962, Cornell University Press, Ithaca, NY

McKechnie, G., editor, Energy Finance, (1983), Euromoney Publications, London EC4

Meyer, M., Markets, 1988, W. W. Norton & Company, New York & London.

Montgomery, J., Span Magazine, Nov. 1986, p. 15

Morse, E. L., et al., Joan Pearce, Ed., The Third Oil Shock, 1983, Routledge & Kegan Paul Ltd., London WCD1E 7DD

New York Mercantile Exchange, "NYMEX Energy Complex", booklet, 1992 New York Mercantile Exchange, Four World Trade Center New York, NY 10048

Petroleum Finance Company, World Petroleum Markets, 1988, World Bank Tech. Paper No. 92

Rafferty, G. P., The Emergence of the Energy Futures Market, 1985, Cambridge Energy Forum, Cambridge MA 02138

Razavi, H. and Fesharaki, F., Fundamentals of Petroleum Trading, 1991, Prager Publishers,NewYork.

Terzian, P., OPEC: The Inside Story, 1985, Zed Press, London N19DN

Yergin, D., et al., The Future of Oil Prices: The perils of prophecy, 1984, Cambridge Energy Research Associates, Cambridge, MA 02138 and Arthur Anderson & Co., Chicago, IL 60602

Yergin, D., The Prize, 1991, Simon & Schuster, New York 10020

APPENDIX III-A
TYPICAL CRUDE OIL ASSAYS

TABLE 3A-1
DUBAI (FATEH), DUBAI, U.A.E.

Crude

Gravity, API: 31.05
Sulfur, wt%: 2.0
Kin. vis., 37.8 cSt: 6.7
Con. carbon, wt%: 4.62
V/Ni, ppm: 42/14
Pour Pt. C: -9
Asphaltenes, wt%: 1.70

Debutanized Naphtha

Range, C: 15-145
Yield, vol%: 16.4
Gravity, API:
Sulfur, ppm: 193
Mercaptans, ppm: 325
Paraffins, vol%: 68.4
Naphthenes, vol%: 23.1
Aromatics, vol%: 8.5
RON, clear: 55.6

Kerosene

Range, C: 145-230
Yield, vol%: 15.0
Sulfur, wt%: 0.25
P/N/A, vol%: 50.9/29.9/19.2
Smoke pt. mm: 22
Analine pt. C: 56.0
Diesel Index: 62.1
Cetane Index: 56.0

Gas Oil

Range, C: 175-400
Yield, vol%: 39.9
Gravity, API 34.1
Sulfur, wt%: 1.43
Kin. vis., cST @ 20C: 3.02
           @ 40C: 1.99
Analine pt., C: 60.8
Diesel Index: 54.6
Cetane Index: 51.0

Residue

Range, C: 350+
Yield, vol%: 45.0
Gravity, API 14.75
Sulfur, wt%: 3.0
Pour pt., C: +19
Con. Carbon, wt%: 9.1
Vni, ppm: 85/27
Kin. Vis. CSt @ 50C: 351
              @ @ 100C: 31.0

Courtesy of Oil & Gas Journal

TABLE 3A-2
EKOFISK, NORWAY NORTH SEA

(Seal Sands, Teeside, U.K. pipeline terminal)

Crude

Gravity, : 43.4
Sulfur, wt%: 0.14
Pour pt., F: 10
Vis., SUS @ 70F: 36.1
RVP, psig: 5.0
Yield, C1—C1, wt%: 6.43

Naphtha

Range, F: 176-302
Yield, vol%: 20.2
Gravity, API: 57.9
Sulfur, ppm: 276
Mercaptans, ppm 276:
P/N/A, vol%: 53.1/36.8/10.1

Light Gas Oil

Range, F: 400-662
Yield, vol%: 23.4
Gravity, API: 38.0
Sulfur, wt%: 0.07
Pour pt, F: 200.5
Aromatics, vol%: 17
Analine pt, F: 200.5
Diesel Index: 62
Vis, cSt @ 100F: 3.01

Residue

Range, F: 700+
Yield, vol%: 25.2
Gravity, API:20.9
Sulfur, wt%: 0.40
Pour pt, F: 97
Con. Carbon, wt%: 4.3
V/Ni/Fe,ppm: 0.4/5.7/1.2
Kin. Vis.cSt @ 50C:

Gasoline

Range, F: ibp -176
Yield, vol%: 11.9
RON, clear: 66

Kerosene

Range, F: 302-400
Yield, vol%: 10.6
Gravity, API: 48.7
Sulfur, wt%: 0.002
Mercaptans, ppm: 182
Smoke pt, F: -85
Freezing pt, F: -85
Analine pt, : 137

Heavy Gas Oil

Range, F: 707-977
Yield, vol%: 13.2
Gravity, API: 25.0
Sulfur, wt%: 0.025
Pour pt, F: 97
Con. carbon, wt%: 0.047
Analine pt, F: 162
V/Ni/Fe, ppm: 0.8/0.3/0.3
Vis, cSt @ 122F: 34.05

Note:
Seven fields contribute to the Eskofisk blend which t h i s assay represents. They are Ekofisk, West Ekofisk, Cod (pure condensate), Tor, Eldfisk, Edda, and Albuskjell.

Condensate is not included in the gasoline and naphha blends listed.

TABLE 3A-3
ABU AL BU KHOOSH, ABU DHABI, U.A.E.
(Storage Tanker Offshore Producing Area, CFP)

Crude

Gravity, API: 31.6

Sulfur, wt%: 2.0

Pour pt., C: -12 (-24.5F)

Kin. vis., 37.8 cSt: 6.7

Con. carbon, wt%: 4.3

RVP, psi: 3.5

V/Ni, ppm: 13/7

IBP -15C yield, wt%: 0.59
                vol%: 0.69

Light Naphtha

Range, C: 15-65

Yield, wt%: 2.74
      vol%: 3.50

Gravity, API: 88.7

Sulfur, wt%: 0.03

Mercaptans, ppm: 325

Paraffins, vol%: 95

Naphthenes, vol%: 4.4

Aromatics, vol%: 0.06

RON, clear: 74

RON +3ml TEL: 94.5

Heavy Naphtha

Range, C: 65-165

Yield, wt%: 14-45
      vol%: 16.83

Gravity, API: 59.6

Sulfur, wt%: 0.05

Mercaptans, ppm: 276

P/N/A, vol%: 68.3/18.5/13.2

Kerosene

Range, C: 165-230

Yield, wt%: 11.52
      vol%: 12.53

Gravity, API: 46.25

Sulfur, wt%: 0.16

Mercaptans, ppm: 132

P/N/A, vol%: 55.5/21.8/22.7

Freezing pt, C: -52.5

Analine pt, C: 56.4

Diesel index: 61.7

Smoke pt. mm: 23

TABLE 3A-3 (continued)
ABU AL BU KHOOSH, ABU DHABI, U.A.E.
(Storage Tanker Offshore Producing Area, CFP)

Light Gas Oil

Range, C: 230-300

Yield, wt%: 12.5
      vol%: 12.96

Gravity, API: 37.6

Sulfur, wt%: 0.85

Pour pt, C: -27

Cloud pt, C: -19

Analine pt, C: 65

Diesel index: 55.9

Kin. vis. cSt @ 37.8C: 2.63

Heavy Gas Oil

Range, C: 300-375

Yield, wt%: 13.83
      vol%: 13.83

Gravity, API: 30.3

Sulfur, wt%: 1.9

Pour pt, C: +2

Cloud pt, C: +7

Analine pt, C: 72.2

Diesel index: 49.1

Kin. vis. cSt @ 37.8C: 6.92

Distillate

Range, C: 375-550

Yield, wt%: 26.63
      vol%: 24.95

Gravity, API: 21.4

Sulfur, wt%: 2.67

Pour pt, C: +26

Cloud pt, : +28

Con. carbon, wt%: 0.7

Kin. vis. cSt @ 50C: 40.96

Residue

Range, C: 375+

Yield, wt%: 45.37
      vol%: 40.69

Gravity, API: 14.9

Sulfur, wt%: 3.4

Pour pt, C: +21

Asphaltenes, wt%: 3.1

V/Ni, ppm: 30/14

Kin. vis. cSt @ 50C: 435
           @ 98.9C: 40.3

Residue

Range, C: 550+

Yield, wt%: 18.77
      vol%: 15.7

Gravity, API: 4.3

Sulfur, wt%: 4.4

Pour pt, C: +50

Asphaltenes, wt%: 9.2

V/Ni, ppm: 67/35

Kin. vis. cSt @ 98.9C: 1,860

Pen. @ 25C: 125

Courtesy Of Oil & Gas Journal

TABLE 3A-4
PRODUCT COMPARISON FOR EKOFISK AND DUBAI CRUDE OILS

	Ekofisk				Dubai			
	Gasoline Mix							
Component	BPSD	RVP	RON	MON	BPSD	RVP	RON	MON
nC4	1,762	52.0	93.6	90.1	974	52.0	93.6	90.1
S.R. Gasoline	5,950	5.0	66.0	65.0	1,779	9.0	74.0	72.0
Reformate	8,368	3.0	95.0	86.4	6,686	4.0	95.0	86.4
Alkylate	2,272	6.0	94.0	91.4	3,051	6.0	94.0	91.4
FCC Gasoline	5,034	7.5	94.6	81.9	8,275	7.5	93.3	80.9
Nat. Gasoline	906	14.0	70.0	69.0	3,913	14.0	70.0	69.0
Blend	24,293	10.0	RON + MON/2 = 83.5		24,697	10.0	RON + MON/2 = 85.1	

	Jet Fuel Mix							
Component	BPSD	P.Pt.	Sulf.	P.Pt.	BPSD	P.Pt.	Sulf.	P.Pt.
Atm Kero.	4638	-85	0.002	137.0	143.1	-63.5	0.16	133.5

	Diesel Mix							
Component	BPSD	P.Pt.	Sulf.	P.Pt.	BPSD	P.Pt.	Sulf.	P.Pt.
Atm. Kero.	4,638	-85	0.002	137.0	143.1	-63.5	0.16	133.5
Diesel Cut	11,700	0°F	0.070	162.0	6480	-16.0	0.85	149.0
Trtd. GO	0	—	—	—	3000	32.6	0.40	167.0
Ligt. GO	1,238	+15	0.350	70.0	0	—	—	—
Trtd. LCO	0	—	—	—	3245	17.0	0.40	85.0
Blend	13,600	0°F	0.093		17,558	-1°F	-.0.50*	

**Sulfur is the active constraint, 4.865 BPSD of high grade jet fuel has to be downgraded to offset the high sulfur level of the Diesel component.

CHAPTER IV

CASH FLOW

The use of money is all the advantage there is in having it.

—Benjamin Franklin

Before cash flow can be discussed, it is necessary to establish what operations are considered to be cash transactions. Cash includes ALL expenditures, both capital and operating expenses, and all income. The total value of each transaction is recognized at the time that an exchange of values occurs. The exchange of values is usually goods or services for cash. Cash can be represented by currency, check, bank draft, money order, or electronic transfer of funds. The net cash flow of a corporation, a division, an area, individual property or project, is the net sum, positive or negative, of all of the individual items of income and expense relating to the particular entity. The total of all of the inflow and outflow of related funds each year, over the life of an investment, is referred to as the cash flow stream produced by the capitalized investment.

Many people confuse cash flow with accounting profits. On an annual before tax basis, net cash flow may be identified as:

$$\text{NCF} = \frac{\text{Net}}{\text{Annual}} \quad - \quad \frac{\text{Net}}{\text{Annual}}$$

NCF = Net Annual Revenue - Net Annual Expenditures

This is not the way the accounting profession handles the matter, however. An accountant will start with the same "dollars in" and "dollars out," but the accountant then adjusts these numbers in two important ways. First, the accountants, who prefer working on an accrual basis, show profit as it is "earned" rather than when the customer and the company get around to paying their respective bills. Secondly, accountants split cash outflows into current expenses, which are deducted when calculating profit, and capital expenditures, of which a portion is deducted as annual depreciation, depletion and amortization (DD&A) charge against profit before taxes. As a result of these procedures the accountants' figures for profit include some cash flow items and exclude others, and are reduced by DD&A charges, which are not cash flows at all. These "non-cash charges" are discussed more fully in Chapter VIII.

Accountants develop the cash flow statement recognizing the applicable income tax rate which, of course, must be recognized as one of several important categories of company expense. It should be recognized, however, that income taxes are assessed against the company as a whole, and not against any of its subdivisions or individual projects. In fact, it is essentially impossible to allocate the company's tax burden properly to all of its contributing activities. For this reason many major companies prefer to operate their upstream business on a before federal income tax cash flow basis, rather than attempting to deduct a hypothetical annual tax for each individual project or company unit. This is not universally true throughout the industry, however. Alternatively, some companies approximate the income tax effect for each project by applying the maximum tax rate to the taxable income generated by each project.

Cash flow is extremely important to any commercial enterprise. A business can go bankrupt, no matter how profitable the company books may indicate it to be, if the firm doesn't have sufficient cash to pay its bills. A frequent saying in financial circles is that "Happiness is a positive cash flow."

Cash flow for a typical Exploration and Production venture can be determined annually or for the life of a project by the following process:

+ **GROSS REVENUE = (PRODUCED VOLUME OF HYDROCARBONS) x (PRICE)**
- **ROYALTY = (FRACTION) x (GROSS REVENUE)**

= **NET REVENUE**

- **DIRECT OPERATING EXPENSES**
- **INDIRECT EXPENSES (OVERHEAD)**
- **OPERATING TAXES (VAT, SALES, PROPERTY, ETC.)**

= **OPERATING CASH INCOME BEFORE INCOME TAX (OCIBT)**

- **CAPITAL EXPENDITURES**

= **NET CASH FLOW BEFORE INCOME TAX (NCF BT)**

- **INCOME TAX**

= **NET CASH FLOW AFTER INCOME TAX (NCF AT)**

. . . (DISCOUNT CASH FLOW AT HURDLE RATE) . . .

= **PRESENT VALUE NET CASH FLOW AFTER INCOME TAX (PV NCF AT)**

CASH FLOW DIAGRAMS

The Cash Flow Diagram is simply a graphical sketch representing the timing and direction of the monetary transfers. The diagram begins with a horizontal line, called the time line. The time line represents the duration of the financial exercise, and is divided into compounding periods.

■ EXAMPLE 4-1
■ CASH FLOW COMPARISONS

Consider a situation in which a sum of money is paid out at time zero. A different sum is received six years later. It may be more, due to receiving interest on the use of the money during the six years. This would be diagramed as follows:

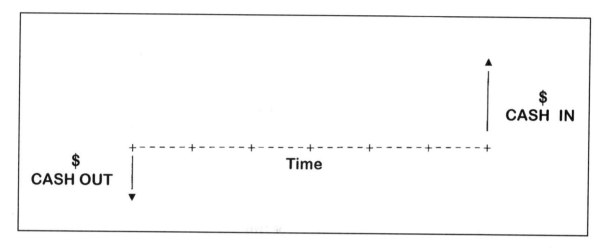

The exchange of monies in an exercise is depicted by the vertical arrows. Money received is represented by an arrow pointing upward from the point in time when the transaction occurs; money paid out is represented by an arrow pointing down. Sometimes money is received and paid out within the same time frame. This may be handled either as a single arrow representing the net of the transactions, or by showing both the positive and negative arrows as a help in keeping track of all the activity. This might be diagramed as follows:

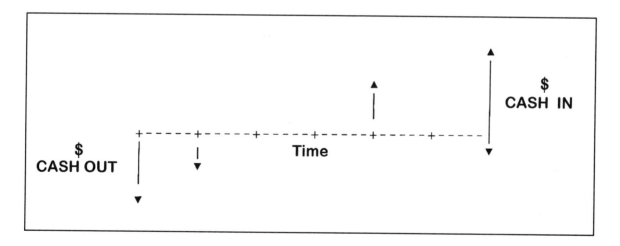

The arrow pointing up at the right hand end of the diagram indicates the money received at the close of the transaction even though there were also disbursements made at that time. The closing net cash is the algebraic sum of the two.

Every completed cash flow diagram must include at least one cash flow, or disbursement, in each direction. Note, however, that cash flows corresponding to the accrual of interest are not represented by specific arrows in cash flow diagraming, unless interest is actually paid by the element being evaluated.

The lengths of the arrows, positive or negative, may be drawn to scale representing the magnitude of the cash flows for the period. Even a rough "not to scale" sketch can be very useful in tracking cash inflows and outflows.

Direct Operating Expense

Both fixed and variable cost items are important in a realistic projection of project expense. Some of the expense forecast may be on a monthly per-barrel or per-well basis. Other significant items of expense such as individual well repair or production-stimulation treatments occur less frequently. These are projected on a statistical basis over the life of the property. For example, it may be assumed that each well will be repaired on the average of once every five years. This type of estimating may give unexpected bumps in the cash flow forecast. This is probably a realistic approach over a long number of years. The difficulty, however, is in forecasting the exact year of the "bump."

Operating Cost Data generally must be developed from historical records for the property or from nearby similar operations. Operating costs are frequently reduced to ratios of monthly costs on a per-well or per-daily-barrel basis for each month of operation. The cost of repairing and working over individual wells will show up on the individual account, and thus must be included in the forecast of operating expenses for cash flow analysis purposes. In the U.S.A. all oil and gas production books for tax purposes must be kept on an individual property basis. Elsewhere in the world the cost accounting unit are specified by the terms of the contract with the host government.

TABLE 4-1
**ELEMENTS OF
DIRECT OPERATING
EXPENSE - ONSHORE**

Elements	Basis
Pumper wages, benefits	Hourly rate
Pumper transportation	Mileage rate
Purchased power, fuel, water	Metered volume
Field power, fuel, water	Metered vol. or allocation
Treating chemicals	Warehouse charge
Small tools, supplies	Warehouse charge
Teaming, trucking, heavy equipment	Hourly rate incl. operator
Gas gathering, compression	Metered volume
Salt water disposal	Tested volume
Roustabout gang wages, benefits	Hourly rate
Roads, bridges, docks	Direct account
Flow lines, tank batteries	Direct account
Wells—pulling, cleanout tbg and csg repair plug back zone or gas shut-off	Individual well account Hourly rate + mileage
Outside labor and equipment Service co. personnel, equip. Service co. deliveries Contractor services	Contr's rate + schedule
Crop damage Pipeline Gauging	Negotiated agreement Measured volume

126

Offshore operations involve quite different direct operating cost items. These would include the costs of workboats subdivided into the bareboat lease, crew costs, and fuel. There would still be warehouse charges, but docking costs might very well substitute for lease road maintenance. Table 4-2 outlines the major items of offshore operating expense.

TABLE 4-2
TYPICAL DIRECT OPERATING EXPENSES - OFFSHORE

- SUPPLY BOATS
- HELICOPTERS
- STANDBY VESSELS
- DOCKING CHARGES
- SHORE BASE EXPENSE
- UNDERWATER INSPECTIONS
    —PLATFORMS AND PIPELINES
- COMMUNICATIONS AND DATA TRANSMISSION
- PERSONNEL
    —PROCESS OPERATORS
    —PROCESS MAINTENANCE
- WAREHOUSE
- PIPELINE TARIFFS
- FUEL
- EQUIPMENT STANDBY
    —WIRELINE
    —CEMENTING PUMPS

Indirect Expenses (Overhead)

The general rule for charging costs directly to an operation is that those charges must be for work physically performed at the project site, or if not on site they must be performed specifically and exclusively for that operation. All other costs which may be incurred at a distant location for a number of different operations are considered indirect operating costs or "overhead." These include a prorated portion of supervisory and administrative expense covering the individual property, which is also shared with other operations under the same district or organizational unit. These are summarized in Table 4-3 and listed in column (i) of the Table 4-6 spreadsheet.

The specific method of allocating these expenses is arbitrary. Specific rules for allocating must be established to maintain uniformity from month to month. Such rules, for example, charging a percentage on top of certain direct costs, may work in many cases. This percentage could be determined as a percent of direct operating costs plus a percent of capital expenditures required for the project, where the percent of capital would occur in those years when capital is spent, whereas the percent of direct operating costs would be charged each year of the project. If this procedure is followed, the percents used for each of these factors should be reviewed annually to determine the appropriate values. This is done by identifying the total corporate overhead allocated to ongoing projects and that allocated to new capital projects and dividing these amounts by the respective total operating costs and capital expenditures. There will also be situations where such a simplified procedure will no longer be justified, for example, a full per-well allocation of district office overhead at the end of the economic life of an individual oilfield.

Functions which benefit more than a single project (overhead Administration G+A) costs which cannot be charged directly to a specific job.

127

TABLE 4-3
INDIRECT EXPENSES (OVERHEAD)

ELEMENTS

- OFFICE EXPENSES, INCLUDING RENT AND UTILITIES
- LEASE SUPERVISION WAGES, BENEFITS
- ENGINEERING SALARIES, BENEFITS
- CLERICAL, ACCOUNTING WAGES, BENEFITS
- TOOLROOM, WAREHOUSE, SHOP WAGES, BENEFITS
- MOTOR POOL EXPENSE, NOT RECOVERED AS DIRECT CHARGE, ON A MILEAGE BASIS
- MANAGEMENT SALARIES, BENEFITS
- SERVICES
- EMPLOYEE RELATIONS
- PUBLIC AFFAIRS
- INSURANCE

Changes in the current operations of the organizational unit such as the start-up of a new drilling program or the phasing down of development drilling are also situations that call for appropriate changes in the allocation of expenses throughout the unit or district.

EXAMPLE 4-2
INDIRECT OPERATING COST (OVERHEAD) FORECAST

In developing a cash flow forecast for the economic evaluation of an investment opportunity, indirect operating costs (overhead) should be included along with the forecast of direct operating costs (DOC). Able Oil Company's economics department has determined that for such evaluations, overhead can be estimated as a fraction of capital expenditures plus a fraction of direct operating costs. A recent study determined that the appropriate fractions would be 9% of capital and 11% of DOC. You need to forecast overhead for a 10-year project that will require an initial capital investment of $1,250,000 and direct operating costs of $100,000 each year for the life of the project.

To the right are spreadsheet columns to develop a forecast of overhead for the 10-year life of the project being analyzed.

YEAR	CAPITAL INVESTMENT	DIRECT OP. COST	INDIRECT OPERATING COST (OVERHEAD) DUE TO CAPITAL (9%)	DUE TO DOC (11%)	TOTAL
1	$1,250,000	$100,000	$112,500	$11,000	$123,500
2		$100,000		$11,000	$11,000
3		$100,000		$11,000	$11,000
4		$100,000		$11,000	$11,000
5		$100,000		$11,000	$11,000
6		$100,000		$11,000	$11,000
7		$100,000		$11,000	$11,000
8		$100,000		$11,000	$11,000
9		$100,000		$11,000	$11,000
10		$100,000		$11,000	$11,000
TOTAL	$1,250,000	$1,000,000	$112,500	$110,000	$222,500

It is common practice in North America for the operator of a joint venture to charge the joint account a fixed rate of overhead for its services in addition to the direct operating expenses incurred. Commonly a per well rate is established for drilling wells, which is applicable only during the drilling phase of the operation, and a significantly lower rate applicable for that well after production begins. Both of these rates are agreed upon prior to execution of the joint venture agreement and included in that agreement. Recognizing that these overhead rates may be in effect for a very long period of time, there is usually a provision in the agreement to adjust the rates for inflation. The Council of Petroleum Accountants Societies (COPAS) annually issues the adjustment factor for the year. The accounting firm of Ernst & Young LLP annually surveys operators in the U.S. and publishes the results in a booklet, Ernst & Young's Fixed Rate Overhead Survey. The results are broken down by producing area and well depth. Table 4-4 is a sample of data from the 2007-2008 survey.

TABLE 4-4
ERNST & YOUNG'S 2007-2008 FIXED-RATE OVERHEAD SURVEY
($/well/month)

Texas

DEPTH IN FEET	Monthly Drilling Wells		Monthly Producing Wells	
	Average	Median	Average	Median
0 – 5,000	$4,282	$3,615	$443	$400
5,001 – 10,000	$6,834	$5,921	$717	$613
10,001 – 15,000	$6,949	$7,206	$778	$769
15,001 – 20,000	$9,168	$9,004	$828	$790
20,000+	$9,158	$9,158	$916	$916
NO DEPTH LIMIT	$7,805	$8,329	$750	$736

Source: Ernst & Young LLP Fixed Rate Overhead Survey ©2007 EYGM Ltd. (www.ey.com)

SPREADSHEETS

Spreadsheets (or worksheets as they are sometimes called) are a basic tool for conducting economic analysis. Physically these consist of sheets with columns (vertical) and rows (horizontal). On these the analyst labels the rows and columns and starts posting data which may be past actuals and forecast futures. A spot or "address" of a particular column and row is known as a cell. Depending upon the project being studied the cell values can conveniently be added, or subtracted from one another, or multiplied or divided by values in other cells. These related cells are usually located conveniently along side or close by, in order to facilitate the analysis. A typical spreadsheet may require 30 columns or more. Excel and Lotus 1-2-3 are popular personal computer spreadsheet software.

Economic analysts coming from an academic background in petroleum engineering generally like to adapt the left-hand most column to listing the years of production. Analysts from the financial side of things prefer to list years as the top row to facilitate annual comparisons much as some financial statements are presented. Either way works.

The most common source of income from an oil and gas producing property is from the sale of crude oil, natural gas, natural gas liquids and sulfur. To begin to project cash flows, it is necessary to build a year by year projection of the production and sale of each of these products. Most contracts permit the royalty-free use of natural gas on the individual property where it is required in the producing operation. Accordingly, there is an important distinction between production and sales. It is sales revenues that should be included in the cash flow spreadsheets.

If the property produces primarily crude oil, sales may be limited by pipeline capacity or by state or provincial restrictions on production. In these situations, the projection of production and sales is a simple matter in the early years until the producing characteristics change, due to the normal depletion of the reservoir. Thereafter, forecasting the income from the producing stream becomes more complex.

PETROLEUM INDUSTRY CASH FLOW SPREADSHEETS

The first step in developing a cash flow forecast of an oil or gas producing operation is the preparation of a year-by-year projection of the future sales revenues from each of the produced streams. The projection of production and sales is a simple matter, at least in the early years, until the producing characteristics change, perhaps due to the normal depletion of the reservoir. Thereafter, forecasting the income from that producing stream becomes more complex. The production forecast may come from a reservoir engineer's computer, or from a decline curve analysis, or even a simple intuitive approximation. Format #1 is a sample spreadsheet format for estimating the cash flow before interest and taxes (BIT) for a property that is primarily oil productive.

TABLE 4-5a
SPREADSHEET FORMAT #1

Forecast of Oil Production and Net Revenue (Before Income Taxes).

(1)	(2)	(3)	(4)	(5)	(6)	(7)	(8)	(9)	(10)	(11)
Year	WI Share Investment ($)	Total Annual Gross Production (bbls)	Oil Price ($/bbl)	Total Annual Gross Revenue ($)	WI Share Revenue ($)	WI Royalty ($)	WI Share Net Revenue ($)	WI Share Operating Expenses ($)	WI Share Net Cash Flow ($)	WI Share Cum. Net Cash Flow ($)

Column
(2) = (Total Investment)* W.I.
(5) = (3) * (4)
(6) = (5)* W.I.
(7) = (6)* (Royalty Fraction)
(8) = (6) - (7)
(9) = (Total Op. Exp.)* W.I.
(10) = (8) - (9) - (2)

The cash flow analysis of natural gas production is slightly more complex due to the additional income projections for a series of product sales including the gas, NIL, and sometimes sulfur. A typical cash flow spreadsheet for this type of cash flow is shown as Format #2.

TABLE 4-5b
SPREADSHEET FORMAT #2

Forecast of Natural Gas Production, By-Products and Net Revenue (Before Income Taxes).

(1)	(2)	(3)	(4)	(5)	(6)	(7)	(8)	(9)	(10)	(11)	(12)	(13)	(14)	(15)
Year	Investment ($)	Gross Pipeline Gas Sales (MMCF)	Gas Price ($/MCF)	Gross Gas Revenue ($)	Gross Liquids Production (bbl)	Liquids Price ($/bbl)	Gross Liquids Revenue ($)	Total Gross Revenue ($)	Royalty ($)	Operating Expense Wells ($)	Plant ($)	Compr. ($)	Net Cash Flow ($)	Cum. NCF ($)

Column
(5) = (3) * (4)
(8) = (6) * (7)
(9) = (5) + (8)
(10) = (9)* (Royalty Fraction)
(14) = (9)-(10)-(11)-(12)-(13)-(2)

Example of a Concession Type Contract

A concession type of contract, also known as a lease or license, allows the contractor to explore for hydrocarbons in a specified area for a limited period of time. If a commercial discovery is made, the contractor has the right to produce hydrocarbons therefrom over an additional specified period of time. In exchange for this right, the contractor agrees to pay royalty to the mineral owner on the hydrocarbons produced and sold and to pay income tax on the profits to the government. Table 4-6 shows a forecast cash flow for the first eight years of a producing operation under a concession type of contract. In this format, the investments are shown in the left-hand-most group of columns, followed by revenue and then expenses. This is a convenient format for evaluating the return on a project's investments.

It should be remembered that individual items of cash flow may be either positive or negative. For most oilfield investments, the cash flows in the early months or years are negative. "Payout" is the point in time when the cumulative net cash flow for the particular project being analyzed is zero or breaks even after which the cash flow remains positive until such time other major capital investments occur such as infill drilling or secondary recovery operations, which may again result in negative cash flows. Additional funding may be required at this point with its attendant payout period.

Investments

The following is a list of several important negative items of Cash Flow that are classed as "Investment," which constitute cash items:

The largest initial development costs on the typical concession, as depicted in Table 4-6, involve expenditures for drilling. The actual costs of drilling and completing wells are available from accounting records at a much later date. Therefore, most cash flow projections as in Table 4-6 must rely on estimates. The actual costs of drilling can be very important, however, in developing improved estimates of the costs of future drilling in the vicinity. Such

131

items as tank battery and other lease production facilities are generally included with the costs of the first well on the lease. All these data must be incorporated in the investment and expense portions of the net cash flow projections.

In Table 4-6, the cash flow investment items are divided into several columns because each has a distinct tax treatment, at least in the U.S.A.

TABLE 4-6
SPREADSHEET OF NET CASH FLOW FOR A CONCESSION

	Investment				Revenue		
Year	Intangible Drilling Cost	Tangible Drilling Cost	Other Tangible Cost		Gross Revenue	Royalty	Net Revenue After Royalty
(a)	(b)	(c)	(d)		(e)	(f)	(g)
1	$95,000	$35,000	$5,000				
2					$175,000	$21,875	$153,125
3					$175,000	$21,875	$153,125
4					$170,000	$21,250	$148,750
5					$165,000	$20,625	$144,375
6					$160,000	$20,000	$140,000
7					$155,000	$19,375	$135,625
8					$150,000	$18,750	$131,250
TOTAL	$95,000	$35,000	$5,000		$1,150,000	$143,750	$1,006,250

	Expenses				Cash Flow		
Year	Direct Operating Expense	Overhead	Operating Taxes (7% of net)		NCF BFIT	Income Tax (45%)	NCF AFIT
(a)	(h)	(i)	(j)		(p)	(o)	(q)
1					$(135,000)	$(35,061)	$(99,939)
2	$27,000	$3,900	$10,719		$111,506	$43,205	$68,302
3	$25,500	$4,100	$10,719		$112,806	$45,049	$67,757
4	$25,500	$4,100	$10,413		$108,738	$44,118	$64,620
5	$25,500	$4,000	$10,106		$104,769	$42,974	$61,794
6	$25,500	$4,000	$9,800		$100,700	$43,709	$56,991
7	$25,500	$4,000	$9,494		$96,631	$41,878	$54,754
8	$25,500	$4,000	$9,188		$92,563	$40,850	$51,713
TOTAL	$180,000	$28,100	$70,438		$592,713	$266,721	$325,992

	Income Tax Calculations					Present Value	
Year	Depreciation Schedule*	Depreciation	Intangible Write-Off	Taxable Net Inc.	Income Tax (45%)	Discount Factor (10%)	PV NCF AFIT (10%)
(a)	(k)	(l)	(m)	(n)	(o)	(r)	(s)
1	14.29%	$5,714	$72,200	$(77,914)	$(35,061)	0.9535	$(95,288)
2	24.49%	$9,796	$5,700	$96,010	$43,205	0.8668	$59,203
3	17.49%	$6,997	$5,700	$100,109	$45,049	0.7880	$53,392
4	12.49%	$4,998	$5,700	$98,040	$44,118	0.7164	$46,920
5	8.92%	$3,570	$5,700	$95,499	$42,974	0.6512	$40,242
6	8.92%	$3,570		$97,130	$43,709	0.5920	$33,740
7	8.92%	$3,570		$93,061	$41,878	0.5382	$29,469
8	4.46%	$1,785		$90,778	$40,850	0.4893	$25,302
TOTAL	100.00%	$40,000	$95,000	$592,713	$266,721		$192,350

* 7-year depreciation schedule (DDb / SL)
** 70% Written off immediately, balance 5-year (SL)

COLUMN:

(b) Intangible Drilling Costs, (IDC's), are separated from the tangible or hardware cash outlays for drilling and completion because IDC's may be taken as a tax deduction in the year of expenditure (only partially in the U.S.). IDC's, for the most part, represent the payments to the drilling contractor for his services, wire line services, cement, etc.

(c) The remaining well investment, or Tangible Drilling Cost, which is depreciable along with other lease equipment, is segregated since depreciation on these amounts must be considered for income tax limitations. Depreciation of this investment item is utilized in the tax calculation only and is not otherwise considered as a cash expense item.

(d) Other Depreciable Investment items include all capital items associated with the operation that are not classified as lease and well investment. An example might be company vehicles that also serve other activities. In such instances only the proportional amount applicable to the property or unit being served should be included.

Note: Tangible (Depreciable) Capital assets have a useful life greater than one year and a recognizable salvage value.

Revenue

The income portions of cash flow are also known as "Revenue" and include columns (e) through (g) in Table 4-6.

COLUMN:

(e) Gross revenue is the product of oil and/or gas volume times the unit price received for each volume of that product.

(f) Royalty is the fraction of gross revenue paid to the owner of those minerals according to the contractual relationship between the parties.

(g) Revenue remaining after the deduction of royalty is used by the project to pay expenses, repay investments, and provide a profit to the producer.

Expenses

The various categories of expense also constitute items of negative cash flow. These occupy columns (h) through (j) in Table 4-6.

COLUMN:

(h) Direct operating costs include all ongoing expenses necessary to conduct the project being evaluated. The various categories of expense constitute items of negative cash flow.

(i) Overhead includes all indirect costs which are allocated to each operating project and includes general supervision, management and services provided at the corporate level, as previously discussed.

(j) Operating taxes include all taxes, other than income taxes, which may be imposed at the project level commonly called excise taxes, such as sales, VAT, property, or production taxes. The example is based upon a production tax of 7% of revenue after royalty.

Income Tax

COLUMN:

(k) The depreciation schedule used in this example is the one developed by the U.S. tax authority (IRS) to be applied to tangible capital expenditures of the oil industry to determine taxable income. It is based upon a combination of double declining balance and straight line depreciation methods using a life of seven years. Mid-year convention is used, allowing only a half-year depreciation for the first year, with the final half-year depreciation taken in year eight.

(o) The other major item of government take is the income tax listed in column (o). Income tax is determined as a separate stand alone calculation which is not a part of the cash flow. To determine the amount of income tax payable, it is necessary to calculate "Taxable Net Income" column (n), which is revenue after royalty (g) minus expenses (h) through (j) and allowable tax deductions (l) and (m). In this example, Income Tax (o) is computed as 45 percent of Taxable Net Income (n). Since there was no revenue in year 1, the taxable net income was negative reflecting depreciation and intangible write-off for that year. The resulting income tax is also negative (-$35,061) and would be used to offset income tax due from other profitable projects. Income tax (o) is a cash flow item and is copied in the Net Cash Flow section.

Net Cash Flow

COLUMN:

(p) Net Cash Flow Before Federal Income Tax (NCF BIT): the algebraic sum of each year's cash flow before federal income tax yields column (p). This is column (g), less the sum of columns (b) through (d) and (h) through (j). Net cash flow before taxes is an important indication of how the operation is progressing from a business standpoint.

(q) Net Cash Flow After Federal Income Tax (NCF AFIT) is NCF BIT (p) minus Income Tax (o). This reflects the true cash flow of the property over the eight year period.

(s) Discounted Net Cash Flow After Federal Income Tax (PV NCFAFIT) was computed with annual mid-year discount factors at 10% (r). Columns (r) and (s) are included here only for completeness. They will be dealt with in much more detail in the following chapter. Net cash flow and present value net cash flow can either be positive (as shown in this example) or negative, which would indicate a loss.

In Table 4-6, royalty at a simple 12 1/2 percent of gross production income [column(e)] is shown separately as column (f). However, in western Canada, where the government is both the royalty owner and the taxing authority, the petroleum industry usually lumps the two together under the term, "Government Take." The North American petroleum industry differs from the rest of the world in its handling and proliferation of the outside share. The practice of including state and local operating taxes as part of the direct operating portion of the cash flow calculation as is done in column (j) of Table 4-6 is quite common.

PSC: *Divides production between gov't + contractor after allowing a portion for cost recovery, income tax + royalty*

Example Of A Production Sharing Contract

Outside of North America, the private-sector oil companies deal with the host governments in a wide variety of arrangements. Figure 4-1 diagrams one such arrangement in the form of a production sharing agreement. This type of agreement originated in Indonesia in 1966. This spreadsheet develops an after-tax net cash flow comparable to the concession example in Table 4-6. Table 4-7 portrays the results of a typical production sharing agreement.

FIGURE 4-1
PRODUCTION SHARING CONTRACT

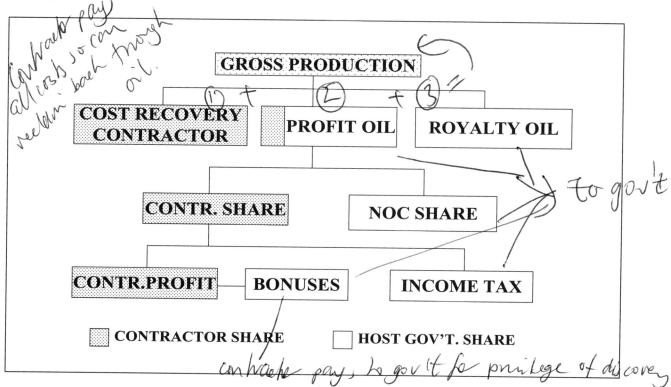

Contractor pays all cost, so can reclaim back through oil.

to gov't

contractor pays, to gov't for privilege of discovery

In the typical production-sharing contract, recovery of costs is handled as a "Cost Oil" account. This accumulates the operating company's exploration, development and operating expenditures for later reimbursement from production income, if and when it occurs. Column (bb) shows the investment schedule over the first two of the five-year period shown.

Column (cc) lists the schedule of bonuses paid to the host government. These are a signature bonus upon signing in year one and a production level bonus paid in year three when the daily producing rate reaches the previously agreed level. Bonuses are not recoverable through the cost oil account, but are paid from the operating company's funds. If commercial production is established, they reduce the operator's profit, otherwise they increase its loss. Even though bonuses may not be recoverable under the contract, they are usually considered a normal business expense for income tax purposes.

Column (dd) includes expenditures other than for development of direct operating costs that are recoverable from cost oil. This might include expenses for training local personnel as required in the agreement. Column (ee), on the other hand, recognizes expenditures that are necessary for the start-

135

up of the operation. Column (ff) recognizes working capital required for the timely payment of freight, shipping and other bills. Working capital is not reimbursable from cost oil on the basis that the funds will eventually work their way through the other accounts back to the operating company.

There is usually a limit to the amount of cost recovery that can be claimed each year. The limitation is commonly a fraction, less than 100%, of the gross revenue after deducting royalty. With a limit on the amount of cost oil that can be claimed by the contractor each year, the government will realize income from profit oil as soon as production begins. Such a limitation also automatically prevents cost recovery from starting before oil production (revenue) begins, because a fraction of zero is zero. This limitation usually delays the recovery of funds expended by the contractor, but eventually the total expenditures will be recovered. As this limit is applied, the costs that the contractor can recover each year is either the limitation fraction times the net revenue or actual unrecovered costs which ever is less. This is shown graphically in Figure 4-2 where the actual expenditures are represented by the area beneath the curved line and the limit by the horizontal line at 40%. Any costs which exceed the annual limit are carried forward until such time as they can be recovered, as represented by the area covered by square shading. Once all past costs have been recovered, the annual cost recovery equals the actual expenditures for that year.

FIGURE 4-2
MECHANICS OF PRODUCTION SHARING CONTRACTS

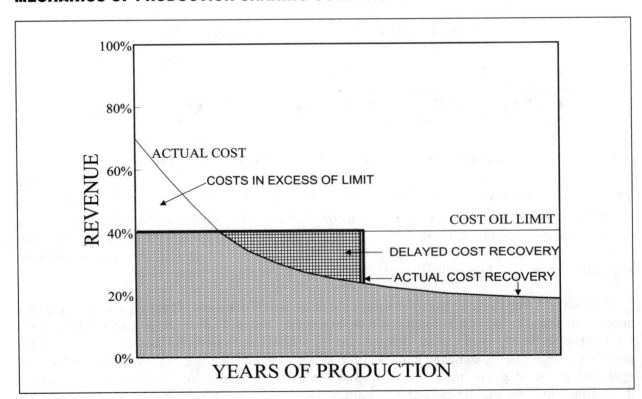

TABLE 4-7
SPREADSHEET FOR A PRODUCTION SHARING CONTRACT

	INVESTMENT					REVENUE	EXPENSES	
Year (a)	Depreciable Dev. Costs (bb)	Bonuses (cc)	Other (dd)	Other than Depreciable (ee)	Working Capital (ff)	Gross Prod. Income (gg)	Direct Oper. Costs (hh)	Overhead (ii)
1	250,000	500,000	10,000	45,000	50,000	0	0	100,000
2	1,250,000		15,000			1,250,000	100,000	100,000
3		500,000				20,000,000	1,000,000	100,000
4						25,000,000	2,000,000	100,000
5						25,000,000	2,000,000	100,000
SUM	1,500,000	1,000,000	25,000	45,000	50,000	71,250,000	5,100,000	500,000

	COST RECOVERY				PROFIT OIL		
Year (a)	Total Recoverable Costs (1) (jj)	Annual Limit 40.00% (kk)	Annual Cost Recovery (ll)	Remaining Unrecovered Costs (mm)	Total (nn)	NOC 85.00% (oo)	Contractor 15.00% (pp)
1	405,000	0	0	405,000	0	0	0
2	1,465,000	500,000	500,000	1,370,000	750,000	637,500	112,500
3	1,100,000	8,000,000	2,470,000	0	17,530,000	14,900,500	2,629,500
4	2,100,000	10,000,000	2,100,000	0	22,900,000	19,465,000	3,435,000
5	2,100,000	10,000,000	2,100,000	0	22,900,000	19,465,000	3,435,000
SUM	7,170,000		7,170,000		64,080,000	54,468,000	9,612,000

	CONTRACTOR CASH FLOW				
Year (a)	Net Cash Flow BFIT (qq)	Income Tax (2) 45.00% (rr)	Net Cash Flow AFIT (ss)	Discount Factor 10.00% (tt)	Present Value Cash Flow (uu)
1	(955,000)	0	(955,000)	0.95346	(910,557)
2	(852,500)	50,625	(903,125)	0.86678	(782,814)
3	3,499,500	1,183,275	2,316,225	0.78799	1,825,152
4	3,435,000	1,545,750	1,889,250	0.71635	1,353,365
5	3,435,000	1,545,750	1,889,250	0.65123	1,230,332
SUM	8,562,000	4,325,400	4,236,600		2,715,478

(1) Excludes bonuses and working capital
(2) Percent of profit oil

The cost oil account is shown in columns (jj) through (mm), with the annual limit as discussed above listed in column (kk). The annual limit for this example is 40% of the values shown in column (gg). Column (ll) is the actual amount of cost recovery allowed each year and it is the lesser of the value for the previous year in column (mm) plus column (jj) or column (kk). The account accumulates until there is production income. Under most agreements, accumulated cost oil account has first call on the production income until the ongoing cost oil comprises only the annual operating costs as listed in column (hh). Some production sharing agreements allow a maximum overhead expense to defray the administrative costs outside the country that directly relate to the project. These expenses might include the costs of arranging for offshore drilling and construction equipment. The category could also include the costs of recruiting and training expatriate personnel for the operation. These expenses are shown as an agreed maximum of $100,000 per year in column (ii).

After the deduction of the accumulated cost oil, the remainder of the production income is designated as "Profit Oil." For the example production sharing agreement (Table 4-7), the split is 85/15 with 85 percent going to the host government/NOC (oo) and 15 percent to the operating company/contractor (pp).

The resulting net cash flow before income tax for the contractor is listed in column (qq). The Host Government applies an income tax rate of 45 percent on the contractor's share of the profit oil which is shown in column (rr).

The contractor's net cash flow after income tax is shown in column (ss). The last two columns (tt) and (uu) show the discounted cash flow (present value) for the operation using a discount rate of 10 percent per year, as was done in Table 4-6 for the concession type project example.

ADDITIONAL CASH FLOW CONSIDERATIONS

Taxation

Taxation of the petroleum industry has been one of its greatest cash flow problems through the years. There are several basic reasons for this. The most equitable tax systems are those that affect the widest possible total of the population. The assumption has been, "what better way to raise revenue than to tax the industry that fuels our transportation and heats our homes?" On the other hand, because companies engaged in the exploration for and the production of oil and gas have to be either large or lucky (preferably both) to sustain the great physical and commercial risks involved, the income and profit figures look so large that the general public sees them as an acceptable target for taxation.

Most governments that levy taxes on the producing sector of the industry have found that taking a share of the oil companies' cash flow, in the form of income and excise taxes after the oil companies have absorbed all the risk, is a lucrative and easily administered source of revenue. The taxing authorities find extreme difficulty in reducing this flow of funds. These income and excise taxes are in addition to the state severance taxes and local ad valorem taxes that have already been discussed.

Taxes, plus economic rents (profits), and royalties on government lands, comprise by far the largest item of oil company expense worldwide. In many countries, the concept of taxation on income derived from oil, gas and mineral production differs significantly from the normal business, accounting or legal concepts of taxable income.

In addition to taxation in the host country where oil and gas are produced, international oil companies are also concerned about income taxes in their home country. Of particular concern to international oil companies is the issue of double taxation of their profits, once in the host country and again in their home country. Many countries have established tax treaties, which minimize this problem. When there is a tax treaty between the host country and home country of the producer, the producer may treat tax payments to the host country as a credit against its home country income tax. The credit is only for income tax actually paid to the host country. If the payment to the host government does not qualify for a tax credit in the home country of the producer, it will be treated as any other business expense and deducted from income to determine taxable income. A tax credit is of more value to a tax payer than a business deduction, so it is important to determine if there is a tax treaty between the countries involved and that the foreign income tax payment is recognized in the home country of the producer. Example 4-3 shows the effect of a tax credit versus tax deduction upon net cash flow.

■ EXAMPLE 4-3
■ TAX CREDIT vs. DEDUCTION ON HOME COUNTRY INCOME TAX

A contractor earns $100 million dollars during a year in a foreign country. Royalty of $10 million dollars is paid to the host country along with a $1 million dollar bonus and the contractor incurs expenses of $15 million during the same year. The host country has a general income tax at the rate of 33% on net income and the contractor's home country also has a general income tax at the rate of 35%.

Determine:

What the contractor's net income will be for that year after paying the required income tax in both the host country and its home country; (a) if there is not a tax treaty between the host country and the home country and (b) if there is a tax treaty between the two countries? The following table shows the effect of income taxation on the Contractor's net income, with income tax payments made to both the

TAX CREDITS vs. TAX DEDUCTIONS (millions of dollars)

	HOST COUNTRY	HOME COUNTRY WITHOUT TAX TREATY	HOME COUNTRY WITH TAX TREATY
GROSS INCOME	$100.000	$100.000	$100.000
ROYALTY	$10.000	$10.000	$10.000
BONUS	$1.000	$1.000	$1.000
EXPENDITURES	$15.000	$15.000	$15.000
HOST COUNTRY			
INCOME TAX RATE	33%		
TAXABLE INCOME	$74.000		
INCOME TAX	$24.420		
HOME COUNTRY			
INCOME TAX RATE		35%	35%
TAX DEDUCTION		$24.420	XXXX
TAXABLE INCOME		$49.580	$74.000
INCOME TAX		$17.353	$25.900
TAX CREDIT		XXXX	$24.420
NET INCOME TAX		$17.353	$1.480
NET INCOME AFTER ALL TAXES		$32.227	$48.100

host government in the country where the operation is located and the government in the contractor's home country.

It can be seen from the bottom line that the contractor will have a greater net income after payment of all applicable income taxes if the foreign income tax paid can be treated as a tax credit on its home country income tax rather than as a business deduction. In the absence of a tax treaty, the income tax paid to the host country is only a normal business deduction for purposes of calculating the home country income tax. This reduces the net income after income tax to the contractor by an amount equal to one minus the home country tax rate times the foreign income tax payment compared to treating the foreign income tax as a credit against home country income taxes. The presence of tax treaties are very important in the international oil business.

Effects of Inflation

Conventional net cash flow analysis are prepared in terms of money. Unfortunately, the buying power of money doesn't remain constant. In an inflationary environment, the value of a given amount of money gradually declines. Each month it takes more Argentine Pesos, or Brazilian Cruizeros, to buy a case of beer. In the oil business long payout projects with low percentage returns may not earn enough to repay the initial investment, when measured in terms of actual purchasing power.

Inflation is that economic phenomena which causes an increase in the average level of prices. When prices decline, which has occurred only rarely since the 1930's, it is called deflation, recession, or depression. The measure for the total, or average, amount of inflation experienced by a country is generally referred to as "Gross Domestic Product Deflator" or may be measured by the "fixed Weighted Price Index for Gross Domestic Product". There are two basic causes of inflation, that being changes in money supply which is controlled by governments and elected officials and non-monetary forces. Since the major factor affecting inflation is public policy, inflation is a country by country phenomena and can vary significantly from country to country at any given time. The two most important non-monetary factors which effect inflation are supply and demand for money and technology. Supply and demand for money will generally reflect the condition of a country's economy and can both increase and decrease the rate of inflation. Technology almost always has a negative effect on inflation as improving technology generally reduces the cost of goods and services. Interest rates always include a component of inflation, so they will move similar to changes in the rate of inflation. Inflation is expressed as the change in prices from one year to the next, so it can properly be handled as an exponential function in a fashion similar to compound interest.

Figure 4-3 shows the history of inflation in the United States from 1980 through 2001. During this period inflation ranged from a low of about 2% per year to over 10% per year. You should note the years between 1973 and 1982. These years of abnormally high inflation in the U.S. were caused primarily by increases in oil prices, with the peaks of 1974 and 1981 reflecting the two OPEC oil price shocks of 1973 and 1979. Most other countries of the world also experienced higher inflation during this period. Currently most countries of the world are experiencing lower inflation rates.

FIGURE 4-3
U.S. INFLATION

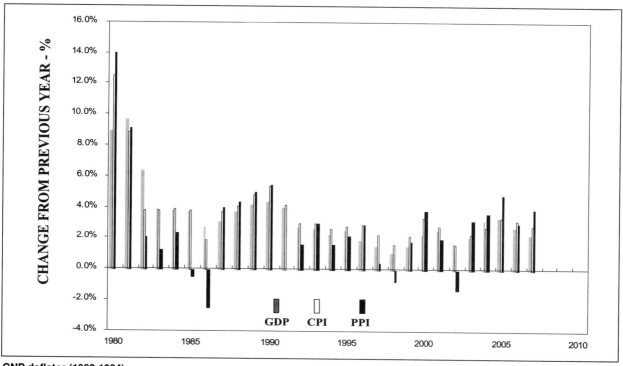

GNP deflator (1980-1984)

Net cash flow projections should be indexed to provide forecasts in terms of actual buying power rather than current dollars. The purpose of this indexing is to eliminate or, at least, reduce the economic risk due to inflation. This indexing procedure involves multiplying each year's forecast of prices, or costs by pre-determined escalation factors so as to represent the monetary values of those costs and prices in the years ahead. The exercise is rendered more complex by the fact that crude oil prices and the costs of drilling, development and production characteristically do not escalate at the same rate. It is also noteworthy that the higher the inflation rates, the greater the uncertainty in their reliability. Nevertheless, inflation is a very serious matter to be dealt with in economic analysis and indexing is the best approach toward reducing this business risk.

In some countries contract terms are indexed according to some standard government report of price levels. Brazil and Israel are examples. Contracts may be fully indexed, or partially indexed. For example, a 10 percent increase in general inflation as shown by the government figures might call for a 7 percent increase in the partial indexing of a contract.

Not all costs will escalate at the same rate as general inflation. Some costs will go up faster than inflation while others will lag inflation. Figure 4-4 is a plot of the Nelson-Farrar Yearly Refining Construction Indexes published by the Oil and Gas Journal, plus the labor component and the components for materials and equipment. It can be seen that the N-F index and its subsets have performed quite differently over the period of this graph, indicating that not all costs have increased equally. Over the period from 1926 to 1980 labor costs went up at the rate of about 6.8% per year while equipment and materials increased at about 5.7% per year. Since 1980 those costs have increased at the rate of 3.0% and 2.0% per year respectively. This performance is not surprising as technology has probably had a greater effect upon the cost of materials and equipment than on the cost of labor. The increase in fringe benefits to employees has also contributed to this difference. Therefore when incorporating the effect of inflation into a cash flow it may be necessary to escalate different items in the cash flow at different rates. This will provide a better forecast of future costs in terms of money of the day when the money will actually be spent. Whether the inflation observed prior to 1980 or that since 1980 is what can be expected in the future is a matter of speculation. But all of these factors must be considered when making a forecast of future inflation.

FIGURE 4-4
NELSON - FARRAR / YEARLY REFINING CONSTRUCTION INDEXES

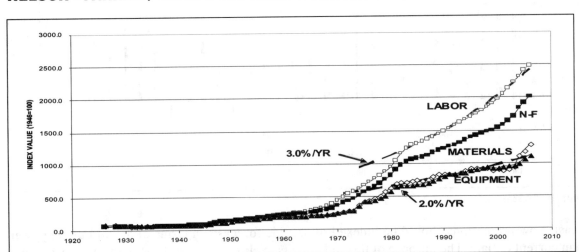

The usual approach to handling the problem of inflation in a cash flow projection is to convert the spreadsheet to one of buying power estimates (both revenues and expenses) expressed in "current dollars," or "Money of the day" as the Europeans refer to it (see Example 4-4). Depreciation (which will be discussed in Chapter VIII) is an exception to the current dollars handling since it relates to a portion of the fixed original investment cost. This yields a somewhat larger taxable income projection than would be the situation in the "constant dollar" case, which ignores the effects of inflation. The government also benefits from inflation from the real value of current tax deductions such as depreciation, depletion and amortization which will also be discussed in subsequent chapters. These unintended effects on taxes may not detract from the reasonableness of the economic analysis, however.

The spreadsheet at this stage of development yields a year by year forecast of after tax cash flow. To convert this to a forecast of future buying power it is necessary to apply stepwise indexing according to some generally recognized forecast. This set of inflation factors may be the same as those employed for adjusting the revenue and expense cash flows. In the petroleum industry, however, history has shown that each of the three sets of indexing factors is likely to differ somewhat one from the other. In other words, changes in the price of crude oil are not immediately reflected in the costs of drilling, or offshore construction. The resulting annual cash flow stream can then be measured by the various economic "yardsticks," described in Chapter V, employing discount factors that have not otherwise been modified for inflation. If future buying power is not considered, the yardsticks which involve present value can give quite misleading results according to Davidson (loc. cit.). For example, capital intensive projects may look unduly favorable because the tax liability is understated.

The shortcoming of the foregoing procedure, which is rigorous in concept, lies in the reliability of the several indexing forecasts required. In many analytical situations it is expedient to simply compose a short term discount factor which recognizes 1) the uninflated cost of money, say 3 percent, 2) a risk return of, say 3 percent, plus 3) an inflation factor of 5 percent per year. This suggests a "hurdle rate" discount factor of 11 percent applied to the net cash flow.

The relatively short-term effects of inflation on the budgeting process, discussed in Chapter IX, can often be handled with a relatively simple set of annualized indices since capital projects are generally completed within a year or two. The long-life nature of petroleum production with ongoing sales and operating costs imposes problems of a much more complex nature. Over a span of years continuing inflation takes on a cumulative compounding effect, much like the compounding of interest on a financial debt. That is, each year's devaluation of monetary value must be weighted according to its time of occurrence, and becomes a multiple of the previous year's factor, or index.

Modeling Escalation of Prices Under Inflation

Not all prices move at the same rate of inflation. Some move faster, some slower, and some may even move opposite to inflation. It may be better to refer to these price changes under the more general heading of price escalations. This recognizes that various components of an economic valuation may be subject to different escalators, some of which may go as up. Recent movements in the pricing of crude oil is a good example.

In the first instance, if p represents the price at the beginning of year one, then at the end of this first period we will have that price plus the inflation for the period: $p + p$ II or $p(1 + II)$. The second period will start with $p(1 + II)$ so that at the end of the second period we will have:

$$p(1 + I_I) + p * I_I(1 + I_I) = p(1 + I_I)(1 + I_I) = p(1 + I_I)^2.$$

The basic mathematical equation for compound inflation at a constant rate over a span of periods is thus written as follows:

$$p_t = p_i(1 + I_I)^t \tag{4.1}$$

Where:

p_i = average price, or annual cost, for initial year
p_t = average price, or average cost, for year, "t"
I_I = effective annual inflation rate, fraction per year
= $[(p_2 / p_1) - 1]$

$t = a + b + c$

The mathematical statement, Equation 4.1, can be used to account for inflation in the cost of goods and services sold. It has the same format as the equation for compound interest which will be introduced in the Chapter V. It does imply that the rate of inflation is forecast to be constant in the future. If a forecast is desired incorporating future changes in the inflation rate Equation 4.1 can be modified to accommodate such changes as shown in Equation 4.2.

$$p_t = p_i(1 + I_{Ia})^a(1 + I_{Ib})^b(1 + I_{Ic})^c \qquad (4.2)$$

Where:

I_{Ia}, I_{Ib}, I_{Ic} are the inflation rates in effect for time periods a, b, and c respectively, and $t = a + b + c$.

When forecasting the escalation of crude oil prices as well as goods and services it may be convenient to make the projection with reference to an initial forecast of general inflation. One may predict, however, that one service (perhaps drilling) will escalate X% faster than general inflation, while another (possibly tanker rates) may escalate Y% slower. This is done by imposing another concurrent series of escalation factors, I_{Ex}, I_{Ey}, I_{Ez}, etc.

These two escalator sets, I_I and I_E, are then combined in a series composed of:

$$p_{tx} = p_{ix} (1+I_{Ex})^t (1+I_I)^t = p_{ix} [(1+I_{Ex})(1+I_I)]^t$$

$$p_{ty} = p_{iy} (1+I_{Ey})^t (1+I_I)^t = p_{iy} [(1+I_{Ey})(1+I_I)]^t \qquad (4.3)$$

$$p_{tz} = p_{iz} (1+I_{Ez})^t (1+I_I)^t = p_{iz} [(1+I_{Ez})(1+I_I)]^t$$

Where:

p_{tx}, p_{ty}, p_{tz} = average price/cost of items x, y, and z in year "t"

p_{ix}, p_{iy}, p_{iz} = average price/cost of items x, y, and z in initial year

I_I = general rate of inflation, fraction per year

I_{Ex}, I_{Ey}, I_{Ez} = effective annual escalation rate of items x, y, and z above the general rate of inflation, fraction per year

which can be adapted to the specific circumstances. If a cost is expected to increase at a rate less than inflation, this can be handled easily in Equation 4.3 by using negative values for I_{Ex}, I_{Ey}, and I_{Ez}.

Note: Values of I_E, I_I and t should be chosen in accordance with the foregoing.

Example 3-3, in Chapter III, showed how an equation similar to Equation 4.3 can be used to prepare a schedule of crude oil prices over the period of time encompassed by an economic evaluation.

Economic Analyses Involving Inflation

When dealing with future expenditures in an inflationary economic climate, there are several terms and concepts that should be understood. When money is actually spent at some future date, the cost at that time will reflect all escalation and inflation that has occurred up to that point in time. The expenditure at the future date is considered to be made in "current currency" or "money of the day." However, economists find it convenient to view future and past expenditures in terms of "constant currency," which means it is on the basis of the value of that currency at some particular point in time. This assumes the values include all inflation and escalation up to that base time period, but then is frozen at that time with no further escalation or inflation. Thus it is possible to compare values for different years, excluding the effect of inflation on those values. Economic evaluations can make good use of the concept of "constant currency" in that most estimates of future costs are based upon current or past actual experience with that of a similar expenditure. Once these estimates are made, it is a simple matter to escalate the costs to the future date of expenditure so they will be in terms of "currency of the day," as needed for the economic evaluation.

Another aspect of the economic analysis started in Example 3-3, which dealt with the effects of inflation on crude oil prices, is the forecast of expenditures, both operating expenses and capital investments under inflation.

■ EXAMPLE 4-4
FORECAST OF EXPENDITURES WITH INFLATION

Operating costs are estimated to be $175,000 per year in terms of last year's dollars and it is estimated that in terms of last year's dollars, it will remain at that level for the next 10 years. A capital investment of $1,000,000 is anticipated for year 1 and an additional investment of $100,000 in year 6. The estimate of both capital investments is also in terms of last year's dollars. Inflation is forecast to be at the rate of 6 percent per year for the next 5 years and then to drop to a rate of 5 percent per year thereafter. Prepare a schedule of expenditures in "dollars of the day" when the money will be spent.

If:

E_C = Operating expenses in "constant dollars"

E_D = Operating expenses in "dollars of the day"

t = Elapsed time between "constant dollar" year and "dollar of the day" year

I_I = Inflation rate

Then:
$$E_D = E_C (1+I_I)^t \qquad (4.4)$$
(Applied in a manner similar to shown in Eq. 4.1 and 4.2)

If:

C_C = Capital investment in "constant dollars"

C_D = Capital investment in "dollars of the day"

Then:
$$C_D = C_C (1+I_I)^t \qquad (4.5)$$
(Applied in a manner similar to shown in Eq. 4.1 and 4.2)

Years from Base Year, t	Constant Dollars		Dollars of the Day			
	Op. Exp. E_C ($)	Cap. Inv. C_C ($)	Inflation I_I (%/yr)	Cum. Inflation *	OP. Exp. E_D ($)	Cap. Inv. C_D ($)
0	(base year for "constant dollars")			1.0000		
1	175,000	1,000,000	0.06	1.0600	185,500	1,060,000
2	175,000		0.06	1.1236	196,630	
3	175,000		0.06	1.1910	208,428	
4	175,000		0.06	1.2625	220,933	
5	175,000		0.06	1.3382	234,189	
6	175,000	100,000	0.05	1.4051	245,899	140,514
7	175,000		0.05	1.4754	258,194	
8	175,000		0.05	1.5492	271,104	
9	175,000		0.05	1.6266	284,659	
10	175,000		0.05	1.7080	298,892	
Total	1,750,000	1,100,000			2,404,428	1,200,514

*Cumulative Inflation Factor for Year 8 = $(1.06)^5 (1.05)^3 = 1.5492$

The effect of technological improvements resulting in reduced costs can be included very easily while making a forecast in the manner discussed in Example 4-4. This can be done by adjusting the constant dollar costs to reflect the cost reductions anticipated before applying the cumulative effect of inflation. This will be shown in Example 4-5.

EXAMPLE 4-5
FORECAST OF EXPENDITURES WITH INFLATION AND TECHNOLOGICAL COST REDUCTIONS

Using the same data and inflation equations used in Example 4-4 one additional factor will be added; that is, assuming that in the absence of inflation we would expect both operating costs and capital costs to decline at the rate of 2% per year due to technological improvements and increased productivity. This can also be handled as an exponential effect since the future costs are expected to go down each year by 2% of the previous year's costs. Written as equations that can be included in a spreadsheet analysis:

$$E_{cf} = E_c(1-PI)^t \tag{4.6}$$

$$C_{cf} = C_c(1-PI)^t \tag{4.7}$$

Where:

E_{cf} = Operating Expenses in "Constant dollars" at future date

C_{cf} = Capital Investment in "Constant dollars" at future date

PI = Annual productivity improvement (cost reduction)

Future operating expenses and capital expenditures, in terms of the value of money when these expenditures are to be made, can be determined as was done in Example 4-4. This is done by inflating the forecast values after first including the effect of technological improvements anticipated to reduce costs. This is shown in the following table.

Years from Base Years, t base yr.	Constant Dollars		Dollars of the Day			
	Op. Exp.* E_{Cf} ($)	Cap. Inv.* C_{Cf} ($)	Inflation I_I (%/yr)	Cum. Inflation	OP. Exp. E_D ($)	Cap. Inv. C_D ($)
	$175,000	$1,000,000	(yr. 0)	1.0000		
		$100,000	(yr. 0)			
1	$171,500	$980,000	0.06	1.0600	$181,790	$1,038,800
2	$168,070		0.06	1.1236	$188,843	
3	$164,709		0.06	1.1910	$196,171	
4	$161,414		0.06	1.2625	$203,782	
5	$158,186		0.06	1.3382	$211,689	
6	$155,022	$88,584	0.05	1.4051	$217,828	$124,473
7	$151,922		0.05	1.4754	$224,145	
8	$148,884		0.05	1.5492	$230,645	
9	$145,906		0.05	1.6266	$237,334	
10	$142,988		0.05	1.7080	$244,216	
Total	$1,568,601	$1,068,584			$2,136,442	$1,163,273

* Future costs reduced 2% each year for anticipated technological cost reductions

REFERENCES

Black, S. J.: Oil and Gas Investment Evaluation, 1983, Lafayette, LA 70503.

Bierman, Harold, Jr., Financial Management and Inflation, 1981, The Free Press, Macmillan, New York

Davidson, L. B.: "Investment Evaluation Under Conditions of Inflation," JPT, p. 1183-1189, Oct. 1975.

Dept. of Energy, Development of the Oil and Gas Resources of the United Kingdom, 1987, H. M. Stationery Office, London.

McKechnie, Gordon, Editor: Energy Finance, 1983, Euromoney Publications, London EC4.

Park, William R.: Cost Engineering Analysis, 1973, John Wiley & Sons, New York.

Stermole, F. J.: Economic Evaluation and Investment Decision Methods, 1980, Investment Evaluation Corporation, Golden, CO 80401.

Thompson, R. S. and Wright, J. D.: Oil Property Evaluation, 1983, Thompson-Wright Associates, Golden, CO 80402.

ECONOMIC DECISION TOOLS

Just as surely as water runs downhill ...
capital will seek the best investment opportunities.

Analysis of the various alternatives to a given investment decision may show important differences with respect to costs, profits, savings, project lives, tax considerations, effects of escalation and inflation, and in general, the risks and uncertainties associated with the projects under consideration.

THE FUNDAMENTAL BASES OF ECONOMIC DECISION

All economic analyses are based on the premise that an investor choosing between two cash flow streams of equal risk should prefer:

1. The larger monetary benefits, (i.e., profit), over the smaller ones

2. The earlier benefits (quick return) over the later ones.

The normal approach to investment decision making is to apply one or more of the measures of economic performance, or "Economic Yardsticks," which are discussed in this chapter. However, an economic analysis is only as good as the data that go into it. Because of the effects of risk and uncertainty, which include escalation and inflation, as well as geological, technical engineering, and political risk, it is not possible to develop evaluation techniques that are exact and infallible in dealing with future events. By employing one or more of the economic yardsticks that are presented in this chapter, one should be able to do a consistently better job of economic decision-making than can be achieved without their application.

There are many situations in the petroleum industry to which economic analyses are regularly applied. Some of these include:

- Establishing the costs of borrowed funds, both short and long term (hurdle rate).
- Establishing the economic feasibility of an investment or rate acceleration opportunity (screening).
- Weighing the relative merits of several investment prospects when funds are not available for all of them (ranking).
- Choosing among mutually exclusive alternatives.

- Evaluating purchase proposals from several different suppliers selling the same (or similar) equipment or services in order to select the best one.
- Buy or lease decisions.
- Determining a value, or price, for buying or selling producing properties.
- Replacement of existing equipment or service.

Most business decisions require choosing between several alternative courses of action. In many cases business decisions are made intuitively. Sloan J. Black (loc. cit.) employs a sieve analogy to decision making. He points out that some decisions are so compelling, such as handling a blowout, that no analysis is necessary. This process is likened to a very large grid screen that catches all of this kind of expenditure requirements immediately out of the hopper of available cash. Everything else falls through to the next screen which then catches the most obvious very rapid payout projects, or those that are immediately necessary to comply with a governmental regulatory order. These investment opportunities are disposed of from the second screen. As the screen grids get progressively smaller the expenditure decisions for the remaining opportunities become more difficult. When funds are limited various decision making tools are needed in order to choose the best alternatives.

One way to approach this selection problem is to subject all expenditures to a standard economic yardstick as a screening devise. Those that fail to meet the minimum criteria would be discarded from further consideration. Then, possibly with a different yardstick better suited to ranking, in order from the most to the least desirable, all of the remaining opportunities are listed.

Any major decision involving substantial amounts of money, and which may have significant impact on the short and long term profitability of the organization unit, should routinely utilize the added insight that the economic decision tools bring to bear on the decision.

Approaches to Economic Quantification

Several analytical methods are described and demonstrated in this chapter. Most of these involve some recognition of the time value of money. One should become familiar with as many of the evaluation techniques as possible. This will yield benefits in being able to communicate more effectively with other analysts within the petroleum and banking industries, and even within one's own company. The basic advantages and disadvantages of the various evaluation methods are described later in this chapter.

ECONOMIC YARDSTICKS

The purpose of using economic yardsticks in evaluating investment opportunities is to establish consistent criteria for determining which investments are desirable and which ones are not. Yardsticks used in this manner will be considered "screening" yardsticks. Before applying screening criteria, investments should be divided into two broad categories. The first group are those investments which are NECESSARY for continuation of the business. Investments required to meet safety and environmental standards or those mandated by law require no economic justification, as those expenditures are necessary to remain in business. So these types of investments might be considered to have infinite profitability. However, there is an alternative to these expenditures which might be considered if the cost is too high. That would be to close the business or that portion of the business requiring the expenditure. Follow-up expenditures for ongoing projects may also be included in this category under a similar philosophy to that just presented.

The remainder of investments being evaluated should be considered to be DISCRETIONARY. It is the discretionary investment opportunities which will be measured with the economic yardsticks. The first measurement is done with screening criteria which identify those investments which meet or exceed a predetermined standard and are therefore investments which would be "desirable." The desirable investments may be undertaken or subjected to further evaluation. The "undesirable" investments, which do not meet the standard, are rejected unless something can be done to improve their profitability.

If sufficient funds are available to undertake all "desirable" investments, then no further economic criteria need be applied. However if, as is frequently the case, the "desirable" investment opportunities exceed the funds available, it will then be necessary to prioritize the investment opportunities so that the better ones will be undertaken with the limited available funds. A ranking yardstick will be used for this purpose so that the investment opportunities can be put in order of most desirable to least desirable. Then it is just a matter of going down the list, until the available funds are exhausted.

A number of screening and ranking criteria will be presented, because no single yardstick matches the goals of every organization. It will be necessary for each organization to select the one or combination of criteria which most closely match their established goals.

Return on Investment (ROI)

Probably the oldest, and still a very popular economic yardstick which is widely employed in the oil industry, is "return on investment," or ROI. This is simply the total net income which is attributable to a particular investment divided by the total investment or cost outlay for the project. Here, 'net income' means gross revenue minus royalty minus operating expenses. Many oil companies require an ROI of at least 2.0 for any project. This yardstick is completely independent of time which limits the usefulness of ROI as a single criterion for investment decisions, and so other yardsticks which incorporate time factors should be employed along with the ROI parameter.

Profit to Investment ratio (PIR) is another form of expressing the same yardstick. Whereas, ROI divides the total net income by the investment, PIR first subtracts the investment from the net income, which is net cash flow, before dividing by the investment. Thus, PIR is always equal to ROI minus one.

A further variant of the basic ROI is Return on Maximum Cash Out-of-Pocket. This economic yardstick recognizes that in some types of projects income can begin before the project is finally completed and all of the cash required to complete it does not have to be on hand before the project can commence. None of these is superior to the other. However, one should know that these yardsticks exist and are widely used in the industry. It is also interesting to note that to a degree Return on Maximum Cash Out-of-Pocket recognizes the factor of time prior to the payout of the project.

Unfortunately, perhaps due to their widespread use through the years, the exact definition of these yardsticks varies among different organizations and writers even though they employ the same names for the ratios. For the purposes of this text the following will apply:

$$\text{Return on Investment, ROI} = \frac{\text{Cumulative Net Income}}{\text{Total Investment}}$$

$$\text{Profit to Investment Ratio, PIR} = \frac{\text{Cumulative Net Income} - \text{Total Investment}}{\text{Total Investment}}$$

$$= \frac{\text{Net Cash Flow}}{\text{Total Investment}} = \text{ROI} - 1$$

$$\text{Maximum Out of Pocket ROI} = \frac{\text{Cumulative Net Cash Flow}}{\text{Maximum Negative Cash Position}}$$

Although the net cash flow curve shown in Figure 5-1 suffers somewhat from not being adaptable to tabulation along with other data it does have several distinct advantages as a tool of comparison. For example:

1. Variations in the type and projection of cash flow histories such as the differences in a gas processing plant and a drilling venture are immediately apparent

2. Net out-of-pocket cash flow and payout are depicted

3. The curves come to a discrete point of termination when the economic limit or life of the project is achieved

FIGURE 5-1
CASH FLOW COMPARISONS

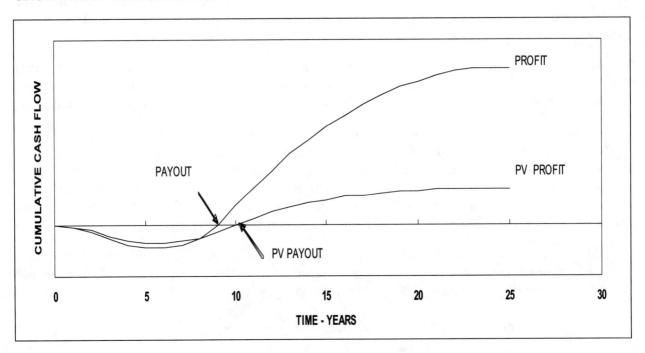

Payout Period

Payout period is the time lapse from the initial expenditure on the project until its cumulative net cash flow becomes positive. Stated otherwise, payout occurs at the time that the net cash position curve crosses the zero axis, as shown in Figure 5-1. The same plot shows the maximum out-of-pocket cash requirement, which is the maximum negative cumulative cash flow value before payout.

Payout period is one of the oldest of the several economic yardsticks still in popular use in the oil industry. Traditionally it is undiscounted and is sometimes referred to as "Payback Period."

Payout Period by itself is a simple measure of capital turnover. For some projects Payout provides some rough measure of risk by indicating how long the investment capital is exposed. One significant weakness of Payout lies in the fact that it disregards the cash flow which occurs after payout. Neither does it recognize any positive and negative cash flows during the payout period itself.

A useful variation of payout time is payout volume. This is the volume of hydrocarbons which must be produced to payout the investment and it is reported in barrels, cubic feet, barrels of oil equivalent, or what ever volume measurement is appropriate. Occasionally present value payout time is desired. This is the time required to recover not only the initial investment, but earn interest on that investment. Everything that has already been said about payout applies equally to present value payout.

Minimum Bailout Period

A variant of payout period is known as the minimum bailout. This is the minimum time period at which the cumulative net cash flow plus the appropriate early salvage value equal the original investment. Minimum Bailout Period can sometimes be employed as a useful decision tool in choosing between several project alternatives, particularly if one or more of them is likely to provide early indication of success or failure soon after the initial phases of field development. Assessment of the "early salvage value" is critical to determination of the minimum bailout period. It is the fair market value of any salvageable equipment or materials, less salvage and abandonment costs.

Financial Ratios

Several of the oldest economic decision tools which are still widely employed in the industry are the investment ratios. They provide indices of economic efficiency resulting from a unit of investment. The investment ratios are essentially independent of project life and the pattern of cash flow. Those yardsticks which involve the time value of money are described later in this chapter.

Book Profit

Annual book profit is obtained by subtracting expenses, taxes, DD&A, non-operating adjustments and other write-offs (see Chapter VIII) from operating revenue. The computation of income tax and DD&A make this a rather subjective economic yardstick, although it may have appeal to some managers who may even attempt to apply this on a project basis.

Book Rate of Return

The accounting profession employs a Book, or Accounting Rate of Return which is used primarily as one of the financial ratios in annual reports. Book Rate of Return is a measure of the accounting (current dollar) earnings of a project as a percentage of the book value of the assets in each period evaluated. It is computed as:

$$\text{Book ROR} = \frac{\text{Book Profit for the Year}}{\text{Capital employed at beginning of the year}}$$

Book rates of return will likely vary from period to period because they are strongly influenced by depreciation schedules. This renders the results noncomparable from period to period, and may have little relationship to the rate of return achieved over the life of the project. This ratio is frequently referred to as ROCE (return on capital employed) or ROACE (return on average capital employed). The denominator of this ratio is original cost of investments minus cumulative DD&A on those investments and may be the average of successive year's balance sheet values.

Stockholder's Rate of Return

The average stockholder's rate of return is obtained by dividing total book profit for the year by the year's average stockholder invested capital. The invested capital for each year is taken as the sum of the original investment attributable to equity and any subsequent investments made through that year, less the cumulative DD&A written off on those investments up to that time. The denominator in this case is then the mean of the invested capital figures for the beginning and end of the year. The Stockholders' Rate of Return has a similar appeal, with the same strengths and shortcomings of book profit.

$$\text{Stockholders' Rate of Return} = \frac{\text{Book Profit for the Year}}{\text{Average Shareholders' Investment}}$$

THE TIME VALUE OF MONEY—PRESENT VALUE CONCEPTS

Time is an all important factor in the earning power of an investment. A dollar received today is worth more than a dollar to be received sometime in the future. The value of money to be received, or to be paid out, is directly related to the timing of the receipt or disbursement. This is a principle that must be applied to economic evaluations in order to relate the impact of varied timing in the cash flow to enable meaningful comparisons. So simply; time is equal to money.

The handling of inflationary effects on buying power is quite independent of the present and future value concepts, as such. The problems and techniques of dealing with inflation are discussed in detail in chapters III, IV, and VIII.

Time = Money.
Interest is the rent paid to use someone else's money

Inflation is rise in price, not due to any change to the product.

[handwritten margin notes, top:] Taking future money × by 1.01% (1 + 1%)
1000 × 1.01 = 990.90 + this becomes Net present value. You have discounted the future amount 1000 = 990.90 now in future (opportunity cost) Divide by 1 + discount rate. 1%. Discount factor.

EXAMPLE 5-1
TIME VALUE COMPARISONS

Suppose one has been offered the following alternative options:

1. For an investment of $5,000, you are promised $800 per year for four years and $10,900 at the end of the fifth year.

2. For an investment of $5,000, you are promised $15,000 at the end of five years.

The first alternative indicates a total gain of $9,100, and the second a yield of $10,000, both over the same period of five years. It would appear that the second proposal is $900 better than the first. However, the simple comparison of the total gain to be obtained overlooks other important differences of the two alternatives, including timing of the gain.

First, it assumes that both alternatives involve the same risk. Between the time one enters into the investment and the time at which the project is finally completed many things may happen to affect the "promised" returns. The quick analysis also overlooks inflation which may diminish the value of the money during the contract period. The simple analysis also ignores the matter of how accessible the money is during the time involved. If the funds are locked into the investment and one must forego the use of the money or the income produced for some period of time (referred to as opportunity costs), one should consider that deprivation. These are a few of the factors involved in what is commonly referred to as the "time value of money."

Now again compare profits from the two propositions just mentioned. The first produces income early on, so one regains access and therefore use of the money sooner. Assuming steady inflation, or the political chances of expropriation, the sooner one regains his funds the less exposure there is to their loss or reduction in value on exchange. Since the payback of the first option begins at an earlier time, the amount of funds at risk declines each year. The first alternative now intuitively seems the best. Intuition is often not satisfactory, however. In this chapter several analytical methods will be investigated which serve to quantify the time value of various types of investment propositions.Calculations and the conclusion of this example are presented later in this chapter in the section on Net Present Value.

Project Life and Cash Flow Patterns with Time

Project life, or the span of time during which a potential undertaking will continually produce an income stream, is also an important economic consideration. Two or more potential projects may have quite different projected periods of income during which the invested capital remains at risk.

Oil industry investments may vary quite widely with regard to the time at which they generate a positive income stream. Timing becomes a particular problem with projects having long lead times (the time span between the first expenditures and investments and the first income). The North Sea and North Slope are examples in the recent past. The Beaufort Sea, South China Sea, Santa Maria Basin, Hibernia, North Sea above the 62nd parallel, the Athabasca oil sands, Colorado oil shale, and the entire field of enhanced oil recovery (EOR) are more current and near future examples of long lead time undertakings. Unintended delay in startup of a project can significantly affect the project economics as demonstrated in Figure 6-12, of the next chapter, and its related commentary.

[handwritten margin notes, bottom:] In 10 years time $\frac{1000}{(1.01)^{10}} = 905.29$ this is exactly equal to 1000 in 10 years time we have applied a discount rate

155

Some capital outlays in the petroleum producing industry such as those for gasoline plantsand pipelines generate a relatively uniform income throughout their useful lives. Most exploration and producing ventures are characterized by relatively long lives and a high degree of irrevocability. Producing oil and gas wells usually provide relatively high initial incomes that may hold steady for several years under proration or pipeline capacity limitations and then decline gradually for their remaining productive life. The cash flows from these two types of investment differ appreciably.

Interest and Principal

Interest is the 'rent' that is paid to use someone else's money. In industry a company does not own the money it uses, but acquires those funds from another source; i.e. stockholders, banks, other lenders or in the case of national oil companies, the people of the country. The interest rate is a function of many things, but basically reflects the risk to the lender and economic conditions at the time that the money is acquired. Inflation is a major risk to the lender, as the funds will be repaid with money valued on the date of repayment. If inflation has eroded the value of money over time, the repaid funds will be worth less than what was borrowed. So the duration of the loan will affect interest rates. The credit worthiness of the borrower will also be reflected in interest rates as it reveals the probability of payment of interest and repayment of principal at a future date. Thus an interest rate is the composite of many factors which are constantly changing in response to varying economic conditions. Since interest rates are constantly changing, it may be desirable to allow a contract interest to 'float' with going interest rates. This may be accomplished by tying a contract interest rate to some frequently well publicized interest rate; such as LIBOR (London Interbank Borrowing Rate) or U.S. Treasury interest rates. Both of these interest rates are considered risk-free, so a contract rate based on either of these would be higher, reflecting the appropriate risk to the lender.'

The promise of one dollar to be received one year from now is not equivalent to the receipt of that dollar today. A dollar received today can be invested, i.e., put to work. If it is deposited in a savings account earning 7% interest, it will grow to $1.07 (ignoring compounding for the moment) one year from now. Similarly, if $0.93 is deposited in the same account it will grow to $1.00 over the same period. Thus, the promise of a dollar to be received one year from today is equivalent to $0.93 received today when the interest rate is 7%.

The concept of "interest" is key to accounting for the "time value of money." The term "interest" denotes synonymously, either the cost of borrowed money, or the rate of return on an investment by the investor. Interest can be viewed as the monetary rent for the use of the borrowed funds. The amount of the loan or investment involved is referred to as the "principal" and is expressed numerically in the local monetary units such as dollars, pounds sterling, pesos, dinars, or whatever. The loan fee, or "interest" is expressed as a percent and represents the ratio of the interest amount for a period of one year to the amount of the principal. Although always expressed as a percent, interest is always dealt with as a decimal fraction in its mathematical treatment.

Most problems in finance are unchanged regardless of the direction in which the term "interest" is used. If a bank loans $10,000 at 10% interest compounded annually, the banker is receiving a rate of return of 10% per year on his loan. Similarly, the borrower is paying 10% as the rental cost for the borrowed funds. In determining the payments required in future time periods to pay off the loan, the 10% rate of interest is applicable whether one views the matter from the standpoint of the borrower or the lender.

Future Value = Present Value + the interest fee

= Present Value + (i) * (Present Value)

Or :

$$v_f = v_p (1+i)$$

[handwritten: to get the present value in answer, plus the interest fee, rather than just the % of the PV.]

Where :

(5.1)

v_r = Future Value

v_p = Present Value

i = the interest rate expressed as a decimal

Present Value (vp) is defined as the investment required at the time of the evaluation to yield the Future Value (vf) at the selected interest rate. Some writers prefer the term Present Worth which is synonymous with Present Value as employed in this text.

Numerically, using a 10% Annual Interest rate for example: *[handwritten: gives you that in answer, plus the 1%]*

$$v_f = (\$1,000)(1 + 0.1) = \$1,100$$

The interest on $1,000 for one loan period at 10 percent is $100. The interest can be withdrawn and used for other purposes. In such a case it would be known as simple interest, which basically is the rental fee for a one-year period divided by the principal, and expressed as a percent. If a loan extends into additional time periods there may be rental amounts owing on the accrued, or still unpaid interest. Thus, if the interest earned as rent for the first time period is left in the account it too will earn interest during the subsequent time periods.

$$v_{f1} = (\$1,000) (1.10) = \$1,100 \text{ (first Period)}$$

$$v_{f2} = (\$1,100) (1.10) = (\$1,000) (1.10) (1.10) = \$1,210 \text{ (two periods)}$$

$$v_{f3} = (\$1,210) (1.10) = (\$1,000) (1.10) (1.10) (1.10) = \$1,331 \text{ (three period, etc.)}$$

Therefore :

$$v_{fn} = v_p (1+i)^n$$

(5.2)

Where :

n = the number of compounding periods

Adding to the principal the interest earned by the principal itself is referred to as "compounding." Interest paid on the sum of the initial principal and the accumulated interest is called "compound interest." This compounding may be computed at the end of any specified period of time such as a year, month or day. If compounding is done on a yearly (end of period) basis the process is referred to as compounding annually.

The general formula for compounding is:

$$v_f = v_i * FVIF_{in} = v_i (1+i)^n$$

(5.3)

Where:

v_f = future value

= principal + accrued interest at time, n

v_i = the initial principal

$FVIF_{i,n}$ = Future Value Interest Factor for i and n

n = the number of compounding periods

i = nominal rate of interest per period, expressed as a decimal

It should be noted that the units of time in the previous equations must be consistent. If interest is a decimal value per year, then n must be in years. The value of n can be any real number; i.e., positive, negative, integer or fraction. Commonly used FVIFs are listed in Table 5B-2 in the Appendix to this chapter for mid-period payments.

Nominal Interest and Continuous Compounding

Interest may be compounded more frequently than once a year; i.e., quarterly, monthly, daily, etc. When this is done it is called a "nominal" annual interest and is commonly identified as an Annual Percentage Rate or APR. The periodic interest rate is the APR divided by the number of compoundings per year and the "effective" interest is the equivalent annually compounded interest rate.

The effective interest rate that is equivalent to a nominal interest rate compounded more than once a year is easily determined by compounding the periodic interest rate for the number of compounding periods used each year. This is shown in Equation 5.4;

$$\text{Effective interest rate} = \left[(1+ i/m)^m - 1 \right] \tag{5.4}$$

When nominal interest is compounded over a number of years, the future value is determined by using Equation 5.5;

$$v_f = v_p \left[1 + (i/m) \right]^{nm} \tag{5.5}$$

Where:

i = nominal interest rate per year (APR)

m = number of interest compoundings per year

n = number of years

So, for example, money compounded twice each year at a nominal interest rate of 10 percent per year:

$$v_{f1} = (\$1,000)(1+0.05)^2 = \$1,102.50 \text{ (first year)}$$

$$v_{f2} = (\$1,000)(1+0.05)^4 = \$1,215.51 \text{ (two years)}$$

$$v_{f3} = (\$1,000)(1+0.05)^6 = \$1,340.10 \text{ (three years) etc.}$$

Similarly, compounding quarterly at 10 percent per year:

$$v_{f1} = (\$1,000)(1+0.025)^4 = \$1,103.81 \text{ (first year)}$$

$$v_{f2} = (\$1,000)(1+0.025)^8 = \$1,218.40 \text{ (two years)}$$

$$v_{f3} = (\$1,000)(1+0.025)^{12} = \$1,344.89 \text{ (three years) etc.}$$

It may be noted that the more frequent the compounding the greater the future value.

For the case of continuous compounding, which is commonly used within the petroleum producing industry, the relationship between the present and future values becomes:

$$v_f = v_p (e^{jn}) \hspace{4cm} (5.5a)$$

Where "n" is the number of "j" interest periods, i.e., if j is nominal interest per year, and n is the number of years. The constant "e" is the base of natural logarithms, 2.71828.

Then:

$$v_{f1} = (\$1,000)(e^{0.1}) = \$1,105.17 \text{ (first year)}$$

$$v_{f2} = (\$1,000)(e^{0.2}) = \$1,221.40 \text{ (two years)}$$

$$v_{f3} = (\$1,000)(e^{0.3}) = \$1,349.86 \text{ (three years)}$$

for nominal interest of 10 percent per year, compounded continuously.

Summary

$1,000 compounded at a nominal interest rate of 10 percent per year, with the "rent" to be received at the end of the year in each case is shown below in Table 5-1 for several frequencies of compounding.

TABLE 5-1
EFFECT OF FREQUENCY OF COMPOUNDING INTEREST

		One Year	Two Years	Three Years
Frequency:	$(1 + i)^n$ Annual	$1,100	$1,210	$1,331
	$(1 + i)^n$ Semi-Annual	$1,102	$1,216	$1,340
	$(1 + i)^n$ Quarterly	$1,104	$1,218	$1,345
	$(1 + i)^n$ Monthly	$1,105	$1,220	$1,348
	$(1 + i)^n$ Daily	$1,105	$1,221	$1,350
	(e^{in}) Continuous	$1,105	$1,221	$1,350

These examples have all assumed that the interest rate remains constant with time. However, these equations are equally applicable for varying interest rates. For example; if monthly variations in interest rates of 10, 9, 8, 13 and 15 over a five month period were applied to a $5,000 investment (applying Equation 5.5) it would be worth:

$$\text{Compound amount} = \$5,000 \times \left(1 + \frac{0.10}{12}\right) \times \left(1 + \frac{0.09}{12}\right) \times \left(1 + \frac{0.08}{12}\right) \times$$

$$\left(1 + \frac{0.13}{12}\right) \times \left(1 + \frac{0.15}{12}\right)$$

$$= \$5,233.35 \text{ after five months.}$$

Discounted Value of Future Funds

A related process known as "discounting," is the inverse of compounding. The interest is, in effect, deducted in advance. For example, assume that the principal sum is $1000 and the effective interest rate is 10 percent. Under this process the $1000 loan (investment) will grow to $1100 after one year. Under the discounting process, $1000/1.10 or $909.09 is the amount that would grow to $1000 after one year. Thus $909.09 is the Present Value of $1000 to be received one year hence at 10 percent interest, or stated differently, the discounted value of $1000 due in one year at 10 percent is $909.09. Frequently the term 'discount rate' is used when discounting a future value to a present value. This is just another name for 'interest rate' and merely denotes that discounting of a future value to a present value is taking place rather than compounding a present value to determine a future value. So don't be confused by the terminology.

Rearranging Equation 5.2 to solve for present value (vp) yields the lump sum discount Equation 5.6;

$$V_p = V_{fn} * \frac{1}{(1+i)^n} = V_{fn} * (1+i)^{-n} \tag{5.6}$$

$(1+i)^{-n}$ is known as the discount factor, or $PVIF_{i,n}$ (Present Value Interest Factor), with the figures for i and n being specified. The interest rate used to discount a future value to a present value is commonly called the discount rate.

As previously demonstrated, the investment of $1,000 at 10 percent per year for 3 years yields $1,331. So the present value of that amount is:

$$V_p = \$1,331 * \frac{1}{(1.1)^3} = \$1,000$$

The equation for calculating present value of a future income or expenditure can also be written in an exponential form similar to Equation 2.3a. Using parameters similar to Equation 5.5, above, and rearranging Equation 5.5a it is:

$$V_p = V_{ft}\left(e^{-jt}\right) \tag{5.6a}$$

Where:

j = nominal annual interest or discount rate

$= Ln(1 + i)$

t = time (years)

If the proper relationship between "i" and "j" shown above is observed, equations 5.6a and 5.6 will give identical results. Therefore, the choice of either equation is determined by its use, selecting the one that can be employed most easily for a particular application.

If we increase the interest rate to 15%, for a more dramatic illustration, we have the relationship shown in Table 5-2.

TABLE 5-2
INTEREST AND DISCOUNT AT 15% COMPOUNDED ANNUALLY

Year	(1) How $1.00 left at 15% compound interest will grow:	(2) What $1.00 due in the future is worth today at a discount rate of 15%:
0	1.0000	1.0000
1	1.1500	0.8696
2	1.3225	0.7561
3	1.5209	0.6575
4	1.7490	0.5718
5	2.0114	0.4972
6	2.3131	0.4323
7	2.6600	0.3759
8	3.0590	0.3269
9	3.5179	0.2843
10	4.0456	0.2472

The foregoing serves to illustrate the increase or decrease of the cash income inherent in an investment with a 15% interest rate. In column (1) we see that $1.00 grows to $4.05 in ten years @15% compound annual interest. The second column shows the inverse. One dollar loaned out today is worth only 25 cents if we cannot recover it for ten years. Conversely, 25 cents today will grow to $1.00 in ten years if invested at 15 percent over the period. The "Rule of 72" states that the time required for money to double can be approximated by dividing the interest rate in percent into 72, i.e. 72/15 = 4.8 years.

Compounding and discounting can be illustrated more fully by considering the successive years of each process. Assume a principal of $1000 with an effective interest rate of eight percent and no repayment is to be made for two years. Through compounding the first year's interest will be (0.08 x $1000) = $80, plus (0.08 x $1080) = $86.40 for the second year. The repayment of principal and interest at that time would be $1000 + $80 + $86.40 = $1166.40. Under the discounting process $1000/1.08 1/1.08, or $857.34 is the present value of $1000 due two years in the future.

Table 5-3 demonstrates the construction of a discount table.

TABLE 5-3
PRESENT VALUE FACTORS - DISCOUNTED AT 8%

Years hence	Present Value of 1 (Discounted @ 8% rate)
0	1.000
1	$(1.00)/(1+0.08) = (1.00)/(1.08) = 0.926$
2	$(1.00)/(1.08)^2 = (1.00)/(1.17) = 0.857$
3	$(1.00)/(1.08)^3 = (1.00)/(1.26) = 0.794$
4	0.735
5	0.681
6	0.630
7	0.583
8	0.540
9	0.500
10	0.463, etc.

The discounting process is very important in profitability analyses. Discounting computations are carried out with the use of Interest Tables, a hand calculator or computer. Tables are generally published in handbook form consisting of many pages which are necessary to cover the ranges of interest and discount rates and the span of years which may be of concern. Four commonly used simplified interest and discount tables are presented in Appendix V-B. It is often more convenient to let the computer calculate the values as needed, or use a hand-held calculator.

Figure 5-2 shows the Present Value of one unit of money at several discount factors versus time. Present value drops off rapidly with time at the higher interest rates. It is also apparent from this graph that money to be received in the distant future has a reduced present value that may be nearly zero when a high discount rate is used.

FIGURE 5-2
PRESENT WORTH COMPARISONS (Mid-Year Discounting)

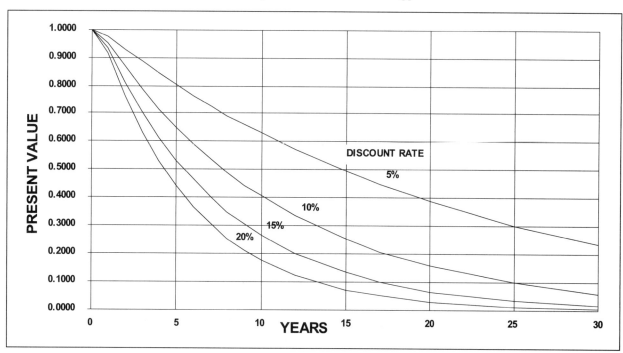

Discounting, being the inverse of compounding, is subject to the same variations in compounding frequency as is the interest computation. Discounting may, as indicated, be performed more often than once a year which yields more severe rates of discount. The resulting present worth factors for income discounted at 10 percent per year with payment at the end of the period for several frequencies is shown in Table 5-4. This is the inverse of the previous Table 5-1.

TABLE 5-4
EFFECT OF FREQUENCY OF END OF PERIOD DISCOUNTING
with payment at end of the year

Frequency:		One Year	Two Years	Three Years
	$PVIF_{10,n}$ Annual	0.9091	0.8264	0.7513
	$PVIF_{10,n}$ Semi Annual	0.9070	0.8227	0.7462
	$PVIF_{10,n}$ Quarterly	0.9060	0.8207	0.7436
	$PVIF_{10,n}$ Monthly	0.9052	0.8194	0.7417
	$PVIF_{10,n}$ Daily	0.9048	0.8187	0.7408
	$PVIF_{10,n}$ Continuous	0.9048	0.8187	0.7408

Time of payment is a separate variable. The resulting present worth discount factors for income realized over a three year period at 10 percent per year, compounded at several frequencies, and assumed to be paid at mid-point would be:

TABLE 5-5
EFFECT OF FREQUENCY OF MID-PERIOD DISCOUNTING
with payment in the middle of year three

End of year, annual	$PVIF_{10,3}$ = 0.7513
Mid-year annual	$PVIF_{10,3}$ = 0.7880
Mid-year monthly	$PVIF_{10,3}$ = 0.7796
Mid-year continuous	$PVIF_{10,3}$ = 0.7788

The above variations, although they appear to be slight, can become significant with the big multipliers of daily barrels of oil encountered in the oil business. This fact becomes particularly pronounced when extended over a long producing life. One should be aware that as the discount rate is increased, the longtime future income receives progressively less weight, or impact, compared to income earned early in the life of a project. For this reason the selection of the precise discount rate to be applied, at least for the early years, will have a profound effect on present or future values which are computed.

In the petroleum producing industry income is normally received in monthly segments rather than appearing all at one time at the end of the year, which is implied by equation (5.1). A better approximation is to assume that all of the income occurs at the middle of the year. This is known as mid-year discounting. Mid-year discounting results in higher present values of future income than end of year discounting since the values are discounted for one-half year less than for the end of year discounting. The assumption that full year's income and expenditures are treated as a single lump sum amount at the mid-point of the year is not only a convenient approximation for the petroleum industry but a reasonably accurate one. Only if a large percentage of annual amounts actually occur at specific times during the year would this not be a reasonable approach. The discount and compound interest tables presented at the end of this chapter are prepared on this basis, discounting from mid-period to the beginning of the time interval and compounding from mid-period to the end of the time interval. This is accounted for simply by subtracting a half-year from the exponent of the equation as is seen in Equations 5.7 and 5.7a and the following example.

For money to be received at the mid-point of the year, and compounded annually:

$$v_f = v_p (1+i)^{(n-0.5)} \qquad (5.7)$$

So, for example, $1,000 invested at 10 percent per year to be received in the middle of the third year:

$$v_f = (\$1,000)(1.1)^{2.5} = \$1,269.06$$

In this case the discount factor, or PVIF is:

$$PVIF_{10,3} = \frac{1}{(1.1)^{2.5}} = 0.78799$$

The present value of $1,000 to be received in the middle of the third year, discounted annually:

$$v_p = (\$1,000)(0.78799) = \$787.99$$

Likewise, the present value of $1,000 to be received at the end of the third year, discounted annually, according to Equation 5.3:

$$v_p = (\$1,000)(0.75131) = \$751.31$$

Continuous compounding provides somewhat different values:

$$v_f = (\$1,000)(e^{0.1 \times 2.5}) = \$1,284.03$$

$$PVIF_{10,3} = e^{-0.25} = 0.77880$$

So that the present value of $1,000 to be received in a lump sum at the middle of the third year discounted continuously:

$$v_p = (\$1,000)(0.77880) = \$778.80$$

If the funds are to be received at the end of the third year with the same 10 percent per year continuous discounting:

$$v_p = (\$1,000)(e^{-0.3}) = \$740.82$$

Oil and gas production income generally occurs at relatively uniform rates and is usually discounted, or compounded, as the case may be, at the mid-point of the year using either annual or continuous discounting.

$$v_f = v_p (1+i)^{(n-0.5)} \tag{5.7}$$

$$v_f = v_p = \left[e^{j(n-0.5)} \right] \tag{5.7a}$$

A further discussion of the mathematical refinements of these relationships is contained in the appendix to this chapter.

Annuities

An annuity is a method of evenly spreading a cost or scheduling the repayment of a debt, with interest, evenly over a specified period of time. An annuity provides for the full repayment of the cost or debt over the specified period of time. Each payment includes repayment of a portion of the total cost plus interest on the balance remaining to be repaid. Figure 5-3 is an illustration of the annuity concept. It shows the single principal amount as a negative bar at time zero and equal annual bars thereafter as repayment. The portion of each constant payment that represents interest and principal is indicated by shading of the annual bars. It is apparent that interest declines and principal repayment increases with each payment.

FIGURE 5-3
ANNUITY (Constant annual payments)

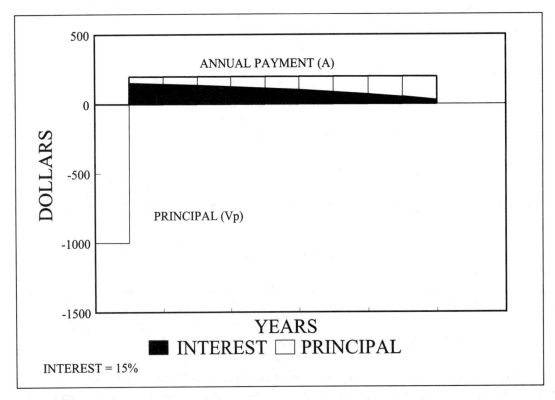

This technique can be used in compounding/discounting. It involves investing or paying equal amounts of money each year, or period and is used, for example, in figuring bond sinking fund payments and in calculating the equivalent average annual capital cost as used in Chapter X. The present value of an annuity is expressed as:

$$PVIFA_{i,n} = \sum \left[(1 + i)^{-n}\right] = \left[1 - (1 + i)^{-n}\right]/i \tag{5.8}$$

Where:

v_p = principal (PV) amount

A = annual payment (year end)

i = annual interest rate, decimal

n = number of years to maturity

$$v_p = A(PVIFA_{i,n}) = (A/i)[1 - (1 + i)^{-n}] \tag{5.9}$$

Or

$$A = \frac{iv_p}{\left[1-(1+i)^{-n}\right]} = \frac{iv_p(1+i)^n}{\left[(1+i)^n - 1\right]} \tag{5.10}$$

When payments (A) are made more frequently than at the end of each year, Equation 5.10 can be altered to match the frequency desired. This is accomplished by multiplying "n" by the desired number of compoundings per year and dividing "i" by that same number of compoundings per year. The formula for monthly mortgage payments, with "vp" being the amount borrowed and "M" the monthly payments, would be:

$$v_p = M(PVIFA_{i/12,12n}) = \{M/(i/12)\}\{1 - [1 + (i/12)]^{-12n}\} \tag{5.11}$$

Or

$$M = \frac{v_p(i/12)}{1-(1+i/12)^{-12n}} \tag{5.12}$$

Average Annual Capital Cost

Average annual capital cost is similar to depreciation of a capital investment, but with interest. It is not used frequently, although it may prove to be a useful concept in some situations. It can be calculated easily using the formula previously presented for determining the present value of an annuity. However, in this case we use the present value of the capital investment as the present value of the annuity and determine the corresponding annual amount that when present valued at the corporate discount rate will be exactly equal to this value. This can be accomplished by rearranging Equation 5.9 to solve for the annual value "A" as shown in Equation 5.9a.

$$A = v_p/(PVIFA_{i,n}) = (v_p * i)/[1 - (1+i)^{-n}] \tag{5.9a}$$

Equation 5.9a can also be used to determine the annual repayment amount required to extinguish a loan of amount (v_p) at an interest rate (i).

PROJECT EVALUATION

Economic evaluation of investment alternatives requires that all opportunities be appraised on the same basis, and further, that the time values of money be properly taken into account. When alternate sources of funding are available with different repayment schedules it becomes difficult or impossible to determine intuitively which source will be the least expensive. Valid comparison is achieved using one of several possible "equivalence" methods involving Present Worth values. These rely on the mathematical techniques of handling interest and discount computations.

Before analyzing a prospective investment, or applying any of the remaining economic yardsticks to be described in this chapter, four items must be determined. These are:

1. Cash flow stream generated by the project

2. Capital expenditure, or net initial outlay

3. Hurdle rate (or a risk adjusted discount rate), specific for the project and firm

4. Project life

Cash flow was presented in Chapter IV. The other three items will be considered in this chapter.

Discounted Cash Flow

The example spreadsheets presented in Chapter IV included columns which showed the annual values discounted at an assumed 10% "hurdle rate." This recognizes the time value of money and shows that the money received early in the life of a project has greater economic value than that received at a later time.

[handwritten: Taking into account alternative investment opportunities & use those % as discount rate.]

Net cash flow (revenue minus expenditures) is the basis of most economic decisions. Discounting these estimates of cash flow over time provides a realistic means of evaluating investments and the likely profits from them in terms of alternative opportunities. Cash flow forecasts, with or without discounting, do encounter difficulty in certain long life types of investment such as basic research or frontier exploration. Nevertheless the discounted cash flow approach to economic yardsticks represents the most consistent method of analysis overall.

Having recognized the concept of the time value of money as an important criterion for economic decision making, it becomes necessary to develop methods of incorporating the concept into simple workable measures of fiscal efficiency. The key parameter in such a measure is the discount factor itself. There are two basic approaches to the application of discounting so as to achieve the requisite time value effects while identifying economically acceptable projects. Each method has its fundamental advantages and disadvantages. One involves discounting a cash flow stream at a predetermined discount rate, called the "hurdle rate," to determine the increase in present value which will be obtained as the result of making the investment being evaluated. The other involves discounting cash flow at multiple interest rates to determine by trial and error which interest rate is equivalent to the return on a simple interest bearing investment of the same value.

Factors Which Influence Interest

The four most important factors that affect interest rates are: (1) the credit rating of the borrower, (2) financial market conditions on the date of borrowing, (3) the length of time to repayment of the debt, and (4) the rate of national inflation. It is because of the inter-play of these four factors that interest rates can differ significantly from time to time, from country to country, and from borrower to borrower. Each of these factors represent a risk to the provider of funds. Anything which increases risk to the lender increases interest rates.

Each of the factors discussed above add a component to the final interest rate such as blocks depicted in Figure 5-4. The starting point of the final interest rate is the rate of interest in the absence of risk and inflation, which is commonly called a "real interest rate". Each factor is added to the real interest rate to arrive at the final rate. Unlike the diagram of Figure 5-4, these components are not equal and change significantly over time. The only value of each component that is important is what it is on the day that the interest rate is determined. The final block, labeled Inc. Tax, has been added to reflect the fact that some interest is not taxed. Where that is the case the last block would be absent.

FIGURE 5-4
COMPONENTS OF INTEREST RATES

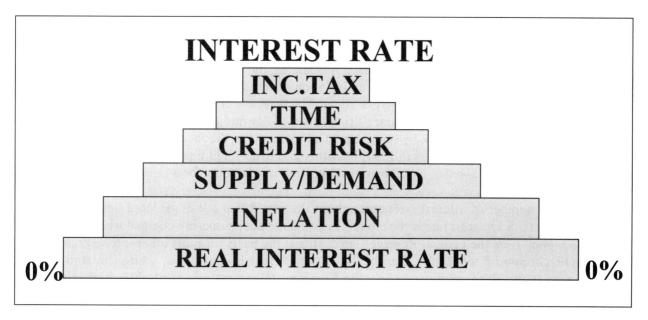

The credit rating of the borrower is a reflection of the credit worthiness of the borrower as assessed by one or more of the organizations that provide this service. This is discussed in detail in Chapter VII—Financing and Ownership of the Oil and Gas Industry and in Appendix VII-A of that chapter.

Financial market conditions continually change, reflecting the general economic conditions of the country, i.e., boom and recession, which in turn affects the supply and demand for money. For money is no less a commodity than wheat, copper, gold, or oil; so interest, which is the cost of using money, goes up and down as the supply and demand for money changes.

The elapsed time until a debt is repaid affects the lender's risk. For this reason longer term debt will usually bear an interest rate which may be several percentage points higher than short term debt. The interest premium for long term debt is not a constant value, but will also respond to the market conditions discussed in the paragraph above. Occasionally financial markets get out of balance for short periods of time, with long term interest rates lower than short term rates.

Inflation in the country where the money is borrowed has probably the greatest impact upon interest rates and may indirectly influence the other three factors as well. The relationship between the nominal (actual) interest rate paid to borrow money and inflation can be expressed as follows:

$$(1 + i) = (1 + i_{const.}) * (1 + I_I) \qquad\qquad (5.13)$$

or

$$i = (1 + i_{const.}) * (1 + I_I) - 1 \qquad\qquad (5.14)$$

where:

i = nominal (current) interest rate, decimal/year

$i_{const.}$ = "real" interest rate at zero inflation, decimal/year

I_I = national inflation rate, decimal/year

Equation 5.14 shows that changes in the rate of inflation contribute directly to the rise and fall of interest rates.

The real risk free interest rate in the U.S., i.e., the interest rate at zero inflation, is between2% and 3%. According to Chicago Researchers Ibbotson Associates, the real return on U.S. Government bonds over the period since 1925 was 1.9%. However, inflation-indexed securities recently issued by the U.S. Treasury Department carried a real interest above 3%. These securities will earn interest at a rate slightly higher than 3% plus the Consumer Price Index (CPI) rate of inflation. The principle of the bonds are also indexed for inflation by semi-annual adjustments of the face amount, also with the CPI. The U.S. first issued these securities in 1997, but other countries including Australia, Britain, Canada, Israel, New Zealand, and Sweden have been offering such investments for a number of years.

There are a number of interest rates throughout the world that are considered benchmarks. The Federal Reserve (U.S.) interest rate is the interest rate that member banks are charged when they borrow short term money from the Federal Reserve Bank. This is the basis for most other interest rates in the U.S. LIBOR (London Inter-Bank Offering Rate) is the British equivalent, while the Bundesbank Discount rate is the German counterpart. The Lombard (German) rate is slightly higher than the Bundesbank Discount rate and sets the upper limit on German money market rates. It is not uncommon for interest rates elsewhere in the world to be tied to one of these interest rates, which is frequently the case in petroleum industry agreements.

Hurdle Rate

The "hurdle rate" discount factor, more properly referred to as the "guideline discount (or interest) rate," is the minimum acceptable rate of return on investments. It is also considered the corporate cost of capital, which can be judged from two points of view. One is the acquisition cost of capital, i.e., the cost of obtaining both debt and equity funds. The other is the opportunity cost of capital, or the marginal rate of return on the least profitable investment to be undertaken. For this reason it is sometimes referred to as the "alternate use of money" factor which comes from its use as a budgetary "culling" tool. At times these values differ, but over the long run they should equilibrate and become nearly the same value. Figure 9-1 shows how the opportunity cost of capital may be estimated from an analysis of an organization's investment opportunities. The hurdle rate used is normally intended as the return which the firm has been able to realize on its investments of a similar nature in previous years. Some authorities disagree, and argue that the discount factor which is used should be the firm's average cost of capital. In most large companies the choice of discount factor is not delegated to the individual analyst but a standard value is established and is used throughout the firm for every economic evaluation.

The hurdle rate should recognize:

1. THE ACQUISITION COST OF CAPITAL,
2. SUITABLE RETURN ON THE INVESTMENT INVOLVED
3. A COMPENSATION FOR CORPORATE RISK, AND
4. INFLATION.

The hurdle rate may be related to the size of the company and the riskiness of the industry in which the firm operates. Within large international corporations, specific hurdle rates may be used for individual countries reflecting the overall economic risk of operating in that country. For example, two exploratory drilling prospects, similar in geologic potential, but one to take place in Texas, the other in a politically and economically troubled country in the third world might call for different hurdle rates due to the differing risks involved. Technical risk, including geological, is not properly accounted for by adjusting the hurdle rate, but should be evaluated more rigorously, as will be explained in the next chapter. Since all nominal interest rates include a component of inflation, it may be necessary to adjust the hurdle rate as inflation varies, even though the hurdle rate in the absence of inflation (real hurdle rate) remains unchanged.

Internal Rate of Return (IRR)

The internal rate of return (IRR) concept determines a discount rate which becomes the economic measure, or "yardstick." IRR measures the effective rate of return earned by an investment as though the money had been loaned at that rate. IRR evaluates the economic benefits over the project's life. It provides a single figure which is easy to communicate to management. Rate of return is intended as a measure of the profitability of a business or project. It is derived from the discounted annual cash flow stream resulting from a unit of invested capital. Rates of return may be determined before tax or after tax. It is important to specify which basis is employed. IRR will be indeterminate if there is no capital expenditure or initial outlay.

The internal rate of return (IRR) of a cash flow is the discount rate at which the present value of the cash flow is zero. The IRR is not directly calculable, but involves a trial-and-error solution. Hand held calculators will do the necessary interpolations, as will personal computer spreadsheet programs. The IRR can be interpolated graphically as shown in Figure 5-5, or mathematically as a simple interpolation between two "trial" points which straddle the actual value. The line connecting the calculated values is not a straight line, so interpolation between calculated values must account for the curvature.

Disant rate is low *[handwritten]* Long deemed. that in company alternative investment opportunity chart return could you get eg 80% 16% etc.

ECONOMICS OF WORLDWIDE PETROLEUM PRODUCTION

FIGURE 5-5
PRESENT VALUE PROFILES

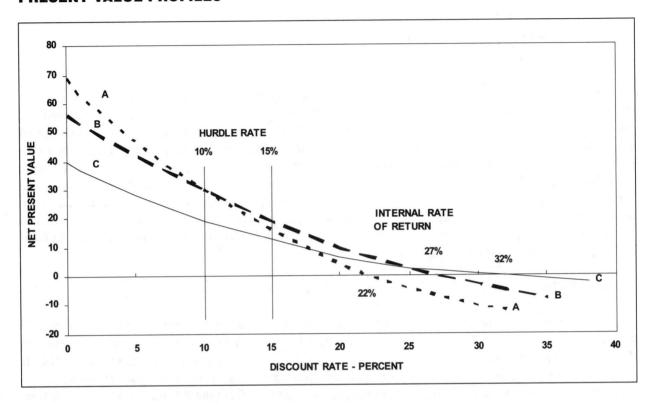

```
DECISION RULE:

If IRR, as a percentage, exceeds the hurdle rate percentage,
    accept the project

If IRR is less than the hurdle rate, reject the project

If IRR = the hurdle rate, management should be indifferent
```

Another way of looking at IRR is to assume that the net cash income from a project is used to repay the project investment plus interest on that investment. The interest rate which would allow the investment to be repaid plus interest precisely over the life of the project would be the IRR. Table 5-6 is a mortgage schedule using a project net cash income to repay the investment plus interest, using the mid-period convention. It can be seen that if the interest rate chosen is such that the investment is just paid off at the end of the project life (unrecovered balance = $0) it meets the definition of IRR because the PV Net Cash Flow is zero.

[handwritten: This is discount rate that which gives no NPV of zero.]

TABLE 5-6
THIS IS HOW INTERNAL RATE OF RETURN (IRR) WORKS

[handwritten: 10,000 / 0.5 / 1-49]

YR	INVESTMENT	NET INCOME	PV* NET INCOME 49%	MID-YEAR INTEREST 49%	MID-YEAR PRINCIPAL PAYMENT	UNRECOVERED BALANCE @ MID-YEAR
0	$16,128	$0	$0	$0	$0	$16,128
1		$10,000	$8,192[1]	$3,559[2]	$6,441	$9,687
2		$8,000	$4,399	$4,746	$3,254	$6,433
3		$6,000	$2,214	$3,152	$2,848	$3,585
4		$4,000	$991	$1,757	$2,243	$1,342
5		$2,000	$332	$658	$1,342	$0
SUM	$16,128	$30,000	$16,128	$13,872	$16,128	

IRR = 49%   *MID-POINT OF YR. DISCOUNTING

[1] $(1+0.49)^{-0.5}*\$10,000 = \$8,192$

[2] $[(1+0.49)^{0.5} -1]*\$16,128 = \$3,559$

[handwritten: 49% IRR]

[handwritten marginal note: raise to power 0.5 a mid yr discounting. So ± a yr. Then there yr is 1.5, then 2.5, then 3.5 etc.]

Internal Rate of Return is quite sensitive to the convention followed in determining present value. Throughout this text the mid-period convention, using a lump sum amount at the middle of the period, has been presented as a good approximation of the continuous nature of oil industry income and expenditures. For this reason it is believed that use of the mid-period convention for determining IRR is also preferable. There are functions in both Microsoft Excel and Lotus 1-2-3 which will automatically determine the IRR of a cash flow. However, these functions are based upon the end-of-period convention, where the income and expenditures over a period of time are approximated as a single lump sum value at the end of the period. This convention is not a good approximation to the way that business is conducted in the oil industry and will yield values for IRR which are less than that determined using the mid-period convention. The difference is small for low values of IRR, but may be 100% or more in error for large values of IRR. Therefore, if a more accurate value for IRR is desired, it will be necessary to create a discounting routine based upon mid-period values. Once this has been done, IRR can easily be determined by using the "Goal Seek" tool in Excel to determine the "trial-and-error" value.

■ EXAMPLE 5-2
INSURED PRODUCTION PAYMENT

- Calculation of IRR -

An insured production payment from a producing property has a guaranteed income stream of $6,000 a year for ten years. It can be acquired for a NIO (Net Initial Outlay) of $25,000.

DETERMINE:

What is the IRR of this potential investment employing values from Appendix VB of this chapter?

SOLUTION:

The constant income stream of $6,000 per year can be treated as an annuity. Using $(PVIFA_{i,n})$ values from Table 5B-3 it is apparent that the IRR is between 20% and 30%.

$$NPV_{20\%} = -25,000 + 6,000 \ (PVIFA_{20,10})$$
$$= -25,000 + 6,000 \ (4.5926) = \$2,555.60$$
$$NPV_{30\%} = -25,000 + 6,000 \ (PVIFA_{30,10})$$
$$= -25,000 + 6,000(3.5249) = -3,850.6$$

Using linear interpolation between the two values:

$$IRR \ = \ 20 + \frac{2,555.60}{2,555.6 - (-3,850.6) \ *} \ * \ 10 = 23.99\%$$

The major portion of an investment normally occurs at the start of a project. Where this pattern is substantially altered by multiple reversals from positive to negative cumulative cash flows there may be multiple IRR's, which can be a serious limitation to the use of this technique One of the principal attributes of IRR lies in the fact that it clearly reflects time values. It also has the advantage of direct comparison with the firm's cost of capital.

IRR has mathematical limitations with the problem of multiple and meaningless values if the cumulative cash flow moves from negative to positive and back to negative. Multiple rates of return occur when there are two or more discount rates for which the NPV of cash outflows equals the NPV of cash inflows. This occurs when the time pattern of net cash flows is such that the sign changes more than once during the time period in which the project is being analyzed.

For example, consider the cash flow diagrams of Figure 5-6.

FIGURE 5-6
PATTERNS OF CASH FLOW

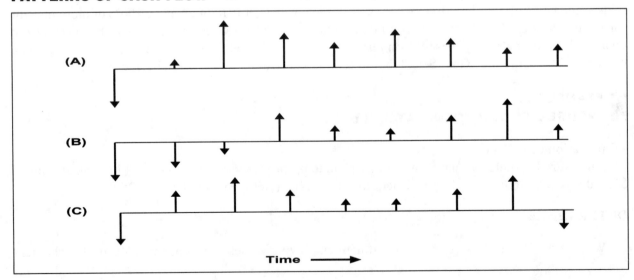

Pattern (C) with its mid-life decline of cash flow is typical of some secondary recovery projects, late infill drilling, field gas compression and other rate acceleration projects, when analyzing the full cycle economics of such ventures. The negative cash flow in the last project year could be due to a significant abandonment expenditure. These unusual cash flow patterns may yield multiple values for IRR, in which cases IRR is not the proper investment criteria to use.

IRR goes by various names within different companies operating in the petroleum industry. These include such labels as Internal Rate of Return (IRR), Rate of Return (ROR), Internal Yield, Discounted Cash Flow Rate of Return (DCFROR) (Mobil), DCFR (Exxon), Earning Power (Shell), Profitability Index, or (PI) (Amoco), AAROR (Phillips). "Profitability Index" is the same term widely used for quite different parameters, i.e. the PV of future cash flows divided by initial investment, and sometimes for IRR. Although all of these bear a common concept they may vary somewhat from company to company in the way they are computed—something to be aware of during negotiations. Pennant-Rae and Emmott (loc. cit.) observe: "Rate of Return—Simple in outline, in detail a statistical minefield."There are functions named IRR in both Excel and Lotus 1-2-3 spreadsheet programs which can be used to calculate this yardstick.

The rate of return, (IRR) approach places heavy emphasis on the cash flow early in the life of a project while being relatively independent of the ultimate return on investment (ROI). The IRR yardstick is better suited as project qualifying parameter than as a ranking tool. IRR with a high minimum standard will quickly isolate those projects with a short payout. Since IRR focuses on the near term it is quite insensitive to risk, and uncertainties which tend to increase with longer project lives.

IRR bears some similarity to "book rate of return" discussed earlier in this chapter, in that it is the rate of return on the unrecovered balance of the investment made. However the similarity is only a casual one, since IRR is the average return over the entire life of an investment while the "book rate of return" is calculated for a single year during the life of the investment. Book rate of return has the further disadvantage in that it uses accounting depreciation to approximate the investment repayment schedule. Book rate of return during the mid-portion of the life of an investment can be used to approximate IRR and does provide a surveillance tool for checking the profitability of an ongoing project.

IRR has become the most frequently used economic decision tool in the petroleum industry. When employed with other of the common economic yardsticks it can be an effective culling, or selection, tool in the budgeting process.

Net Present Value, (NPV)

Another approach, which is older than the IRR and currently regaining favor in the petroleum industry is the net present value (NPV). NPV measures the capital created over and above the company's hurdle rate. It employs a given discount factor (hurdle rate) whereas the IRR seeks to compute the discount rate. When the cash flow of the operation is discounted at the given percentage rate a positive or negative value is computed. The application merely calls for choosing the project or investment with the highest positive value of NPV in preference to the lesser.

Net Present Value (NPV) is the difference between the present value of the cash inflows and the present value of the cash outflows generated by the investment and discounted at the assumed hurdle rate. For NPV to be a useful number, the present value date and the discount rate must be specified. It is most commonly used as a screening criterion.

175

The lower righthand columns in the spreadsheet examples in Chapter IV, Tables 4-6 and 4-7, show present values of the net cash flow discounted at 10 percent per year. These are derived according to the concepts described above.

■ **EXAMPLE 5-1**
TIME VALUE COMPARISON (continued)

It is now possible to quantify the difference between the two alternate investments with their associated different repayment schedules. Using a hurdle rate of 10% and applying mid-year discounting to the annual income of Investment #1 for years 1 through 4 and year-end discounting for the lump sums for each at the end of year 5, the NPV#1 = $4,428 and NPV#2 = $4,314. It is apparent that at a 10% discount rate, Investment #1 has a higher NPV than Investment #2 and would be preferable on the basis of this criteria, since the cost of each investment is the same.

Year	Discount Factor (Mid-Year) 10%	15%	NCF INV. #1	10% PV NCF INV. #1	15% PV NCF INV. #1	NCF INV. #2	10% PV NCF INV. #2	15% PV NCF INV. #2
0	1	1	($5,000)	($5,000)	($5,000)	($5,000)	($5,000)	($5,000)
1	0.953463	0.932505	$800	$763	$746		$0	$0
2	0.866784	0.810874	$800	$693	$649		$0	$0
3	0.787986	0.705108	$800	$630	$564		$0	$0
4	0.716351	0.613137	$800	$573	$491		$0	$0
5	0.620921	0.497177	$10,900	$6,768	$5,419	$15,000	$9,314	$7,458
			$9,100	$4,428	$2,869	$10,000	$4,314	$2,458

(handwritten margin notes: "divided by 10/0 mid year")

■ **EXAMPLE 5-3**
COMPUTATION OF NET PRESENT VALUE (NPV)

An interest in an Australian oil producing property is purchased for $100,000. In its first year of operation it sustains a net operating loss of $25,000. The next year it nets $125,000 and the following year, $60,000 before it is sold for $50,000.

DETERMINE:

The net present value, (NPV), of the project discounted at 10 percent per year (mid-period).

SOLUTION:

Year	Bonus	Sale	Operating NCF	Total NCF	Discount Factor*	NPV$_{10}$
0	$(100)	$ 0	$ 0	$(100)	1.0000	$ (100)
1	0	0	(25)	(25)	0.9535	(24)
2	0	0	125	125	0.8668	108
3	0	0	60	60	0.7880	47
3**	—	50	—	50	0.7513**	38
Totals	$(100)	$50	$160	$ 110		$ 69

As an economic decision tool the NPV method recognizes the time value of money and applies the same weighting to all future income. The net initial outlay (NIO) denotes the purchase price of the equipment to be used for the project. The costs of freight, installation and so forth are included as part of the purchase price. Many potential projects involve the replacement of an old asset with a new set of equipment. If the old equipment can be sold, its salvage value is deducted in arriving at the NIO. There may also be tax consequences which should be recognized in any after tax cash flow projection.

Like IRR, the NPV, with its assumed hurdle rate, has its advantages and disadvantages. One of the principal advantages of NPV stems from its capability of providing a uniform comparison tool which is readily understood. Thus, if all the company's organizational units use a common hurdle rate it is a simple matter at budget time to choose the best of the many projects submitted on the basis of the largest NPV values.

Choosing the appropriate hurdle rate percentage constitutes the principal drawback of the NPV approach. Obviously, the assumed hurdle rate must exceed the company's cost of capital, which will be discussed in Chapter VII. Otherwise the firm might continue to take on projects which would not even pay their way at the bank, let alone help the company to grow. A secondary problem lies in its inability to relate to a unit dollar of investment for projects under comparison. NPV does not, of itself, provide any reference to the magnitude of the investment required to achieve the resulting value. A small or a large investment can produce the same dollar NPV result. An NPV15 of $50,000 can be obtained on a $10 million investment as well as from an initial outlay of a mere $25,000.

DECISION RULE:

If NPV 0, i.e., positive dollar value, accept the project.

If NPV < 0, i.e., negative dollar value, reject the project.

If NPV = 0, management should be indifferent to the project

(at the hurdle rate employed).

■ EXAMPLE 5-4
■ EVALUATION OF EXPLORATION PROGRAMS

An exploration venture has been proceeding for five years at an annual cash outlay of $10 million per year. A commercial discovery has been achieved at the beginning of year 6.

DETERMINE:

Whether the following development of the field would be a profitable undertaking.

SOLUTION:

Assume that the net cash flow from the beginning of exploration is:

The summation of the net cash flows from the true beginning through abandonment in year 20 is +$92 million. From discovery forward it is +$142 million. To answer the basic question as to whether the development would be economic, the past would be ignored as "sunk costs," and year 6 would be treated as though it were year 1. The $50 million of sunk costs would not enter into the decision. It is well to keep in mind, however, that the venture appears better than it actually is in its entirety.

The overall profitability (full cycle) of the entire venture is also pertinent; for example, in evaluating the desirability of undertaking similar ventures in the future. In this case time zero can be set at year 1 of the venture even though that time is now long past. The discounting is then straightforward. A second method, which yields the same result, is to set time zero at the beginning of development in year 6 and bring the past expenditures plus interest forward to time zero. The same effective interest rate should be used in figuring the cash flows prior to time zero as are employed in discounting the flows from the future years of the venture. This is illustrated in the following tabulation:

ADJUSTMENT OF NET CASH FLOW FOR TIME ZERO RESET ($ millions)

NCF Year	Venture Year	Annual NCF	Effective 10% Factor*		Adj. Past NCF	Adj. NCF	PV NCF
	1	- 10	(5)	1.536 FV	- 15.36		
	2	- 10	(4)	1.396 FV	- 13.96		
	3	- 10	(3)	1.269 FV	- 12.69	= -64.03	
	4	- 10	(2)	1.154 FV	- 11.54		
	5	- 10	(1)	1,049 FV	- 10.49		
Time Zero						- 64.03	- 64.03
1	6	- 10	(1)	0.953 PV		- 10	- 9.53
2	7	- 5	(2)	0.867 PV		- 5	- 4.33
3	8	+ 5	(3)	0.788 PV		+ 5	+ 3.94
4	9	+10	(4)	0.716 PV		+ 10	+ 7.16
5	10	+20	(5)	0.651 PV		+ 20	+ 13.03
6	11	+35	(6)	0.592 PV		+ 35	+ 20.72
7	12	+25	(7)	0.538 PV		+ 25	+ 13.46
8	13	+20	(8)	0.489 PV		+ 20	+ 9.79
9	14	+15	(9)	0.445 PV		+ 15	+ 6.67
10	15	+ 8	(10)	0.404 PV		+ 8	+ 3.24
11	16	+ 6	(11)	0.368 PV		+ 6	+ 2.21
12	17	+ 5	(12)	0.334 PV		+ 5	+ 1.67
13	18	+ 4	(13)	0.304 PV		+ 4	+ 1.21
14	19	+ 3	(14)	0.276 PV		+ 3	+ 0.83
15	20	+ 1	(15)	0.251 PV		+ 1	+ 0.25
		92					

NPV$_{10}$ (Full Life)   + $6.27

NPV$_{10}$ (Forward from Time Zero) $70.30

*The numbers in parentheses represent the number of time periods. Factors are from Tables 5B-1 and 5B-2 for mid-year compounding and discounting.

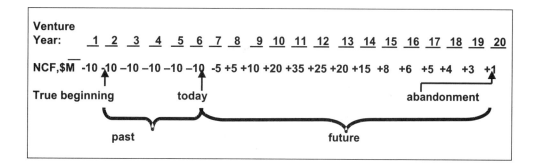

CONCLUSION:

At a hurdle rate of 10% the development project appears to be warranted.

Discounted Return on Investment, (DROI)

DROI is the ratio of a project's NPV to the PV of the total investment required for the project, over its entire operating life. These must be discounted at the same rate. As is the case with other of the frequently used economic yardsticks, DROI is also known as DPR, (Discounted Profit to Investment Ratio), and sometimes as NPVI (Net Present Value Index), and PWPI (Ratio of Net Present Value to Present Worth of Investment).

DROI is an effective measure of capital efficiency which may be viewed as the amount of after-tax NPV generated per dollar of discounted investment. It has the advantages of NPV as an economic yardstick, such as realistic reinvestment rate, not a trial-and-error solution, and no multiple rates. DROI is thus a measure of NPV profit generated in excess of the average, or hurdle rate per dollar of investment. This provides a particularly useful tool for ranking prospects from a group of investment opportunities whose total capital requirements exceed the available funds.

A useful property of DROI is that it is independent of the present value date used in its calculation, since both the numerator and denominator of this ratio are multiplied by the same value. For this reason, the discount rate used is the only auxiliary information required when using DROI as a ranking criterion. Since DROI is independent of PV date, it can also be used to compare projects with different PV dates.

It is obvious that DROI is only useful for comparing projects requiring an investment. If the investment is zero, the denominator of the ratio is zero and DROI is mathematically indeterminate. However, occasionally a significant "one-time" expense is made that has an effect over a long period of time, i.e., a well workover, equipment repair, or plant turnaround. Most of these costs are generally recognized as expense items, but having a lasting effect on the operation. They can be treated just like an investment for economic purposes and included in the denominator of DROI.

$$\text{Discounted ROI, DROI} = \frac{\text{NPV of the Project}}{\text{PV of the Total Project Investment}}$$

Note: When the investment is all at the start of the project, the denominator of DROI equals the total investment. If investments are made after the start of the project, they must be discounted appropriately at the hurdle rate. "NPV of the project" should be after income tax while the "PV of the Total Project Investment" should be as it would be presented in the budget; i.e., before income tax. This apparent discrepancy occurs because the primary use of this ratio is to rank order budget projects, where the budgeted capital expenditure is the total capital expected to be spent to complete the project; i.e. before taking any tax credits. On the other hand, the economic evaluation (NPV) to justify the expenditure should be after all expenditures, including income tax.

■ **EXAMPLE 5-5**
■ **COMPARISON OF METHODS OF PROFITABILITY ANALYSES**

Table 5-7
ILLUSTRATION OF SEVERAL ECONOMIC YARDSTICKS

Time	Major Cash Investment	Annual After Tax Cash Flow	Year End Cumulative Cash Flow	PVIF* 15%	PVIF* 33%
0	-100	-100	-100	1.0000	1.0000
1		50	-50	0.9325	0.8671
2		50	0	0.8109	0.6520
3		50	50	0.7051	0.4902

*Mid-Period Discounting

Interpreting Table 5-7:

$$ROI = 150/100 = 1.5 \ (150\%)$$
$$PIR = 50/100 = 0.5 \ (50\%)$$
$$Payout = 2 \ years$$

NPV at 33% interest:

$$NPV_{33\%} = (-100 * PVIF_{33,0}) + (50 * PVIF_{33,1}) + (50 * PVIF_{33,2}) + (50 * PVIF_{33,3})$$

$$NPV_{33\%} = (-100 * 1.0000) + (50 * 0.8671) + (50 \ \ 0.6520) + (50 * 0.4902)$$

$$NPV_{33\%} = 0.00$$

Therefore:

$$IRR = 33\%$$

In this example, the present worth is zero discounted at 33 percent. Usually, however, it is necessary to compute NPV at two or more values, say 30 and 35 percent, and interpolate, as was done in Example 5-2.

Present Worth at 15% interest:

$$NPV_{15} = (-100 \times 1.0000) + (50 \times 0.9325) + (50 \times 0.8109) + (50 \times 0.7051)$$
$$NPV_{15} = 22.4$$
$$DROI = 22.4 / 100 = \underline{0.224 \ (22.4\%)}$$

Table 5-8 further demonstrates the time value effects of several analytical economic methods discussed. Both Case I and Case II require an initial investment of $50,000.

TABLE 5-8
EFFECT OF CASH FLOW TIMING

Year	Cash Flow $	Cash Flow Discounted @ 10% (End of Year) $	Cash Flow $	Cash Flow Discounted @ 10% (End of Year) $
1	20,000	18,182	6,000	5,455
2	20,000	16,529	6,000	4,959
3	10,000	7,513	6,000	4,508
4	10,000	6,830	6,000	4,098
5	2,000	1,242	6,000	3,726
6			6,000	3,387
7			6,000	3,079
8			6,000	2,799
9			6,000	2,545
10			6,000	2,313
11			6,000	2,103
12			6,000	1,912
13			6,000	1,738
14			6,000	1,580
15			6,000	1,436
16			6,000	1,306
17			6,000	1,187
18			6,000	1,079
19			6,000	981
Total	62,000	50,296	114,000	50,190

	CASE I	CASE II
Investment	50,000	50,000
Net Income	62,000	114,000
Net Profit	12,000	64,000
Profit on Investment	$\dfrac{12,000}{50,000} = 0.24$	$\dfrac{64,000}{50,000} = 1.28$
Payout	3 years	8.33 years
Rate of Return (IRR)	10.30%	10.06%
Life of Project	5 years	19 years
NPV_{10}	$296	$190
$DROI_{10}$	0.0059	0.0038

In Table 5-8 two cases of equal investment, $50,000, have quite different project lives and ultimate profit. Both projects have identical IRR values of 10 percent. Payout of Case I is 3 years compared to 8.33 for Case II, although Profit on Investment (POI) values are just the opposite with the long life Case II being the better. The NPV values are both so small compared to the investments of $50,000 that the DROI values have little meaning.

Acceleration Investments

The development of incremental cash flow streams will be discussed more fully in Chapter X. Rate acceleration projects are a special type of incremental cash flow stream that deserves special attention. These include such activities as infill drilling, field compression of natural gas, well workovers or repairs, facility debottlenecking, etc., which do not recover additional oil or gas, but merely produce it more quickly. However, a savings in direct operating costs may be realized if the producing life is shortened.

FIGURE 5-7
ACCELERATION PROJECT

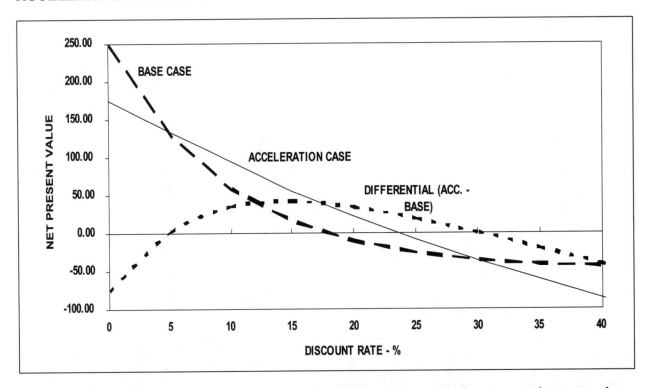

Since additional money is spent for acceleration than would be required to recover the same volume of hydrocarbons, there is no undiscounted payout for the additional expenditure. However, because the additional expenditure accelerates future income, the venture should exhibit a greater net present value. This is the justification for making the added expenditure. The present value profile of incremental change due to acceleration of a project will cross the zero net present value line twice, yielding two theoretical values for internal rate of return (IRR), as is shown in Figure 5-7. Since there is not a unique value for IRR it would appear that IRR is not the proper criteria for evaluating acceleration projects and that one of the other criteria should be used.

WHICH YARDSTICKS TO USE

It was pointed out at the beginning of this chapter that there is no single universally acceptable economic decision tool. Several of these yardsticks, perhaps IRR, NPV, DROI, payout period, and ROI should be computed at the outset before any serious selection of the alternatives is undertaken.

When using an economic yardstick in making investment decisions, it is first necessary to establish what the yardstick will measure. If the investment is to be measured against goals to be achieved, then that must be done with a screening criteria. If it has already been established that the investment opportunity meets or exceeds the established standards, but there are only limited funds available, then a ranking criterion must be used to select the better investments.

The investment screening criteria which have been presented are;
- Net Present Value
- Internal Rate of Return

As screening criteria, both will provide the same information. Each will indicate whether or not the rate of return of the investment opportunity exceeds the hurdle rate. Since Net Present Value is the simplest to calculate and more easily understood, it might be considered to be preferable.

When considering two or more mutually exclusive projects with essentially the same investment requirement, the one with the higher NPV should be selected. In a budgeting situation when considering an array of possible projects with different investment requirements, the NPV of the total package should be maximized within the budgetary limit. The decision as to which projects to undertake and which to reject is complicated by the fact that not all of the projects offering the same present value will require the same capital investment. This was illustrated in Figure 5-5.

The investment ranking criteria which have been presented are;
- Discounted Return On Investment
- Internal Rate of Return
- Payout Time

When these ranking criteria are applied to a set of investment opportunities, each will likely give a different rank order. Therefore, the one or combination of criteria should be used which best reflect the established goals. Discounted return on investment might be more applicable to a large corporation with long term goals, that has sufficient resources to weather short term reverses. Use of DROI to rank investments will maximize the present value net cash flow to the firm and increase its value by the net present value of the investments undertaken. Internal rate of return might be appropriate for a new company with limited resources that needs a rapid return on investments. Payout time may also be appropriate for this type of organization, or in situations where risk increases with time.

Payout time serves an added useful purpose when employed in conjunction with IRR and NPV. Not only does payout time indicate how long a period of time the investment capital is at risk, but it also functions as a rough indicator of liquidity. For instance, if the bond rating agencies have been indicating concern over the firm's ability to service its debt obligations, or if a major cash demand is foreseen in the coming year, a short payout time for current investment opportunities could be an important factor in the decision process.

Before selecting investment criteria, it will be necessary to determine which one or ones will best implement the established goals. It is not uncommon to specify a primary criterion as well as one or more secondary criteria which must be satisfied before an investment opportunity is judged to be acceptable.

ECONOMIC OPTIMIZATION

How profitability of a project can be increased.

As the result of studying the present value concept and the various investment criteria based upon this philosophy, it should be obvious what can be done to improve the economics of an investment opportunity or an ongoing project.

- Reduce capital expenditures
- Reduce operating costs
- Reduce indirect costs (overhead)
- Delay expenditures
- Increase income (production rate)
- Improve value of product
- Accelerate income

Economics will only be improved if the project can be altered without any significant reduction in income or without incurring offsetting costs. Therefore before making any change to a project as suggested in the above list it may be necessary to evaluate the actual impact of such changes.

ESTIMATING PRESENT VALUE UNDER INFLATION

Equation 5.6 can be used to calculate the present value of a future income or expenditure. With the appropriate variables, it is:

$$v_p = v_{ft} (1+ i)^{-t} \qquad\qquad (5.6)$$

where:

v_p = present value of future expenditure or income

t = elapsed time between PV date and expenditure date

v_{ft} = future expenditure or income at end of year "t"

i = effective annual interest or discount rate

The value of future expenditures (v_{ft}), or income at the end of year, "t," is in terms of currency of the day, reflecting the effects of inflation to year,"t."

The value of t can be any real number, positive or negative, integer or fractional. The convention followed is that positive values for t reflect future times while negative values are past times. Mid-year discounting is very common, in that it is a very good approximation for values continuously received over the entire year. This is accomplished by using a value of t equal to an integer plus or minus 0.5 depending upon how the equation is used.

184

Equation 2.3, describing production or income decline, Equation 4.2, describing continuous escalation, and Equation 5.6 for calculating the present value of a future income are all of the same form. They can easily be combined into a unified equation for calculating the present value of a future cashflow stream subject to both a production decline and an inflationary escalation. Equation 5.15 is the resulting equation incorporating mid-year discounting which can be used to calculate the present value of future revenue or cash flow stream.

$$v_p = [R_1(1+i)^{-0.5}][(1-X^t)/(1-X)] \qquad (5.15)$$

where:

$X = (1 - d)(1 + I_I)/(1 + i)$

R_1 = annual revenue or cashflow for the first year

t = number of years of the revenue or cashflow

d = effective decline rate, fraction per year

I_I = constant escalation or inflation rate, fraction per year

i = interest or discount rate, fraction per year

Figure 5-8 is a graph of $[(1 - X_t)/(1 - X)]$ versus X for values of t from 2 to infinity. Normally values of X will be in the range of 0.5 to 0.9, so it can be seen that for all practical purposes, 20 years can be considered an infinite time when considering present value. Figure 5-8 also gives a visual understanding of the interrelation of these three important variables which appear in most economic evaluations.

If Equation 5.15 is used to represent the net income of a cash flow, then it can also be used to determine the internal rate of return (IRR) of that cash flow. To determine the IRR make v_p equal to the present value of the capital investment at the start of the cah flow being evaluated. Then by "trail-and-error" determine the interest rate which discounts future net income to a value equal to the present value of the investment. Alternatively, use the "Goal Seek" tool in Microsoft Excel to determine the interest rate which discounts net income to the present value investment.

■ EXAMPLE 5-6
USE OF A CASH FLOW MODEL

You have been asked to determine the present value net cash flow and Internal Rate of Return (IRR) for a situation where very little information is available. The only value that you can reasonably estimate are the investment required, the net income for the first year, the project life and the expected production decline rate. This is a case where the cash flow model represented by Equation 5.15 is useful. With only the following data the PV NCF can be determined.

Investment at time zero = $300,000

First year Net Income = $100,000

Estimated project life = 15 years

Estimated decline rate for production = 10%/year

Forecast Inflation rate = 5%/year

Hurdle rate = 10%

Using Equation 5.15 to determine the present value net income:

$$X = (1-0.1)(1+0.05)/(1+0.1) = 0.8591$$
$$v_p = [(\$100,000)(1+0.1)^{-0.5}][1-(0.8591)^{15}]/(1-0.8591)$$
$$= \$607,316$$

Then by subtracting the investment at time zero, the PV project net cash flow is estimated to be;

$$PV\ NCF = \$607,316-\$300,000 = \$307,316$$

And;

$$DROI = \$307,316 / \$300,000 = 102.44\%$$

Equation 5.15 can also be used to determine the internal rate of return (IRR). Set vp equal to $300,000, the investment at time zero. Determine the interest rate that satisfies this condition either by "trial-and-error" or using the Excel "Goal Seek" tool. Using the latter procedure, the IRR for the above conditions is 32.65%

FIGURE 5-8
PRESENT VALUE RELATIONSHIPS UNDER INFLATION

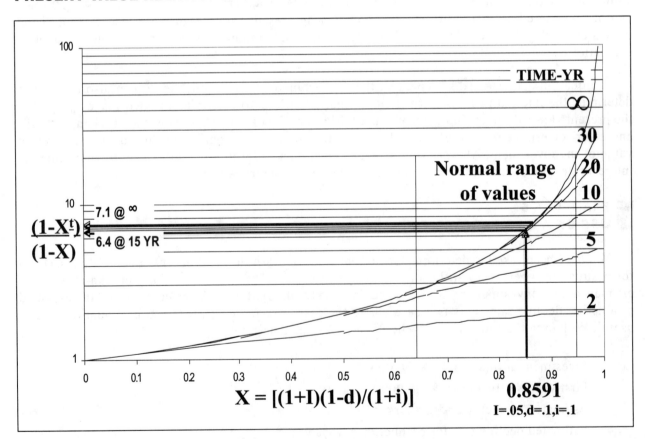

REFERENCES

Arps, J. J., "Profitability of Capital Expenditures for Development Drilling and Producing Property Appraisal," JPT, July 1958, Society of Petroleum Engineers, Richardson, TX 75083-3836

Berry, C. W., "A Wealth Growth Rate Measurement for Capital Projects," Decision Making in the Mining Industry, 1971, Canadian Institute of Mining and Metallurgy

Black, Sloan J., Oilfield Economics, 1983, Prentice and Records, Lafayette, LA. 70503

Blecke, C. J., Financial Analysis for Decision Making, 1966, Prentice Hall, Englewood Cliffs, NJ 07632

Brealey, R., and Myers, S., Principles of Corporate Finance, 2nd Ed., 1984, McGraw-Hill, New York

Campbell, John M., "Petroleum Evaluation for Financial Disclosures," 1982, Campbell Petrolcum Series, Inc., Norman, OK 73069

Capen, E. C., Clapp, R. V., and Phelps, W. W., "Growth Rate—A Rate-of-Return Measure of Investment Efficiency," JPT, May 1976, p. 531-543

Cleland, N.A., Evaluation of Canadian Oil and Gas Properties, 1980, Sproule Associates Limited, Calgary, Alberta

Johnson, Daniel, Oil Company Financial Analysis in Non Technical Language, 1992, Pennwell, Tulsa 74101.

Johnson, Daniel, Petroleum Fiscal Systems and Production Sharing Contracts, 1992, PennWell, Tulsa 74101

Kaitz, Melvin, "Percentage Gain on Investments—An Investment Decision Yardstick," JPT, May 1967, p. 679-685

Kroeger, H. E., Using Discounted Cash Flow Effectively, Dow Jones-Irwin, Homewood, IL 60430

McCray, A. W., Petroleum Evaluations and Economic Decisions, 1975, Prentice Hall, Englewood Cliffs, NJ 07632

Newendorp, Paul. D., Decision Analysis for Petroleum Exploration, 1975, PennWell, Tulsa 74101

Megill, R.E., Exploration Economics, 2nd Ed., 1979, PennWell, Tulsa. OK 74112

Pennant-Rae, R. and Emmott, B, "The Pocket Economist," The Economist, London SW1A 1HG

Phillips, C. E., "The Appreciation of Equity Concepts and its Relationship to Multiple Rates of Return," JPT, Feb. 1965, p.159-163

Seba, R. D., "The Only Investment Selection Criterion You Will Ever Need," Proc. 1987, SPE Hydrocarbon Economics and Evaluation Symposium, p. 173-180.

Silberg, M. and Bronz, F., "Profitability Analysis. Where are we now?" JPT, Jan. 1972, p. 90-100

Stermole, F. J., Economic Evaluation and Investment Decision Methods, 1980, Investment Evaluation Corp., Golden, CO 80401

Steinmetz, Richard, Editor, The Business of Petroleum Exploration, 1992, AAPG, Tulsa 74101.

Theusen, H. G.,Fabrycky, W. J.,and Thuesen, G. J., Engineering Economy, 5th Ed., 1977, Prentice Hall, Englewood Cliffs, NJ 07632

Vichas, R. P., Handbook of Financial Mathematics, Formulas and Tables, 1979, Prentice-Hall, Inc., Englewood Clifts, N.J.

Wooddy, L. D., and Capshaw, T. D., "Investment Evaluation by Present-Value Profile," JPT, June 1960, p. 15

APPENDIX V-A
SOME HELPFUL RULES ABOUT DISCOUNTING

The present value of two or more future values, all at the same point in time, is the sum of their individual present values. In equation form this is;

$$PV\left[\sum_{n=1}^{N}(v_{fn})\right] = \sum_{n=1}^{N}\left[PV(v_{fn})\right] \qquad (5.16)$$

If each future value is multiplied by the same constant value, the constant (K) can be removed from the present value calculation as is shown in the following notation;

$$PV\left[K * v_f\right] = K * PV\left[v_f\right] \qquad (5.17)$$

Sometimes it is convenient to discount first to an intermediate date and then discount from there to the desired PV date. In this case, the discount factors are multiplied together. This can best be understood by reference to Equation 5.18 and Figure 5-9.

$$(1+i)^{-n} = (1+i)^{-a} * (1+i)^{-b} \qquad (5.18)$$

where:
$$n = a + b$$

FIGURE 5-9
SEQUENTIAL DISCOUNTING

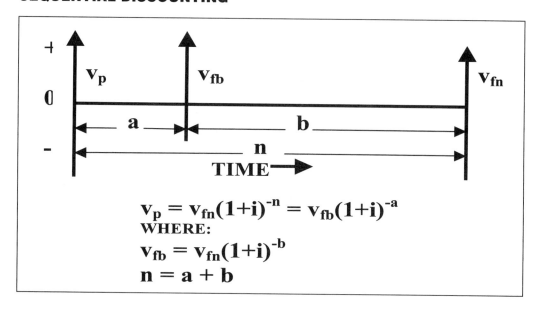

In Figure 5-9 it is seen that future value (v_{fn}) is first discounted "b" years and then discounted another "a" in accordance with Equation 5.18. This technique is useful when it is desired to calculate a present value for a new present value date or when the present value can be more easily calculated at some date other than the prescribed PV date. This can be seen in the equations of Figure 5-9.

Using the principle of sequential discounting, it is then an easy matter to adjust a present value for a change in PV date. This is done by lump sum compounding or lump sum discounting using either Equation 5.2 or Equation 5.6. If the PV date is advanced to the future use Equation 5.2. If the PV date moved back in time, Equation 5.6 is used with its negative exponent, which accounts for a move from right to left on the time line.

APPENDIX VB
COMPOUND AND DISCOUNT TABLES

TABLE 5B-1
PVIF: PRESENT VALUE OF $1 - PAYMENT AT MID-PERIOD

$$PVIF_{i,n} = (1+i)^{-(n-0.50)}$$ (5.19)

DISCOUNT RATE

PERIOD	1.0%	2.0%	3.0%	4.0%	5.0%	7.0%	10.0%	15.0%	20.0%	30.0%	40.0%	50.0%
1	0.9950	0.9901	0.9853	0.9806	0.9759	0.9667	0.9535	0.9325	0.9129	0.8771	0.8452	0.8165
2	0.9852	0.9707	0.9566	0.9429	0.9294	0.9035	0.8668	0.8109	0.7607	0.6747	0.6037	0.5443
3	0.9754	0.9517	0.9288	0.9066	0.8852	0.8444	0.7880	0.7051	0.6339	0.5190	0.4312	0.3629
4	0.9658	0.9330	0.9017	0.8717	0.8430	0.7891	0.7164	0.6131	0.5283	0.3992	0.3080	0.2419
5	0.9562	0.9147	0.8755	0.8382	0.8029	0.7375	0.6512	0.5332	0.4402	0.3071	0.2200	0.1613
6	0.9467	0.8968	0.8500	0.8060	0.7646	0.6893	0.5920	0.4636	0.3669	0.2362	0.1571	0.1075
7	0.9374	0.8792	0.8252	0.7750	0.7282	0.6442	0.5382	0.4031	0.3057	0.1817	0.1122	0.0717
8	0.9281	0.8620	0.8012	0.7452	0.6936	0.6020	0.4893	0.3506	0.2548	0.1398	0.0802	0.0478
9	0.9189	0.8451	0.7778	0.7165	0.6605	0.5626	0.4448	0.3048	0.2123	0.1075	0.0573	0.0319
10	0.9098	0.8285	0.7552	0.6889	0.6291	0.5258	0.4044	0.2651	0.1769	0.0827	0.0409	0.0212
11	0.9008	0.8123	0.7332	0.6624	0.5991	0.4914	0.3676	0.2305	0.1474	0.0636	0.0292	0.0142
12	0.8919	0.7963	0.7118	0.6370	0.5706	0.4593	0.3342	0.2004	0.1229	0.0489	0.0209	0.0094
13	0.8830	0.7807	0.6911	0.6125	0.5434	0.4292	0.3038	0.1743	0.1024	0.0376	0.0149	0.0063
14	0.8743	0.7654	0.6710	0.5889	0.5175	0.4012	0.2762	0.1516	0.0853	0.0290	0.0106	0.0042
15	0.8656	0.7504	0.6514	0.5663	0.4929	0.3749	0.2511	0.1318	0.0711	0.0223	0.0076	0.0028
16	0.8571	0.7357	0.6324	0.5445	0.4694	0.3504	0.2283	0.1146	0.0593	0.0171	0.0054	0.0019
17	0.8486	0.7213	0.6140	0.5235	0.4471	0.3275	0.2075	0.0997	0.0494	0.0132	0.0039	0.0012
18	0.8402	0.7071	0.5961	0.5034	0.4258	0.3060	0.1886	0.0867	0.0411	0.0101	0.0028	0.0008
19	0.8319	0.6933	0.5788	0.4840	0.4055	0.2860	0.1715	0.0754	0.0343	0.0078	0.0020	0.0006
20	0.8236	0.6797	0.5619	0.4654	0.3862	0.2673	0.1559	0.0655	0.0286	0.0060	0.0014	0.0004
21	0.8155	0.6663	0.5456	0.4475	0.3678	0.2498	0.1417	0.0570	0.0238	0.0046	0.0010	0.0002
22	0.8074	0.6533	0.5297	0.4303	0.3503	0.2335	0.1288	0.0495	0.0198	0.0035	0.0007	0.0002
23	0.7994	0.6405	0.5142	0.4138	0.3336	0.2182	0.1171	0.0431	0.0165	0.0027	0.0005	0.0001
24	0.7915	0.6279	0.4993	0.3978	0.3177	0.2039	0.1065	0.0375	0.0138	0.0021	0.0004	0.0001
25	0.7837	0.6156	0.4847	0.3825	0.3026	0.1906	0.0968	0.0326	0.0115	0.0016	0.0003	0.0000
26	0.7759	0.6035	0.4706	0.3678	0.2882	0.1781	0.0880	0.0283	0.0096	0.0012	0.0002	0.0000
27	0.7682	0.5917	0.4569	0.3537	0.2745	0.1665	0.0800	0.0246	0.0080	0.0010	0.0001	0.0000
28	0.7606	0.5801	0.4436	0.3401	0.2614	0.1556	0.0727	0.0214	0.0066	0.0007	0.0001	0.0000
29	0.7531	0.5687	0.4307	0.3270	0.2489	0.1454	0.0661	0.0186	0.0055	0.0006	0.0001	0.0000
30	0.7456	0.5576	0.4181	0.3144	0.2371	0.1359	0.0601	0.0162	0.0046	0.0004	0.0000	0.0000
31	0.7382	0.5466	0.4059	0.3023	0.2258	0.1270	0.0546	0.0141	0.0038	0.0003	0.0000	0.0000
32	0.7309	0.5359	0.3941	0.2907	0.2150	0.1187	0.0497	0.0122	0.0032	0.0003	0.0000	0.0000
33	0.7237	0.5254	0.3826	0.2795	0.2048	0.1109	0.0452	0.0106	0.0027	0.0002	0.0000	0.0000
34	0.7165	0.5151	0.3715	0.2688	0.1951	0.1037	0.0411	0.0093	0.0022	0.0002	0.0000	0.0000
35	0.7094	0.5050	0.3607	0.2584	0.1858	0.0969	0.0373	0.0081	0.0019	0.0001	0.0000	0.0000
36	0.7024	0.4951	0.3502	0.2485	0.1769	0.0905	0.0339	0.0070	0.0015	0.0001	0.0000	0.0000
37	0.6955	0.4854	0.3400	0.2389	0.1685	0.0846	0.0308	0.0061	0.0013	0.0001	0.0000	0.0000
38	0.6886	0.4759	0.3301	0.2297	0.1605	0.0791	0.0280	0.0053	0.0011	0.0001	0.0000	0.0000
39	0.6818	0.4665	0.3205	0.2209	0.1528	0.0739	0.0255	0.0046	0.0009	0.0000	0.0000	0.0000
40	0.6750	0.4574	0.3111	0.2124	0.1456	0.0691	0.0232	0.0040	0.0007	0.0000	0.0000	0.0000
41	0.6683	0.4484	0.3021	0.2042	0.1386	0.0646	0.0211	0.0035	0.0006	0.0000	0.0000	0.0000
42	0.6617	0.4396	0.2933	0.1964	0.1320	0.0603	0.0192	0.0030	0.0005	0.0000	0.0000	0.0000
43	0.6552	0.4310	0.2847	0.1888	0.1257	0.0564	0.0174	0.0026	0.0004	0.0000	0.0000	0.0000
44	0.6487	0.4226	0.2764	0.1816	0.1197	0.0527	0.0158	0.0023	0.0004	0.0000	0.0000	0.0000
45	0.6422	0.4143	0.2684	0.1746	0.1140	0.0493	0.0144	0.0020	0.0003	0.0000	0.0000	0.0000
46	0.6359	0.4062	0.2606	0.1679	0.1086	0.0460	0.0131	0.0017	0.0002	0.0000	0.0000	0.0000
47	0.6296	0.3982	0.2530	0.1614	0.1034	0.0430	0.0119	0.0015	0.0002	0.0000	0.0000	0.0000
48	0.6234	0.3904	0.2456	0.1552	0.0985	0.0402	0.0108	0.0013	0.0002	0.0000	0.0000	0.0000
49	0.6172	0.3827	0.2384	0.1492	0.0938	0.0376	0.0098	0.0011	0.0001	0.0000	0.0000	0.0000
50	0.6111	0.3752	0.2315	0.1435	0.0894	0.0351	0.0089	0.0010	0.0001	0.0000	0.0000	0.0000

TABLE 5B-2
FVIF: FUTURE VALUE OF $1 - PAYMENT AT MID-PERIOD

$$FVIF_{i,n} = (1+i)^{(n-0.50)} = 1 / (PVIF_{i,n}) \tag{5.20}$$

INTEREST RATE

PERIOD	1.0%	2.0%	3.0%	4.0%	5.0%	7.0%	10.0%	15.0%	20.0%	30.0%	40.0%	50.0%
1	1.0050	1.0100	1.0149	1.0198	1.0247	1.0344	1.0488	1.0724	1.0954	1.1402	1.1832	1.2247
2	1.0150	1.0301	1.0453	1.0606	1.0759	1.1068	1.1537	1.2332	1.3145	1.4822	1.6565	1.8371
3	1.0252	1.0508	1.0767	1.1030	1.1297	1.1843	1.2691	1.4182	1.5774	1.9269	2.3191	2.7557
4	1.0354	1.0718	1.1090	1.1471	1.1862	1.2672	1.3960	1.6310	1.8929	2.5050	3.2467	4.1335
5	1.0458	1.0932	1.1423	1.1930	1.2455	1.3559	1.5356	1.8756	2.2715	3.2565	4.5454	6.2003
6	1.0563	1.1151	1.1765	1.2407	1.3078	1.4508	1.6891	2.1569	2.7258	4.2334	6.3636	9.3004
7	1.0668	1.1374	1.2118	1.2904	1.3732	1.5524	1.8580	2.4805	3.2710	5.5034	8.9091	13.951
8	1.0775	1.1601	1.2482	1.3420	1.4418	1.6610	2.0438	2.8526	3.9252	7.1544	12.473	20.926
9	1.0883	1.1833	1.2856	1.3957	1.5139	1.7773	2.2482	3.2804	4.7102	9.3008	17.462	31.389
10	1.0991	1.2070	1.3242	1.4515	1.5896	1.9017	2.4730	3.7725	5.6523	12.091	24.446	47.083
11	1.1101	1.2311	1.3639	1.5096	1.6691	2.0348	2.7203	4.3384	6.7827	15.718	34.225	70.625
12	1.1212	1.2557	1.4048	1.5699	1.7526	2.1773	2.9924	4.9891	8.1392	20.434	47.915	105.94
13	1.1324	1.2809	1.4470	1.6327	1.8402	2.3297	3.2916	5.7375	9.7671	26.564	67.081	158.91
14	1.1438	1.3065	1.4904	1.6980	1.9322	2.4928	3.6208	6.5981	11.721	34.533	93.914	238.36
15	1.1552	1.3326	1.5351	1.7660	2.0288	2.6673	3.9828	7.5878	14.065	44.893	131.48	357.54
16	1.1668	1.3593	1.5812	1.8366	2.1303	2.8540	4.3811	8.7260	16.878	58.361	184.07	536.31
17	1.1784	1.3864	1.6286	1.9101	2.2368	3.0537	4.8192	10.035	20.253	75.869	257.70	804.46
18	1.1902	1.4142	1.6775	1.9865	2.3486	3.2675	5.3012	11.540	24.304	98.630	360.78	1206.7
19	1.2021	1.4425	1.7278	2.0659	2.4661	3.4962	5.8313	13.271	29.164	128.22	505.09	1810.0
20	1.2141	1.4713	1.7796	2.1486	2.5894	3.7410	6.4144	15.262	34.997	166.68	707.13	2715.1
21	1.2263	1.5007	1.8330	2.2345	2.7188	4.0028	7.0559	17.551	41.997	216.69	989.98	4072.6
22	1.2385	1.5307	1.8880	2.3239	2.8548	4.2830	7.7614	20.184	50.396	281.70	1386.0	6108.9
23	1.2509	1.5614	1.9446	2.4169	2.9975	4.5828	8.5376	23.211	60.475	366.21	1940.4	9163.3
24	1.2634	1.5926	2.0030	2.5135	3.1474	4.9036	9.3914	26.693	72.570	476.07	2716.5	13745
25	1.2761	1.6244	2.0631	2.6141	3.3047	5.2469	10.330	30.697	87.084	618.89	3803.1	20617
26	1.2888	1.6569	2.1250	2.7186	3.4700	5.6142	11.364	35.302	104.50	804.55	5324.3	30926
27	1.3017	1.6901	2.1887	2.8274	3.6435	6.0072	12.500	40.597	125.40	1045.9	7454.1	46389
28	1.3147	1.7239	2.2544	2.9405	3.8257	6.4277	13.750	46.686	150.48	1359.7	10436	69584
29	1.3279	1.7583	2.3220	3.0581	4.0169	6.8776	15.125	53.689	180.58	1767.6	14610	104376
30	1.3412	1.7935	2.3917	3.1804	4.2178	7.3590	16.637	61.743	216.69	2297.9	20454	156564
31	1.3546	1.8294	2.4634	3.3076	4.4287	7.8742	18.301	71.004	260.03	2987.3	28636	234846
32	1.3681	1.8660	2.5373	3.4399	4.6501	8.4254	20.131	81.655	312.04	3883.4	40090	352269
33	1.3818	1.9033	2.6134	3.5775	4.8826	9.0151	22.144	93.903	374.45	5048.5	56126	528404
34	1.3956	1.9414	2.6918	3.7206	5.1267	9.6462	24.359	107.99	449.34	6563.0	78576	792606
35	1.4096	1.9802	2.7726	3.8695	5.3831	10.321	26.795	124.19	539.20	8531.9	110006	1188909
36	1.4237	2.0198	2.8558	4.0242	5.6522	11.044	29.474	142.81	647.04	11091	154009	1783363
37	1.4379	2.0602	2.9414	4.1852	5.9348	11.817	32.421	164.24	776.45	14419	215612	2675044
38	1.4523	2.1014	3.0297	4.3526	6.2316	12.644	35.664	188.87	931.74	18745	301857	4012566
39	1.4668	2.1434	3.1206	4.5267	6.5432	13.529	39.230	217.20	1118.1	24368	422600	6018849
40	1.4815	2.1863	3.2142	4.7078	6.8703	14.476	43.153	249.78	1341.7	31678	591640	9028274
41	1.4963	2.2300	3.3106	4.8961	7.2138	15.490	47.468	287.25	1610.1	41182	828296	13542411
42	1.5113	2.2746	3.4099	5.0919	7.5745	16.574	52.215	330.34	1932.1	53536	1159614	20313617
43	1.5264	2.3201	3.5122	5.2956	7.9533	17.734	57.437	379.89	2318.5	69597	1623460	30470425
44	1.5416	2.3665	3.6176	5.5074	8.3509	18.976	63.180	436.87	2782.2	90477	2272844	45705637
45	1.5570	2.4138	3.7261	5.7277	8.7685	20.304	69.498	502.40	3338.6	117619	3181981	68558456
46	1.5726	2.4621	3.8379	5.9569	9.2069	21.725	76.448	577.77	4006.3	152905	4454773	102837684
47	1.5883	2.5114	3.9530	6.1951	9.6672	23.246	84.093	664.43	4807.6	198777	6236683	154256526
48	1.6042	2.5616	4.0716	6.4429	10.151	24.873	92.502	764.10	5769.1	258410	8731356	231384789
49	1.6203	2.6128	4.1938	6.7007	10.658	26.614	101.75	878.71	6922.9	335933	12223898	347077183
50	1.6365	2.6651	4.3196	6.9687	11.191	28.477	111.93	1010.5	8307.5	436713	17113458	520615775

TABLE 5B-3
PVIFA: PRESENT VALUE OF AN ANNUITY OF $1 PER PERIOD
FOR n PERIODS - PAYMENT AT MID-PERIOD

$$PVIF_{i,n} = (1+i)^{0.5}\left[1-(1+i)^{-n}\right]/i \qquad (5.21)$$

DISCOUNT RATE

PERIOD	1.0%	2.0%	3.0%	4.0%	5.0%	7.0%	10.0%	15.0%	20.0%	30.0%	40.0%	50.0%
1	0.9950	0.9901	0.9853	0.9806	0.9759	0.9667	0.9535	0.9325	0.9129	0.8771	0.8452	0.8165
2	1.9802	1.9609	1.9420	1.9234	1.9053	1.8702	1.8202	1.7434	1.6736	1.5517	1.4488	1.3608
3	2.9557	2.9126	2.8707	2.8300	2.7905	2.7146	2.6082	2.4485	2.3075	2.0707	1.8800	1.7237
4	3.9214	3.8456	3.7724	3.7018	3.6335	3.5038	3.3246	3.0616	2.8358	2.4699	2.1880	1.9656
5	4.8776	4.7604	4.6479	4.5400	4.4364	4.2413	3.9758	3.5948	3.2761	2.7770	2.4080	2.1269
6	5.8244	5.6572	5.4978	5.3460	5.2010	4.9305	4.5678	4.0584	3.6429	3.0132	2.5652	2.2344
7	6.7618	6.5364	6.3230	6.1209	5.9293	5.5747	5.1060	4.4616	3.9486	3.1949	2.6774	2.3061
8	7.6898	7.3984	7.1242	6.8661	6.6228	6.1768	5.5953	4.8121	4.2034	3.3347	2.7576	2.3539
9	8.6087	8.2435	7.9020	7.5826	7.2835	6.7394	6.0401	5.1170	4.4157	3.4422	2.8149	2.3858
10	9.5185	9.0720	8.6572	8.2715	7.9124	7.2652	6.4445	5.3820	4.5926	3.5249	2.8558	2.4070
11	10.4193	9.8842	9.3904	8.9340	8.5115	7.7567	6.8121	5.6125	4.7401	3.5885	2.885	2.4212
12	11.3112	10.6806	10.1022	9.5709	9.0821	8.2160	7.1463	5.8130	4.8629	3.6375	2.9059	2.4306
13	12.1943	11.4613	10.7933	10.1834	9.6255	8.6452	7.4501	5.9873	4.9653	3.6751	2.9208	2.4369
14	13.0686	12.2267	11.4643	10.7723	10.1431	9.0464	7.7262	6.1388	5.0506	3.7041	2.9314	2.4411
15	13.9342	12.9771	12.1157	11.3386	10.6360	9.4213	7.9773	6.2706	5.1217	3.7263	2.9390	2.4439
16	14.7913	13.7128	12.7481	11.8831	11.1054	9.7717	8.2056	6.3852	5.1810	3.7435	2.9445	2.4458
17	15.6399	14.4341	13.3622	12.4066	11.5525	10.0992	8.4131	6.4849	5.2303	3.7566	2.9483	2.4470
18	16.4801	15.1412	13.9583	12.9100	11.9783	10.4052	8.6017	6.5715	5.2715	3.7668	2.9511	2.4478
19	17.3119	15.8345	14.5371	13.3940	12.3838	10.6912	8.7732	6.6469	5.3058	3.7746	2.9531	2.4484
20	18.1356	16.5141	15.0990	13.8595	12.7700	10.9585	8.9291	6.7124	5.3344	3.7806	2.9545	2.4488
21	18.9510	17.1805	15.6445	14.3070	13.1378	11.2084	9.0708	6.7694	5.3582	3.7852	2.9555	2.4490
22	19.7584	17.8338	16.1742	14.7373	13.4881	11.4418	9.1997	6.8189	5.3780	3.7888	2.9562	2.4492
23	20.5578	18.4742	16.6884	15.1511	13.8217	11.6600	9.3168	6.8620	5.3945	3.7915	2.9568	2.4493
24	21.3493	19.1021	17.1877	15.5489	14.1394	11.8640	9.4233	6.8995	5.4083	3.7936	2.9571	2.4493
25	22.1330	19.7177	17.6724	15.9315	14.4420	12.0546	9.5201	6.9320	5.4198	3.7952	2.9574	2.4494
26	22.9089	20.3213	18.1430	16.2993	14.7302	12.2327	9.6081	6.9604	5.4294	3.7964	2.9576	2.4494
27	23.6771	20.9129	18.5999	16.6530	15.0046	12.3991	9.6881	6.9850	5.4374	3.7974	2.9577	2.4494
28	24.4377	21.4930	19.0435	16.9931	15.2660	12.5547	9.7608	7.0064	5.4440	3.7981	2.9578	2.4495
29	25.1908	22.0617	19.4742	17.3201	15.5150	12.7001	9.8269	7.0250	5.4495	3.7987	2.9579	2.4495
30	25.9364	22.6193	19.8923	17.6345	15.7521	12.8360	9.8870	7.0412	5.4542	3.7991	2.9579	2.4495
31	26.6747	23.1659	20.2982	17.9368	15.9779	12.9630	9.9417	7.0553	5.4580	3.7995	2.9580	2.4495
32	27.4056	23.7019	20.6923	18.2275	16.1929	13.0817	9.9913	7.0676	5.4612	3.7997	2.9580	2.4495
33	28.1293	24.2273	21.0750	18.5070	16.3977	13.1926	10.0365	7.0782	5.4639	3.7999	2.9580	2.4495
34	28.8458	24.7424	21.4465	18.7758	16.5928	13.2963	10.0776	7.0875	5.4661	3.8001	2.9580	2.4495
35	29.5553	25.2474	21.8071	19.0342	16.7786	13.3932	10.1149	7.0955	5.4680	3.8002	2.9580	2.4495
36	30.2577	25.7425	22.1573	19.2827	16.9555	13.4837	10.1488	7.1025	5.4695	3.8003	2.9580	2.4495
37	30.9531	26.2279	22.4973	19.5217	17.1240	13.5683	10.1797	7.1086	5.4708	3.8004	2.9580	2.4495
38	31.6417	26.7037	22.8274	19.7514	17.2844	13.6474	10.2077	7.1139	5.4719	3.8004	2.9580	2.4495
39	32.3234	27.1703	23.1478	19.9723	17.4373	13.7213	10.2332	7.1185	5.4728	3.8004	2.9580	2.4495
40	32.9985	27.6277	23.4589	20.1847	17.5828	13.7904	10.2564	7.1225	5.4735	3.8005	2.9580	2.4495
41	33.6668	28.0761	23.7610	20.3890	17.7215	13.8550	10.2774	7.1260	5.4741	3.8005	2.9580	2.4495
42	34.3285	28.5157	24.0543	20.5854	17.8535	13.9153	10.2966	7.1290	5.4746	3.8005	2.9580	2.4495
43	34.9836	28.9468	24.3390	20.7742	17.9792	13.9717	10.3140	7.1317	5.4751	3.8005	2.9580	2.4495
44	35.6323	29.3693	24.6154	20.9558	18.0990	14.0244	10.3298	7.1339	5.4754	3.8005	2.9580	2.4495
45	36.2745	29.7836	24.8838	21.1304	18.2130	14.0737	10.3442	7.1359	5.4757	3.8006	2.9580	2.4495
46	36.9104	30.1898	25.1443	21.2983	18.3216	14.1197	10.3573	7.1377	5.4760	3.8006	2.9580	2.4495
47	37.5400	30.5879	25.3973	21.4597	18.4251	14.1627	10.3692	7.1392	5.4762	3.8006	2.9580	2.4495
48	38.1634	30.9783	25.6429	21.6149	18.5236	14.2029	10.3800	7.1405	5.4764	3.8006	2.9580	2.4495
49	38.7805	31.3611	25.8814	21.7641	18.6174	14.2405	10.3898	7.1416	5.4765	3.8006	2.9580	2.4495
50	39.3916	31.7363	26.1129	21.9076	18.7068	14.2756	10.3987	7.1426	5.4766	3.8006	2.9580	2.4495

TABLE 5B-4
FVIFA: SUM OF AN ANNUITY OF $1 PER PERIOD
AT END OF N PERIODS - PAYMENT AT MID-PERIOD

$$FVIFA_{i,n} = (1+i)^{0.5} \left[(1+i)^n - 1 \right] / i = (1+i)^n * (PVIFA_{i,n})$$

(5.22)

INTEREST RATE

PERIOD	1.0%	2.0%	3.0%	4.0%	5.0%	7.0%	10.0%	15.0%	20.0%	30.0%	40.0%	50.0%
1	1.0050	1.0100	1.0149	1.0198	1.0247	1.0344	1.0488	1.0724	1.0954	1.1402	1.1832	1.2247
2	2.0200	2.0401	2.0602	2.0804	2.1006	2.1412	2.2025	2.3056	2.4100	2.6224	2.8397	3.0619
3	3.0452	3.0909	3.1369	3.1834	3.2304	3.3255	3.4716	3.7238	3.9874	4.5493	5.1588	5.8175
4	4.0807	4.1626	4.2459	4.3306	4.4166	4.5927	4.8675	5.3548	5.8803	7.0543	8.4056	9.9511
5	5.1264	5.2558	5.3882	5.5236	5.6621	5.9486	6.4031	7.2304	8.1519	10.3107	12.9510	16.1513
6	6.1827	6.3709	6.5647	6.7643	6.9699	7.3994	8.0922	9.3873	10.8777	14.5441	19.3146	25.4517
7	7.2495	7.5083	7.7765	8.0547	8.3431	8.9518	9.9502	11.8678	14.1487	20.0475	28.2237	39.4023
8	8.3270	8.6684	9.0247	9.3967	9.7849	10.6128	11.9941	14.7204	18.0738	27.2020	40.6964	60.3283
9	9.4153	9.8517	10.3104	10.7924	11.2989	12.3901	14.2423	18.0008	22.7841	36.5027	58.1582	91.7171
10	10.5144	11.0587	11.6346	12.2439	12.8885	14.2918	16.7153	21.7733	28.4363	48.5937	82.6046	138.800
11	11.6245	12.2898	12.9985	13.7534	14.5576	16.3267	19.4357	26.1117	35.2190	64.3120	116.830	209.425
12	12.7458	13.5455	14.4033	15.3234	16.3102	18.5040	22.4280	31.1008	43.3583	84.7458	164.745	315.363
13	13.8782	14.8264	15.8503	16.9561	18.1504	20.8336	25.7196	36.8383	53.1254	111.310	231.826	474.269
14	15.0220	16.1329	17.3407	18.6542	20.0826	23.3264	29.3404	43.4365	64.8459	145.843	325.740	712.628
15	16.1772	17.4655	18.8758	20.4201	22.1114	25.9937	33.3233	51.0243	78.9105	190.736	457.219	1070.17
16	17.3439	18.8248	20.4570	22.2567	24.2417	28.8476	37.7044	59.7503	95.7881	249.097	641.289	1606.48
17	18.5224	20.2112	22.0856	24.1668	26.4785	31.9014	42.5236	69.7853	116.041	324.966	898.988	2410.94
18	19.7126	21.6254	23.7631	26.1533	28.8271	35.1689	47.8248	81.3254	140.345	423.596	1259.77	3617.63
19	20.9147	23.0678	25.4908	28.2192	31.2932	38.6651	53.6561	94.5966	169.509	551.815	1764.86	5427.67
20	22.1288	24.5391	27.2705	30.3678	33.8825	42.4061	60.0705	109.859	204.506	718.499	2471.98	8142.73
21	23.3551	26.0399	29.1035	32.6023	36.6013	46.4089	67.1264	127.410	246.503	935.189	3461.96	12215.3
22	24.5936	27.5706	30.9914	34.9262	39.4561	50.6919	74.8878	147.593	296.899	1216.89	4847.93	18324.2
23	25.8446	29.1320	32.9361	37.3431	42.4536	55.2748	83.4254	170.805	357.375	1583.09	6788.28	27487.5
24	27.1080	30.7246	34.9391	39.8566	45.6010	60.1784	92.8168	197.498	429.945	2059.16	9504.77	41232.5
25	28.3841	32.3490	37.0021	42.4707	48.9057	65.4253	103.1473	228.195	517.029	2678.05	13307.9	61850.0
26	29.6729	34.0059	39.1271	45.1893	52.3757	71.0395	114.5108	263.497	621.531	3482.60	18632.2	92776.3
27	30.9746	35.6960	41.3158	48.0167	56.0192	77.0467	127.0107	304.094	746.932	4528.52	26086.3	139166
28	32.2893	37.4199	43.5701	50.9571	59.8448	83.4743	140.7606	350.780	897.414	5888.22	36521.9	208750
29	33.6172	39.1782	45.8921	54.0152	63.8618	90.3520	155.8854	404.469	1077.99	7655.83	51131.9	313126
30	34.9584	40.9718	48.2838	57.1956	68.0796	97.7110	172.5228	466.212	1294.69	9953.71	71585.8	469690
31	36.3130	42.8011	50.7472	60.5033	72.5082	105.5852	190.8239	537.216	1554.72	12941.0	100221	704536
32	37.6811	44.6671	53.2845	63.9432	77.1583	114.0105	210.9551	618.871	1866.76	16824.4	140311	1056805
33	39.0629	46.5704	55.8979	67.5207	82.0410	123.0257	233.0994	712.774	2241.21	21872.9	196437	1585209
34	40.4585	48.5118	58.5897	71.2414	87.1677	132.6719	257.4581	820.763	2690.54	28435.9	275013	2377815
35	41.8681	50.4919	61.3623	75.1108	92.5508	142.9933	284.2528	944.950	3229.75	36967.8	385019	3566723
36	43.2917	52.5117	64.2181	79.1351	98.2030	154.0373	313.7268	1087.76	3876.79	48059.2	539028	5350086
37	44.7296	54.5719	67.1595	83.3203	104.138	165.8543	346.1483	1252.00	4653.24	62478.1	754640	8025130
38	46.1819	56.6733	70.1892	87.6729	110.369	178.4985	381.8120	1440.87	5584.99	81222.7	1056497	12037696
39	47.6487	58.8167	73.3097	92.1996	116.913	192.0278	421.0420	1658.08	6703.08	105591	1479097	18056546
40	49.1302	61.0030	76.5239	96.9074	123.783	206.5042	464.1950	1907.86	8044.79	137269	2070736	27084820
41	50.6265	63.2330	79.8345	101.803	130.997	221.9939	511.6633	2195.11	9654.85	178451	2899032	40627231
42	52.1377	65.5076	83.2445	106.895	138.571	238.5678	563.8784	2525.45	11586.9	231987	4058646	60940847
43	53.6641	67.8277	86.7567	112.191	146.525	256.3020	621.3151	2905.34	13905.4	301585	5682106	91411272
44	55.2057	70.1942	90.3743	117.699	154.875	275.2775	684.4954	3342.22	16687.6	392061	7954950	137116909
45	56.7628	72.6081	94.1004	123.426	163.644	295.5814	753.9938	3844.62	20026.2	509681	11136931	205675365
46	58.3354	75.0702	97.9383	129.383	172.851	317.3065	830.4419	4422.39	24032.5	662586	15591704	308513049
47	59.9237	77.5815	101.891	135.578	182.518	340.5524	914.5349	5086.82	28840.1	861363	21828387	462769575
48	61.5280	80.1431	105.963	142.021	192.669	365.4254	1007.037	5850.91	34609.2	1119773	30559743	694154364
49	63.1482	82.7559	110.157	148.722	203.327	392.0396	1108.790	6729.62	41532.1	1455706	42783641	1041231547
50	64.7847	85.4210	114.476	155.690	214.518	420.5168	1220.718	7740.14	49839.7	1892418	59897099	1561847322

RISK AND UNCERTAINTY

CHAPTER VI

In this world nothing is certain but death and taxes.

—BENJAMIN FRANKLIN

The petroleum exploration and production industry is characterized as a "risk business." The usual reference is to the geological risk of drilling non-productive wells. With the growing volatility of oil and gas prices, financial risk is also becoming an increasingly important factor. The traditional method of coping with risk has been through methods such as diversification, sheer size, and vertical integration of oil and gas production and downstream refining and marketing.

The terms, risk and uncertainty, are frequently used almost interchangeably in everyday discussion. For our purposes this will be differentiated. Uncertainty will be used to characterize the fact that the eventual outcome of a decision or event is not precisely known, with the degree of uncertainty described by the probability that it will occur. This concept implies that the range of possible outcomes can be determined and that the probability of each occurring can be estimated. This approach to uncertainty permits application of the mathematical theories of probability to the decision-making process.

Risk, on the other hand, denotes that there is a possibility of incurring economic loss or reduced value. High risk ventures are ones with a chance of a large loss, even if the probability of such an occurrence is small. It is possible for a project to be highly uncertain, but have a low risk, if failure would be inconsequential. Assessment of the probability of each anticipated outcome can be the most difficult part of including uncertainty in the decision making process. Such assessments will range from apriori values based upon a complete knowledge of the system, such as casino type games of chance, to those where the probability of each outcome could be anywhere in the range of zero to one. Unfortunately, most situations encountered in the petroleum exploration and production industry are more in the direction of the latter type of probability assessment than the former. Frequently such probabilities are based on no more than a guess or a hunch.

The decision to undertake a project is not only affected by the anticipated gain, but also by the degree of uncertainty of both the timing of events and the ultimate outcome. Various economic yardsticks discussed in the previous chapter must be adjusted to reflect both of these factors either subjectively or objectively. In order to achieve the degree of realism required for investment decision analyses, it is necessary to incorporate allowances for possible alternate outcomes, including failure, in our calculations. This becomes a challenging and often elusive problem.

Depending upon the availability of factual engineering data and our knowledge of the economic as well as the physical environment, the range of probability for each outcome may be very broad or narrow. We will categorize our decision-making situations under the three headings: certainty, uncertainty and risk.

CERTAINTY, UNCERTAINTY AND RISK

The following definitions are employed:

Certainty: Only one possible outcome

Uncertainty: Recognition that more than a single outcome is possible, with each outcome having a finite probability of occurrence

Risk: Possibility of incurring economic loss or reduced economic value

High Risk: The chance of incurring a large loss, even if the probability of doing so is very small.

Outcome: One of the possible events that can take place

Probability: The chance between 0% and 100% that a particular outcome will occur. A probability of 0% means that it is certain NOT to occur, while a probability of 100% means that it is certain TO occur. All other probabilities in this range describe uncertainty.

These and other definitions related to risk, uncertainty and statistics can be found in Appendix VI-A at the end of this chapter.

Types of Risk

There are many varieties of such risks including such items as:

TABLE 6-1
TYPES OF RISK

TECHNICAL	ECONOMIC	POLITICAL
Dry holes	Inflation	Governmental policy
Geological	Oil and gas prices	Government regulations
Engineering	Gambler's ruin	Laws
Storm damage	Interest rates	Nationalization
Earthquake	Environmental	Environmental
Timing	Timing	Timing
	Exchange rate	Exchange rate
	Financing / capital	Financing / capital
	Supply / demand	Taxation
	Operating costs	Export / import
		Personnel

As there may be several of the various risks listed above which have simultaneous impact on a particular project, it may be necessary to weight them mathematically in order to obtain a proper overall approximation. The weighting applied to each of the several risks will depend upon the nature of the project being evaluated. Such risks are generally considered under three broad categories namely, technical, economic and political.

When evaluating exploration and production ventures there are three types of uncertainty which are most significant. These are; (1) uncertainty of occurrence, (2) uncertainty of magnitude and (3) uncertainty of rate of production. During the exploratory phase, uncertainty of occurrence is of major concern and will probably dominate exploratory evaluations, even though the other two uncertainties surely have an impact upon the possible successful outcomes. Once a discovery is made, uncertainty of magnitude, which includes both volume and value, and rate become the dominant uncertainties. These uncertainties will remain throughout the entire producing life of a project, but will diminish with time as producing performance is observed, becoming zero at abandonment.

Political risks involve the uncertainty arising from possible changes in the policies of regulatory authorities and the degree to which such changes may affect the project revenues. Regulatory considerations can be subdivided into fiscal and non-fiscal considerations. The fiscal aspects primarily include continuity in the levels of local and national taxation, exchange controls and limitations on import and export of foreign and local currencies, changes in levels of customs duties on imported equipment and supplies, and possible imposition of locally denominated prices for the production.

Non-fiscal political risks may relate to possible interruptions by regulatory authorities over environmental matters, disagreements over hiring or firing of local personnel, determinations of commerciality, or outright nationalization. Matters such as the provisions for transfer of operatorship to the NOC and the potential for political unrest in the host country also fall under this category.

Risk also affects oil company financing. *"One doesn't ask one's bankers for a loan to play roulette"**—neither does an oil company take on long term debt to drill exploratory wells. Funding these expenditures is usually accomplished with internal company resources. *"At the development stage, it is feasible to call on external financing because one is very likely to make a profit."*

Economic risk also covers a very broad range of potential situations, not the least of which are the present and future levels of oil and gas prices. The physical nature of the project is also highly important. For instance, the principal economic risk associated with an infrastructure project may be confined to the possibility of capital cost overrun and timing of completion, whereas in the case of a depleting asset, such as mining or petroleum production, the prime economic concerns will probably involve drilling and operating costs, inflationary effects and interest rates, as well as the always important product prices and demand over the life of the project add additional risk.

Technical risks involve the operational nature of the project, i.e., has the procedure been employed many times before or is it a brand new technique. Technical risks may include the capability and experience of the engineering talent assigned to the project. In the case of reserve estimation the degree of technical risk may involve the hydrocarbon volumes which actually exist underground and whether the producing rates and ultimate recoveries projected by the engineers will actually be realized. Pricing risk and the other categories listed above are discussed more fully in other sections and chapters.

---

* ALAIN BRION (LOC. CIT.)

Characteristics of Risk and Uncertainty

The first characteristic of uncertainty is that it may be either objective or subjective. Uncertainties that can be accurately calculated ahead of time, as in the case of flipping a coin, are known as objective uncertainties. (Newendorp's second category.) In these cases there is no room for disagreement as to the probability of the outcome.

Describing the odds for rain next Tuesday is not so clear cut, however. This represents a subjective situation. Armed with the same information one forecaster may think the chance of rain is 30% while another weatherman may judge the odds to be about 65%. Neither is wrong. Subjective uncertainty is always open to reassessment on the basis of new information, further study, or by recognizing the opinion of other observers. Most uncertainties are subjective, and this is particularly true of the oil business.

Subjective assessments of uncertainty are likely to change as additional information becomes available. The possible option of spending money for additional information, such as adding another seismic line, or waiting for more production data, in order to improve the risk assessment becomes part of the decision making process.

Most uncertainty quantification involves some degree of personal judgement. This can be due to any of several reasons. There may not be complete or adequate information on the situation; or, the situation may not be repeatable like the coin flip; or it may be too complex to permit a finite answer. The recognition that a certain operation is risky involves a judgement. What may be considered unduly risky by one individual may not be so appraised by another person.

The second important characteristic of uncertainty is its adaptability to quantitative interpretation. This leads to the third characteristic of uncertainty, i.e., the possible choice of accepting or avoiding risk. This characteristic is particularly likely in decisions involving investments in the oil industry. In most cases the operator is not contractually required to develop a project. He has the option of "walking away." To the extent that individual judgements regarding the degree of a particular risk may differ (in magnitude, if not in nature), identical risks may involve dissimilar responses from different investors.

Dealing with Risk and Uncertainty

There are four fundamental approaches to coping with risk and uncertainty: (1) diversification, (2) reduction of exposure, (3) avoidance, and (4) insurance. In the case of geological risk, diversification means participating in the drilling of ten wells instead of putting all of one's resources in a single prospect. This reduces exposure by taking a lesser interest in a greater number of ventures. If the risk is so great or the reward insufficient to justify the risk it may be better to avoid the undertaking completely. Insurance does not reduce risk, but distributes it over time and shares it with others in the same insurance pool.

Vertical integration of crude production and downstream refining has also been an acceptable means of diversification. The benefits of integration have been seriously eroded, however, by the number of host country nationalizations during the 1970's. This greatly increased the dangers of price risk during a time of high market volatility and unpredictability.

198

When uncertainties are great, and where there is no recognized method of quantifying the situation, decision making becomes particularly difficult. As a start it may be helpful to identify the types of occurrences which are contemplated.

TYPES OF OCCURRENCES OR OUTCOME

- Mutually Exclusive
- Independent
- Dependent, Conditional, or Contingent
- Sequential, i.e., must occur in a specific order of events

Many times there is even no experience to be drawn upon. Hunches and rumors, as well as personal optimism or prejudice, all too frequently play their part. Judgments of some kind must be made, the relative desirability of various possible outcomes must be visualized, and the whole process mentally integrated. One is forced either consciously or subconsciously to make assumptions and predictions in arriving at any "intuitive" decision.

Rules of thumb are often relied upon with some appropriate modification recognizing the implied degree of uncertainty. For example, a normally acceptable payout standard of three years might be arbitrarily reduced to two years, as a screening basis for project selection in a high-risk situation. There are more reliable, albeit more complicated, techniques for handling decision-making under uncertainty when adequate data are available.

Risk versus Reward

Investors are willing to accept the risks of the oil business and invest in it is because the potential reward (profit) is sufficient to compensate for the risk of losing their entire investment as the result of failure. The relationship of risk and reward is described in Figure 6-1. The upper line (curved) shows the relationship for unrisked investment analysis, while the lower line (dashed) represents the results of an investment analysis which fully incorporates investment risk. The area above each of the lines represents conditions for an acceptable investment opportunity, while the area below each line represents conditions unfavorable to making investments. Each organization must establish its own criteria in this regard. It can be seen on the graph of unrisked analysis that as risk increases the minimum acceptable rate of return also increases to compensate for the higher risk. The minimum acceptable rate of return rises slowly for small amounts of risk, but eventually rises almost vertically, indicating that at some point there is no reasonable rate of return which would justify accepting the risk of that investment. The "risked" line is flat because that analysis has included a portion of the failure cost as part of the cost of each successful venture.

There are two other interesting concepts also presented on this diagram. The first is the effect of competition for investment opportunities. As indicated by the down pointing arrow, competition reduces the rate of return investors can expect from an investment and may actually push that return into the unacceptable range. If this should occur, the investor would no longer compete for that investment opportunity. This is a common occurrence in mergers, acquisitions, and auctions where many investors are competing for the same investment opportunity.

The other concept presented in Figure 6-1 is the effect of technology on risk and rate of return. It is shown as a diagonal arrow pointing upward, because technology can both reduce risk and increase profitability. Through technology and engineering design the chance of failure can be significantly reduced. Technology and engineering can also reduce costs which would increase return. Some organizations may choose to avoid competition or reduce their competition by concentrating on those situations where they have some sort of a technical advantage. The recent concentration of many major oil companies on deep water drilling is one example of their benefiting from technical and financial advantages.

FIGURE 6-1
RISK vs. REWARD

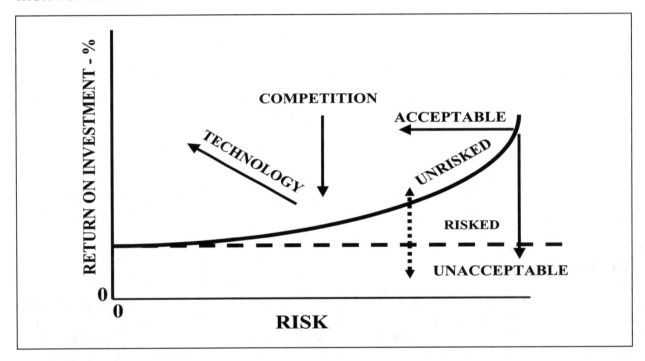

PROBABILITY

Probability is the mathematical study of the uncertainty of events. It is a number between zero and unity, including both end points. The most likely outcome of a particular event under study has the highest probability of occurrence. Higher values are taken to indicate greater likelihood of occurrence. Thus, an event with zero probability never occurs, while one with a probability of one is certain to occur. In the oil business, analysts and managers are frequently forced to assign a numerical probability of a particular outcome on the basis of intuition and judgement due to the lack of sufficiently applicable statistical information.

The characteristics of a single numerical probability under this concept are thus:

1. A positive number between one and zero. A probability of one is considered certain to occur whereas events which cannot occur are assigned probabilities of zero. Uncertain events which may or may not happen have probabilities between one and zero.

2. The sum of probabilities of occurrence of all possible mutually exclusive events equals 1.0 (or 100 percent).

3. It is a quantitative measure of the likelihood of a specific event occurring. The assignment of a probability of 0.8 means that the particular event is four times more likely to occur than if it had an assigned probability of 0.2. In any single trial this does not mean that it will happen. Only with the background of a long series of similar events can that probability be used for the prediction of results.

4. If two events are mutually exclusive, i.e., that one (x) or the other (y), but not both will occur, then:

$$P(x \text{ or } y) = P(x) + P(y) \qquad (6.1)$$

An example of this is the roll of a pair of dice:

P(seven or eleven)

$$= P(7) + P(11)$$
$$= 1/6 + 1/18$$
$$= 2/9 \text{ or } 22.2\%$$

5. If the two events are independent of each other, but it is desired to compute the probability that the two will both occur in any sequence, then:

$$P(x \text{ and } y) = P(x) \ P(y) \qquad (6.2)$$

Or with two rolls of the dice coming up a seven both times:

P(seven and seven)

$$= P(7) * P(7)$$
$$= 1/6 * 1/6$$
$$= 1/36 \text{ or } 2.8\%$$

6. The probability that a certain event will occur which is contingent on another event having taken place first is similar to sequential events. The probability of a contingent event occurring is the probability of the first independent event times the independent probability of the contingent event. This will be dealt with in more detail in the subsequent section on Decision Trees, but the following is a simple example of a contingent probability.

The probability of drawing 2 aces from a normal deck of 52 cards, without replacing the first one is;

$$P(A_1, A_2) = P(A_2, A_1) = P(A_1) * P(A_2)$$

Where: $P(A_2, A_1)$ is the statistical notation for a contingent probability of event A2 occurring if event A1 has already occurred.

So the probability of drawing a second ace (i.e., a pair of aces) from a deck of 52 cards is;

$$P(A_2, A_1) = (4/52) * (3/51) = 0.0045$$

Handling Probabilities

Recognizing uncertain situations is only the first step. It is then necessary to "quantify" the risk and uncertainty in order to deal with it effectively. Quantifying risk means determining all of the possible values an uncertain variable can take, and determining the relative likelihood of each value. The applicability of these classic definitions of risk and uncertainty to actual industrial situations depends in large part upon the level of available knowledge regarding the specific element and its environment being forecast. The size and detailed knowledge of the sample, or environment, e.g., geological region, upon which a forecast is to be based is critical to its accuracy.

Incredible as it may seem, typically just a few cubic centimeters of core sample are taken to the laboratory for analysis. The results of porosity and permeability determinations are then applied to define a whole reservoir of millions of barrels of crude oil.

It is an unfortunate human tendency to be more proud of our ability to estimate exact probabilities of expected values than we really should be. In most physical situations, a reasonably realistic range of outcomes can almost always be determined. Even one completely unfamiliar with the petroleum industry could come up with some range of possible offshore platform costs for 1,000 feet of water. The individual might merely resort to making the range extremely large, say between $1 million and $1 billion. Then, it would be quite difficult to establish meaningful probabilities for each outcome, so the estimate would be considered to have a high degree of uncertainty.

An experienced petroleum engineer could, without question, reduce the range of possible outcomes, to say, a range of between 250 to 350 million dollars. He may also be able to draw on experience and professional judgment in order to estimate that $310 million is the best estimate with a 50 percent chance of happening, and that outcomes of $250 million and $350 million each have a 25 percent chance of occurrence. Or more simply, the engineer might say that $310 million is the most likely with $250 million and $350 million having equal chances of occurring.

The engineering ability of providing this type of estimate varies considerably with the nature and situation surrounding the element to be forecast. The cost of an offset well onshore in an established field would be far easier to estimate with a high probability of reliability than say, the reserve size of a new discovery in a remote new frontier area of the arctic. The degree to which uncertain elements can be subdivided into discrete elements affects the type of analysis which is best employed in your decision-making evaluation.

In most real life situations, experimentation is not a viable approach to determining probability. Oilwell drilling is a good case in point. A great deal might be known about the prospect, but the only way to know whether the specific location will, in fact, be productive of oil or gas is by drilling. After the well is drilled and completed the uncertainty of occurrence of hydrocarbons is gone. Prior to the drilling, however, there is no mathematical formula from which one can determine the uncertainty associated with the possible outcomes. Uncertainty has to be estimated using the best information available at the particular point in time.

A more basic tool of probability theory involves the analysis of ranges of values from a large number of trials to describe variables that cannot be adequately quantified by single numerical estimates. For example, the determination of the greatest, the least, and most likely values will often more adequately quantify a wide-ranging variable than will the simple average value. This type of situation frequently occurs regarding future expenditures of either time or money.

Actuarial tables used by the life insurance industry to establish premiums by the age of the insured or for payout of annuities are good examples of empirical probability estimates. These tables reflect the expected life span of males and females for a given locality as a function of the insured current age and are constructed by observing the actual lifetime of millions of individuals. Such tables are updated from time to time as new data suggest changes in lifetime due to changes in diet, advances in medicine, etc.

A large percentage of probabilities in the petroleum industry fall into the last category of subjective. For even though observations and experiments can be conducted to obtain data for the situation being considered, the sample taken is unlikely to be completely representative of the case being evaluated. As an example; just because "W" large, "X" medium, "Y" marginal, and "Z" dry hydrocarbon accumulations may have been discovered in a given basin, there is no assurance that future discoveries will occur at the same frequency. Usually the large fields are discovered first.

Subjective estimates of probability are a numerical statement of a personal opinion on the likelihood of the occurrence of an event. It is based upon the knowledge, experience, personal preferences, and judgement of the estimator. Such values are strongly influenced by the experience, or lack thereof, and bias of the individual making them. Since these factors will cause probability estimates to differ among people, or professional groups, it is usually desirable to get a "second opinion." This second independent approach may take the form of a report by a committee convened for such purpose, an informal poll of associates, or consensus between management and technical personnel as a project is subjected to management review. In any event, group participation in the determination of subjective probabilities is often the best procedure for such determinations.

When making subjective probability estimates one should not overlook the possibility of obtaining additional information to improve that estimate. Significant information might be obtained from additional seismic data, cores, well tests, etc. However, acquisition of further data will usually delay a decision and that may not be an acceptable alternative.

Subjective probability estimates are not unique, as they will vary with the personal bias and ability of the individual making the estimate. In most cases it is a simple number representing a complex problem. Therefore a commonly used technique is to arrive at a probability of occurrence through group discussion, getting independent input from a number of individuals. This will involve different backgrounds and points of view, all of which are relevant to the situation being evaluated. This may be done on a formal basis at a conference with the end result being a probability which is the average of values submitted by each participant. More commonly it is based upon an informal survey of peers and/or input from various levels of management as a recommendation moves up within an organization for approval.

Subjective estimates of probability are a numerical statement of personal opinion on the likelihood of the occurrence of an event. It is based upon the knowledge, experience, and personal preferences of the estimator. Such values are influenced by the experience, or lack thereof, and bias of the individual making them.

Frequency distributions may change, or evolve over time, particularly in the case of petroleum exploration. Figure 6-2 shows the distribution of fields discovered in Western Canada's sedimentary basins during two successive time periods. Each bar represents the number of fields discovered in each size class, with the maximum size of each class shown at the base of each bar. The bottom portion of each bar (light shading) represents the number of fields discovered up until 1979 while the top portion represents the discoveries between 1979 and 1988. If an analytical model had been developed for this area using the discoveries up to 1979 as the basis for investments after that date, it certainly would not have been representative of what was eventually discovered. The larger fields are almost always discovered first with progressively smaller fields discovered as a specific geological area is more fully developed. In this case the only significant discoveries were fields of less than 10 million barrels, where previously there had been a one billion barrel field plus 28 fields between 100 million and one billion barrels.

FIGURE 6-2
OIL FIELD SIZE DISTRIBUTION VARIATION WITH TIME
(Western Canada)

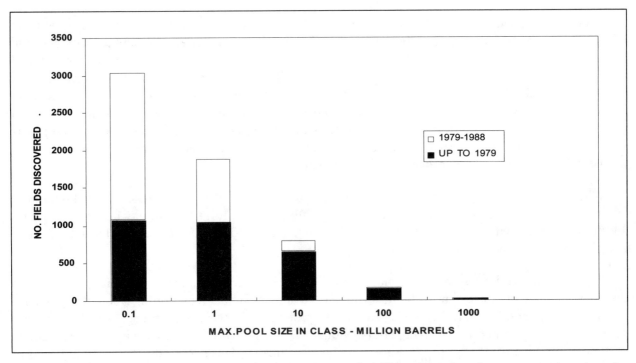

Another technique frequently used in estimating probabilities is to break an uncertainty into its component parts. It may be easier to estimate the probability of occurrence of the smaller parts either through analysis of empirical data or experimentation, where that would not be possible for the overall occurrence being evaluated. A good example of this is the estimation of the probability of occurrence of a commercial hydrocarbon accumulation for a particular prospect as illustrated in the following Example 6-1. Recognizing that certain conditions must be present for there to be a commercial hydrocarbon accumulation, these are first identified and then the probability of each estimated independently. The probability of occurrence of the commercial hydrocarbon accumulation is then the product of each of the individual independent component probabilities. This is much like a chain composed of individual links. If one link fails, the chain breaks. So if the probability of any component part is zero, then the total probability is zero. The following example may help in understanding this concept.

■ EXAMPLE 6-1
■ HYDROCARBON OCCURRENCE MODEL

The existence of a commercial hydrocarbon reservoir is dependent upon a number of physical factors, all of which must be present. The following tabulation represents the individual probabilities that each will occur in a given geologic situation.

Event	Probability Estimate
Existence of Source Rock	0.95
Sufficient Level of Maturity of Source Rock	0.75
Migration Path to Reservoir	0.50
Reservoir Rock with Porosity and Permeability	0.50
Existence and Persistence of Seal	0.75
Combined Probability of Occurrence of a Commercial Reservoir (P_S)	**0.13**

Determine:

What is the weighted probability of the occurrence of a commercial hydrocarbon trap under the conditions outlined?

Solution:

In accordance with Equation 6.2:

$$P_S = 0.95 \times 0.75 \times 0.50 \times 0.50 \times 0.75 = 0.1336$$

Before a commercial hydrocarbon reservoir can exist to be discovered, all of the items listed in the previous example must exist. By breaking the probability of occurrence of a commercial hydrocarbon reservoir into its component parts, it should be possible to estimate the probability of each part on the basis of existing data, experience, or measured data, which would lead to a more realistic estimate of the combined probability of occurrence of a commercial hydrocarbon reservoir.

The absence of one or more of these factors will result in a "dry hole." An offshore North Slope of Alaska prospect that was drilled several years ago as a joint venture of a number of major oil companies is a good example of this. Apparently all of the parameters of the example except one were found to exist. However, apparently several million years ago the seal was broken and all of the hydrocarbons had leaked out, resulting in a dry hole that cost the companies over one hundred million dollars.

No matter what technique is used to arrive at a subjective probability, it is not possible to completely overcome the weakness of estimating; which is the fact that it is made by people on the basis of their feelings. Estimation of probability is seldom easy. Even though it is the weakest link in decision analysis, it is a very important part of the process of decision making.

Frequency Distributions

The estimation of probability is seldom easy and even though it is the weakest link in decision analysis, it is a very important part of the process of decision making. The classic way of describing the risk involved in an investment is by means of a probability distribution of its anticipated performance. This is sometimes referred to as a "risk profile"and is a prime tool of the statistical analysts who are fortunate enough to have very large masses of data to work with and analyze. Some typical examples of these frequency distributions,or risk profiles, are shown in Figure 6-3. By definition, the area under a probability distribution curve equals unity, i.e., 100% probability, as this represents a mutually exclusive alternative condition and the area represents all possible occurrences.

All of these distribution shapes use a set of arguments to specify a range of the actual values and distribution of the possibilities. The normal distribution, for example, uses a mean value and a standard deviation as its arguments. The mean defines the value, or point, around which the bell curve will be centered and the standard deviation is a measure of dispersion of values around that mean value.

In a normal distribution curve (Figure 6-3e) the peak is at the "mode," which is coincident with the "mean" and the "median." How useful it may be is often determined by the shape of the rest of the distribution curve. More often than not the curve will be skewed in some manner such as shown in Figure 6-4.

FIGURE 6-3
TYPES OF PROBABILITY DISTRIBUTIONS

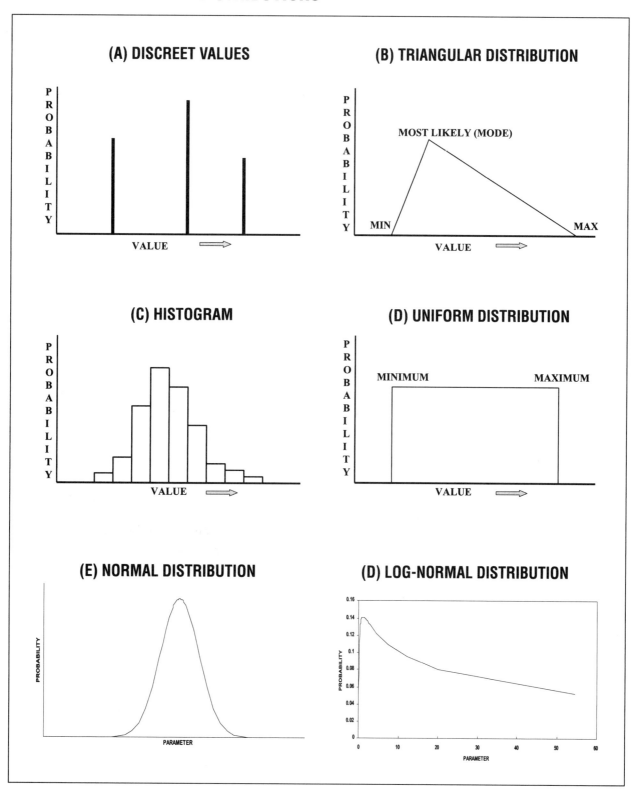

FIGURE 6-4
TYPICAL FREQUENCY DISTRIBUTION

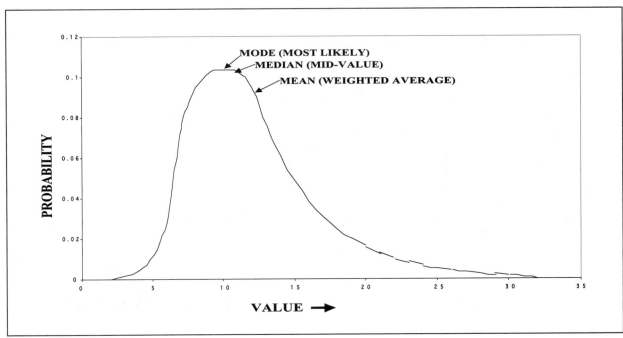

The log normal distribution (Figure 6-3f) describes a random variable which exhibits small likelihood of large numerical values and a large likelihood of smaller numerical values. This type of curve or distribution is often representative of formation permeabilities as determined from core samples in the laboratory and the distribution of field sizes (reserves) within a geologic unit.

As a general observation of the nature of frequency distributions, a quick reading of Figure 6-5 would indicate a much greater degree of certainty in the oil recovery estimate depicted for Project A than for Project B. This is true because Project B exhibits a much wider dispersion of possible outcomes, or the greater standard deviation of the two, although both would report the same expected value reserve. Project B has a probability of lower downside values which, in some cases might lead to financial difficulty, but it also has a chance of greater values on the upside. Project A might be preferable to some decision makers in order to avoid the potential downside of B. This is an example of Risk Aversion.

FIGURE 6-5
POSSIBLE RECOVERY OUTCOME FOR TWO PROJECTS

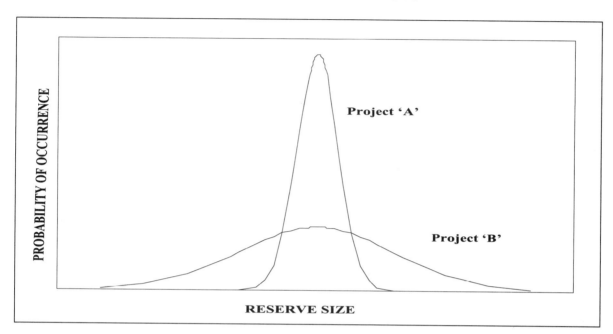

Whereas plots of normal frequency distribution (the bell shaped curves) are very helpful conceptually, the cumulative frequency type of plot is the more useful in working with these kinds of data. Figure 6-6 shows a typical relationship for assessing the cumulative frequency distribution of pay thickness data in a typical geologic basin superimposed on the frequency distribution histogram for those data.

FIGURE 6-6
PLOT OF CUMULATIVE PAY THICKNESS DATA

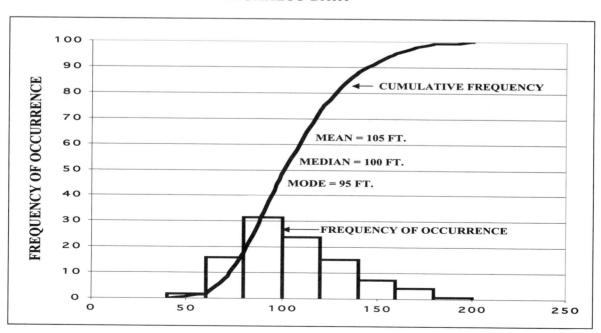

In this manner of plotting, the 50 percentile (median) value is easily read as 100 feet. This is the 50-50 value, i.e., the chances of the pay thickness being higher or lower than 100 feet are equal. It is considered to be the most realistic average to use when dealing with a wide range of random numbers which are key to the analysis at hand. While the mode, median and mean values are not identical for this example, they are close.

Example 6-2 illustrates the numerical derivation of Mean, Median and Mode of a set of core analysis data from the Niabrara Formation in the Denver-Julesburg Basin of Colorado covering the depth interval from 4807 to 4850 feet.

▎ EXAMPLE 6-2
STATISTICAL ANALYSIS OF CORE DATA

DEPTH (FEET)	PERME-ABILITY (MD)	POROSITY (%)	DEPTH (FEET)	PERME-ABILITY (MD)	POROSITY (%)	DEPTH (FEET)	PERME-ABILITY (MD)	POROSITY (%)
4807.5	25	17.0	4807.5	25	17.0	4843.5	882	15.1
4809.5	221	19.1	4821.5	98	16.9	4825.5	260	15.1
4811.5	275	23.3	4813.5	108	23/3	4833.5	610	15.5
4813.5	108	23.3	4849.5	139	20.5	4827.5	312	15.6
4815.5	290	17.2	4819.5	167	16.2	4845.5	407	15.7
4817.5	278	15.9	4809.5	221	19.1	4817.5	278	15.9
4819.5	167	16.2	4825.5	260	15.1	4819.5	167	16.2
4821.5	98	16.9	4823.5	266	20.3	4831.5	275	16.4
4823.5	266	20.3	4831.5	275	16.4	4839.5	339	16.8
4825.5	260	15.1	4811.5	275	23.3	4821.5	98	16.9
4827.5	312	15.6	4817.5	278	15.9	4807.5	25	17.0
4829.5	395	19.4	4815.5	290	17.2	4815.5	290	17.2
4831.5	275	16.4	4827.5	312	15.6	4837.5	597	17.7
4833.5	610	15.5	4841.5	332	18.0	4847.5	479	17.8
4835.5	535	18.3	4839.5	339	16.8	4841.5	332	18.0
4837.5	597	17.7	4829.5	395	19.4	4835.5	535	18.3
4839.5	339	16.8	4845.5	407	15.7	4809.5	221	19.1
4841.5	332	18.0	4847.5	479	17.8	4829.5	395	19.4
4843.5	882	15.1	4835.5	535	18.3	4823.5	266	20.3
4845.5	407	15.7	4837.5	597	17.7	4849.5	139	20.5
4847.5	479	17.8	4833.5	610	15.5	4813.5	108	23.3
4849.5	139	20.5	4843.5	882	15.1	4811.5	275	23.3

SOLUTION

	PERMEABILITY (md)	POROSITY (%)
MEAN	331.4	17.8
MEDIAN	278-290	17.0-17.2
MODE	200-300	15.0-16.0
VARIANCE	39,026.4	5.65
STANDARD DEVIATION	197.6	2.38

RISK ANALYSIS

If a quantitative assessment of probabilities can be derived from historical data or personal experience (i.e., subjective judgment), the uncertainties can be more effectively handled through the various techniques of risk analysis.

Expected Value Concepts

The discipline of decision analysis seeks to handle risk and uncertainty in a quantitative manner. One must understand from the outset, however, that numerical approaches leading to precise "solutions" may often be misconstrued to imply a degree of certainty and precision which just is not there.

The Expected Value approach is a method of combining quantitative estimates of probability, i.e., "uncertainty" of each of the alternative elements that comprise an investment opportunity. This enables a risk-adjusted decision criterion, known as Expected Value, or sometimes Expected Monetary Value (EMV) to be determined. These terms are widely used in the petroleum industry when applying the concept of Expected Value to the assessment of projects in a risk environment.

Expected Value combines quantitative probabilities (estimates) with each alternative. The parameter is computed as the sum of the mathematical product of the probability of each outcome times the value of that outcome for all the possible outcomes. It is appropriate to use expected values for projects that have frequent repetition. Over many "repeats" the expected return will provide a more reliable indication of the actual return. For individual, or infrequently repeated undertakings, the use of expected value is not so reliable. Nevertheless, we in industry must work with what data we have, and expected value calculations can at least serve as a worthwhile guide to highlight, and steer one away from potentially unprofitable projects, and to indicate possible improvements in the decision process.

In principle, the expected value concept adds an additional cost to successful projects that represents a portion of the company's total failure cost. Thus the 'acceptable' investment opportunities, based upon expected investment criteria, will not only yield a rate of return which is above the company's minimum acceptable value, but they will also generate sufficient additional cash flow to pay for all of the failed investment opportunities. Sometimes expected investment criteria are also referred to a fully risk discounted values. This concept is represented by the horizontal dashed line on the graph of Figure 6-1. Expected investment criteria are discussed more fully later in this chapter in a section titled "Investment Criteria With Risk and Uncertainty."

Expected Value analyses requires the identification of two or more outcomes for each alternative. However, the outcomes identified must include ALL possible outcomes for the alternative being evaluated. Each of the possible outcomes must have some finite likelihood of occurring, but none can be certain of happening. The assigned probabilities must be proportional to the likelihood of that individual event's occurrence, and the sum of all such probabilities must add up to one.

If one lets R's (R_1, R_2, R_3,...) represent the possible numerical outcomes of a decision situation, and P's (P_1, P_2, P_3,...) represent the corresponding probability that each of the R's will occur, then in general:

$$\text{Expected value} = (R_1 * P_1) + (R_2 * P_2) + (R_3 * P_1) + ... \tag{6.3}$$

211

(handwritten annotations at top): $E(P|S)=V$ $Y=(V)X-CC(1-x)$

(handwritten): $E(C/F)=C$

Where:

(handwritten box): $\left(y=(V+C)X-C\right)$ Simplified

(handwritten): $P_S = 1-P_f = x$

$$P_1 + P_2 + P_3 + \ldots = 1, \text{ since, something must occur.} \tag{6.4}$$

(handwritten): $EV=y$

The R's can be either positive or negative, and may occur in various types of combinations, depending on the project, with the number of R's dependent upon the number of possible outcomes. The expected value can be viewed mathematically as the average, or mean, outcome (value) anticipated from a large number of ventures of similar type. EV can be expressed in barrels, acres, feet of pay, or any unit, including dollars, or even a discounted present worth of those quantities. This technique thus becomes the basis for risk-adjustment in decisions.

■ EXAMPLE 6-3
EXPECTED VALUE OF DRILLING A WELL

A drilling prospect is proposed. It is estimated that the dry hole cost (failure cost) of this well, after income tax, would be $500,000 and there is a probability of 60% that this will occur. Three successful outcomes are considered. The conditional value of each outcome and the probability of each occurring is; $1,000 (5%), $750,000 (20%), and 2,000,000 (15%). The expected value of drilling this well can be determined from the following "pay-off table".

Result	Conditional Value		Probability		Expected Value
Dry hole:	-$500,000(R1)	*	0.60(P1)	=	-$300,000
Producer:	$1,000(R2)	*	0.05(P2)	=	$50
	$750,000(R3)	*	0.20(P3)	=	$150,000
	$2,000,000(R4)	*	0.15(P4)	=	$300,000
Expected value of drilling			1.00		$150,050

(handwritten annotations in table): 60%, 5%, 20%, (6%), 100%)

The sum of the probability factors must equal 1.00 or 100%. The fact that the expected value of drilling this well (EMV) is positive indicates that given enough repetitions of this type of project, the probable fiscal benefits of these projects would outweigh the cost of failures.

The expected value assuming that only successful outcomes occur, can be determined by adding up the expected values of only the successful outcomes and dividing by the total probability of success, which is one minus the probability of a dry hole or 40%. So the expected value conditioned only on a successful outcome would be;

[E (P | S)] = ($50 + $150,000 + $300,000) / 0.4 = $1,125,125

If only the two alternatives of success and failure are considered, a simpler form of Equation 6.3 can be written. This equation may be particularly useful when using a model to represent all of the successful outcomes. In Equation 6.5 the expected value of success as determined from a probabilistic model would be represented by the first term and the expected cost of failure by the other term.

$$\text{Expected Net Present Value} = [\, E(P \mid S)\,] \, P_S - [\, E\,(C \mid F)\,]\,(1-P_S) \tag{6.5}$$

Where:

P_s = Probability of Success = $(1 - P_f)$

P_f = Probability of Failure = $(1 - P_s)$

$E(P|S)$ = Expected profit conditioned on success

$E(C|F)$ = Expected cost of failure

While the Expected Values concept is the fundamental basis for decision making under uncertainty, we should also note that the EXPECTED VALUE is NOT the VALUE ANTICIPATED, or counted upon from a single venture.

One more important aspect of statistical theory which applies to all Expected Value analyses should be examined before we proceed further. This is the point that a series of investments cannot result, in the long run, in consequences which are any better than the mean, or average, of the series of individual projects. In other words a few very good projects must inevitably be balanced by a large number of poor ones if the theories are to hold. If one is to use Expected Values to sell a project, one must also recognize that everything can't be "above average." The analyst must be careful to fully define the pertinent set of outcomes and apply consistently realistic probability factors and economic values to each.

Even though there are a number of events which have a probability of occurrence, only one event will actually occur. The value of the event which does actually occur will usually not be the expected value. Only if the operations is repeated a great number of times will the average value of the events approximate the statistically calculated Expected Value.

As an example consider the flipping of a coin. Statistically a head should come up half of the time, but on any flip of the coin it will be either heads or tails, not a half a head and half a tail. However, if you flip that coin 100 times you should get approximately 50 heads and 50 tails.

These points are of extreme importance since the Expected Values which result from these studies and analyses are used so extensively in accepting or rejecting investment opportunities in the petroleum industry. These techniques do not reduce the uncertainty in the investment decision. They simply express the degree of ignorance concerning the decision consequences in a quantitative manner while providing the benefit of furnishing a single numerical value for the decision maker. The principal shortcoming of Expected Value as a "stand alone" decision tool in economic analysis is that it fails to give adequate weight to the chance of very large financial loss when the probability distribution for a profitable project resembles Project B in Figure 6-5.

The decision analysis process under uncertainty can be summarized in five steps.

(1) Define all possible outcomes which might occur as a result of making the decision being considered. Neglect of any possible outcome may lead to a faulty decision.

(2) Evaluate the conditional profit or loss of each outcome independently, conditioned upon the premise that each outcome is the one that occurs.

(3) Determine the probability of occurrence of each of the outcomes being considered. The sum of these probabilities must equal 1.0.

(4) Compute the expected (weighted average) value of all possible outcomes.

(5) Sum expected value of all outcomes.

(6) Select the decision alternative which provides the greatest expected monetary value.

■ EXAMPLE 6-4
DETERMINATION OF OPTIMUM BREAKWATER HEIGHT

Let us assume a situation involving the design and construction of a shore base to serve an offshore producing operation. The closest landfall is an exposed continuous shoreline requiring dredging and extensive breakwater construction. The height and extent of the breakwater is obviously cost related and its effectiveness is sensitive to wave height and direction.

The operations offshore are stockpiled for storm periods of three days duration. After that the expense of using helicopter supply and alternate more distant ports mounts rapidly. Weather Bureau records show that normal storm duration, and ensuing rough seas, spans several days depending upon the severity of the storm as indicated by the maximum wave height.

During periods of bad weather when the new facility would not be usable the monetary cost to the operation has been determined according to the data shown in Table 6-2 along with the relevant storm frequency information for the area. There is also an increasing cost of the alternate supply arrangements which is related to the severity of the storm which is shown in the same table.

TABLE 6-2
PROBABILITY AND COST INFORMATION FOR DETERMINING OPTIMUM BREAKWATER HEIGHT

Wave Height Above Mean Sea Level, x	Number of days per year sea exceeds x feet for 3+ days	Probability of wave height (x)	Added Cost of Shut-Down	Initial Cost to build Breakwater x ft. high
A	B	C	D	E
0	36	0.098	0	0
3	72	0.197	$135,000	$500,000
6	194	0.533	$150,000	$700,000
9	45	0.123	$160,000	$750,000
12	15	0.041	$165,000	$850,000
15+	3	0.008	$180,000	$925,000
	365	1.000		

The operator has an unexpired primary term of sixteen years on the offshore concession and it is estimated that the producing fields will not be depleted prior to the expiration date. The cost of down-time expected for the port facility under study will vary with storm wave height above mean sea level, in the absence of a breakwater, or wave height above breakwater height if one is installed, as shown in Column D. The extent of down-time increases with the severity of the storms due to the added time required for the sea conditions to return to normal.

Figure 6-7 depicts the several wave-height situations which this example seeks to evaluate.

Examples of the calculations for individual breakwater heights are shown for two of the alternatives.

FIGURE 6-7
DETERMINATION OF OPTIMUM BREAKWATER HEIGHT \ (Example 6-4)

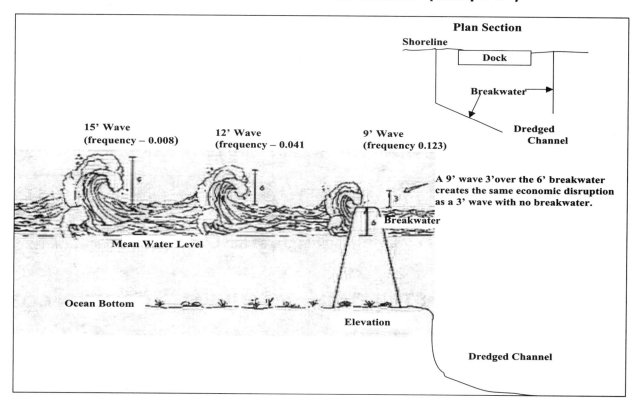

Six Foot Breakwater:

There are three storm categories of Column A that can exceed the six foot wall. Their frequency of occurrence is shown in Columns B and C. The shutdown costs for these three situations are represented by the top three items of Column D.

Annual Cost of Investment Capital @12% =(0.1355)($700,000)	= $94,850
Expected Annual Shutdown Cost=(0.123)($135,000)+	
(0.041)($150,000) + (0.008)($160,000)	= $24,035
	$118,885

The value (0.1355) discounts the investment, Column E, at 12% and spreads it over 16 years to provide an annual capital cost figure. It is the reciprocal of PVIFA from Equation (5.21) of Appendix VB, Table 5B-3. The value, (0.123), Table 6-2 Column C is the frequency, (i.e., 45 days per year) that nine foot waves will occur. These exceed the six foot high breakwater by three feet, costing $135,000 per shutdown. Correspondingly, is the frequency of twelve foot waves (0 0.41) which go over the six foot breakwater by six feet at a cost of $150,000 per shutdown and 15+ foot waves (0.008) at a cost of $160,000.

Nine Foot Breakwater:

There are only two categories of Column A which exceed the nine foot wall. The shutdown costs are $135M and $150M.

Annual Cost of Investment Capital @12% =
(0.1355)($750,000) = $101,625

Expected Annual Shutdown Cost =
(0.041)($135,000)+ (0.008)($150,000) = <u>$ 6,735</u>
$108,360

The costs developed for varying the height of the breakwater are summarized in the following.

SUMMARY OF CONSTRUCTION AND ANNUAL SHUTDOWN COSTS

Breakwater Height (Feet)	Annual Investment Charge	Expected Annual Shutdown Cost	Total Expected Annual Cost
0	$0	$134,430	$134,430
3	67,750	98,285	166,035
6	94,850	24,035	118,885
9	101,625	6,735	108,360 (Min)
12	115,175	1,080	116,255
15+	125,338	0	125,338

Under the assumptions employed, construction of the contemplated breakwater is justified. A breakwater constructed nine feet above mean sea level will be the optimum and its cost is less than $134,430, the expected annual shut-down cost in the absence of a breakwater.

Probability Weighting of Reserve Estimates

Another frequent application of the Expected Value concept is the probability weighting of oil and gas reserve estimates. Reserve estimates are generally categorized as: Proven, Probable, and Possible according to the definitions listed in the Appendix to Chapter II. The "proved"category may be further subdivided into: Proved producing, Proved developed nonproducing, Proved behind pipe (meaning that the formation has been penetrated and found to be productive but then cased off), and Proved undeveloped. Table 6-3 lists a range of commonly applied Expected Value probability factors pertaining to these classifications.

TABLE 6-3
PROBABILITY FACTORS FOR VARIOUS RESERVE CLASSIFICATIONS

Proved producing	0.90 to 1.00
Proved developed nonproducing	0.80 to 0.95
Proved behind pipe	0.70 to 0.95
Proved undeveloped	0.60 to 0.90
Probable	0.35 to 0.60
Possible	0.00 to 0.10

The appropriate single probability number from these ranges is used as the multiplier with the normal unweighted recovery estimate to obtain the expected value, or probability weighted reserve figure. This is sometimes referred to in the industry as "risk weighted reserve." An economic analysis will often include reserves in several of the categories which are then added as their weighted, or expected value totals to obtain a probability weighted sum for the entire company, or project. These totals are generally employed in comparing one alternative investment with another. Probability weighted reserves may also be incorporated in oil company annual reports for the same purpose of comparing companies. It should be borne in mind, however, that the degree of uncertainty associated with the estimation of reserves changes with time as implied in Figure 6-11. These are statistical comparisons, and the actual ultimate recoveries, may differ quite substantially from the earlier estimates.

The Mineral Management Service, the government agency responsible for oil production on U.S. Government lands, relates reserves to both geological and economic uncertainty. Figure 6-8 shows the relationship of reserve identification and their degree of uncertainty.

Probability weighted reserves adjustments tend to introduce a downward bias in the analysis which could result in reduced expected profits, if the probability factors are applied too conservatively.

FIGURE 6-8
MINERAL MANAGEMENT SERVICE PETROLEUM RESOURCE

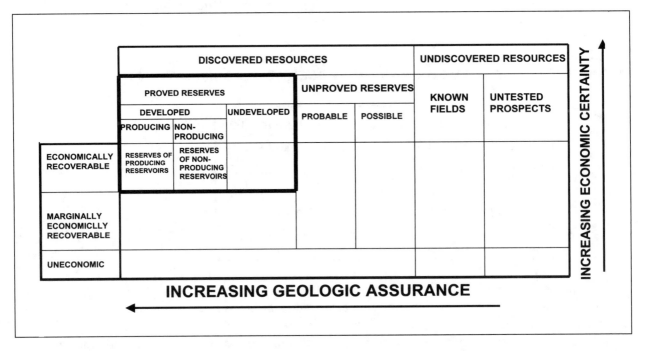

Proved and unproved reserves shown in the upper left portion of the diagram meet the Society of Petroleum Engineers' and the World Petroleum Congress's definition of reserves as presented in Appendix II-B. Only "proved reserves," shown in the box, are disclosed in oil company annual reports.

EXAMPLE 6-5
PROBABILITY WEIGHTED RESERVES

Onxy oil company is involved in a potential sellout or merger and has engaged the services of a consultant to establish a "third party" evaluation of its reserve base. Since the company has a wide portfolio of exploration and development at different stages of processing, a probability weighted evaluation in terms of barrels of oil equivalent is appropriate.

The following is presented:

	Total Company {boe}	Probability	Probability Weighted Reserve (boe)
Proved Producing	600,000	1.00	600,000
Proved developed non-producing (waiting on pipeline connection)	50,000	0.90	45,000
Proved behind pipe	125,000	0.75	93,750
Proved undeveloped (within 1 kilometer of a producing well)	Basin A 200,000	0.50	100,000
	Basin B 50,000	0.40	20,000
Probable	500,000	0.35	175,000
Possible (2,000 hectares of exploratory leases)	2,000,000	0.03	60,000
			1,093,750

The term, barrels of oil equivalent (boe), seeks to relate the heating value of a cubic foot of natural gas to that of a barrel of crude oil (both at surface conditions of temperature and pressure) and report the total volume as barrels. This criterion is frequently convenient in appraising prospects which may ultimately yield either oil or gas. Depending upon hydrocarbon composition, this ratio is generally in the order of six thousand cubic feet per barrel of oil.

Note: An alternative boe ratio is sometimes employed using the relative current economic values of natural gas and crude oil. This is not illustrated here.

Risk Capacity

Another useful concept is the determination of how much risk a project can stand and still yield a breakeven expected net present value. Application of this concept enables the computation of the minimum probability of success that a project must have in order to achieve a zero net present value. This is not a method of determining the probability of success of a venture, but it is a method for judging the acceptability of a risky project.

This concept is based upon the fundamental premise of applying risk and uncertainty to investment decisions, as presented earlier in this chapter. That is, that the expected value of a venture is the sum of the product of the net present value of each possible outcome and the probability of that outcome occurring, where by definition the sum of all probabilities must equal unity (1.0). If only two possible occurrences are considered; i.e., success and failure, a simple equation can represent the above concept. This is Equation (6.5) for an Expected Net Present Value of Zero.

$$P_R(NPV) - (1 - P_R)(DHC) = 0 \tag{6.6}$$

Where:

P_R = Risk Capacity
NPV = net present value of entire successful project
DHC = all costs which would be incurred before the project is determined to be a failure

Algebraic manipulation of the above equation produces the following equation for risk capacity.

$$P_R = \frac{DHC}{NPV + DHC} \tag{6.7}$$

When using this equation there are a number of rules which must be followed.

(1) If the net present value of success (NPV) is after income tax, the cost of failure (DHC) must also be after income tax.

(2) If this method is applied to a wildcat well, the DHC would be the dry hole cost of that single well, but the NPV would be the net present value of the entire discovery made by the successful wildcat well, including all follow-up development.

(3) If this method is applied to an exploration program, the DHC should include the total cost of failure (all required expenditures as discussed in the section entitled Failure).

Once the risk capacity (PR) has been determined, it can be used in judging whether or not the risky project should be undertaken. This requires an independent assessment of the probability of success of the project (Ps). If that independently determined probability of success is greater than the calculated risk capacity, then the expected net present value of the entire venture is positive and the venture is acceptable. If it is not then the venture should not be undertaken. Another way of looking at risk capacity is that it indicates the number of failures a successful project can support. As an example if one in (1/PR) wildcat wells are successful then the program will generate a zero net present value.

Risk Capacity criteria;
     If:

          $P_S > P_R$: Acceptable Project
          $P_S = P_R$: Indifferent
          $P_S < P_R$: Unacceptable

The concept of "risk capacity" is demonstrated by the chart of Figure 6-9. This is a graph of expected present value net cash flow as a function of probability of success, which is a plot of Equation 6.5 for values of probability of success between zero and one. NPV is the expected PV profit conditioned on success and DHCis the expected cost of failure. The point that this straight line crosses zero expected PVNCFis the minimum probability of success necessary to just break even, which by definition is "risk capacity". This graph also explains the decision criteria previously presented. It shows that a probability of success less than the "risk capacity" yields a negative expected NPV, so it is unacceptable, and one that is greater than the "risk capacity" yields an acceptable positive expected NPV.

FIGURE 6-9
RISK CAPACITY

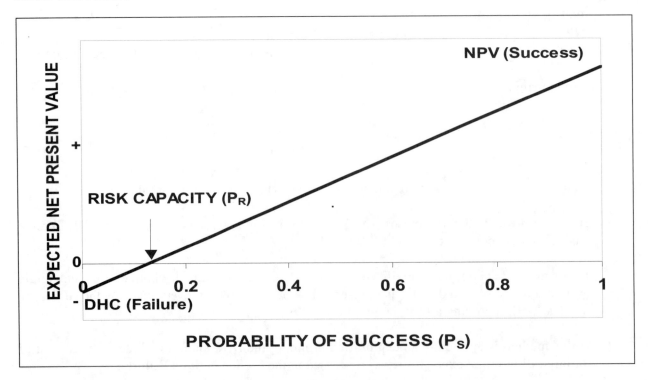

EXPLORATION AND PRODUCTION PROBABILISTIC ANALYSIS

The triangular frequency distribution (Figure 6-3b) has been found to be a very useful tool for applying risk and uncertainty to petroleum exploration and production evaluations. Two of its greatest advantages are that it is simple to use and requires a minimum of work to apply. It can be used to approximate a normal distribution or a log normal distribution, both of which are frequently encountered in petroleum exploration and production ventures.

The triangular distribution, as shown in Figure 6-10, requires the determination of only most likely, minimum and maximum values. Theoretically there is a zero probability of the minimum value or less occurring. The probability of greater values occurring increases up to the most likely value, which by definition has the highest probability of occurrence. As values increase further, the probability of each decreases to a value of zero at the maximum value. The probability of occurrence of values greater than the maximum is also zero. Practically speaking it is commonly assumed that there is a 5% chance of values occurring below the minimum and a 5% chance of values occurring above the maximum value. Thus the values within the triangle should encompass about 90% of occurrences or about 1.6 standard deviations of data around the mean or expected value. The best estimate that can be made is usually considered the most likely case. The least successful case that can be economically justified would represent the minimum case, and the most optimistic case the maximum. It is unnecessary to assign probabilities to the minimum, maximum, and most likely cases because the mean or expected value is precisely the numerical average of the minimum, maximum, and most likely values, irrespective of the skewness of the distribution. Thus the triangular distribution represents a continuous distribution of values, ranging between the minimum and maximum values, including ALL possible outcomes in this range. The area within the triangle has a unit value (1.0), as it represents all possible outcomes for a given occurrence. Mathematical properties of a triangular distribution are presented below.

EXPECTED VALUE (MEAN) = [MINIMUM + MODE + MAXIMUM] / 3

PROBABILITY OF MOST LIKELY VALUE (MODE) = 2 / (MAX - MIN)

MEDIAN: IF; MODE (MIN. + MAX.) / 2

      THEN; MEDIAN = MIN. + [(MAX - MIN) (MODE - MIN) / 2]$^{0.5}$

MEDIAN: IF; MODE (MIN. + MAX.) / 2

      THEN; MEDIAN = MAX. - [(MAX - MIN) (MAX - MODE) / 2]$^{0.5}$

VARIANCE: $\sigma 2$ = [(MAX - MIN) 2 - (MAX - MODE) (MODE - MIN)] / 18

STANDARD DEVIATION: = (VARIANCE)$^{0.5}$

Figure 6-10 shows two probabilistic models, with success represented by triangular distributions similar to the shape shown in Figure 6-3b. There is a single value for failure.

Failure

Failure may occur for a myriad of causes from human carelessness to natural catastrophes. The most common cause of failure in the E&P business is geologic failure. For countless reasons the prospect turns out to be dry when finally drilled. No matter how good the geologic evidence appears to be, only the drill bit will determine the success or failure of a geologic prospect. Every prospect has

221

a probability of geologic failure which may be fairly high. While exploration failure does not persist after a discovery is made, geologic uncertainty will persist throughout the producing life of a prospect affecting producing rate and ultimate recovery. The exact reserves discovered will not be known until the last barrel of oil or cubic foot of gas is produced.

Mechanical and technical failure must also be considered. This includes minor failures that would merely add to the drilling cost all the way to a well blowout or natural disaster, such as a storm or earthquake, that would completely destroy the project. Mechanical and technical failures can occur at anytime during the life of a project. Sound engineering and technological achievements may reduce risk, but the cost of total elimination may be excessive.

Political failure is also of real concern due to the very long producing life of a typical oil field. Many things of a political nature may occur which would seriously effect project economics. Tax rates and even entire tax systems may change. Political parties and forms of government may shift. Political philosophies may vary and contracts can be voided.

Recognition of the failures just described, plus others that may be unique to a particular project, is necessary when estimating, calculating, hypothecating, or guessing a numerical value for the probability that a given project will fail. The best source of such a value would be experience with similar projects or under similar conditions. In Diagram A of Figure 6-10, failure is represented by a single probability at zero ultimate recovery.

The last item to consider when assessing failure is the cost that will be incurred if a failure occurs. That should include all expenditures made to the time that the prospect is abandoned as a failure, including exploration, drilling, and irrevocable commitments. These costs should be present valued at the appropriate hurdle rate to some meaningful date. If possible the cost of failure should be after income tax effects to be comparable to the value of success. In Diagram B of Figure 6-10, failure is represented by a single probability with a negative net present value representing the total cost of failure. Failure includes all outcomes which cannot be considered successful.

At some point in time it will be necessary to decide if a project is a failure or a success. In making this decision, past expenditures are regarded as "SUNK COSTS": i.e. not considered as a cost in the decision process. However, income tax credits which may be available as a result of the expenditure of those "sunk costs" must be recognized. Additionally, any salvage value which may be realized by immediate abandonment should be added to the tax credits. Such tax credits and salvage value are equivalent to future cash income and should be included in determining if a project is a failure or a success. If the present value of net cash flow after income tax of development (success) exceeds the present value of the tax credits generated by the "sunk costs" plus salvage value, then the project should be developed and it is considered a successful outcome. However if this is not so, the project will be declared a failure with the acceptance of the immediate tax credits and salvage value to partially offset the cost of failure. Under most income tax systems, the project must be physically abandoned and title to the project surrendered before cost of failure tax credits can be claimed.

Value of Success

Once all of the factors related to failure have been considered, attention should be directed to the success side of the equation. While this is a lot more pleasant task, it may be even more difficult than that required to appraise the failure side of the evaluation. In most cases geologic failure will occur early in a project, requiring a single decision to abandon the prospect. While the probability of this event may be nebulous, the cost of failure can reasonably be identified. However, when considering the successful outcomes, they are almost limitless, their probability of occurrence is in great doubt, and much of it will occur a long time in the future.

When considering the range of possible outcomes, the minimum value is the easiest to determine. This is the value of the minimum volume of hydrocarbons which must be discovered to economically justify development. Discovery of any volume less than this will be considered a dry hole or geologic failure. In the case of a single well, this is the smallest volume of hydrocarbons which would justify completion of the well, treating the drilling cost to the casing point as a "sunk cost." For offshore operations, this would be the minimum volume which would justify the installation of a platform. For other types of projects, this would be the minimum volume necessary to justify further development. The present value cash flow generated by the minimum case must at least equal the present value of any tax credits available on the failure cost and the salvage value of immediate abandonment, as discussed in the last paragraph under the previous section "Failure".

In the normal course of evaluating a project, there is one successful outcome that appears most likely to occur. The most likely or best estimate, is generally the case developed from the best information available when the evaluation is being made. This case will be extensively reviewed and may strongly influence management decisions about the project.

One necessary condition which must be met in determining expected values is that ALL possible outcomes have been considered. To meet this condition it will be necessary to determine how big the prospect might be. It may be possible to determine this from geologic evidence, from an appropriate analog, experience, or intuition. In any event, the maximum value should be large enough so that it would rarely be exceeded. The maximum case is the most optimistic case that can be imagined. It might be assumed that the hydrocarbon column is the thickest ever observed in a particular horizon or geologic setting, or the largest field ever discovered in a geologic basin, or that the trap is filled to its spill point. The value used for the maximum case must be realistic, but it is also important that it be large enough to include a value which may only rarely be exceeded.

Once all possible successful outcomes have been identified, an economic evaluation of each must be made. This entails forecasting production rates, future oil and gas prices, capital, operating expenses, and taxes. It must include all costs and payments required by contracts or agreements. The resulting cash flow over the entire project life of each possible successful outcome is then discounted at the same hurdle rate and to the corresponding date used for determining the present value of failure.

The most difficult task when evaluating the success side of the equation is determining the probability of occurrence for each successful event. The total probability of occurrence of all successful outcomes is one (1.0) minus the probability of failure. Therefore, the probability of all successful outcomes is limited by the value assigned to the probability of failure. It should be noted that the probability of success is not the probability of the most likely discovery volume, but it is the probability that the project will not fail. It is also the probability of developing the expected successful volume.

The Model

Figure 6-10(A) is a probabilistic model for the volume of hydrocarbons discovered. The ultimate recovery or reserve additions are plotted on the X-axis and the probability of occurrence is plotted on the Y-axis. The area under the triangle represents the probability of success while the height of the bar at the origin represents the probability of failure. Since only two outcomes are considered, i.e., success and failure, the probability of failure is merely one minus the probability of success or vice versa. The triangles in Figure 6-10 are only drawn schematically to show that the "most likely" case has the highest probability of occurrence of all successful outcomes. No specific probability value should be inferred for the "most likely" case. This demonstrates an important property of the E&P triangular probability distribution model which is:

> The only probability value which must be determined to use theE&P Triangular Probability Distribution Model is the probability of failure. ALL successful outcomes are represented by the triangular distribution, with the sum of their probabilities being equal to one minus the probability of failure. Mathematically then, the area under the triangle is also equal to one minus the probability of failure. There is no particular significance to an absolute value of the probability of the most likely outcome, other than it is greater than that for any other successful outcome.

Using the properties of a triangular distribution:

$$\text{Expected Ultimate Recovery} = P_s\,(A + B + C)\,/\,3 \tag{6.8}$$

$$\text{Expected Net Present Value} = [\,P_s\,(E + F + G)\,/\,3\,] - [\,D\,(1 - P_s)\,] \tag{6.9}$$

where:

P_s = probability of success = $(1 - P_f)$

P_f = probability of failure

A = minimum volume to justify completion / development

B = most likely / best estimate volume

C = maximum / very optimistic volume

D = cost of failure (dry hole cost)

E = net present value of minimum volume discovery

F = net present value of most likely volume discovery

G = net present value of maximum volume discovery

Figure 6-10(B) is an economic probabilistic E&P model. It has many of the characteristics of the ultimate recovery model, however, the X-axis represents the net present value associated with volumes discovered as presented in the volume model. On the X-axis, D represents the cost of failure, which in the case of a single well would be the dry hole cost, and for a larger project would be the expenditures up to the decision point of not developing. The cost of failure is a negative value, because this is an expenditure with no income. However, point E is also shown as a negative value (loss). This represents the case where development is undertaken because sufficient volume is found to justify completion of the project, treating all costs up to that point as "sunk costs." Under these conditions the additional investments to complete the well or undertake development are profitable, but due to the sunk costs the overall project is unprofitable. The net present value for the most likely (F) and maximum (G) cases are both positive.

FIGURE 6-10
EXPLORATION & PRODUCTION PROBABILISTIC MODEL

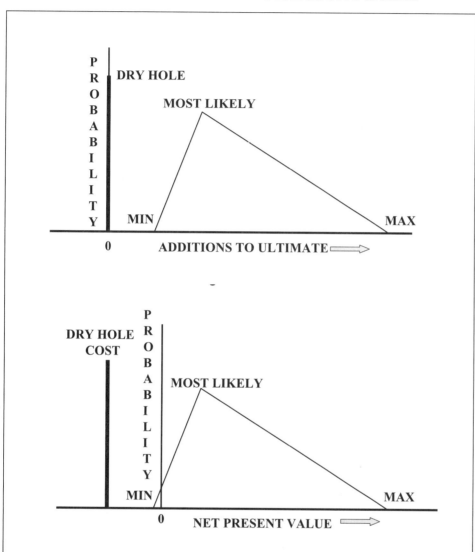

There are a number of reasons for using this probabilistic model when considering risk and uncertainty in exploration and production ventures. It is a relatively simple model requiring the identification of only four easily recognized and well understood outcomes and their evaluation. The probability that the venture will fail is the only probability that must be determined. And once values have been determined for each of the four outcomes, the expected values can easily be calculated.

CENTRAL LIMIT THEOREM

Central Limit Theorem (CLT) states that as the sample size is increased there is a greater tendency for the values to fall in the middle of the distribution and the distribution approaches "normal." The central limit theorem deals with the shape of the distribution curves. In simple terms this merely says that the larger the number of readings the more nearly the curve of frequency distribution will approach the classic uniform bell shape.

225

Figure 6-11 presents another frequently encountered situation in petroleum engineering in which the degree of uncertainty declines throughout the field life from new discovery to economic limit and abandonment. The only exact figure on ultimate recovery is its cumulative production at abandonment.

FIGURE 6-11
PROGRESSIVE REDUCTION IN UNCERTAINTY

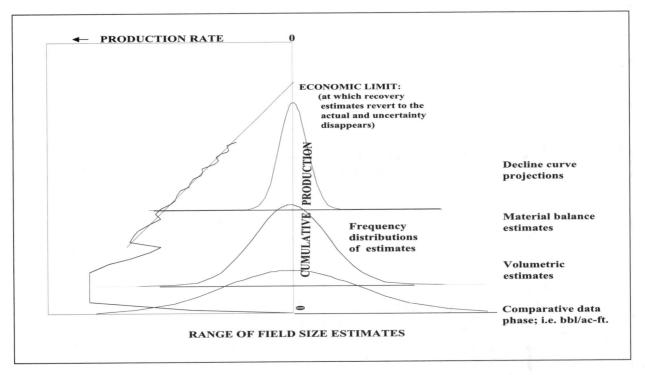

TIMING RISKS

Wilson and Pearson (loc. cit.) present an interesting example of the importance of timing and engineering risk. Figure 6-12 shows an actual case of the effect of time delay in initiating water injection in a unitized operation, coupled with a poor performance prediction for a unitized field wide waterflood operation. The authors report that the initially predicted return of 15 percent in fact produced a rate of return of less than six percent.

FIGURE 6-12
EFFECTS OF TIMING RISK

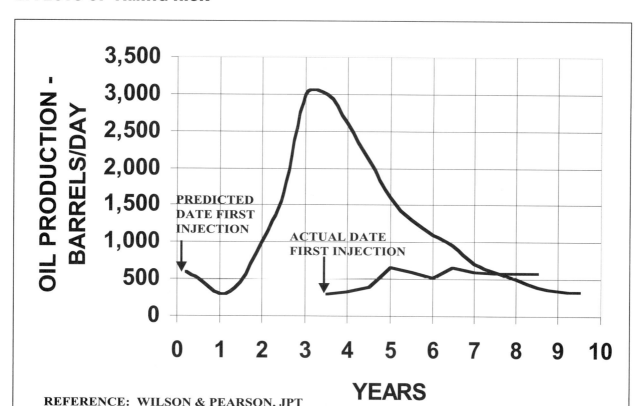

SENSITIVITY ANALYSIS

A common method of handling uncertainty (i.e., where the range of outcomes is known but the probability of each outcome is not) is known as, "sensitivity analysis." Sensitivity analysis goes beyond simple indicators of uncertainty exposure through the development of multiple case comparisons to explore and demonstrate the significance of alternative forecasts of uncertain elements in the economic evaluation. These are designed to cover the ranges of possibilities in each of the factors whose sensitivity is studied. It is possible through sensitivity analyses to develop a great deal of information from relatively straight forward analyses repeated on an optimistic,pessimistic and most likely estimates of the key variables.

The simplest sensitivity analysis relies on those responsible for the development of the case under study merely to select alternative forecasts in any combination that they regard as optimistic or pessimistic. The difficulty here stems from the fact that people who are uncertain about the possible outcome of a particular event usually have no real perception as to the degree, or level, of their uncertainty. If the basis for optimism and pessimism is clearly set out and explained, such sensitivities can be meaningful and quite useful. The sensitivity analysis becomes more useful, however, when particular elements of the forecasts of input data are identified and assumed to change in specific ways and to a specific extent. Elements included in sensitivity analyses typically include items such as the required investment, operating costs, 1,000 reserve size, producing rates,prices, etc. Combinations of elements also are sometimes used.

While sensitivity analysis shows how changes in the physical or commercial input elements affect the outcome of a project, the technique does not concern itself with evaluating the probabilities that such changes actually will, or might take place. Sensitivity analysis methods are useful in the decision-making process to the extent that reasonable values, i.e., within a realistic range of possibilities are chosen. For example, it makes little sense to show sensitivity to a 50 percent increase in investment if experience indicates that such investments are almost always estimated within a plus or minus 20 percent range. On the other hand, reporting a sensitivity to large changes in reserves, if the variation is clearly a potential outcome of exploration activity in a new and highly speculative area, could be quite useful.

There is a rudimentary art in selection of the criteria for sensitivity checks as well as in their presentation. Sensitivity analyses that are well chosen make it possible for management to concentrate its attention on those factors that have the greatest impact on the forecast outcome, and at the same time they can confidently disregard those that have only minor effect. Sensitivity analysis is commonly employed when awareness of specific certainties is inherent, but quantified probability projections are difficult or impossible.

Tornado Diagrams

A tornado diagram displays the sensitivity of Net Present Value to changes in the variables that are involved in computation of NPV for a particular project or investment. In a tornado diagram a solid bar indicates the range of NPV values for each parameter as it is individually varied from its minimum value to its maximum value. The minimum value for each parameter is not its absolute minimum value, but a value which will be exceeded 95% of the time. Likewise the maximum value for each parameter is not is absolute maximum value, but a value which will be exceeded no more than 5% of the time. Thus, there is a 90% probability that the correct value will be included in the interval from minimum to maximum. This interval represents approximately 1.6 standard deviations. The minimum and maximum values of each parameter are shown at the end of each bar to indicate the range evaluated. The bar itself will usually include values both less than as well as greater than the expected NPV, which is shown on the diagram. The parameters are ordered on the diagram in descending order of sensitivity. The top bar will be the longest bar representing the parameter that most greatly influences NPV and the bottom the shortest bar representing the parameter which least affects the value of NPV. The funnel shape of bars, looking much like a tornado, gives rise to the name of this graphical presentation.

FIGURE 6-13
TORNADO DIAGRAM

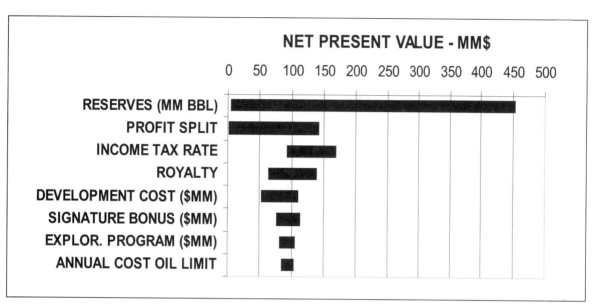

The following table shows the minimum, median and maximum values (from the point of view of the contractor) used in developing the tornado diagram of Figure 6-13. The net present value determined using the median value for each of the parameters shown was $94.61 million dollars.

	MINIMUM VALUE	MEDIAN VALUE	MAXIMUM VALUE
RESERVES (MMB)	50	100	300
PROFIT SPLIT (Gov. /Contr.)	90%/10%	80%/20%	75%/25%
INC. TAX RATE		35%	0%
ROYALTY	25%	15%	0%
DEV. COST ($MM)	$500	$321	$250
SIGNATURE BONUS ($MM)	$75	$50	$25
EXPL. PROGRAM ($MM)	$80	$50	$25
ANNUAL COST OIL LIMIT	30%	40%	60%

Decision Alternatives

Graphical presentations, such as Figure 6-14, can be helpful in charting a course through a decision tree's option nodes. The grid comprises a vertical axis, both positive and negative, of NPV at the preferred hurdle rate. The horizontal axis represents the spectrum of probabilities of success varying from zero (or nil) to 100 percent, (or certainty). Figure 6-14 represents a drill or farm-out decision situation. Three possible alternatives are plotted, with each represented by a straight line across the probability spectrum.

Data for all decision alternatives plotted in accordance with the format of Figure 6-14 will always follow a straight line. Therefore the entire range of probabilities can be evaluated by drawing straight

229

lines from the conditional values for failure (0%) to the corresponding conditional values for 100% success (100%).

In this example, the value of failure (0%) is the cost of a dry hole. The conditional value of success (100%) is the expected value of all of the successful outcomes (excluding any outcomes that are not successful), as explained under Expected Value Concepts.

FIGURE 6-14
SENSITIVITY OF DECISION TO PROBABILITY OF SUCCESS

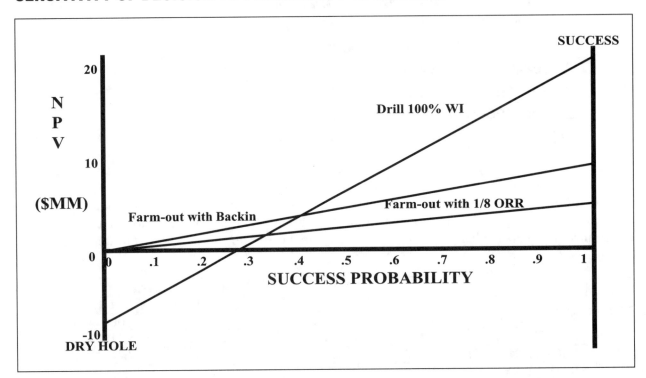

The first alternative considered is to drill the well "straight up" with 100 percent working interest. In this situation engineers estimate that the cost of a dry hole after tax would be $10MM. The best that can be expected from a fully successful well is $20MM after tax. The straight line between the two points is labeled "Drill 100% WI." The other two alternatives are to: 1) farm-out for a 1/8th ORR; or 2) farm-out (see Chapter VII) for 1/16 ORR until payout with the farmee's option at that point to convert, or "back-in," to a 40% working interest. The two farmout alternatives start at zero NPV for a dry hole since there is no initial gain or loss for the party farming-out. The 1/8th ORR provides a positive expected NPV increasing uniformly as the probability of success moves to 100 percent, as does the 1/16 ORR with back in option.

From this presentation (Figure 6-14) it is apparent that for this set of conditions, there is no probability of success for which the straight 1/8 ORR is the best alternative, so it can be eliminated from further consideration. Comparison of the farmout with "back in" alternative with the 100% WI drilling alternative indicates that the farmout is preferable to drilling "straight up" when the probability of success is in the lower range. However, when the probability of success exceeds 50%, drilling is preferable to the 1/16 ORR with back in case.

The diagram of Figure 6-14 can also be used to evaluate the sensitivity of decision among alternatives to the assessment of probability of success. Figure 6-14 would indicate that the decision to farmout is correct if the probability of success is less than about 50% and it should be drilled as a 100% WI venture if the probability of success is assessed to be greater than 50%.

DECISION TREES

It should be recognized that in many decision-making situations, future choices in a train of events are inevitably affected by the actions taken in the immediate present. For cases in which consideration of the consequences of a series of decisions is important, and the likelihood, or probability of the subsequent events is known, the use of decision trees for risk analysis can be very helpful.

Decision trees are an excellent way of breaking down a highly complex decision problem into a series of simpler ones. Decision trees are constructed by diagramming all of the decision options and subsequent chance events associated with the particular alternatives. This allows the user to react to a series of "what if" questions as well as to envision the range of possible outcomes of the project under study. Decision Trees also provide a useful vehicle for examining the value of additional information to the decision process.

The approach is best suited to the evaluation of projects or programs involving a succession of alternative decisions in which:

1. each decision has a limited number of outcomes

2. the probability of occurrence of each outcome can be determined, and

3. each decision is ultimately dependent upon the economic consequences of future decisions and outcomes.

Each decision tree is built for only one decision even though it may forecast other subsequent decisions which will be required. In exploration, for example, a typical sequence of operations and related decisions might involve taking a lease, undertaking geophysical work, deciding to drill a well, deciding to drop the acreage, whether to farmout or drill a confirmation well, and so forth. At each of these branching points the several possible outcomes are indicated. Starting at the left of the tree with the initial decision alternatives each branch depicts a possible outcome which is given a value or "payoff."

Two types of node symbols are normally employed:

■ = decision node

● = chance node

The nodes of the tree from which the branches spread are all either decision nodes or chance nodes. The branches that emanate from a decision node represent alternative courses of action between which the decision maker must choose. The branches leaving the chance nodes represent chance events, or outcomes over which the decision maker has no control. The probability associated with each possible outcome from a chance node is written on the branch. Chance events are usually the work of Nature such as weather, or geology, or possible political outcomes.

FIGURE 6-15
SIMPLE DECISION TREE

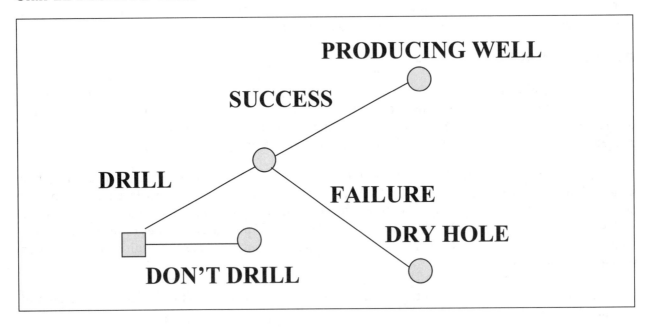

EXAMPLE 6-6
DECISION TREE ANALYSIS

Consider an exploratory drilling venture as depicted in Figure 6-15 by way of illustration. If next, we assume a 65 percent probability of finding no production, the variables with respect to expenditure and outcome may be incorporated, as shown in Figure 6-16:

FIGURE 6-16
DECISION TREE WITH PROBABILITIES OF OCCURRENCE

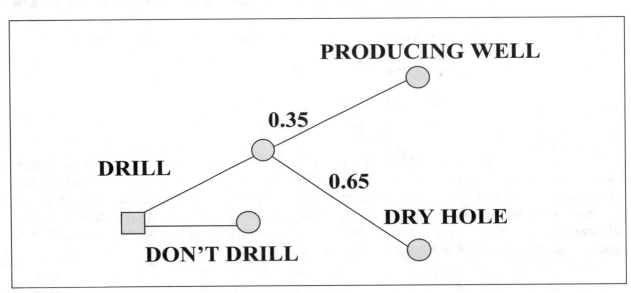

The analysis is further complicated, as is shown in Figure 6-17, by the realization that not all successful wildcats are equally profitable. For this example there is a 75% chance of the well being commercial, with a 25% chance that it will be marginal. A commercial well would have a conditional value (NPV) of $3MM, a marginal well $0.5MM, and the dry hole cost of $0.2MM. All of these are after income tax values.

Thus, the drilling of the prospect can have only one of three possible outcomes. Any one excludes the others (mutually exclusive), and one of the events must happen. The outcomes at this stage, therefore, comprise a collectively exhaustive, mutually exclusive set, and the total probability that one of the three will occur is unity.

The most likely result is always the one having the highest probability of occurrence, or in this case, a dry hole. If the decision maker relied only on the most likely outcome, he would have to reject the project forthwith. On the other hand, a business decision would be made easier if excellent producing wells had a higher relative weighting, in economic terms, than marginal producers and dry holes. Expected value is clearly the preferred approach to decision making in these circumstances if it can be properly utilized.

The decision tree is evaluated using what is known as the "Rollback" procedure. This procedure calls for starting at the tips of the branches and working back toward the starting, or initial node of the tree, employing these two rules:

1. If the node is a decision node take the maximum profit or minimum cost from the adjacent nodes upstream to the right.

2. If the node is a chance node compute the expected value of that node from the "rolled back" values of the adjacent upstream nodes.

As each node is evaluated, working downstream from the right, the Expected Value calculated for Rules 1 and 2 can be determined and posted on the main tree. Thus by working backward down the tree, many of the alternatives may be eliminated quite quickly from further consideration.

The resulting assessment of uncertainty can be shown diagrammatically in Figure 6-18.

FIGURE 6-17
DECISION TREE WITH ECONOMIC CONSEQUENCES

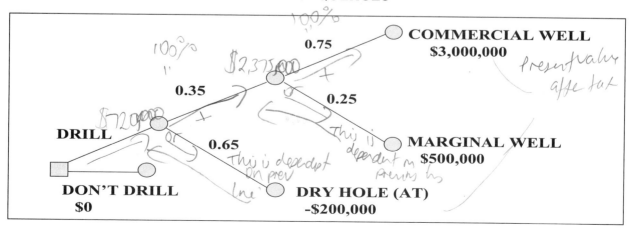

233

The risk-weighted economic value of the drilling decision can also be posted on the diagram. These would be derived from the following value assessment:

Outcome	Probability
P: Productive and Commercial	(0.35 * 0.75) = 0.263
M: Productive but Marginal	(0.35 * .025) = 0.087
D: Dry Hole	(0.65 * 1.00) = 0.650
	1.000

Under the outlined conditions of uncertainty and expected value assignment, the risk-weighted or mean expected value of the particular decision to drill can be calculated as follows:

Outcome	Probability	Economic Consequence	Expected Values, $MM
P	0.26	$ 3.0 MM	(0.263 * 3.0) = 0.789
M	0.09	+ 0.5	(0.087 * 0.5) = 0.043
D	0.65	- 0.2	(0.65 * - 0.2) = - 0.130
			Expected value of drilling = $0.702

Now the question arises as to the value and meaning of the expected NPV of $702,000 calculated for this example. Using criteria of Chapter V, this investment opportunity is acceptable because it is positive, indicating that the rate of return exceeds the hurdle rate, and all risks have been considered. The resulting value, however, is no better than the assumptions that went into its derivation. The successful use of the Expected Value technique as an index of investment quality depends upon the validity of both the probability figures and the expenditure estimates for each of the alternative investments. Risk adjustment does not eliminate uncertainty. It merely expresses its effect in a useful fashion which oftentimes proves helpful in decision making. To further enhance the utility of decision tree diagrams, values can be posted at each node, as has been done in Figure 6-18.

Continuing our discussion of decision tree diagrams, let's next consider adding a farmout option to the drilling decision, in which the decision maker could come back in for a quarter working interest after payout of the drilling and completion cost. This is diagramed in Figure 6-18:

FIGURE 6-18
DECISION TREE WITH FARMOUT OPTION

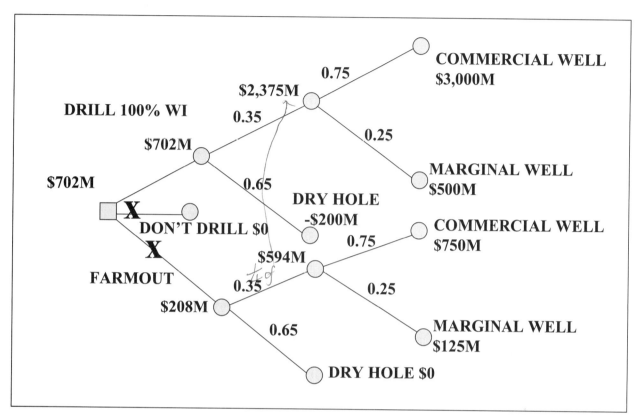

Conclusion: The drilling option is better (EMV + $702 M) vs. the farmout (EMV + $208 M) although both are positive in this case.

Interpretation of Decision Trees

The Decision Tree method of analysis can be further enhanced by scaling the branches in terms of time. For example, in the case of a secondary recovery project, discounting all values back to a common time point for comparison can be helpful.

Several other analytical techniques based on decision trees are documented in the literature. One is Nature's Tree in which only the chance nodes from the basic tree are included and are replotted separately. From this tree it is relatively simple to ascertain what the maximum, or minimum outcomes due to chance alone will be. This method facilitates the computation of any series of chance events that may be dispersed through the basic tree.

The decision tree can also be used to test decision alternatives for the sensitivity of the probability of occurrence and conditional value of any or all of the outcomes anticipated by the tree. This requires the resolving of the decision tree for a range of values or selected discrete values for the parameters being investigated.

Another interesting variation of Decision Tree analyses involves the inclusion of the effects of expenditures for additional information, such as additional seismic surveys, or special laboratory work for an enhanced oil recovery project. The added costs, project delays, and changes in Expected Values, either positive or negative, can be incorporated in the diagram employing the same techniques already described. These costs are carried to the end of all branches to the right of such expenditures and are included in the conditional value of each subsequent outcome.

Decision trees do not eliminate risk, they merely aid in assessing it. One drawback to the decision tree methodology is the implicit assumption that maximizing expected NPV is the ideal decision criterion. This ignores the fact that most companies are to a greater or lesser extent, risk averse. The fact that a decision problem is formalized as a diagram helps, of itself, to clarify the important issues and to demonstrate the interrelationship of the decision process.

The decision tree examples presented up to this point have only included a single decision node. However as noted at the beginning of this section, decision trees are not limited to a single decision node. A tree may include as many decision nodes as the user wishes, representing all future decisions expected to be made. Although each tree is prepared to assist in making only a single decision, which is the decision represented by the decision node at the extreme left side of the diagram. All decision nodes to the right of the prime decision node represent future anticipated decisions. They are dependent upon the prime decision and/or the outcome at certain chance nodes and will not be made until something else occurs. For example, it may be anticipated that a decision to drill a well will be made only after obtaining additional seismic data. The decision will be highly dependent upon that information. A single dry hole may not completely condemn a prospect. It may be desirable to delay making the decision to drill a second wildcat well until the data provided by the first dry hole can be fully analyzed.

Not only can the decision tree be used to evaluate alternatives by including and weighing the various outcomes which may result from those alternatives, but once it has been prepared it does provide a plan for carrying out the decision. The decision tree is also an excellent surveillance tool for monitoring the results of outcomes represented by the chance nodes and for actually making the future decisions that were anticipated when the tree was initially drawn. As new information is obtained the tree should be updated and altered when dictated by the new information. As subsequent decisions are made and outcomes become known, the nodes to the left of these points become moot and those branches emanating from them which were not followed should be eliminated. It may also be necessary to add more branches and nodes to the diagram as the result of additional information obtained.

By way of summary, we can observe that any sequence of decisions, no matter how complicated, can be analyzed by the decision tree method. Contingencies and various alternatives are defined and analyzed in a consistent manner. Almost any number of assumptions can be incorporated. However, the more assumptions, the larger the number of cases, and the laws of diminishing returns limits the practicality of making the tree too extensive. How extensive to make the tree is, of course, a matter of judgment on the part of the analyst.

The procedure to be followed in using a decision tree as a tool in the decision process can be summarized as follows:

1. Identify all future decisions which will follow the current decision considering all of the current choices being contemplated.

2. Identify all future outcomes resulting from the current and all future decisions.

3. Draw all branches and nodes identified in steps 1 and 2. Work from left to right on the diagram.

4. It may be necessary to iterate through steps 1, 2, and 3 several times to identify all future decisions and outcomes.

5. Determine the conditional values of all chance outcomes identified in step 2.

6. Determine the probability of occurrence of all outcomes identified in step 2.

7. Calculate the expected value at all chance nodes.

8. Select the best alternative at all decision nodes. (That is, the decision alternative (branch) which will yield the greatest expected monetary value.)

9. For trees with multiple decision nodes it will be necessary to iterate through steps 7 and 8 for each decision node in solving the decision tree from right to left; i.e., from the most distant future time back to the present.

10. Select the best alternative for the first decision. (This is the decision for which the tree was constructed.)

RELATING DECISION POINTS TO COINCIDE WITH CHANGES IN UNCERTAINTY

In any drilling or production project in the oil business, the degree of uncertainty changes with time, or the project's progress. A good example of adapting to these changes in the degree of risk is the frequency of exploratory drilling deals which employ the "casing point" as the decision time. When this point is reached during drilling, the various parties must opt to reduce or increase their financial commitment, either to continue the operation, or plug the well. Casing point is the stage, in the physical operation, when drilling has reached the target depth, logs have been run, and core samples are taken and analyzed. The greatest part of the exploratory and drilling risk is history at this point. If the decision is to proceed to run casing, and attempt a commercial completion, it is still possible that it is a dry hole, but the likelihood of failure is greatly reduced.

Other examples of coping with exploratory drilling risk might include dry hole and bottom hole contributions. These types of arrangements are undertaken when two operators share an exploration prospect which is bisected by a lease or concession boundary. The results of drilling a test well in such a situation is of value to both parties. In order to avoid the unfortunate event of both operators drilling nearby dry holes, it is common for one to drill, and the other to agree to pay the drilling party a stipulated amount if the well fails to produce. The non-drilling party has full access to the rig floor and all of the geological data from the well. If the well is successful the non-drilling party pays nothing, takes the information, and proceeds to drill on his side of the boundary hoping for the same happy result.

A variation of the dry hole contribution is the bottom hole contribution. This is generally undertaken on a more regional basis in which the drilling of a well primarily for geologic information will be of value to various parties of interest. They agree to pay the drilling party a certain sum if and when his well reaches the agreed depth and the logs and other data are delivered.

Analyses of these various types of risk related deals can generally be incorporated in decision tree analyses.

ECONOMIC RISKS ASSOCIATED WITH MARGINAL FIELD DEVELOPMENT

Marginal oil and gas fields comprise an important risk problem to the petroleum industry. A marginal field is one which is capable of yielding an economic return to the company if all the technical and operational assumptions and analyses prove minimally correct. The economic viability of these fields is highly dependent upon the price of oil throughout their productive life. Once developed, however, it may prove best to continue production, rather than to shut them in, or actually abandon them.

In addition to the oil pricing risk there are a number of physical and engineering risks to be recognized in marginal field analyses. Principal among these is field size. Marginal field undertakings do not permit any large amount of delineation drilling so the productive limits remain uncertain. This increases the need for accurate predictions of the productivity of the wells and their ability to attain and maintain the required level of production at least until payout. In order to accomplish this the operator needs a good understanding of the reservoir drive mechanism and its adaptability to artificial lift and secondary recovery. Extended well testing (EWT) of several months duration can be helpful in assessing these important unknowns but this delays full development and can adversely affect the project's economics.

The complexities of marginal field development complicate their economic analysis. The most effective approach to handling them is by means of a carefully designed sensitivity analysis as described earlier in this chapter.

STOCHASTIC MODELING

Another much more involved approach to dealing with uncertainty is the "Monte Carlo" simulation technique. This approach seeks to relate the numerous factors interacting simultaneously with one another, in the evaluation of ventures, whose outcome is determined by a series of distinct variables.

The technique requires a mathematical model of the situation under study which recognizes all of the pertinent factors, even though only ranges, rather than specific values can be assigned to a number of the parameters. It is particularly important in developing the model to identify the types and degree of dependency among any, or all, of the variables. The concept of simulation permits recognition, or consideration of a statistical combination of all the possible values within the range of each random variable.

This is accomplished by randomly choosing a value for each parameter from its cumulative frequency plot, such as shown in Figure 6-20. This process would be repeated maybe several thousand times and the results averaged to determine the expected value. In effect, the simulation approach is as if one ran hundreds, or perhaps thousands of "what-if" computer spreadsheet analyses in order to find the overall most probable answer.

The procedure produces a simulated probability distribution for the outcome from which the decision maker can gain a reasonable idea of what is likely to occur if he pursues an assumed, or modeled, course of action. The outcome of the analysis may be viewed as a stochastic sample of the real system.

"Random numbers" are employed to perform the sampling process for each specific variable in a mathematical model. Random numbers occur with equal likelihood as each one has an equal probability of occurring, but without any identifiable pattern. They are dimensionless, positive digits. A classic way of generating random numbers is by drawing samples from a hat. One might employ a sequence of say, two-digit random numbers e.g., 00, 01, 02, ..., 98, 99. Any of these two-digit numbers is equally likely to occur in any sequence and there is no pattern in the sequence in which they appear. Random numbers are generated from irrational numbers such as, e, or . There are functions in Excel and Lotus 123 which can be used to obtain the desired random numbers as well as books of tables such as those published by the Rand Corporation (loc. Cit.).

The Monte Carlo simulation process is described as follows;

1. Establish a formula (model) for the value to be determined. The parameters used in the formula should be independent. If any are not independent, their interdependencies must be recognized either in the formula or combined into a single parameter for purposes of Monte Carlo simulation.

2. Define a probability distribution for each independent parameter in the formula of step #1.

3. Determine the cumulative probability distribution (0 to1) for each independent parameter of step #1 over its expected range of value.

4. Randomly and independently select a value of cumulative probability for each independent parameter in the formula of step #1.

5. Determine the value of each independent parameter of step #1 which corresponds with the randomly chosen cumulative probability for that parameter.

6. Calculate the desired result using the formula of step #1 with the independent parameter values determined in step #5.

7. Repeat step #4, #5, and #6 (500 to 5000 times).

8. The "Expected Value" of the situation being analyzed is the average of all calculations made in step #6.

Figure 6-19 after the work of Rodney Schmidt (loc. cit.) endeavors to show, in simplified terms, that a series of independent probabilities are combined to form a single overall probability distribution for a contemplated exploration and development program in Egypt. The general provisions of that country's production sharing type of contract were assumed. Note that there is a significant probability of monetary loss from the project even though the expected outcome would be $53 million profit. Figure 6-19 determined from Monte Carlo simulation demonstrates a frequency similar to the triangular distribution model presented in Figure 6-10(B).

FIGURE 6-19
PROBABLE CASH FLOW FROM SIMULATION ANALYSIS (Egyptian Project)

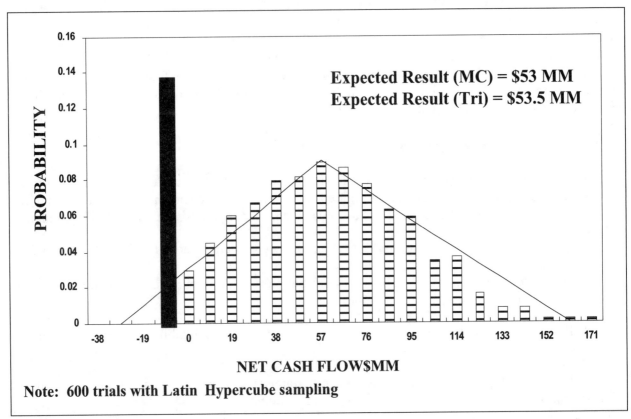

The strong bar on the left represents the specific circumstance of a $3 million signature bonus plus a $14 million initial exploration commitment. There is approximately a 13% probability of the program being a total loss to the contracting company.

Figure 6-20 is a generalized triangular probability distribution and its associated cumulative probability function similar to that presented in Figure 6-10.

FIGURE 6-20
TRIANGULAR PROBABILISTIC MODEL

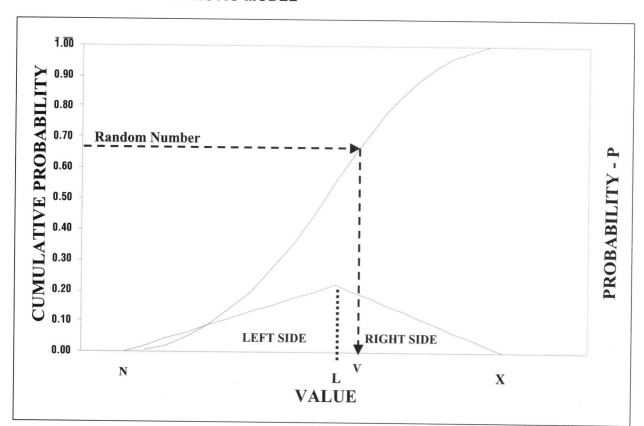

The three triangular values; i.e. minimum, most likely, and maximum have been identified as N, L, and X respectively. As is shown in Figure 6-20, with the dashed lines and arrows, a parameter value (V) is randomly chosen. This is done by determining one that corresponds to a randomly chosen value of cumulative probability in the range of 0 to 1 using the cumulative probability versus value function for that specific parameter. When a triangular probability model is used to describe probability distribution in a Monte Carlo simulation, it is convenient to use equations relating the parameter value to cumulative probability for this model. Two equations are required for this relationship because of the functional change that occurs at the apex of the triangle. Therefore, one equation applies to values between the minimum and most likely values and another is valid in the region between the most likely and maximum values. These are identified as the left and right side of the triangle respectively and are presented as Equations 6.10 and 6.11. The cumulative probability at the apex of the triangle separating the two sides of the triangle is shown as Equation 6.12.

Left side of triangle: $(v < v_{ml})$

$$v = v_{min} + [P_{cum}(v_{max} - v_{min})(v_{ml} - v_{min})]^{1/2} \qquad (6.10)$$

Right side of triangle: $(v > v_{ml})$

$$v = v_{max} - [(1 - P_{cum})(v_{max} - v_{min})(v_{max} - v_{ml})]\}^{1/2} \qquad (6.11)$$

241

Where:

$$P_{cum} = \text{Cumulative probability at a value } v$$

$$P_{cum}(@v_{ml}) = (v_{ml}-v_{min})/(v_{max}-v_{min}) \qquad (6.12)$$

$$v \quad = \text{Parameter value}$$

$$v_{min} = \text{Minimum value}$$

$$v_{ml} \quad = \text{Most likely value}$$

$$v_{max} = \text{Maximum value}$$

To determine the parameter value associated with a randomly chosen cumulative probability value in a Monte Carlo simulation, use the appropriate equation 6.10 or 6.11 selected with an "IF" statement or function, using Equation 6.12 as the selection criterion. The following example should assist the reader in understanding the details of performing a Monte Carlo simulation.

EXAMPLE 6-7
MONTE CARLO SIMULATION

When faced with an economic decision, the present value of the cash flow expected to occur as a result of that decision must be determined. Very little is known about the expected future performance of the project and there is significant uncertainty of the basic economic parameters. In this situation, a mathematical model which incorporates basic economic parameters is probably the best tool to use. It was decided that the model represented by Equation 5.15 would be satisfactory for the purpose since it recognizes declining production, which is characteristic of the cash flow expected, the duration of the cash flow, and inflation of future hydrocarbon values and costs. Future cash flows are to be discounted to a present value using the corporate hurdle rate of 10%.

$$v_p = [R_1(1+i)^{-0.5}][(1-X^t)/(1-X)] \qquad (5.15)$$

Where:

$$X = (1-d)(1+I_I)/(1+i)$$

$$R_1 = \text{annual revenue or cash flow for the first year}$$

$$t \quad = \text{number of years of the revenue or cash flow}$$

$$d \quad = \text{effective decline rate, decimal per year}$$

$$I_1 = \text{constant escalation or inflation rate, decimal per year}$$

$$i \quad = \text{interest or discount rate, decimal per year}$$

There is uncertainty in the initial value of the cash flow, decline rate, inflation rate, and the future time over which the cash flow will be received. These uncertainties can be properly recognized by determining the expected value of the present value net cash flow through Monte Carlo simulation.

To apply the Monte Carlo method, it is necessary to establish a probability distribution for each of the independent uncertain parameters. A triangular distribution for each of the uncertain parameters in Equation 5.15 is believed to be appropriate. The values chosen for the minimum, most likely, and maximum values of each parameter are shown graphically in Figure 6-21. The cumulative probability over its entire range of value from 0 to 1, corresponding to the range of value expected for each uncertain parameter (minimum to maximum value) is also shown for each in Figure 6-21.

FIGURE 6-21
PROBABILITY DISTRIBUTIONS

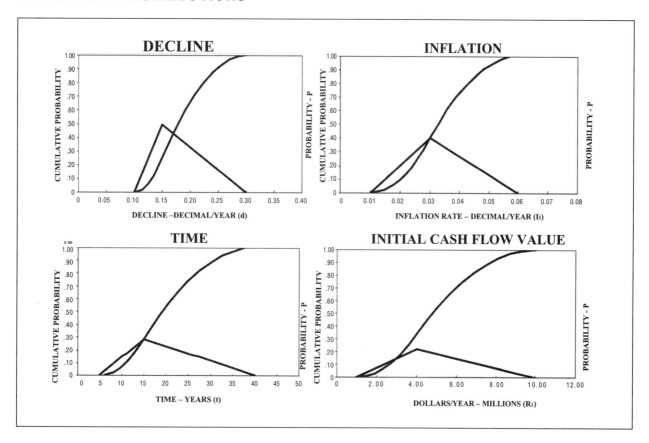

Equation 5.15 was recalculated 5,000 times with each uncertain parameter randomly chosen for each calculation according to its cumulative probability function. The minimum, most likely, and maximum values used for each of the uncertain parameters are listed in Table 6-4 along with the result of these 5,000 calculations.

TABLE 6-4
SUMMARY OF MONTE CARLO MODEL VALUES

(margin note, handwritten): Hurdle rates 10%

	DECLINE (d)	INFLATION (II)	TIME OF CASH FLOW (t-yr)	INITIAL CF VALUE X (R1-$MM)	X (Eq. 5.15)	PV CASH FLOW ($MM)
Minimum(N)	0.10	0.01	5	$1.000	0.8264	$3.3751
Most Likely (L)	0.15	0.03	15	$4.000	0.7960	$18.0782
Maximum(X)	0.30	0.06	40	$10.000	0.6746	$29.2963
Expected Value (Triangular distribution)	0.1833	0.0333	20.000	$5.000	0.7672	$20.375
Expected Value (Monte Carlo - 5,000 calculations)	0.1835	0.0331	20.066	$4.973	0.7664	$20.443
$P_{cum}(@V_{ml})$	0.2500	0.4000	0.2857	0.333		
Hurdle rate =	10%					

Thus the expected value net cash flow for the circumstance being analysed using Monte Carlo simulation is $20.442 million. This is not much different than $20.374 million determined just using the expected values from the distribution of each of the uncertain parameters.

PORTFOLIO ANALYSIS

Portfolio analysis is a procedure to identify the group of investment opportunities which will yield the best return on the limited financial resources available, consistent with the company's goals and objectives. Such an analysis must fully recognize the risks, uncertainties, and interdependencies of each investment and choose that group of projects for which no increase in value is possible without greater risk and no reduction in risk is possible with out loss of value. Such an analysis includes much of the material previously covered in this chapter; e.g. uncertainty, risk, and expected value, and may involve the use of Monte Carlo simulation. Further explanation of portfolio analysis is beyond the scope of this book, but for those interested, further explanation can be obtained from one or more references at the end of this chapter by the following authors; Ball and Savage, Bernstein, Brashear et al, Cairns, Gutleber et al, Newendorp, and Orman et al.

INVESTMENT CRITERIA WITH RISK AND UNCERTAINTY

Hurdle Rate

One other common approach to dealing with risk in general, and political risk in particular involves subjectively increasing the hurdle rate or adding a few percentage points to the minimum acceptable DROI and rate of return which the company uses for selecting investment opportunities. This may be the easiest way of dealing with these problems, but not necessarily the best way to do it. When ever possible, risk considerations should be separated from the fundamental transaction analysis in order to provide a better understanding of the true uncertainty and incorporated in the analysis through the use of "Expected Values" and "expected" economic criteria, using a normal hurdle rate to make investment decisions. It may be helpful to reread the discussion of Figure 6-1.

Expected Decision Criteria

The same investment decision criteria discussed in Chapter V can be used under conditions of risk and uncertainty, but they must be applied to cash flows which have incorporated the principles of risk and uncertainty as discussed in this chapter. Expected Present Value Net Cash Flow can be determined by using either Equation 6.3 or Equation 6.5, where the values used in either of those equations are expected or conditional present value net cash flow numbers and the cost of failure is considered a positive number.

$$\text{Expected PV NCF} = [E(P|S)]P_s - [E(C|F)](1-P_s) \tag{6.5}$$

To ascertain the Expected DROI, the Expected Investment must be determined. It is the weighted average of the cost of failure and the investment required for the successful outcomes, with probability of occurrence being the weighting factor. An equation similar to Equation 6.5 can be used for this purpose, but the terms are additive since both represent expenditures. As was discussed in Chapter V in the section titled 'Discounted Return on Investment (DROI),' the expected NPV determined with Equation 6.5 should be after income tax while the expected investment calculated using Equation 6.13 should be before income tax. This is acknowledged in Equation 6.13 by dividing the expected cost of failure, which is normally on an after tax basis, by one minus the income tax rate. This conforms to the usual practice of budgeting only the actual dry hole cost for wildcat wells.

$$\text{Expected Investment} = [E|(I|S)]P_s + \{[E(C|F)]/(1-T_R)\}(1-P_s) \tag{6.13}$$

$$\text{Expected DROI(EDROI)} = [\text{Expected PVNCF (AFIT)}]/[\text{Expected Investment (BFIT)}] \tag{6.14}$$

Where:

P_s = Probability of Success = $(1 - P_f)$

P_f = Probability of Failure = $(1 - P_s)$

$E(P|S)$ = Expected profit conditioned on success (AFIT)

$E(C|F)$ = Expected cost of failure (AFIT)

$E|(I|S)$ = Expected Investment for successful outcomes (BFIT)

T_R = Income Tax Rate

BFIT = Before income tax

AFIT = After income tax

To determine the Expected Internal Rate Of Return (EIRR) where risk and uncertainty have been considered, it is necessary to create an expected cash flow. This can be done by combining the cash flow of all cases contemplated, both successes and failures. The contribution of each cash flow to the combined cash flow should be in proportion to the probability that each cash flow will actually occur. These will be the same probabilities used in the previous calculations. The Expected IRR is then determined for the combined cash flow as discussed in Chapter V.

OBSERVATIONS ON DECISION ANALYSIS

Use of the decision analysis methods presented in this chapter forces the consideration of all possible outcomes which may result from a given decision. This should eliminate most future surprises and provides a procedure to quantify the risk and uncertainty which is inherently involved in all decisions. It also provides a procedure for evaluating the sensitivity of the ecision to various outcomes and is a rigorous, unambiguous procedure for comparing dissimilar projects. Additionally, the various techniques presented can also be used as a convenient way to communicate judgements about risk and uncertainty. Since this is a means to analyze extremely complex options it also provides a way to focus attention on key uncertainties.

Even though decision analysis is a very powerful tool, there may be some common misunderstandings about it. It does not provide any new information, it merely forces the recognition of information that may have been ignored or considered irrelevant. Decision analysis does not reduce risk, but merely recognizes the risk that is there and enables the decision maker to decide if the benefits outweigh the risk. If they do not then the decision should be to "walk away" from the investment opportunity. So, use of these techniques will not eliminate the need for judgment in making a decision, but it will provide information to assist in making that judgement. Certainly decision analysis has never found a barrel of oil or a cubic foot of gas, but it may have made the difference between success or bankruptcy by preventing investments that had greater risk than reward. Decision analysis certainly will not make your life easier. Quite the contrary, it increases the amount of work required. But if it is done rigorously and consistently, it can make the difference between a successful organization and the other kind.

RISK MANAGEMENT

Risk management does not reduce the probability of an adverse event nor does it eliminate those things which may cause loss. However, it can reduce or eliminate the loss which would be incurred when the adverse event occurs. It is like seat belts in an automobile. They won't prevent an accident nor reduce the probability of one occurring, but if an accident does occur it may prevent or reduce injury to the person wearing the seat belt. Risk management like the seat belt - - - just being there is not enough, it must be used to have an effect. As a further example; there is no way to prevent a severe storm, but proper risk management can prevent or reduce the loss which might be incurred when that severe storm strikes a specific installation.

Risk management will then be defined as preloss planning. It involves identifying all events which might occur that would cause a loss and evaluating each potential loss. The potential loss must include cost of repair or replacement, legal costs, lost or deferred income, and lost future income in the event that repair or replacement are not possible.

There are three primary ways to manage risk;

AVOIDANCE - Evaluation of some risks may indicate that the potential loss is greater than the potential gain. Such ventures should be avoided, if possible. This may require "walking away" from some high risk investment opportunities or terminating certain ventures which have unfavorable risk/reward relationships.

REDUCTION - There may be certain actions which can be taken to reduce loss, if an adverse event occurs. These actions almost always require the expenditure of money and could cost more than the loss they are designed to reduce or eliminate. If that is the case then it may be necessary to terminate the venture, reduce the scope of the loss prevention, or consciously decide to accept the loss if the adverse event occurs. Actions to reduce loss may require modifications or improvements in engineering designs or the strengthening of structural components to withstand greater forces. It may also require extensive training to prepare operating personnel to handle unusual conditions, remote control of operations, or temporary abandonment of operations when threatened with possible disaster.

TRANSFER - It may be possible to transfer a loss to a third party or to spread the loss over a period of time to reduce its adverse impact. A waiver of claims may accomplish the first object for a limited number of losses and insurance for loss accomplishes both of these objectives. However, insurance coverage of a loss should not be considered "something for nothing." Eventually you will pay for that loss, since insurance premiums are designed to cover all losses plus administrative costs and a profit to the insurer. Increased losses will be reflected in increased premiums, so the best that can be hoped for is to recover a high percentage of premiums paid over a long period of time.

The only true transfer of risk is for a third party to assume all or a part of the risk of loss. In the oil business this can be accomplished by a "farmout," where a third party assumes all of the risk of a dry hole or fragmentation of the working interest where several parties proportionately share the risk of loss from a joint venture. Dryhole and bottom hole contributions by third parties also transfer some of the risk of drilling a well.

Risk management requires the preparation of an plan of action, if the adverse event occurs, to minimize the loss and to facilitate rapid restoration of the damage. That plan may also include measures to be taken in advance of a loss which will minimize the loss or reduce the impact of the adverse event. Once the plan has been prepared, it must be implemented to reduce the risks that have been identified. This may involve the installation of safety equipment and the implementation of specific procedures to be followed in the event of an impending disaster. It may also require modifications, or repair and maintenance of facilities to withstand abnormal conditions. But planning and implementation are not enough. Conditions can change making it necessary to continually monitor operations so that changes can be identified. This will assure that the plan and implementation continue to accomplish their desired objectives.

Insurance Coverage

Our consideration of risk in the petroleum industry should also include recognition and discussion of the other major segment of commercial enterprise which is built on risk. We are referring to the insurance business of the Western World. In the insurance industry the business is built around the firm, or "carrier," being paid a fee to take over certain specified risks. Whether or not the insurance company can stay in business depends upon how well its analysts can handle the probabilities of loss, and set their rates and sales accordingly.

Insurance rates are set, where possible, on the observation of losses due to the occurrence for which the insurance is being written. Where empirical data is not available, such as for a new risk or one which has never before been covered by insurance, insurers must rely upon their best subjective estimates. Insurance companies may also spread their risk by selling part of the coverage to other insurance companies or form pools of insurance. Lloyd's of London has probably the largest insurance pools and will insure against virtually any risk. It is not uncommon for insurance companies to change their premiums to reflect actual experience of claims paid. If there is a high incidence of loss due to a particular cause, the premiums for that insurance will no doubt be increased.

247

In the oil business, the survival of the firm is somewhat similarly dependent upon a proper assessment of risk and related costs of replacing produced reserves and maintaining producing rates. Both industries tend to grow into very large commercial enterprises, better to cope with the constant problem of risk. Both also follow a pattern of "laying off" part of their larger risk packages through reinsuring by farming out, or taking on partners.

As with any business, there are a number of basic types of insurance that are required for petroleum operations, some by force of law. It is not surprising that the two risk industries find a bit of common ground. Appendix VI-C of this chapter includes brief summaries of several key types of insurance coverage which are unique to the petroleum business.

Self Insuring - Advantages and Disadvantages

Which coverages a company should purchase, and to what extent, versus assuming the risk (i.e., self insuring) against a particular "peril" is one of the fascinating exercises in risk management. This is true of any business, but the oil business which itself is so risk-oriented, adds another dimension to the game. Most of what we have presented in this chapter is just as applicable to the analysis and decisions in these insurance matters as for the various operating examples in the oilfield. There is no catalogue of fundamental advantages and disadvantages of self insuring. Insurance costs money, and is nice to have, when disaster strikes. Insurance rates change, and specific coverages change, so one in the oil business has to continually update the buy or not-to-buy equation.

Oil and Gas Derivatives

One of the greatest risks facing the petroleum producing industry is the price that will be received for oil and gas when it is produced. The other side of that coin is the price risk to the consumer. The individual consumer is little concerned over this risk, but major consumers such as the electric companies and airlines have a great concern with this major cost of their business. Oil and gas derivatives have been developed to provide "insurance" against wide swings in price. But as is the case for any insurance there is a cost (premium) to be paid. In this case the cost is the price of the option, either to buy or sell the underlying asset, that is purchased to manage price risk.

Derivative contracts give one party a claim on an underlying asset (or its cash value) at some point in the future, and bind a counterparty to meet a corresponding liability. It might bind both parties equally, or offer one party an option to exercise it, or not. It might provide for assets or obligations to be swapped. It could be a complex derivative combining several elements. Futures, or a contract to buy or deliver a specified volume of oil or gas at a specified future date for a specified price were discussed in Chapter III. Options traded on commodity exchanges were also presented in Chapter III as a tool for managing the risk of price. Any perceived value that a put or call option develops above its numerical, or intrinsic, strike price is known as time value. As the expiration date of the option approaches, time value elapses and only the intrinsic value remains. Each passing day erodes the option's time value. This works in favor of the sellers and against the buyers. Actually, most options, both calls and puts, finally expire worthless.

REFERENCES

Ball, Ben C., Jr., and Savage, Sam L., "Portfolio Thinking: Beyond Optimization", Petroleum Engineer International, May 1999, Page 54ff

Bernstein, Peter L., Against the Gods — The Remarkable Story of Risk, 1996, John Wiley & Sons, Inc.,New York, N.Y.

Brashear, J.P., Becker, A.B., and Faulder, D.D., "Where Have All The Profits Gone", Journal of Petroleum Technology, June 2001, Page 20ff

Brion, A., "Oil—a risk business," Petroleum Review, April, 1969, p. 189 ff.

Campbell, John M. Oil Property Evaluation, 1959, Prentice-Hall,Englewood Cliffs, NJ 07632

Campbell, John M. "Petroleum Evaluation for Financial Disclosures," 1983, Campbell Petroleum Series,
Norman, OK 73069

Carins, R.J.R., "Approach to Exploration Risk Analysis", Journal of Petroleum Technology, May 1996, Page 438

Garb, F. A., "Assessing Risk in Estimating Hydrocarbon Reserves and in Evaluating Hydrocarbon-Producing Properties," JPT Vol. 40, No. 6, June 1988, pp 765-778.

Grayson, C. J., Jr, "Decisions Under Uncertainty," 1960, Harvard Business School, Boston

Gutleber, D.S., Heiberger, E.M., and Morris, T.D., "Simulation Analysis for Integrated Evaluation of Technical and Commerical Risk", Journal of Petroleum Technology, December 1995, Page 1062ff

Haley, C. W., and Schall, L. D., The Theory of Financial Decisions, 1979, McGraw-Hill, New York

Magill, R. E., An Introduction to Risk Analysis, 2nd Ed., 1984, PennWell, Tulsa, OK 74101

McCray, Arthur W., Petroleum Evaluations and Economic Decisions, 1975, Prentice Hall, Englewood Cliffs, NJ 07632

Newendorp, Paul. D. And Schuyler, John, "Decision Analysis for Petroleum Exploration (2nd Edition),"
2000, Planning Press, Aurora, Colorado

Orman, M.M. And Duggan, T.E., "Applying Modern Portfolio Theory to Upstream Investment Decision Making", Journal of Petroleum Technology, March 1999, Page 50ff

Rand Corporation, "A million random digits with 100,000 normal deviates," 1955, Free Press, Glenco, IL.

Roebuck, I. Field, Jr. "Economic Analysis of Petroleum Ventures," 1979, Institute for Energy Development, Tulsa, OK 74145

Rose, P. R.,"Dealing with Risk and Uncertainty in Exploration: How Can We Improve?," AAPG Bulletin, 1, No.1, Jan. 1987,pp 1-16.

Schmidt, Rodney, "Assessing Project Risk," Boulder, CO, November 9-10, 1989, The Petroleum Finance Company, Washington, D. C.

Schwettmann, M. W., "Santa Rita," The Texas State Historical Assn., 1943, Austin, TX.

Silbergh, M. and Brons, F. "Profitability Analysis—Where Are We Now?" JPT, Vol. 24, No.1, Jan. 1972, p.261

Theusen, H. G., Fabrycky, W. J. and Theusen, G. J., Engineering Economy, 1977, Prentice-Hall, Englewood Cliffs, NJ 07632

Thompson, R. S. and Wright, J. D., "Oil Property Evaluation," 1983, Thompson-Wright Associates, Golden, CO 80402

Watson, J. F., and Bringham, E. F., Essentials of Managerial Finance, 1974, The Dryden Press

Willis, R. J., Computer Models for Business Decisions, 1987, John Wiley & Sons, New York,

Wilson, W. W. and Pearson, A. J.,"How to Determine the Market Value of Secondary Recovery Reserves," JPT, Aug. 1962, pp 829- 833.

APPENDIX VI-A

DEFINITIONS

Certainty means that one and only one outcome can occur to the exclusion of all others.

Conditional Value is the value of an outcome if that outcome and only that outcome occurs.

Decision Alternative is an option or choice available to the decision maker.

Dependent Event (Conditional Event or Sequential Event) is an event which can only occur if another specified event has already occurred.

EMV is an acronym for Expected Monetary Value.

Event is one particular outcome in the list of all possible things that can occur as the result of a decision under uncertainty.

Expected Value is the weighted average value of all possible outcomes which may occur as the result of a specific decision. The value of each outcome is weighted by its probability of occurrence.

High Risk indicates that if a loss or reduced economic value occurs it will be a large value, not that there is a high probability of that occurring.

Independent Event is an event which if it occurs will in no way affect another or is affected by the occurrence of other events.

Mean Value is the weighted average value of the random variables where the weighting factors are the probabilities of occurrence.

Median Value is the value for which there is a 0.5 probability of random values being less than or greater than that value.

Mode Value is the value that has the highest probability of occurrence within a given set of outcomes.

Most Likely is the value that has the highest probability of occurrence within a given set of outcomes. This is sometimes also called the "best estimate."

Mutually Exclusive Event is an event which if it occurs precludes the occurrence of all other possible outcomes.

Outcome is a state of nature or an event that could occur if a given decision alternative is accepted.

Probability is the likelihood of something happening or the long run average fraction of times that an event will occur.'

Risk indicates that there is a probability of incurring an economic loss or reduced value.

Sample Space is a set or list of all things that can occur as the result of an experiment, chance phenomenon or decision under uncertainty.

Standard Deviation is a measure of the dispersion of data around the mean value. Mathematically it is the square root of the variance.

Uncertainty means that more than single outcome may occur and that there is a finite probability of each outcome happening.

Variance a measure of the degree of spread of data on either side of the mean value. Mathematically it is mean of the squared deviations about the mean value of a set of data.

APPENDIX VI-A

THE STORY OF SANTA RITA NO. 1

One of the great oilfield stories tells of the drilling of the first oil discovery well on the University Lands in West Texas which blew in on May 28, 1923. The University Lands now represent the world's largest university endowment, essentially all of it coming from the oil and gas royalty income from the original University of Texas land grants dating from the days of the Texas Republic.

This is the story of the famous "Santa Rita" well, duly christened with the aid of a Catholic priest in New York City, who admonished a group of his parishioners who were about to invest in the drilling venture that they should invoke the aid of Santa Rita, Saint of the Impossible. Before Frank Pickrell, the well's promoter, left New York two Catholic women gave him a sealed envelope containing a red rose. They requested him to climb to the top of the derrick—to the crown block—take the rose from the envelope, crumble it in his hands, and sprinkle the dried petals over the structure. As they drifted downward over beam, upright and girder, Pickrell was to say, according to instructions, "I Christen thee Santa Rita." This Pickrell did. (Texas Historical Assn.)

The story has more application for our purposes than just the derivation of the discovery well's interesting name. The Santa Rita well was the first Permian discovery in West Texas at a time when a very eminent geologist announced with great fanfare that he would personally drink all the oil ever found in the barren wasteland of West Texas.

(Santa Rita demonstrates maximum uncertainty of discovery in a brand new undrilled geological province.)

The well was spudded after dark on a lease which expired at midnight, a few hours later. The spudding also had to be witnessed by two people—hard to find in sparsely populated West Texas even in daylight. Both the spudding and witnessing were accomplished and the necessary affidavits signed in San Angelo, the county seat, ninety miles away just before the expiration of the lease.

(Santa Rita dealt with the certainty of the provisions in the lease documents.)

The location selected for the drilling of the well by the geologist, Hugh Tucker, was three miles southwest of the railroad tracks at the way station of Best. The ancient water-well drilling rig that had been acquired for the operation was in such poor condition that it nearly disintegrated while getting the rig off the rail car. With the lease about to expire, the decision was made to simply drag the rig off the railway right of way and spud the well.

(Santa Rita demonstrated a high degree of physical risk with respect to the state of the equipment employed.)

Santa Rita was the discovery well in the giant Big Lake Field of Reagan County, Texas at a cable tool TD of 3055 feet. The original location was eventually drilled during the development of the field— and was its first dry hole.

APPENDIX VI-C

SPECIALIZED INSURANCE COVERAGES FOR THE OIL AND GAS PRODUCING INDUSTRY:

Well Blowout

This insurance indemnifies the operator of the well and other owners of interest for the cost of putting out a fire and regaining control of an oil or gas well blowout. The insured is also reimbursed for actual cost and expense to redrill or recomplete the well after it is brought under control. Costs of removing or cleaning up seepage, pollution, or contamination are covered along with indemnification for sums paid for bodily injury or property damage caused directly or indirectly by seepage, pollution or contamination.

This important insurance is frequently purchased as a disaster type of coverage with some minimum and maximum limit. The minimum is usually dictated by the amount of loss which the company feels it can reasonably afford to show on its financial statements for any particular year. The maximum coverage is normally determined by the cost of the insurance in the higher ranges of exposure which the company feels it may encounter. For example, a company which engages in exploratory drilling only for shallow gas need hardly pay the premium for the same type of blowout coverage which a prudent company will carry for drilling in the Tuscaloosa formation of Louisiana or the Devonian reef country of Alberta.

Catastrophe

The oil business being the risk business that it is, even the largest of the major oil companies carry some type of blanket coverage to protect against the worst business setbacks, such as the catastrophic loss of a major facility or refinery due to fire. These are usually "all risk" types of policies which might have a million dollar deductible provision for a smaller independent, and a five to ten million dollar threshold for the largest of the majors.

Payout of Production Payment

Production payment payout coverage permits insuring that an oil payment, as described in Chapter III, will fully pay out to its owners. If the wells will not produce the volume stipulated in the production payment contract the insurance company will make up the difference. This coverage can also be written in dollars, instead of barrels, so as to protect against any decline in the price of oil.

Loss of Production

Usually an accident, which may be included in one or more of the previously described insurance coverage, will result in the loss of production for some period of time. Insurance coverage is available to reimburse the operator for this loss of production until production can again be resumed.

FINANCING AND OWNERSHIP OF THE OIL AND GAS INDUSTRY

CHAPTER VII

The primary source of financing for the petroleum industry is the sale of oil and gas from successful projects.

The ownership rights to recoverable oil and gas in the ground vary significantly in various parts of the world. In by far the greatest number of countries the petroleum resources are owned by governments. The notable exceptions are the fee lands comprising most of the U.S. parts of Canada, areas of Germany and South Africa where the oil, gas and mineral rights are the property of the fee owner of the land. This type of mineral ownership derives from the terms of the original land grants in colonial North America. The landowner, in these situations, has the right, if he chooses, to sell off or convey to others his mineral rights, or to sell the surface and retain all or part of the mineral (oil and gas) rights.

It is common practice in both the fee land and government land to have someone other than the owner himself undertake the exploration, development and production of petroleum. The exploration and production (E&P) costs may be borne or shared in a number of different ways. The historical way of handling the relationship between the mineral owner and the explorer/producer calls for the owner to retain a royalty share of any petroleum found, produced and sold or removed from the property. The royalty is normally an agreed percentage of the value of the production. It is a fundamental premise that the mineral owner does not pay any of the cost of exploration, development or production. He may take his royalty share in kind at the wellhead, or he may opt to let the operator sell the royalty share along with his working interest share and pay the appropriate portion of the revenue to the royalty owner in cash.

On public lands royalties are paid to the host government generally on the basis of a fixed percentage of the oil produced. Royalties represent the most significant part of the government's share from most concession agreements. Governments like them for a number of reasons, not the least of which is that they begin to accrue with the first barrel of production. The other forms of government share must await project payout (which can be slow particularly if the contracting oil company carries the national oil company's NOC's development costs), or the turning of a profit which will yield the government an income tax. The royalty provision thus provides an early cash flow to the host government. Royalty taken in kind by the host government provides a cost-free source of crude oil for the local market.

253

Royalties on government lands are generally in the range of 10 to 20 percent. Some crown i.e., government lands in Alberta carry royalties above 40 percent. Some U.S. government lands on the offshore continental shelf were awarded on the basis of the highest royalty bid, and recently royalty has been reduced on deep water tracts to encourage development in this high cost area.

Royalties are usually flat, or constant for the life of the production. There is a growing tendency to specify sliding scale royalties as a means of encouraging the development and production of marginal fields. These sliding scales may call for a royalty as low as five percent for low, barely economical rates of production to as much as 20 percent for large rates of production from giant fields. Norway was one of the first producing countries to apply sliding scale royalties on a national basis. Some countries have varied their royalty requirements offshore on the basis of water depth.

On the fee lands of North America the Working Interest is highly subject to sale in whole or in part with a myriad of types of oilfield deals. Indirectly these deals generally provide means of coping with the high costs of oilfield operations coupled with the deemed degree of geological and fiscal risk involved. This type of oilfield "wheeling dealing" is not as prevalent outside of North America although it is becoming more common. A recent report indicated that an offshore tract in the Netherlands North Sea had undergone some fifteen changes in ownership of the working interest.

Acquisition of exploration and production rights on fee lands usually involves leasing oil and gas rights from the land or mineral interest owner. In the case of government lands throughout the world a form of exploration and production contract is negotiated between the host government and the operating oil company or companies. The initial stipulation in these contracts is usually a signature bonus which is similar in effect to the lease purchase price in the fee land situation. Beyond the initial signature bonus or lease purchase price the terms and conditions of an E&P contract on government lands characteristically differ significantly from the fee lands conditions. Contracts for E&P operations on both fee and government lands may provide for a royalty payment, however.

The two most common types of E&P contracts between host governments and the oil companies are concessions and production sharing. The spreadsheet deriving cash flow from an Indonesian producing property included in Chapter IV was of this latter production sharing type. On a statistical basis the concession type of agreement, which has greater similarity to the fee land arrangements, is the more common worldwide. Both types of agreements are popular, however, and are discussed in detail in Chapter XI.

Most government contracts make it clear that the contracting company, vis-a-vis the fee land working interest owner, has no ownership of the petroleum in the ground. His rights to the oil and gas occur only after it reaches the surface. This distinction is primarily of political importance in the developing countries of the world.

It is not possible to detail the almost infinite types of transactions to which fee land working interest may be subjected. A large number of these, however, involve farmouts in which the working interest owner deeds a specific portion of his interest to another party while retaining an Overriding Royalty Interest, (ORRI). The term "royalty" is used here to indicate that the owner of the override does not have to pay any of the development or producing costs of the operation. The fraction of the ORRI which is retained is determined by negotiation. The principal considerations are the working interest (farmout) owner's need for funds to undertake development; or his lack of interest in the property due perhaps to discouraging drilling results in the area; and the farmee's perception of the geological risk involved.

Another frequent provision of these oilfield deals is the inclusion of a reversion clause. A reversion clause typically will provide that after certain conditions, such as the accumulation of sufficient revenue to pay for the drilling and completion of a well, have been satisfied then the division of ownership in the working interest will change to some other previously agreed split. Reversionary clauses are also frequently included in E&P contracts on government lands.

The fact that ownership in an oil and gas property may be subject to so much change during its productive life adds appreciably to the complexity of economic evaluation. Appendix VII-B of this chapter on Divisions of Interest should simplify the understanding of some of the most frequent of these transactions. The petroleum industry is one of the most capital intensive in the entire private sector. The industry's demand for funds seems almost insatiable. As a consequence the oil and gas industry is the darling of the banking and financial community.

COST OF ENTRY INTO THE OIL BUSINESS

The petroleum industry is unique in many interesting ways. One of these is the exceedingly large number of small "independent" players in the game. Mr. Peter Ellis Jones (loc. cit.) explains this phenomenon as follows:

"The cost of entry in the United States may be very small. For example, a geologist with a few years' working experience may have developed an enthusiasm for a particular prospect in which he is unable to interest his company. It would not be unusual for him to resign from his employer and set up a new oil company the following Monday, working from home or a small rented office. Having developed information on the particular lease in which he is interested he may be able to obtain an option on it or to purchase the lease very cheaply in return for a commitment to the landowner to drill on it. He may then approach, to quote a simple example, three investors each wishing for tax reasons to participate in drilling a well; he agrees with them that each will pay one third of the cost of the well to earn a 25% interest. The geologist retains a 25% interest in return for his efforts in developing the prospect and managing the operation... Alternatively, if it is a large prospect and he cannot find the money to take a position in the lease in the first place he may introduce the prospect to one or more oil companies, retaining for his efforts a small royalty interest which involves him in no expenditure but gives him a valuable and negotiable right to a small percentage of the oil produced if the well should prove successful. Most independent oil companies in the U.S. originated in this way, the geologists and promoters of the more successful bringing in outside shareholders' money when the basic value of the company has been established.

"In the U.S., Canada and many European countries, particularly Germany, the Netherlands and the U.K., entry into independent oil marketing is also easy and can be very cheap at the lowest level. For instance, it would be quite possible for someone to set up as an independent heating oil distributor by buying one or two second-hand road tankers or even leasing them, by inserting an advertisement in the Yellow Pages or local free circulation newspapers and perhaps distributing hand-bills to local households, by persuading a member of his family to take orders telephoned by customers to his home and by buying his heating oil ex-rack from the depot of a major company. If he can persuade his supplier to extend monthly credit terms to him and if he can receive payment on delivery he may be able to operate on negligible or even negative working capital...

"Operating on a shoestring such as this it is unlikely that the new independent will get rich quickly, but there are certainly many oil distributors whose initial capital was less than 20,000 pounds and which are providing their proprietors with a comfortable living. It would certainly be possible to establish a viable independent oil distributorship... for an investment less than the 300,000 pounds it would require to establish a MacDonalds or similar franchised fast food restaurant."

Whereas the upstream, i.e., exploration and production, and the product distribution phases of the oil business with their low cost of entry continue to attract small independent entrepreneurs, the refining sector lacks such appeal. A modern refinery is more analogous to other heavy industry such as a steel mill.

SOURCES OF FUNDS

Even with the tremendous need for cash to cover its massive outlays for exploration, drilling, and production facilities the oil business has relied most heavily on internal financing from retained earnings. This has been true for both the National Oil Companies (NOC's) and the private sector. In some measure this may have been due to the rather unique nature of the business, and to an incomplete understanding within the financial and banking community of the petroleum industry's very high degree of risk, most particularly with respect to exploratory drilling. This need for the large amounts of money required to undertake enormous development projects is at least part of the reason behind many recent mega mergers in the petroleum industry.

The industry does enjoy, and actually employs, an unusual variety of funding arrangements. Depending upon the size of the enterprise these may vary all the way from one-well drilling deals to multi, multi-million dollar financing deals between major oil companies and large pension funds, insurance companies and other pools of "big money" (generally in the form of co-venturer undertakings). Every working day hundreds, perhaps thousands, of farmout-farmin deals are struck by the North American petroleum industry. More and more of these types of deals are occurring outside North America as the "independents" seek out opportunities within the larger overseas land holdings of the major companies.

It is at the development drilling and facility installation stage that the NOC, or its host government, is likely to be faced with a dilemma on financing. In many contracts the NOC may opt to have its development costs carried by the foreign contracting company as a loan against future production. Where this is not possible the NOC may look to its government for the necessary funds which, in turn, may come from the World Bank. (See Appendix VII-A at the end of this chapter.)

Overrides

The sale of overriding royalties is a common and less complicated means of raising capital within the producing industry. In these transactions the owner of a working interest in a producing property sells a stipulated share in future production for an agreed compensation. This may be in dollars, in exchange for the obligation to drill on the property, retaining a fraction of the working interest in any production which may result, or occasionally the exchange of leasehold interests in other properties.

Overriding royalties do not have to bear operating expenses. They are usually expressed as a percentage of the working interest, or a fraction, e.g., a 1/8th ORR (overriding royalty). Most overrides are the result of farmout agreements rather than a sale for cash. Australia provides the exception where cash transactions seem to predominate.

Production Payments

Sales of production payments are another favorite form of financing acquisitions and drilling and production activities. A production payment transaction is a means whereby the owner or potential purchaser of a producing property can raise immediate capital for various pursuits by selling, at fair market value, a portion of his interest in a property, or group of properties. The sale consists of conveying ownership in a specified number of barrels of future production, or a fixed amount of future income (the "primary sum"), plus accruals equivalent to the interest on the unamortized amount of the primary sum, plus certain production related taxes and nonproducing expenses.

In a petroleum production payment the purchaser acquires from the transaction a stipulated percentage (75% in this example) of the hydrocarbons in, under or produced from the dedicated interest until the specified number of dollars (or cumulative production) is received by the purchasing party. Figure 7-1 illustrates a typical example of features common to all production payments. There can be a number of variations to the basic approach.

FIGURE 7-1
PRODUCTION PAYMENT

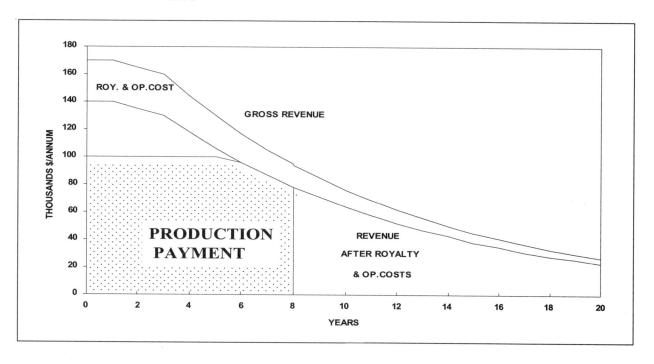

The percentage dedication is normally a constant figure, at or below a level which leaves a margin of the production which will cover royalty, operating, and perhaps capital costs. The dedication is self extinguishing upon receipt of the stipulated amount, and is payable only out of the dedicated hydrocarbons, and there is no recourse if the production from the dedicated properties should fail to produce a sufficient volume of hydrocarbons to satisfy the dedication. This method of funding is analogous to the future sale of agricultural crops such as Egypt's future sale of several years of its entire cotton crop as a means of raising money for its wars on the Sinai. As with crop insurance, surety may be purchased so that the production payment will be paid in its entirety as described in Chapter VI.

When a production payment loan is provided by a bank or banks, they may charge a fee or fees to cover their out-of-pocket expenses of the transaction. This may be for a consultants confirmation of the property's reserves, legal examination of title, or other services required to make the loan. Payment of these fees may be handled as a lump sum payment at the closing of the loan or spread out over the loan repayment period as a monthly add-on or as a higher interest rate. If it comes out of the production payment dedication, it correspondingly lengthens the payout of the obligation.

Production payments are usually sold to a nominally capitalized entity ("the grantee"), often owned by an institution such as the pension funds or insurance companies mentioned earlier in this chapter. The grantee pays an amount equal to the primary sum for the production payment, obtains a 100% bank or institutional financing of the purchase price, and mortgages the production payment to secure and liquidate the loan. As such, the funding becomes a non-recourse loan to the producing company. The grantee's incentive and reward is a share in the interest payments, or sometimes a fixed fee provided for in the non-producing expense addons to the primary sum. If the grantee is a taxable organization, the opportunity to utilize the depletion allowance without the usual geological risk can be a prime inducement for this type of transaction.

Production payment non-recourse financing may also be provided by non-financial institutions that are willing to take more risk than traditional financial institutions. Of course in exchange for taking the greater risk, these organizations will expect a greater return on their investment, i.e., the loan. Terms commonly involve the granting of a limited term overriding royalty interest in the property in exchange for the loan. The override percentage is set to allow the producer to retain sufficient operating income from the property to pay the basic royalty on production as well as normal operating costs plus a small profit for unexpected contingencies. Typically the override might be in the range of 50% to 60%. The override would remain in effect until the lender had earned a specified rate of return (IRR) on the amount of the loan. The specified rate of return may be in the range of 15% to 25%. Alternatively the lender may specify that a fixed margin in terms of $/BBL of the ORRI be dedicated to paying the rate-of-return to the lender, with the balance of the sale price being used to pay off the loan. With either method the ORRI would be terminated when the specified conditions are satisfied.

Production payments are designed to pay out before the property is depleted. In typical transactions to limit the risk to the lender, payout of principal and interest is planned to occur with production of no more than 50% of the property's reserves. Since only income from the specified property(s) is used to repay the loan and interest, there may be a penalty for prepayment of the loan out of other corporate funds, if it is allowed at all. The production payment owner has an economic interest in production and may pay a proportionate share of any production taxes as well as be entitled to depletion allowance for income tax purposes. If the property is depleted before the production payment is satisfied, both the financial and non-financial institutions will treat it as overriding royalty for tax purposes.

Farmouts and Non-Consent Provisions

A farmout is a conditional transfer of rights to explore for and develop any hydrocarbons discovered in exchange for a commitment to drill a well(s) and/or conduct other exploratory activities. Title to those rights is not perfected until the party taking the farmout fulfills the commitment made. In exchange for this commitment the party taking the farmout is entitled to a share of the profit generated by a success. Farmouts and non-consent, or stand-out provisions comprise an indirect form of funding. A party in interest, other than the original working interest owner, undertakes to fund the project in exchange for a stipulated interest in future production, if any. The initial working interest owner benefits by reducing his dry hole risk to zero. In return he must forego the upside benefits of that early portion of future production when the wells are found to be commercial. Farmouts, as well as stand-outs, are means of realigning the risks and rewards of investment among the parties.

Frequently, after the party taking the farmout has recovered their costs from a successful discovery, a joint venture is created between the parties with a predetermined division of ownership. A joint venture is a pooling of interests, with each party having an undivided interest in the operation; i.e. providing their share of costs and receiving their share of the benefits. The formation of such a joint venture is at the election of the party granting the farmout. When a joint venture is formed as part of a farmout, the original party pays no part of the cost of the committed activity, as that cost was recovered before the formation of the joint venture, but must pay their share of all costs incurred after that. During the payout period, the party granting the farmout normally receives a small overriding royalty, as was discussed in an earlier section. A further discussion of joint ventures can be found in chapter XI.

The Production Sharing Contract (PSC), which is a popular arrangement between governments and contractors in many parts of the world, is a form of farmout. Under a PSC the government farms-out its right to explore for and produce what hydrocarbons may be discovered in a designated area of their country to a contractor in exchange for a share of the profit generated by any discovery. The Government transfers the exploration risk to the contractor and also requires the contractor to finance both exploration and development of the designated tract. The contractor is allowed to recover his costs out of gross production from a successful discovery and earns a share of the profit from a successful venture.

Farmins form the opposite of farmouts depending from which side the transaction is viewed. Each deal is individually negotiated, and there is wide variation in the terms and conditions. It is fairly common in the Rocky Mountain region of the U.S. to see conditions under which a party farms in to a fairly limited amount of acreage, perhaps three or four drillsites, in return for agreeing to pay one-third of the cost of drilling and completing one well with the understanding that the party farming-in will receive one-quarter working interest in the tract after payout of the first well, if, of course, it produces. These deals are known as "third for a quarter" transactions. Comparable conditions have evolved in many other producing areas.

Independent producers have traditionally sought access via farmout to acreage held by companies with large holdings. Majors also frequently farmin, i.e., take farmouts from other operators, when they detect geological opportunities which the initial acreage holder may have overlooked. The net result is a significant increase in the number of wells drilled within a given time frame. Inevitably, this finds more oil, to the benefit of all parties concerned.

Stand-out, or non-consent penalty provisions, can also be looked upon as a means of funding oilfield projects. These provisions exist in most unit and partnership agreements. Non-consent provisions allow owners of working interest who do not wish to participate in a particular project, such as additional drilling, secondary or tertiary recovery, or gasoline plant construction to opt out of the initial investment. The participating parties then receive a stipulated 200, to as much as, 500 percent of the non-participating owners share of the resulting income from the investment before the non-consent party comes back in for his original working interest.

There are an infinite number of variations to farmout deals practiced in North America. The following two examples illustrate typical arrangements.

EXAMPLE 7-1
ALTERNATE MEANS OF FINANCING

A well has been proposed that will cost $450,000 to drill and equip for production. An economic evaluation of this well indicates that it would yield a net cash flow over its 11 year life of $350,000, which when discounted at10%would be $160,000 with a DROI of 35.6%. Because of the low profitability of this well, two alternative ways of financing the drilling of this well have been considered.

One alternative is to sell a 25% working interest in the well to generate some cash up front, plus getting another party to share in the drilling and completion cost of the well.

Another alternative to be considered would be to farm-out the drilling of the well, retain a 5% overriding royalty until payout and then back-in to a 50% working interest for the remainder of the producing life of the well. This alternative would completely eliminate the need to invest any money in the drilling and completion of the well.

Which alternative is the most desirable?

A complete analysis of the present value net cash flow of these three alternatives indicate that the farmout alternative with the back-in to a 50% joint venture yields almost the same present value net cash flow over the life of the well as would be earned by spending the full $450,000 to drill the well. It was also determined that if the 25% working interest could be sold for $40,000, the PV NCF of that case would be the same as for the other two cases evaluated. So from an economic point of view, these three alternatives would be almost identical. This performance is shown in the following graph of cumulative present value net cash versus time for these three alternatives to financing the drilling of the proposed well.

FIGURE 7-2 (FOR EXAMPLE 7-1)
ALTERNATE MEANS OF FINANCING (PV Cumulative net cash flow)

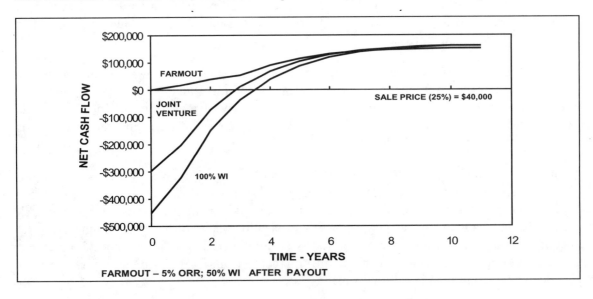

So the choice of method to finance drilling the proposed well will not be determined solely upon economics, but upon external factors, such as risk, availability of funds, and company resources.

EXAMPLE 7-2
COMPUTATION OF OUTSIDE SHARE — IN A DRILLING DEAL

Mr. R. A. Smith owns the surface and the minerals under two sections of land (1280 acres) in Hill County, Texas. Smith agreed to lease his oil and gas minerals to John Able of Midland for a period of five years or as long as commercial production may continue. The deal included a cash bonus of $20 per acre with a 3/16ths (18.75%) royalty, plus a $2.00 per acre annual delay rental. The lease was dated February 15, 200X.

After obtaining the lease, Able proceeded to evaluate the acreage as a drilling prospect and to raise necessary funds. The local crude oil purchaser's posting for Strawn production was $15 per barrel at the wellhead. During 200Y after an analysis of a geological and geophysical report costing $50,000 and paying delay rentals of $2,560 on February 1, 200Y, Able decided to drill a 10,000 ft. test well estimated to cost $750,000 to the casing point. This investment was too much for Able to handle alone, and so the company conveyed a 25% working interest to Baker Oil Company of Corpus Christi, Texas, and Charlie Drilling Company of San Antonio for $250,000 each.

This provided a cash receivable to Able of $500,000 a major part of which accrued as Able's share to Charlie Drilling, who agreed to turnkey the well to the casing point for $600,000. Before making assignment to Baker and Charlie, Able conveyed a 1/16th (6.25%) overriding royalty to his favorite charity, the Hill County Memorial Hospital. The participants then proceeded to drill the initial test well, the Smith #1. The following table summarizes the resulting ownership interest.

OWNERSHIP INTERESTS

Party	Working Interest	Landowner Royalty	Overriding Royalty	WI Net Revenue Interest	Total Revenue Interest
Landowner	0%	18.75%	0%	0%	18.75%
Hill County Hospital	0%	0%	6.25%	0%	6.25%
Able	50%	0%	0%	37.50%	37.50%
Baker	25%	0%	0%	18.75%	18.75%
Charlie	25%	0%	0%	18.75%	18.75%
	100%	18.75%	6.25%	75.00%	100.00%

Able was designated "operator" and thus assigned overall responsibility for conducting the drilling, testing and the management of the property. The other two working interest owners, Baker and Charlie, executed the Operating Agreement, which recorded Able's rights and obligations as operator. The landowner, Smith, and the Hill County Memorial Hospital do not have working interest and thus do not become parties to the operating agreement. Neither, of course, do they have any direct voice in the conduct of the drilling and production operations. All parties including the "Outside Share" owners, i.e., the hospital and the landowner, became parties to the Stipulation of Interest document, and eventually to the Division Order for the property.

Smith #1 was drilled to the authorized TD (total depth) and was found to be productive from the objective Strawn Sand. Casing was set on bottom and the well completed through 20 ft. of perforations with 4 shots per foot and fraced with 25,000 pounds of 20-40 sand in gelled water and swabbed back. Tubing and pump were run for a total completion cost of $350,000. The completed well potentialed at a daily rate of 56 barrels of 38.9 API gravity oil, 80 mcf of gas and 28 barrels of brackish water. Pumping unit, gun barrel, lease stock tanks, separator and lease flow lines were installed at a cost of $110,000, and the well placed on production. The posted price of the produced crude oil is $15.00 per barrel. There is no pipeline connection, and so production is trucked at a cost of 25 cents per barrel. The gas is flared due to lack of a pipeline connection. Costs to operate the pumping unit, dispose of the salt water, and the payroll for a contract pumper to handle the daily operation is $2,500 per month including his small tools and supplies.

The initial investment allocation and the first year's revenues and operating costs are calculated as follows:

Total Initial Investment = ($25,600 + $50,000) G&G + $2,560 delay rentals + $600,000 drilling + $350,000 compl. + $110,000 surface facilities = $1,138,160

Total Annual Operating Costs = $2,500/mo. 12 = $30,000

Total Annual Revenue = 56 BOPD * 365 $14.75/bbl. = $301,490

INVESTMENT AND OPERATING COSTS

Party	Working Interest	Initial Well Investment	Operating Cost/Yr.
Landowner	0	0	0
Hill County Hospital	0	0	0
Able	50%	$608,160*	$15,000
Baker	25%	265,000	7,500
Charlie	25%	265,000	7,500
	100%	$1,138,160	$30,000

*Includes Able's lease and G&G costs prior to farmout

DIVISION OF REVENUE

Party	Total Revenue Interest	Total Revenue First Year of Production	Net Cash Flow First Year of Production
Landowner	18.75%	$ 56,529	$ 56,529
Hill County Hospital	6.25%	18,843	18,843
Able	37.50%	113,059	98,059
Baker	18.75%	56,529	49,029
Charlie	18.75%	56,529	49,029
	100.00%	$301,490	$271,490

Thus, on a before tax basis, the promoter, Able, after selling off one half of his working interest for $500,000 and giving a 1/16th of 8/8ths ORR (overriding royalty) to the Hill County Hospital will require 6.2 years to pay off his outlay of $608,160 assuming constant production and oil price. Baker and Charlie should each pay out their $250,000 entrance fee, plus their one quarter shares of the drilling and completion costs, about six months earlier, if production and price hold at the initial rates.

All the foregoing ignores the outside share, "government take" consisting of ad valorem, income and severance taxes to the county, state and federal governments, as due. These must also be paid by each individual party to the stipulation agreement as assessed by the various jurisdictions.

Unfortunately, most exploratory drilling ventures result in dry holes. If Smith #1 were drilled and logged to the casing point, then plugged and abandoned as a dry hole the out of pocket cash flow for the participants, Able, Baker and Charlie, might appear as follows:

POSSIBLE OUTCOME OF SMITH NO. 1 AS A DRY HOLE

	Bonus	Delay Rental	G&G	Drilling	WI Purchase	Net
Landowner	$25,600	$2,560	0	0	0	$28,160
Hill County Hospital	0	0	0	0	0	0
Able	-$25,600	-$2,560	-$50,000	-$300,000	+$500,000	+$121,840
Baker	0	0	0	-$150,000	-$250,000	-$400,000
Charlie	0	0	0	-$150,000	-$250,000	-$400,000
	0	0	-$50,000	-$600,000	0	-$650,000

Two important comments:

1) Charlie Drilling's situation requires further analysis since he would receive the $600,000 turnkey drilling money. Out of this he, of course, would have labor expenses for his drilling crew, as well as the fuel and maintenance costs for the rig.

2) Able is shown to have pocketed $121,840 even for a dry hole. This is by no means an unusual occurrence in drilling deal promotions of this type. It all depends upon the terms of the transactions, which are highly dependent upon Able's ability as a salesman, and Baker and Charlie's subjective assessment of the probability of Smith No. 1 being a commercial discovery well.

Bank Financing

The major banking institutions of the world are an important source of funds to the oil and gas industry. A number of the largest banks in North America make a specialty of oil and gas loans. Bank loans tend to be more flexible, but are generally more costly than other forms of financing. Most companies consequently rely on the banking institutions for short term borrowings. A frequent first use of new long term debt in the industry is to pay off the more expensive short term bank loans that haven't been amortized out of current earnings.

Unsecured short term bank loans by definition are payable in one year or less. Such loans are typically of several different types:

Line of Credit: This is usually an informal agreement between the bank and its customer. The specified amount is intended as a ceiling of the credit available at any one time, and is based on the bank's assessment of the credit worthiness of the borrower. The line of credit, for which a small fee is charged, is subject to adjustment or renewal as the conditions warrant. Business practice not withstanding—there is no legal obligation on the part of the bank to actually extend the promised credit. This form of financing is analogous to the bank credit cards that many individuals carry.

Revolving Credit Agreement: This involves an actual legal commitment on the part of the bank. A fee, normally a fraction of a percent, is usually charged on the unused portion of the total credit. The portion which is drawn down is, of course, also subject to the agreed interest charges. A revolving credit agreement may extend beyond one year, thus technically disqualifying it as a short term transaction.

Transaction Loan: This type of loan is undertaken when a firm has need of funds for a specific purpose or project. An example might be the company's participation in the construction of a new gasoline plant. The principal determinant of the loan and its terms is the bank's evaluation of the cash flows which the project is expected to generate.

Dedicated COI Payment: Dedicated cash operating income (COI) payment loans are the most common type of borrowing, particularly to the independent sector of the oil and gas industry. This type of arrangement is also known as a production payments interest (PPI). The dedicated COI payment loan is repaid from a "dedicated" portion of the COI. Since income from a producing property varies on a monthly basis, the total periodic payment also varies, and there is no set time period under which the loan must be repaid. Normally, the dedicated portion is a percentage of the COI, after a retainer has been taken out. Interest is paid first from the dedicated monies, then the remaining funds are used to repay the principal. If the dedicated monies are not sufficient to pay the interest charges, then the remaining amount of interest is compounded.

Leasing as a Means of Financing *Fees are tax deductable*

A variety of equipment is necessary for the normal operation of oil and gas properties. This equipment provides a specific function. Most of it is permanently installed at a given location and remains there until either it is worn out, becomes obsolete, or no longer needed. Normally in the oil industry this equipment is purchased and owned by the operator of the property. However, ownership of the equipment by the operator is not necessary, because the operator is only interested in acquiring use of that equipment to provide a given function for a specific period of time. If use of the equipment for a specific period of time can be acquired by means other than outright purchase at a lesser cost, then alternatives to purchase should be considered.

Equipment leasing, to obtain the use of an asset without purchasing it, should be considered as a financing method. Leasing replaces the need for corporate funds up front by committing to pay specified rental payments over a designated period of time. A small amount of up-front capital for installation of the equipment may be required, but this is only a fraction of the capital that would be required for outright purchase. Not only is the capital cost spread over time, but the rental payments are generally considered operating expenses rather than capital.

The equipment lessor may choose to finance the equipment rather than pay for it out of their funds, and it is not unlikely that part of the collateral considered by the lender will be the signed equipment lease contract. Therefore, the lessor may be using the credit rating of the lessee to obtain a more favorable loan. Leasing is usually a more expensive means of financing than outright borrowing for general corporate purposes.

Leasing may be desirable for other reasons: (1) it is usually considered an off-balance sheet item, unless it represents a significant commitment for the company, (2) enables timely replacement if equipment becomes obsolete, and (3) enables short term use of equipment without concern for its disposal when it is no longer needed. There may be situations where a required special service can only be obtained through leasing the equipment or perhaps the equipment just isn't for sale. There also may be tax advantages to leasing as well as political considerations that should be included in arriving at a decision as to whether to lease rather than purchase equipment.

There are two general types of leases. One is called a "payout lease" or "capital lease," in which the total rental payments equal or exceed the total cost of the equipment. The total rental payments for the other type of lease are less than the cost of the equipment, so they are called "non-payout leases" or "operating leases." The type of lease will be determined by the length of the lease and periodic payments required. The payout lease may also include terms for purchase of the equipment after payout or a reduction in payments at that time. In any event, it is important for the lessor to know what type of lease is being executed, to be assured of the most favorable terms.

The evaluation of leasing versus purchasing equipment requires the determination of the total cost of both alternatives and selection of the alternative which has the lowest cost over the period of time that the equipment is needed. The most difficult part of the evaluation will be determining the total cost of each alternative. Leasing costs include all periodic payments required plus any initial fees. The cost must also include any maintenance and service cost not provided by the lessor as well as insurance and other costs which will be incurred by the lessee as a result of using the equipment. The purchase evaluation should include the normal initial capital expenditure, operating and maintenance costs plus a salvage value for the equipment at the end of its service life. The service life of the purchased equipment should coincide with the term of the lease to provide a valid basis for comparison.

Comparison of lease and purchase alternatives is normally done on the basis of net present value of total cost over the service life of the equipment, discounted at the appropriate cost of capital. The alternative selected will be the one which enables the use of the equipment at the lowest cost.

The following simplified example may clarify some of the factors involved in choosing between alternatives of purchasing or leasing equipment. For such an evaluation to be valid the purchased and leased equipment must provide the same service, although they need not be identical, over the same period of time.

EXAMPLE 7-3
LEASE vs PURCHASE

A special pumping unit is required for a well, but it is only forecast to be needed for five years. The cost to purchase that pump is $100,000 and annual operating and maintenance is forecast to be $5,000 per year. It is estimated that at the end of five years the pump will have a salvage value of $60,000. The same pump can be leased for $13,000 per year and the owner will provide the maintenance that is estimated to run $1,000 per year. The lessor also requires an initial fee of $10,000. Using a cost of capital of 10%, determine whether the pumping unit should be purchased or leased.

	PURCHASE EQUIPMENT			LEASE EQUIPMENT			
Year	Investment	Operating & Maint. Cost	Mid-Year Present Value @ 10%	Initial Fee	Rental Payments	Operating Cost	Mid-Year Present Value @ 10%
0	$100,000		$100,000	$10,000			$10,000
1		$5,000	$4,767		$13,000	$4,000	$16,209
2		$5,000	$4,334		$13,000	$4,000	$14,735
3		$5,000	$3,940		$13,000	$4,000	$13,396
4		$5,000	$3,582		$13,000	$4,000	$12,178
5		$5,000	$3,256		$13,000	$4,000	$11,071
5	($60,000)		($37,255)*				
	PRESENT VALUE COST		$82,624				$77,589

*$60,000 salvage value taken at end of year 5

The obvious decision for the previous example would be to lease rather than buy, because the total present value of that alternative should yield the lower cost. However, this evaluation was made on a before tax basis and the income tax consequences; i.e., depreciation of tangible capital and tax credits for expenditures, should be included in such an evaluation. For this particular example the tax effects of both alternatives were about equal. This will not always be the case, which in a large part may be a function of the individual tax payers position. So an after tax analysis may be preferable.

Project Financing

Limited recourse funding (project financing) of specific projects, such as the construction of a natural gasoline plant, has been available through the commercial banking industry in the U.S. for some time. Limited recourse financing has been used successfully overseas for a number of years in the mining industry. The first major application of limited recourse project financing in the petroleum industry was in 1972 for the development of the giant Forties field in the U.K. sector of the North Sea. This is described in Figure 7-3.A project financing loan must be made to a legal entity. This may be a company, a person, or partnership with no assets except for those related directly to the project itself. In the case of Forties a "paper" company known as NOREX was established for the purpose.

FASB 47 (March 1981) sets out the project finance concept:

FIGURE 7-3
FORTIES FIELD – PROJECT FINANCING

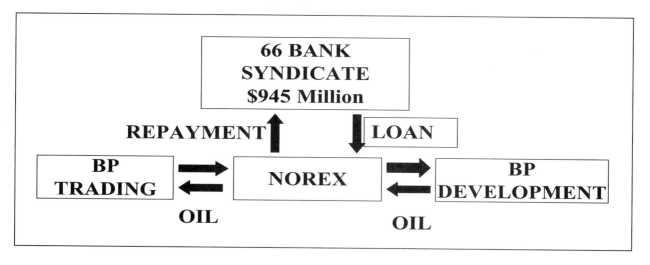

1. Banks accepted the risk that the oil reserves might be deficient.

2. Banks were protected, however, on three other risks:

   A. Possible failure to complete the project.

      BP guaranteed that BP Development would carry out the plan.

   B. Possibility of early depletion.

      BP guaranteed that any difference between market value of the crude oil produced and amount needed to service the loan which would be paid into a reclaim account which would be drawn upon if production declined.

C. Force Majeure

      If it were impossible to produce any oil before 1978, and if a 44 percent recovery factor would have been sufficient to repay the loan, then NOREX could draw from a "ratification account" guaranteed by BP.

--The banks had no recourse against BP if the reserves were insufficient.

--The loan did not show on BP's balance sheet. BP Trading's promise to repurchase the oil from NOREX was recorded as a deferred liability against future oil delivery.

The financing of a major capital project in which the lender looks principally to the cash flows and earnings of the project as the source of funds for repayment and to the assets of the project as collateral for the loan. The general credit of the project entity is usually not a significant factor, either because the entity is a corporation without other assets or because the financing is without direct recourse to the owner(s) of the equity.

The appeal of this type of financing to the operator in reducing or avoiding the impact on his balance sheet is several fold:

1. Maintains flexibility for future borrowing requirements.

2. Avoids committal of a disproportionate amount of borrowing capacity to a single project.

3. Reduces effects upon the borrower's credit rating.

4. Facilitates compliance with existing loan covenants.

5. Isolates the security, and avoids possible impact on the operating company of cross-default provisions.

The security for these project financing loans is the future cash flow of the project rather than the borrower's commitment to repay from his own balance sheet, if need be. Frequently, however, the banks may require some financial recourse to the operator during the initial stages of the project to ensure that the construction work is completed, any cost overruns are paid, and production is commenced and built up to some agreed initial target.

It is often a major challenge for NOC's to establish the criteria necessary to convince commercial lenders to provide non-, or limited-recourse, financing for oilfield development. Although the World Bank is precluded from lending to other than sovereign governments, they have been able to participate in some carefully segmented co-financed projects while taking on certain of the high risk aspects such as the foreign exchange guarantees for the project.

Oilfield Equipment Export Purchase Loans

Many countries, particularly the so called Newly Industrialized Countries (NIC's) extend cheap credit to their private sector industries for the manufacture and export of goods and services as a means of encouraging local industry. This type of financing is quite universally available to NOC's for those portions of a development project.

Bonds and Debentures

Bonds and debentures are in effect formal IOU's (I owe you) in which the borrower or "issuer" commits to repay the total amount borrowed on a fixed date. The issuer also agrees to pay interest on the debt at a fixed rate, usually in semi-annual payments. In the language of the financial community the period of time that the money is held by the borrower is called the "term." The interest rate is called the "coupon" even though it is no longer the custom to actually attach coupons to the certificates which would be clipped off and cashed as the means of collecting the interest due. The coupon rate varies with the degree of financial risk which the borrowing carries, and with the general level of interest rates, at the time the bond is first issued. The total amount to be repaid on the maturity date is referred to as the "face value," "par value," or "principal amount."

The bond certificate spells out all of the stipulations listed above. The certificate may also indicate what items of collateral such as certain leases or equipment are pledged as security for the indebtedness. Most bonds issued by operators in the oil industry offer only the "full faith and credit" of the issuing company. These obligations are known under the generic term of "debentures." Technically, government bonds fall in this category.

Bonds are issued in various denominations, usually multiples of $1000. As negotiable instruments, most bonds are bought and sold many times before maturity. Bonds can be bought and sold on the country's security exchanges or over-the-counter. Their market prices fluctuate both with supply and demand and with changes in the prime interest rates. When a bond trades in the open market in North America its price is quoted as a percent of its par, or face value. Accordingly, a bond selling for its par value of $1000 would be quoted in the financial press as 100. If the bond is selling at a discount, at say, $980 it would appear as 98. A bond trading at a premium and quoted at 104 7/8 means an actual price of $1048.75.

New bonds are issued at rates reflecting the general state of the economy as well as the prospects of the issuing company. As interest rates change such as the prime lending rate of the major banks, then the investor requires that the return, called "yield," from the bonds already trading on the open market must also adjust in order to remain competitive for the investor's dollar. Since the annual interest which the bond will pay throughout its term is fixed, the adjustment in yield is made through the bond's actual trading price.

A bond's current yield is found by dividing its coupon (annual interest) rate by its trading price. For a bond bought at par the current yield is its coupon rate. If the bond is purchased at any other price, its yield will differ from the coupon rate. The price of a bond and its yield always move in opposite directions, as illustrated below for a $1,000 bond paying $100 annual interest:

$$\text{Bond price at par: } \frac{\$100}{\$1,000} = 10.00\%$$

$$\text{As bond price rises its yield declines: } \frac{\$100}{\$1,100} = 9.09\%$$

$$\text{When the bond price declines, its yield rises: } \frac{\$100}{\$900} = 11.11\%$$

Bond prices do not generally swing as widely in price as do the "common" or "ordinary (U.K.)" stocks.

In addition to considerations of "current yield," the investor is interested in "yield to maturity." Here again, at par the bond's yield to maturity is its coupon rate. If it is purchased at a premium it must be recognized that a portion of that premium must be amortized as a deduction from the coupon rate each year to its maturity in determining its true yield. A bond purchased at a discount, of course, has the opposite and more favorable effect. These calculations are important in pricing new issues, since the potential investor will not be interested in buying the new bonds if older ones from the same company are available in the market at more favorable terms. The yield to maturity may be determined by calculating the internal rate of return (IRR) of the bond.

Bond evaluation calls for the summation of the present values of the principal (par value) and the annual interest receipts (annuity value), using semi-annual discounting. Altering Equations 5.6 and 5.9:

$$\text{PV of principal} = \frac{v_f}{[1+(i/2)]^n} \tag{7.1}$$

$$\text{PV of annuity portion} = I_s * \left[\frac{1-[1+(i/2)]^{-n}}{(i/2)} \right] \tag{7.2}$$

Combining Equations 7.1 and 7.2 yields the following equation for determining the current value of a bond as a function of the current interest rate, the coupon interest rate on the bond, and its face value.

$$\text{Current bond value} = v_f\{[1+(i/2)]^{-n}[1-(i_c/i)]+(i_c/i)\} \tag{7.3}$$

Where:

   v_f = face, or par value of the bond, \$, usually \$1,000

   i = annual interest rate, decimal (current yield)

   n = number of semi-annual interest payments remaining to maturity

   I_s = semi-annual interest payment, \$ (coupon value) = $v_f(i_c/2)$

   i_c = coupon interest rate, decimal

These computations are relatively simple. An example of bond pricing is given in Example 7-4.

EXAMPLE 7-4
ILLUSTRATION OF BOND PRICING

Assume an existing bond with an 8% coupon with interest paid semiannually. The prevailing interest rate is ten percent per year and the bond matures in ten years from now.

FIGURE 7-4 (FOR EXAMPLE 7-4)
CASH FLOW DIAGRAM

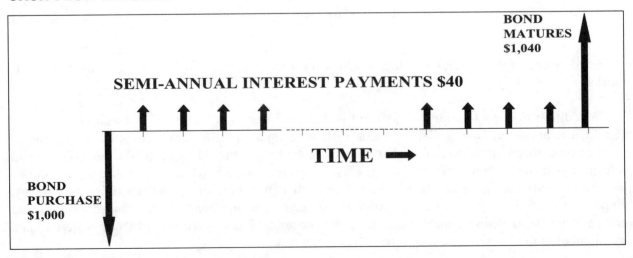

DETERMINE

The value of the bond 1) if the prevailing interest rate rises to 10%,, 2) if the prevailing interest rate returns to 8%, and 3) if it should drop to 6% per year.

SOLUTION:

(1) If the prevailing interest rises to 10%

^{V}Bond = [PV of interest stream annuity] +
[PV of $1,000 returned in 10 years]
i.e., 20 semi-annual payments

Using Equations 5.6 and 5.9 from Chapter V.

^{V}Bond = 40 PVFIA.05,20 + 1,000 PVIF.05,20
= 40 * 12.4622 + 1,000 * 0.3769
= $875.39

(2) If the prevailing interest rate drops back to 8%:

^{V}Bond = (40 PVIFA$_{.04,20}$) + (1,000 PVIF$_{.04,20}$)
= 40 * 13.5903 + 1,000 * 0.4564
= $1,000

(3) If the prevailing interest rate falls further to 6%:
^{V}Bond = 40 PVIFA$_{.03,20}$ + 1,000 PVIF.03,20
= 40 * 14.8775 + 1,000 * 0.5537
= $1,148.80

Still another factor in evaluating bonds and debentures lies in the possible privilege of the issuer, as stated on the face of the certificate, to "call" the bond, i.e., to redeem it ahead of its maturity date. Most bonds are not subject to call, however, until a specified number of years have passed. The call option gives the issuer an important escape mechanism whereby if interest rates should decline substantially after the new issue the borrower can call them in and presumably replace them with another issue of lesser coupon. This can sometimes mean a multi-million dollar saving to the issuer. From the investor's stand point an early call, particularly if he paid a premium to purchase the bond, can result in a substantial loss when his only return is the face value of the security. If there is reason to believe that a company intends to exercise its option to call a bond, this may actually limit the premium price of a bond. Its market value would be no more than the present value of interest payments to the first call date plus the present value of the call price, all discounted at the prevailing "yield to maturity." Thus at the call date, the market price and call price would be the same.

Oil Company Mergers

It is not uncommon to merge two or more oil companies as a means of acquiring the cash assets or healthy cash flow of the acquired company by a cash-short but perhaps more aggressive firm. In this

271

sense "merger" may be viewed as a means of funding. Such activities may be very sensitive to dilution of shareholder's earnings and control in one or the other firm.

Drilling Funds

A drilling fund is a collection of private investors brought together by a promoter to funnel their investments through an operator into exploration for oil and gas. These funds normally operate as partnerships with the investors as limited partners with the general, or managing partner, receiving a fee for his services in managing the fund.

Production Funds, Development Drilling Funds, and Well Completion Funds are variations on the same theme designed to offer the investor progressively lesser degrees of risk.

OIL COMPANY CAPITAL STRUCTURE IN THE PRIVATE SECTOR

Capital Structure refers to the combination of securities which a company employs in the private sector to raise and maintain the capital assets required to conduct its business. There are a wide assortment of securities available within the two major financial categories: debt and equity.

Debt may take the form of mortgage bonds, debentures, or long term notes. Debt carries an obligation by the company to repay the principal by a specified date plus a fixed rate of interest paid normally every six months. Default on these obligations can place the company in bankruptcy. Unlike stockholders, who own the company, bondholders do not share in its profits and losses. If the company is dissolved, debts must be paid before the shareholders receive any of the company's assets. Thus bonds are better protected than equity and are referred to as "senior securities."

Equity takes two basic forms of company stock: common and preferred. Common or ordinary stock represents the basic ownership of the company. Holders of common stock elect the Board of Directors which runs the company. Equity owners share the company's profit or loss after the obligations to all of the other types of investors have been satisfied.

Equity capital in the form of common shares may be raised either by private subscription or placement, or through a public offering employing a financial institution or syndicate as underwriters. This latter route usually requires prior registration with a governmental securities exchange commission having jurisdiction in the particular time and place of the stock offering.

Preferred stock is a form of equity. Preferred dividends come ahead of any payment to the holders of the common stock of the company. Preferred stock has sometimes been characterized as "a bond with a tax disadvantage" since the cost to the company of the dividends is not deductible for tax purposes. Preferred stock is generally more expensive than bonds or debentures in terms of the costs of raising new capital, but less costly than common equity.

Any default in paying preferred stock dividends does not represent a situation which can lead to bankruptcy. Failure to pay dividends on preferred stock may, however, sometimes permit the holders of this type of equity to acquire certain rights which may give them equal voice in the management of the company along side the directors elected by its common shareholders.

Typical Capital Structure of a private sector company is shown in the pie chart of Figure 7-5:

FIGURE 7-5
OIL COMPANY CAPITAL STRUCTURE

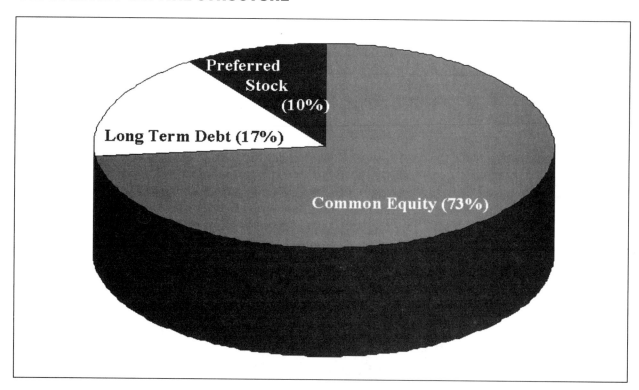

In other words, capital structure which is detailed on its balance sheet includes the company's equity plus all of its long term obligations. Long-term obligations are those items of debt, e.g., bonds and debentures with maturity of one year or more. Obligations of less than a year are included in the balance sheet as Current Liabilities.

The capital structure for a typical oil company consisting principally of debentures, rated AA by Moodys, and common stock might be as shown in Table 7-1:

Both capital surplus and earned surplus are included as part of the equity of the common shareholders. Since the company is owned by the common shareholders, the firm's earnings also belong to them. When earnings are retained in the business rather than paid out as dividends they constitute an addition to the company's equity just as though new money had been raised by the sale of additional shares, although no new, or added voting rights go with it.

TABLE 7-1
TYPICAL OIL COMPANY CAPITAL STRUCTURE

DEBT		
Debenture 12.5%'s due 2020 (Moodys AA)	$48,000,000	
Long Term Notes	1,150,000	$ 49,150,000
COMMON EQUITY		
Capital Stock ($15 par value) 4,900,000 shares	$ 73,500,000	
Capital Surplus	37,437,000	
Earned Surplus	135,624,000	$ 246,561,000
Total Capitalization		$ 295,711,000
Debt/Equity Ratio = 20%		

The degree to which a business firm is funded by loans, or debt, as distinct from shareholder's equity is known as "leverage" in the U.S., and "gearing" in the U.K. Leverage is usually expressed as the "debt equity ratio" in percent. In the illustration in Table 7-1 the ratio is 20 percent. There is considerable variation in the typical capital structures of companies engaged in the several sectors of the oil business. Figure 7-6 shows some examples.

FIGURE 7-6
REPRESENTATIVE CAPITAL STRUCTURES
(Various Segments of the Petroleum Industry)

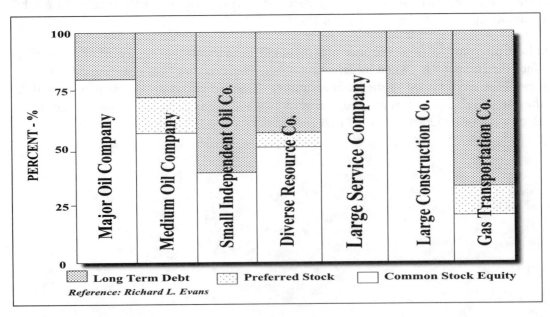

Reference: Richard L. Evans

274

The debt equity ratio for companies in the petroleum industry is directly related to each company's participation in its various segments. Figure 7-7 shows typical debt equity ratios for investments in various parts of the industry from exploration to refining and distribution. It is not surprising that exploration in the upstream sector is almost totally financed by equity or internally generated funds, as this is the most risky part of our business. The downstream sector, being the least risky part of the industry, is financed by a significant amount of long term debt.

FIGURE 7-7
TYPICAL DEBT / EQUITY RATIOS
for oil and gas projects

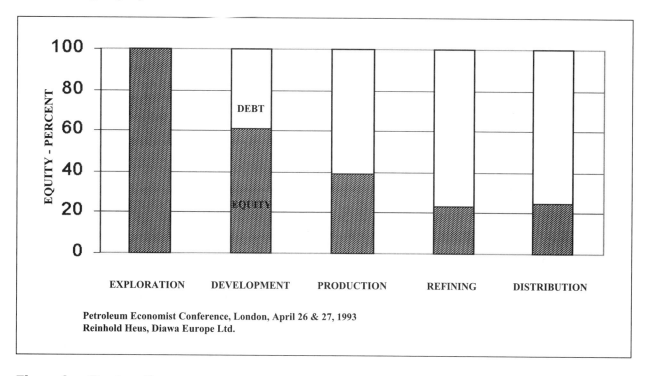

Petroleum Economist Conference, London, April 26 & 27, 1993
Reinhold Heus, Diawa Europe Ltd.

Financing Exploration

The funding of oil exploration with its high degree of geological and economic risk occupies a unique place in corporate capital structure. The level of exploration risk can vary considerably from company to company. Most major companies undertake a mix of high risk and low risk exploration. This is generally possible because a portion of their exploration activity follows successful endeavors and may border on stepout extensions.

Smaller independents may choose to concentrate their available funds "elephant hunting" in high risk-high reward areas, or to focus on low risk bread-and-butter plays in established producing areas. These latter areas generally afford early cash flow opportunities, if not large ultimate reserves, due to the normal proximity to transportation. This group typically likes to brag that their geophysicist has a special, unique and secret interpretive technique which has led to the company's higher than average drilling success rate. There does seem to be a high order of commonality nonetheless.

The capital structure of the high reward, but consequent delayed cash flow, group of independent oilmen differs significantly from the bread-and-butter group. A third factor affecting the capital structure is the degree of diversification in terms of the number of wells drilled; the capital commitment per well; and the geological and geographical spread of the company's exploration activity.

The company's risk profile is largely determined by its exploration philosophy and diversification, and its relevant reliance on cash flow to fund its program as well as service its debt. In the early 1980's with easy bank credit and ever increasing oil prices there was a tendency among the independents to increase their debt load, an unfortunate portion of which went into exploration. The number of active independents in North America has declined in recent years due to being over-leveraged and the consequent inability to fund both debt servicing and ongoing exploration commitments out of current cash flow in the face of declining oil prices.

COST OF CAPITAL

Estimating a company's private sector cost of capital is important due to its relationship to investment decisions. The money or capital required to establish, build and operate a business does not come "for free." This is true whether one considers its capital stock, its bonded indebtedness, or its short-term bank borrowings. This must obviously be paid for by the company as part of the cost of doing business. The specific means by which this cost of capital is computed
 is an area of serious controversy among financial experts. Capital costs in the public sector are essentially undefinable.

A large part of the difficulty in determining the cost of capital stems from two basic conditions: the connotations of business risk, or uncertainty, and the variety of arrangements that are, or can be, developed in the financial markets of the world. Uncertainty, (covered in Chapter VI), for the current example can be assumed to be confined to business risk. This pertains to
variations in the future returns from various investment opportunities.

Institutional practices in estimating the cost of capital which vary all the way from assuming pure equity, to assuming pure debt are not unusual. A portion of the company's capital structure, which may be significant, is comprised of the firm's retained earnings. Another source is the non-cash flow fromDD&A.Since both of these internal sources comprise a return from previous investment they are usually not included in the firm's current cost of capital calculations. The real world of finance is full of a myriad of borderline cases between debt and equity such as variable income bonds and preferred stock, although equity, long term debt, and short term bank financing are the principal means of finance for the private sector. For illustrative purposes we have been considering a capital structure which is reasonably representative of the forms with which the petroleum industry is accustomed to dealing.

The cost of capital is the standard against which prospective investments are compared. Most NOC's resort to using a hurdle rate based on a general average, or consensus, from the private sector. The criterion for accepting or rejecting a capital expenditure, as described in Chapter V, calls for approving the investment if, either:

1. The net present value of future cash is greater than zero, or

2. The internal rate of return, as a percent is greater than the firm's cost of capital.

These decision rules lead to one definition of the cost of capital which states that: "A company's cost of capital is the rate, expressed as a percentage, which must be earned in order to satisfy the combined required rates of return of the firm's investors." As a corollary then, the cost of capital becomes the minimum rate of return which the investors will accept.

The situation is not as simple as this definition implies. When a public company issues securities, the entire amount paid by the investors does not flow through to the issuing firm. Instead, the investment bankers who initially sell the securities generally receive their compensation by retaining a previously agreed upon portion of the receipts from sale of the securities and the issuing firm receives the balance.

U.S. investment bankers who underwrite stock in an Initial Public Offering (IPO) routinely charge a fee of about 7%. On amounts over $80 million or so, the fee may be negotiated slightly lower. When public companies return to sell more stock, investment banks charge about 5%. These costs must be recognized as part of the cost of capital. Thus, if the banker keeps 7% of the initial sales and the stock sells for one hundred dollars a share, the issuing corporation would net only ninety-three dollars. If the investor purchases the stock expecting to receive a ten percent return on his money, or ten dollars on the one hundred dollar investment, the company must earn 10.75% (ten dollars divided by ninety three dollars) on its equity in order to keep the investor from disposing of his ownership and taking his money elsewhere.

An issuing company is typically involved with several different types of investors such as its debenture holders and preferred stockholders, as well as the owners of its common stock. The cost of each of these types of capital will vary due to their fundamental nature. It is usual to attempt a weighted average cost of capital for the corporation. This cost of capital value is often used for a number of purposes including derivation of the firm's hurdle rate as described in Chapter V. The weighted average cost of capital is normally computed on an after tax basis.

The weighted average cost of capital can be calculated as follows:

$$C_{RW} = \frac{C_{RD}\ (\ 1-T_R\ )\ C_D + C_{RE}\ C_E + C_{RP}\ C_P}{C_D + C_E + C_P} \tag{7.4}$$

Where:

T_R = taxrate

C_{RW} = weighted average cost of capital

C_{RD} = cost of debt, %

C_{RE} = cost of common stock equity, %

C_{RP} = cost of preferred stock, %

C_D = market value of debt

C_E = market value of co E mmon stock equity

C_P = market value of preferred P stock

It should be noted that a common name for the term C_{RW} of Equation 7.4 is WACC or Weighted Average Cost of Capital. It is understood that WACC is an after income tax number, which is why it is necessary to include a term in that equation for the income tax rate. Generally interest paid is considered a normal business expense and is deducted along with the other costs before calculating that portion of a company's profit that is subject to income tax (taxable income). Therefore, by multiplying the annual interest cost in Equation 7.4 by $(1-T_R)$, that term becomes the effective after tax cost of debt. On the other hand, dividends on both common and preferred stock are paid out of the company profits after the payment of income tax, so those costs have already been adjusted for the payment of income tax.

Securities are of little value without earnings on the investment involved. Let's look for a moment at the earnings which the investors will probably require to keep their money in the particular enterprise. The financial experts for the firm after "testing the market" may have determined the following as minimum acceptable rates of return to the investing public:

TABLE 7-2
TYPICAL COST OF VARIOUS SOURCES OF CAPITAL

	Probable Return to Investor	Maintenance Cost to Company before tax	Effective Cost to Company after tax (50% rate)
Debt Capital	8%	3%	5.5%
Preferred Stock	7%	1%	7.5%
Common Stock	11%	2%	12%

A good part of the difference in relative costs to the company of debt and equity stems from the fact that interest payments and support costs are tax deductible expenses for the company, while dividends are not. Therefore, debt is normally much cheaper. The 11% probable return to the investor is based upon the fact that the average return to investors of U.S. stock funds over the past 35 years has been 10.84%. The return to stockholders is the composite of dividends plus the long term increase in value. A significant portion of this return is through capital gains.

EXAMPLE 7-5
CALCULATION OF WEIGHTED AVERAGE COST OF CAPITAL

If it is assumed that the pie chart of Figure 7-5 represents the capital structure of a company that has common stock, preferred stock, and bonds outstanding, and the costs of each class of capital is as presented in Table 7-2, then the Weighted Average Cost of Capital for that company could be calculated using Equation 7.4. The following table is the solution to Equation 7.4 which indicates that the WACC or hurdle rate for the company should be about 10.5%.

Source of Capital		After Tax Cost		Fraction Total Capital		Weighted Cost
Long Term Debt	=	5.5%	x	0.17	=	0.935%
Preferred Stock	=	7.5%	x	0.10	=	0.750%
Common Stock	=	12.0%	x	0.73	=	8.760%
				WACC	=	10.445%

Whether most corporations would rather raise new capital by issuing additional debt or new equity depends upon a number of factors and will vary from time to time. The most critical factor is probably the current market (cost) for each type of security issue. The more shareholder's equity they can show on their balance sheet, the better their debt/equity ratio and the more money they can borrow at reasonable rates of interest. Shareholders as a group are generally opposed to diluting their ownership, and thus their share of profits, by the issue of additional shares of equity. Another factor is that corporations can't be forced into bankruptcy if they don't make dividend payments, as they can be for failure to make the stipulated payments on debt.

A brief word is in order regarding the supply and demand for funds, as well as the effects of inflation on the cost of each of these sources of capital. In principle, as the demand for money in the overall economy changes relative to the supply, investors will typically alter the rate of return which they require. When funds are in short supply relative to demand, short-term interest rates can be expected to rise. While short-term rates are directly determined by supply-demand factors, long-term rates must anticipate supply-demand relationships for the future life of the interest-bearing securities.

As an added factor, if inflation is expected to reduce the purchasing power of the dollar, investors again require a higher return (interest) to compensate for this anticipated loss. During inflationary periods borrowers are generally willing to pay much higher amounts to obtain money. They anticipate repaying their lenders with cheaper money. The borrowers also have reasonable expectation that operating expenses will go up but that they probably will also be able to raise prices sufficiently to cover the increased cost of the borrowed money. It must be kept in mind, however, that being caught in recessionary times with long-term high interest rate borrowings can result in severe financial strain on the firm.

REFERENCES

Brealey, R, and Myers, S., *Principles of Corporate Finance*, 2nd Ed., (1984) McGraw-Hill, New York, pp.504-507

Campbell, John M., (John M. Campbell Company, Norman, OK) *Oil Property Evaluation,* 1982, Prentice-Hall, Inc. Englewood Cliffs, NJ 07632

Childs, John F., (Irving Trust Company,New York, N.Y.) *Long Term Financing: 1964*, Prentice-Hall, Inc. Englewood Cliffs, NJ 07632

Ellis Jones, Peter, *OIL,A practical guide to the economics of world petroleum*, Woodward=Faulkner, Cambridge CB2 1QY

Gallum,Rebecca A. and Stevenson,John W., 1983, *Fundamentals of Oil & Gas Accounting,* PennWell, Tulsa, OK

Hindle, T., *Pocket Banker,* 1985, Blackwell and The Economist, London SW1A 1 HG

Little, J. B. and Rhodes, L., *Understanding Wall Street*, 1980, Liberty, Cockneysville, MD 21030

McKechnie, G., et al., *Energy Finance,* (1983), Euromoney Publications, London EC4

Pennant, R, and Emmott, B., *The Pocket Economist,* 1983, Robertson and The Economist, London SW1A 1HG

Philippatos, G. C., *Essentials of Financial Management—Text and Cases:* 1974, Holden-Day, Inc., San Francisco 94111

Petty, J. W., Keown, A.J., Scott, D.F.Jr., and Martin, J.D., *Basic Financial Management*, 2nd Ed., 1982, Prentice-Hall, Inc., Englewood Cliffs, NJ 07632

Stamper, F. A., *A Handbook for Texas Abstracters and Title Men,* Primitive Press, Bellaire, TX 77401

APPENDIX VII-A

THE FINANCIAL COMMUNITY, AND HOW IT OPERATES

The financial communities in each western country exist for two basic reasons. First is the important matter of bringing both investor and entrepreneur together. In this "primary market" corporations sell their stocks and bonds directly to the investing public, thereby obtaining the money needed for expansion.

The "secondary market" occurs after a company has gone public and its shares are traded. The secondary market provides the investor with an opportunity to buy or sell shares at any point in time. The stock prices in the secondary market rise and fall according to supply and demand for the particular stock.

The financial communities comprise a rather large group of specialized activities but it should be pointed out that precise delineations of these specialties of individual firms is difficult, if not impossible, because of the extensive overlapping of many of their activities.

World Bank

The World Bank group consists of three institutions which generally share the same staff. These are the International Bank for Reconstruction and Development (IBRD), the International Development Association (IDA), and the International Finance Corporation (IFC). The IBRD, established in 1945, has the primary purpose of providing loans to developing countries for productive projects. The IDA is a legally separate loan fund administered by the staff of the IBRD. It was established in 1960 to furnish credits to the poorest developing countries on much easier terms than those of conventional IBRD loans. The IFC, founded in 1956, supplements the activities of the IBRD through loans and assistance designed specifically to encourage the growth of private enterprises in less-developed countries. The three institutions are owned by the governments of the 132 countries that subscribe their capital. The head office of all three is on H Street NW in Washington, D.C.

The World Bank group is the largest provider of long-term funds for petroleum development in developing countries. The World Bank can only lend to sovereign governments. By definition, its petroleum project specific loans must have recourse to the NOC's local government. Typically, World Bank loans have a five year grace period and are repayable in up to 20 years.

Commercial Banks

The financial institutions with which we are all most familiar are commercial banks. Commercial banks deal with the business community at large, handling loans, checking and savings accounts and other similar services. When demands reach or exceed the capability of a smaller institution it goes to a larger banking institution with whom it has a "correspondent" relationship. Very large transactions which may be negotiated by one bank are frequently laid off to several correspondent banks, each taking a share of the deal and also spreading the risk. There is great variation in the size of commercial banks from small institutions serving a local community to some of the largest of commercial banks serving businesses world wide.

Investment Banks

The key players in the primary market are the investment banks as they are known in North America, or merchant banks in Europe. These firms, such as Morgan Stanley, Solomon Brothers and Bankers Trust in New York; Warburg, Morgan Greenfell, and Hill, Samuel in London; and Dominion Securities, and Wood Gundy in Toronto are examples. The principal specialty of investment banks is to raise the capital that client businesses need for long term growth. Investment banks guide the client company into the marketplace and generally assist in raising the necessary funds. Their compensation is normally in the form of a fee for the transaction rather than as interest on loans and fees for routine services which are the principal sources of income for the commercial banks. When all the requisite transactions are accomplished, a large check is normally delivered, with due ceremony to the client company's chief financial officer, less a sizable commission to the investment banker.

The investment banker must consider a number of factors such as the general economic conditions, the investment market environment at the particular moment including any competing issues on the same day or week; as well as the client company's financial condition, history of earnings and business prospects. The investment banker is expected to recommend an offering price which has to be agreed to by the company's financial management.

We will assume for our present purposes that the investment banker and the company have done their homework and concluded that a new stock issue is preferable to a debenture offering. Such is not always the case, of course. The choice of means of financing is largely a marketing problem. It may be that there has been a recent run-up in the market price of the company's stock and it is judged to be an opportune time to issue new equity both as a means of raising needed capital and at the same time improving the firm's debt/equity ratio.

The investment banker agrees to "underwrite" the new stock issue by buying all the shares for resale at a previously agreed price per share. In the case of very large issues the investment banker will normally want to spread the risk of actually being able to move the stock out into the market by inviting other financial houses to join in an underwriting group, or "syndicate." The syndicate will oftentimes bring in other security dealers to join in selling the new issue at a set price. The issuing company, i.e., the borrower, pays all of the under writing costs so that the buyer can acquire the new stock free of any commissions. After the initial sale the price of the shares is determined by supply and demand in the secondary market.

Secondary Market, The Brokerage Houses

The principal players in the trading of stocks after their initial issue are the registered representatives, or the "broker/dealers." These are such firms as Merrill Lynch, PaineWebber, and many others with "seats" on one or more of the major exchanges, and whose employees actually execute a client's order on the trading floor or in the OTC market in the U.S.. Wood, Mackenzie in the U.K., and RBC Dominion Securities, and Wood Gundy in Canada are familiar names performing similar roles in their respective countries. The broker, in his role as middleman or agent of the customer, is compensated for this services through a commission on the value of the sale or purchase. A dealer, on the other hand, acts as a principal in a transaction buying from and selling securities to the customer.

Stock Exchanges

The most popular stocks are "listed" on one or more exchanges throughout the western world. To be listed, which greatly facilitates trading and exposure to the widest possible investing public, requires that the stock issue meet certain requirements regarding the number of outstanding shares, the frequency and proportion of the issued stock which actually trades, and the financial condition and history of the company. The requirements for listing on the New York Stock Exchange (NYSE) are more rigorous than they are for the American Exchange (AMEX), or the other exchanges around the U.S. stocks are occasionally delisted by the exchanges for failing to maintain their listing qualifications.

Brokerage houses must purchase their seats on the various exchanges. Since the number of seats are limited, they are bought and sold frequently for very high prices. Some brokerage houses specialize in doing their commission business with individual customers. These are the better known "retail houses." Many of the seats on the exchanges are held by the "institutional houses" which deal essentially only in large blocks of stock for the major pension funds, insurance companies and other institutional investors.

Many stocks, particularly those of new smaller companies, are not listed on any exchange and are said to be traded "over the counter." This OTC, or "Third Market," is also a very active part of the financial community. Stocks listed on the NYSE and AMEX are also often traded in very large blocks in the OTC market.

Stock Specialist

If a dealer, with a seat on the NYSE or AMEX, brings an order for a stock trade to the designated location on the trading floor of the exchange and there are no takers the order (for not over 2000 shares) is filled by the "specialist" for the listed stock. This assures an "orderly market" for each listed stock so that there will always be a trade available. The Toronto, Pacific Coast, and other exchanges do not designate specialists and operate more like the OTC markets in this regard.

When a specialist trades for his own account, it is said he is acting as a "dealer," in much the same manner as the dealers in the OTC market.

Bonds and Debentures

Nearly 80 percent of all new corporate financing in recent years has been through the sale of bonds and debentures. Consequently the financial community devotes a major part of its activity to the trading of debt securities. So long as interest is deductible for tax purposes the high amount of activity in the bond markets will probably continue.

Bond trading occupies a significant part of the activity on the major exchanges. It is not unusual for 25 percent of the trading on the NYSE on a given day to be in corporate bonds. On the American exchange the proportion of trading in listed corporate bonds is much lower than on the NYSE. The majority of bond transactions is by institutional investors and is done through OTC (unlisted) market exchanges.

To better understand bonds and debentures you have to understand their special vocabulary. The following are the most common terms encountered when dealing with bonds and debentures.

ACCRUED INTEREST: The interest accumulated on a bond since issue date or the last interest payment. The buyer of the bond pays the market price of the bond plus accrued interest, which is payable to the seller.

BASIS POINT: One one-hundredth of one percent. One hundred basis points equals one percent.

BEARER BOND: A bond which does not have the owner's name registered on the books of the issuer and whose proceeds (principal and/or interest) are payable to the holder.

BOND: A written, interest-bearing certificate of debt with a promise to pay on a specific date, generally paying interest semi-annually.

BOOK ENTRY: Securities not represented by a certificate with the ownership recorded in a customer's account with a Trust Company, broker or bank.

CORPORATE BOND: Evidence of debt by a corporation. Differs from a municipal bond in various ways, but particularly in taxability of interest.

COUPON: The part of the bearer bond certificate which shows each interest payment due by date and amount. The coupons are clipped as they come due and are presented by the holder for payment of interest.

COUPON RATE: The annual rate of interest the borrower promises to pay to the bondholder.

CREDIT RATING: The designation used by investors' services for relative indications of quality.

DEBENTURE: An obligation secured by the general credit of the issuer rather than being backed by a specific lien on property.

GENERAL OBLIGATION BONDS: Bonds secured by the pledge of a municipal issuer's full faith and credit, including unlimited taxing power.

JUNK BONDS: Bonds with a low credit rating.

LIQUIDITY: The ability to convert a security into cash promptly with minimum risk of principal.

MUNICIPAL BONDS: Bonds issued by domestic public agencies, authorities and other governmental entities below the level of the federal government.

POINT: A point is worth $10 since bond prices are quoted as a percentage of $1,000 maturity value.

PRICE: Bond prices are generally quoted either in terms of a certain percent of maturity value or in terms of yield to maturity.

PRINCIPAL: The face or par value of an instrument, exclusive of accrued interest.

REVENUE BOND: A bond secured by revenues from the operations of a specific public enterprise such as bridges, toll roads, or utility systems. These have no claim on the borrower's taxable resources unless otherwise specified in the bond indenture.

SPREAD: Term indicating the difference between two different figures or percentages. It also represents the difference between the bid and asked prices of a quote.

YIELD: The expected rate of annual return on an investment, expressed as a percentage (1) Current or Income Yield is obtained by dividing the current dollar income by the current purchase or cost price of the security, (2) Yield to Maturity is the interest rate to be earned on the purchase price of the bond until the face value is paid at the maturity date.

Bond Ratings

Not all bonds are equally attractive to the investor. Howwell known and managed the issuing company is, and its probability of bankruptcy over the long life of these instruments, are important considerations regarding the relative attractiveness of a bond to the market. There are specialized organizations within the financial community who make a business of rating bonds from these standpoints.

In the U.S., Moody's Investor Service Incorporated, and Standard and Poor's Corporation are the two best known who perform this service. U.S. companies Fitch Investors Service L.P., Duff & Phelps, and A. M. Best, along with IBCA (London) also rate the credit worthiness of bond issuers. In Canada, both the Dominion Bond Rating Service and the Canada Bond Rating Service perform much the same function. The nomenclature of the two companies bond ratings varies slightly, but A-ratings are always better than B's in both systems. The important consideration is that the borrower is forced to pay a higher rate of interest on a BB bond than one rated in the A range. Triple A is the highest rating in Standard and Poor's system. It is reserved for U.S. government borrowings and the blue chip companies of the Fortune 500 list. These high ratings are highly cherished by management since they mean lower cost of capital to those that can borrow under the select rating of AAA.

Table 7A-1 has been included to show how two bond rating services, i.e., Standard&Poor's Corporation and Moody's Investors Service, rate corporate bonds according to their assessment of risk.

TABLE 7A-1
BOND RATING SYSTEMS

S&P RATING	MOODY'S RATING	EXPLANATION		
AAA	Aaa	Highest quality		
AA+	Aa1			
AA	Aa2	High quality		
AA-	Aa3			
A+	A1			
A	A2	Upper medium grade		
A-	A3			
BBB+	Baa1		↑	
BBB	Baa2	Medium grade		
BBB-	Baa3			Investment
BB+	Ba1			Grade
BB	Ba2	Moderately speculative		Speculative
BB-	Ba3			Grade
B+	B1		↓	
B	B2	Speculative		
B-	B3			
CCC+	Caa1			
CCC	Caa2	Highly speculative		
CCC-	Caa3			
CC	Ca	Poor quality		
C	C			
D	N.E.	Lowest quality		

N.E.: No equivalent
Source: Baron's Dictionary of Finance and Investment Terms

The Money Market

Money moves quickly through the financial community. Much of this is in the form of money market instruments which are characterized by their short maturities and complete liquidity. The amounts involved can vary from several thousands of dollars to many millions. Money market instruments are bought and sold between investment banks, commercial banks, brokerage houses, institutional investors, and large commercial and industrial companies; in fact, any part of the financial community that has a large intermittent need for cash. Examples might be for the distribution of a dividend or to pay suppliers upon delivery, and those firms that don't want to hold any substantial amount of non-interest earning cash even over a weekend are heavy users of money market instruments.

The money market is not a single homogeneous institution. Rather, the money market is a combination of individual trading activities each with its own characteristics. The money market's principal forms of exchange are treasury bills and notes, markets for federal funds, repurchase agreements ("repos"), certificates of deposit, or CD's, Euro dollars, and commercial paper. These are essentially non-equity fixed income securities, maturing within one year, which are issued, traded and redeemed. Most money market trading is done "over the phone" with major corporations acting as both borrowers and lenders. As lenders they seek to use any temporary surplus of funds to turn a profit while maintaining liquidity. As short-term borrowers, (usually days or weeks), they issue a variety of financial obligations (usually as notes, or day-to-day "call" loans) with a wide range of terms and conditions.

Futures Markets

Trading in crude oil and product futures has become a large and active segment of the financial community in recent years. This has been due primarily to the growing volatility of the spot markets in crude oil. Essentially all of the oil companies in the U.S. that are engaged in refining and marketing, as well as the crude oil traders, are currently making use of the futures markets. Trades of averaged 1.1 million barrels a day in 1987 across the board of the NYMEX, New York Mercantile Exchange, which is the principal North American market for trading in crude oil futures. Crude oil futures are also traded in London on the International Petroleum Exchange and the SIMEX in Singapore.

By judicious buying and selling in both the spot and futures markets for crude oil, gasoline, and heating oil along with forward contracts in these products, it is possible to hedge against sudden price disruptions, minimize stock buildups for refinery turnarounds, and handle the seasonal variations in demands for the respective products. Typical futures contracts are written in 1,000 barrel units at a fixed price for delivery at a specified time several months in the future. These are known as paper barrels in contrast to wet barrels which represent actual inventory of hydrocarbon crude oil, or product, in storage facilities or afloat in transit.

Foreign Exchange Dealers

An important activity of the financial community in any country has to do with the exchange of the money of various countries in order to permit the conduct of overseas commerce. Associated with this is the speculation in foreign currency futures which is important to maintaining liquidity in the financial system.

Forward contracts on a currency, such as pounds Sterling, are important. These contracts involve buying the currency, such as Sterling or U.S. dollars, for future delivery at the price fixed when the contract was made.

Swaps is another technique which is often employed (although it is a bit more complex and frequently requires the development of a degree of fortuitous special situations). As a result there is great variation and latitude of swaps agreements.

There are four general types of swaps:

1. An exchange of one foreign currency for another at some specific date in the future.

2. An exchange of long term obligation, such as a fixed rate mortgage, for a short term obligation which may be advantageous to a firm with a high short range cash flow.

3. Crude oil futures swaps as described in Chapter III.

4. Exchanges of financial obligations in two different countries where two or more firms may have cash flows and debt obligations which transcend the national boundaries.

The following example will illustrate one such transaction.

EXAMPLE 7-6
SWAPS AGREEMENT

Sorghum Products, S.A. of Buenos Aires has a $10 million U.S. dollar obligation for specialized equipment and technology payable within one year with 10% interest to the account of Sorghum's affiliate with the Barclay's Bank at their Cayman Islands branch.

The parent of Sorghum is concerned that the local government may impose foreign currency restrictions which would preclude, or seriously delay, paying off the unsecured obligation.

Able Oil Company, a U.S. firm, has need of Argentine pesos to fund a drilling program in that country.

SOLUTION:

Able Oil, Sorghum, S.A., and its parent company arrange a swaps deal whereby Able takes over the Sorghum S.A. obligation, in this case discounted at nine percent from par, and will make the required payments to the Barclay's Bank.

Similarly, Sorghum, S.A. will make pesos available at Able's call in Argentina over the next twelve months at the "parallel," i.e., unregulated exchange rate on the previous day's closing reported by Banco Boston in their main U.S. office. Sorghum is to pay any related bank charges in Argentina.

Other Service Institutions

Timely information is essential to the operation of the financial community. In the past the old Edison "dome" stock market ticker and the ticker tape parades of the twenties were once considered synonymous with Wall Street. Nowadays necessary trading information is relayed by computer networks.

Charles H. Dow, one of the founders of the Dow Jones Co., was the first editor of the Wall Street Journal. Through the years the Wall Street Journal has been, and still is, an important factor in keeping the U.S. financial community informed. The Financial Times in London and the Toronto Globe and Mail have been similarly effective in their respective countries. Specialized printing houses comprise an important part of the service sector of the financial community in every country. For example, the usual development of the required prospectus for a new security issue involves a long session of (several days to a few weeks) intermittent duration, with lawyers and representatives of all the firms and institutions directly concerned with working out all of the terms and conditions. Typically at the close of each day the draft of the prospectus and other necessary documents is sent to the printer, who overnight prints an updated version for distribution to the meeting participants the next morning. The printer may also be requested to undertake courier distribution of each morning's new printing to others, who do not actually sit in the meeting, for their review and timely comment.

Auditors

There are a number of major auditing firms which compose an important segment of the financial community. In recent years these companies, most of which are partnerships, have tended to diversify into a number of management advisory pursuits in addition to their prime activity of third party financial umpires.

The original 'Big Eight' of international accounting firms has been reduced to the 'Big Four' as a result of recent mergers and the closing of Arthur Andersen LLP. They are in order of size:

- PRICE WATERHOUSE COOPERS LLP

- KPMG LLP

- ERNST & YOUNG LLP

- DELOITTE TOUCHE TOHMATSU LLP

With the exception of KPMG, most of the firms originated in the U.K. or the U.S. and now operate worldwide.

Arbitrageur

There are several brokerage firms that specialize in arbitrage. The term refers to the simultaneous buying and selling of either 1) the same security in different marketplaces, frequently in different time zones, in order to profit from a disparity in market prices, or 2) two different securities which have a close relationship, such as convertibility of one into the other, so as to take advantage of the variation in their prices. These operations are generally referred to as "non-risk" arbitrage.

There may also be considerable arbitrage activity surrounding corporate mergers, tender offers, recapitalizations and divestitures. This is known as "risk arbitrage." The petroleum industry has seen an unusual amount of this type of arbitrage action recently in connection with the large number of mergers, buyouts, royalty trust spinoffs, and the like. In these situations the arbitrageur must continually assess the probability that the contemplated exchange of securities will actually take place. The arbitrage opportunity obviously terminates when the deal closes. Up to that time the day to day spread in the prices of the stocks involved is a direct gauge of the likelihood that the transaction will actually be finalized.

Security Analysts

A large number of people employed in and around the financial community are engaged in the detailed day to day study of the equities and senior securities of corporations and assessment of the investment potential of these securities. Some investment analysts specialize in particular industrial groups such as the drilling and service company sector of the petroleum industry. Others may study only a small handful of individual companies, keeping close track of their exploration successes and failures, management changes, past and projected future earnings and, in fact, everything related to those particular firms from a business standpoint. The results of certain analysts' work is published in "market letter" reports which are widely read by investors.

The security analysts have their own professional societies, the largest and best known being the New York Society of Security Analysts. This group meets for lunch every working day of the year to hear a presentation from one of the listed companies regarding its current and future business situation. These presentations usually involve the chief financial officer and his top management, all of whom take the opportunity to present their company in the best possible light to a group who have great influence in determining the popularity of the company's securities in the marketplace.

Investor Relations

Many of the larger oil companies and other industrial firms have a senior member of their management team designated as the liaison between the company and the financial community. These people must have at their fingertips an intimate knowledge of the company's past history and current activity on a full corporate-wide basis. They serve as a focal point for answering inquiries from the large number of investment analysts regarding the company, its policies and prospects. The investor relations managers are also in a key position to keep management advised of pending developments within the financial community which may affect the company.

Securities and Exchange Commissions

Each western country has regulatory bodies firmly established to monitor the activities of their financial communities to see that all of the rules and regulations which are designed to protect the investor are followed. In the U.S. the SEC was set up in 1934 following the crash of 1929. In the U.K. a public limited company (plc) registered under the Companies Acts is the near equivalent of an SEC-registered publicly traded company in the U.S. In France the Commission des Operations de Bourse (COB) is essentially equivalent to the American SEC.

Among the specific responsibilities of these governmental agencies is the precise review and approval of the prospectus which the dealer must give to every buyer or potential buyer of a new security offering. These regulatory bodies may also be involved in licensing many of the various firms and positions required to conduct the financial processes. They are also responsible for policing the financial community against trading irregularities and insider abuses.

The Growing Lack of Distinction of Specialties within the Financial Community

Although we have listed the numerous specialties within the financial community, it must be recognized that the situation is not static. At present there is a growing tendency toward building large conglomerates of financial institutions which handle a broad range of activities through investment banking, brokerage and other commercial activities. Examples are Sears and American Express, referred to as 'non-banks', who have been moving into such areas as brokerage services and real estate. Other firms, formerly devoted primarily to stock brokerage such as Merrill Lynch are also diversifying into banking, and so on.

APPENDIX VII-B

DIVISIONS OF INTEREST OILFIELD DEAL STRUCTURES

The drilling of Smith Number 1 which was reviewed in this chapter represents only a single episode of a typical North American drilling venture. There are literally thousands of variations on the theme of changing up the ownership of the mineral interest in order to accomplish certain objectives.

Petroleum exploration and production involve a number of distinctive features as a commercial enterprise. Many of these are discussed in Chapter XI under the heading of "What makes the oil business unique as a commercial enterprise." Several of the more pertinent which relate to oilfield deal structures are repeated here for ease of reference.

1. Petroleum is an extractive industry—thus, there is a financial relationship between the landowner, (the owner of the resource), and the lessor, who undertakes at his entire expense, to explore and produce the resource while providing the minerals owner his proportionate share of any proceeds.

   It is a fundamental concept of most commercial enterprise that any eventual profit is shared on the same basis as the original investment in the project. In the extractive industries, where substantial royalties may exist, the royalty owner shares in any revenue while the owners of working interest pay all costs and investments and share only in the net income after royalty, costs and investments. Producing properties are normally plugged and abandoned when the working interest owner's net income from the declining production no longer pays the cost of the operation and any unamortized investment. Thus an abnormally high royalty will force an earlier abandonment of the producing stream.

2. Petroleum is a risk industry—so that participants in the business may wish to reduce their interest in one venture in order to participate in others as a means of spreading their risk. This is very similar to the insurance business where the carrier of a substantial policy will often "lay off a portion of the risk" by disposing of part of his ownership to others.

3. Petroleum is an extremely capital intensive industry—A particular oil and gas development venture can easily be so costly that a working interest owner may dispose of a share of his interest to another party who will take on his proportionate part of the financial burden in the expectation of a similar share of the rewards.

4. Petroleum producing wells may yield a variety of commercial streams—these include crude oil, casinghead gas, natural gas, condensate, natural gas liquids, and sulfur.

5. Petroleum may be produced from a number of separate sources, or strata, underlying the same tract—The ownership and Division of Interest may vary with the different sources.

All of these factors lead to possible Divisions of Interest of one type or another. The list of types of deals which result from the infinite combinations of the factors listed above is essentially endless. Thousands of such deals are made each day throughout the world of petroleum. The greatest number, however, are consummated in North America and particularly in the U.S. More and more are developing outside of North America as the smaller independent oil companies venture abroad.

The mineral interest owner of a tract of land is normally the landowner, unless the ownership of the minerals has somehow been severed from the surface ownership of the tract. The owner of the minerals is generally shielded from risk. His deal is usually set from the start with the signing of the oil and gas lease. The lease agreement sets out the royalty interest retained by the mineral owner in all future revenues derived from the production of oil, gas and other minerals from the lease. Traditionally the mineral owner's royalty is fixed at 1/8th of the gross revenue at the wellhead, although larger royalty fractions are not unusual.

A significant variation from the fixed royalty is the Sliding Scale Royalty. Although Sliding Scale Royalties are uncommon in North America they are not at all unusual with host governments overseas. The economic rational for Sliding Scale Royalties is the general observation that very large rates of production from a well, or groups of wells, are much more profitable on a per barrel basis than a small producing rate. On this basis the royalty owner may retain a larger royalty on very large rates of production and progressively smaller royalties on lesser rates of production. Sliding Scale Royalties are usually defined in a series of bracketed rates of production rather than as a continuous mathematical formula.

The remaining portion of the mineral interest—after the mineral owner's royalty is called Working Interest, WI. Figure 7B-1.

FIGURE 7B-1
SINGLE WORKING INTEREST

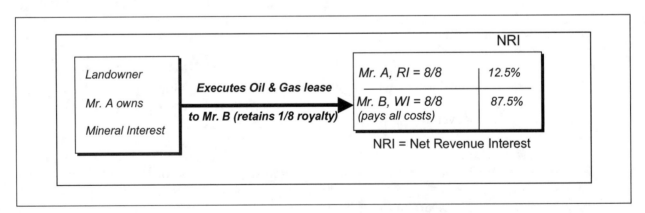

Working Interest is highly subject to oil deals. These may be to reduce risk (2 above) and/or reduce capital requirements (3 above). It should also be recognized that the royalty interest may be subdivided and portions sold off to other parties. As mentioned previously, however, royalties are essentially secure from the major concerns of risk and capital that are so all important to the Working Interest owners.

OIL DEALS INVOLVING CHANGES IN WORKING INTEREST TO REDUCE CAPITAL REQUIREMENTS

Figure 7B-2 continues the dealing on the ownership interests of the subject tract of land. Here Mr. B, the lessee and Working Interest owner, sells 1/2 of his 7/8ths interest in the minerals (Net Revenue Interest, NRI) to Mr. C for cash.

FIGURE 7B-2
SALE OF PARTIAL WORKING INTEREST

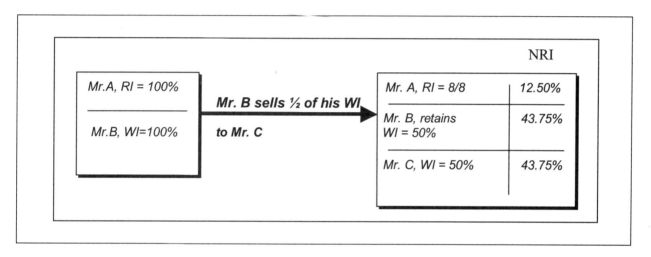

The dealing may continue: in fact, it normally does in some fashion. In Figure 7B-3 we see that Mr. B in the hope of speeding up the process toward drilling a well on the tract has hired Mr. D, a consulting geologist to work up a drilling proposal that can be shown to other prospective purchasers of interest in a test well. The geologist may desire cash for his work but in this case settles for a 1/8th Overriding Royalty on Mr. B's share of the Working Interest. Many geologists will request more cash and a smaller override for their services since in the example shown in Figure 7B-3 a dry hole would leave him with no compensation for his efforts. On the other hand the overriding royalty interest, ORRI, from a good commercial well producing over a number of years might yield income far in excess of the geologist's normal per diem fees.

FIGURE 7B-3
OVERRIDING ROYALTY INTEREST AND PRODUCTION PAYMENT

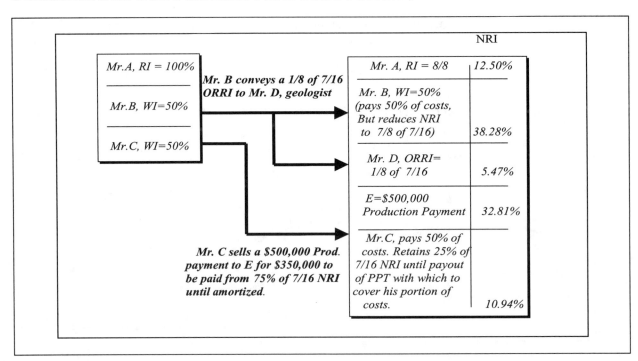

Meanwhile Mr. C, who seems more interested in dealing than immediate drilling, has sold a $500,000 Production Payment to E for $350,000 to be paid out of 75% Mr. C's 7/16ths if and when production is established. This provides Mr. C with funds for his share of the drilling of the test well.

OIL DEALS INVOLVING CHANGES IN WORKING INTEREST TO CUT RISK

The simplest way to lay off risk is merely to sell a portion of the working interest, perhaps with a "promote." The promote transfers a disproportionate part of the risk to the purchaser. A portion of the promote, which is the part of the purchase price in excess of the actual apportioned out-of-pocket expenses, is sometimes looked upon as compensation for "working up" the original prospect. The principal role of the promote, however, is to transfer risk.

Figure 7B-4 might be redrawn to show this type of transaction. If Mr. B had spent $100,000 in acquiring the lease including the lease bonus and other expenses, he might pass half of this on to Mr. C for his share of the costs to date plus a promote of say, another $100,000. Mr. C in this case would be willing to pay this extra amount to "get in on the deal" if, for instance, there had been other recent favorable drilling in the vicinity.

In Figure 7B-4 we see that the Net Revenue Interest of the parties has not changed even though Mr. B has promoted the deal in such a way as to recover his out-of-pocket expenditures to date.

FIGURE 7B-4
SALE OF PARTIAL WORKING INTEREST AT A PREMIUM

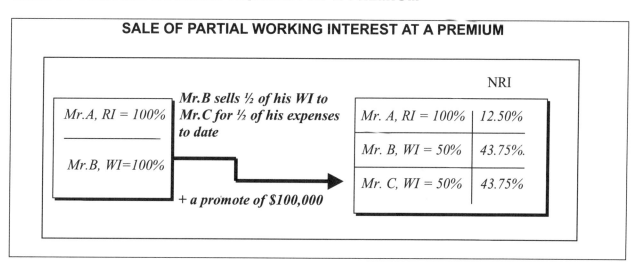

Transactions involving spreading the risk (incurring costs) prior to, and during the drilling of the initial exploratory well, typically will contain specific conditions for the sharing of costs, etc. to the casing point. Once a well is down and electric logs run, interpretation of the logs, cores, cuttings and drillstem tests can give a fair, if not complete, indication of the presence or absence of hydrocarbons. This stage in the drilling operation is known as the "Casing Point" since the next operation typically, if the indications look favorable, would be to run the "long string" of casing and attempt completion of the well.

If the indications at the Casing Point are unfavorable for a commercial completion the well normally is plugged and abandoned. Once the owners of interest in the well have "seen the sand" at the Casing Point, found it favorable, and decided to proceed, the prospect of completing a successful commercial well is greatly increased. Many oilfield deals and divisions of interest are hinged on this point in time because of the tremendous reduction in risk from this time forward.

If the decision is to run pipe, an entirely different set of pre-agreed financial conditions covers the expenditures for the casing and completion of the well.

Typical Third-for-a-Quarter Oilfield Deal—Free Well Agreement

A very common arrangement of this type in fairly heavily drilled areas such as the northern Rocky Mountains is the "Third for a Quarter Deal." This type of agreement calls for the drilling interests to take on a disproportionate share of the drilling risk to the Casing Point in return for a lesser share of the net revenue interest if the well produces. Various ratios of risk to revenue are employed. In the Third-for-a-Quarter deal the ratio is 1/3 of the drilling obligation for an eventual 1/4 of the Working Interest revenue. (21.88% of the Net Revenue Interest before expenses)

A simple example in Figure 7B-5, using three other oilmen (Messrs. X, Y and Z), for ease of understanding, would be a case in which a wildcatter, Mr. B, acquires a drillable lease. He arranges for Messrs. X, Y and Z each to take a third Working Interest to Casing Point. Mr. B, the wildcatter, is thus relieved of all drilling expenses to the Casing Point. Thereafter, Mr. B has the option to come back in for a quarter working interest. At this point the wildcatter and his three promoted partners would each own equal quarter Working Interests in the venture.

FIGURE 7B-5

LEVERAGED SALE OF WORKING INTEREST

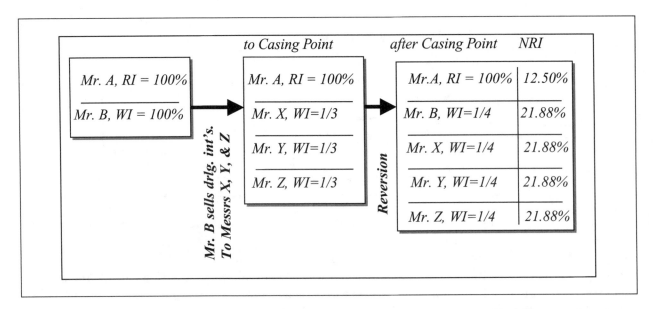

This assumes, of course, that the indications at Casing point favor a completion attempt. If the indications at casing point are discouraging and the well is abandoned as a dry hole, the wildcatter, Mr. B, has succeeded in transferring the initial drilling risk to his three partners and he is "home free." In actual practice the Third-for-a-Quarter deal may not involve transferring all of the working interest to other parties prior to Casing Point. The wildcatter might be bold enough to keep some part of the unpromoted Working Interest for his own account—or, perhaps he couldn't find enough new participants to take up all the thirds.

Carried and Net Profit Interests

Other techniques for transferring risk which may overlap some of those already mentioned are the "Carried Working Interest" and a "Net Profits Interest." The Carried Interest is generally for a limited segment of the drilling, completion and production process. It denotes that the "Carried" party is relieved of the investment costs and expenses during that stage. Being "Carried to the Tanks" merely means that the other Working Interest owners foot the entire bill for drilling, completion and installation of production equipment including the typical lease storage tanks. The "Carried" party is thus free of all of these costs as well as the risk that the venture may prove nonproductive, as so many do. The Carry agreement will usually stipulate that at the end of the Carry the owner of that interest will revert to a normal Working Interest participation, sharing both costs and revenues.

A Net Profits Interest, NPI, which is a form of "carried interest," is generally of much longer duration. These are arrangements whereby the owner of a NPI is relieved of all costs of drilling and production, as well as the risk of the venture being nonproductive, and participates only in any net profits of the venture. Net Profit arrangements are typically for the life of the lease.

296

Farmout/Farmin

By far the most popular means of "laying off" risk (to resort to the terminology of the insurance industry) is by farming out a portion of the Working Interest to another oilfield operator in return for some service and retaining a ORRI. Farmouts and Farmins are the inverse of one another depending upon which interest is being evaluated. The simplest case is shown in Figure 7B-6

FIGURE 7B-6
FARMOUT / FARMIN

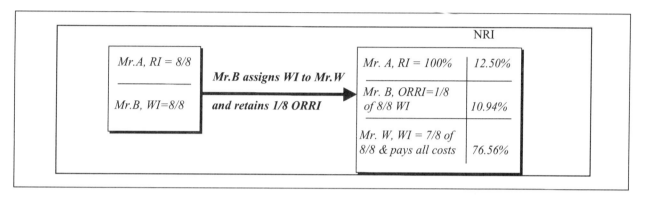

It frequently happens that Mr. B who owns the lease on a tract may have cooled to its productive possibilities. Mr. W, on the other hand, may be highly optimistic about its potential and be quite willing to take on the risk of drilling the prospect at his expense. Mr. B doesn't want to forego entirely the possible outcome of his earlier efforts in acquiring the lease. And so, Mr. B may farmout the tract to Mr. W retaining an override, ORRI. The magnitude of the override is entirely the result of negotiation between the two parties. Typically ORRI's may vary anywhere from 1/32nd to as much as 50 percent depending upon the relative enthusiasm of each party regarding the prospectivity of the tract.

Farmins may be undertaken when a major oil company has assembled a large block of nonproducing acreage beyond its current exploratory drilling budget limitations. Mr. Y, an independent producer, may agree to farmin on several tracts in the block and agree to drill at his entire expense within a certain time limit. The agreement will further provide that Mr. Y will earn a negotiated, say 50 percent, of the Working Interest in all of the tracts in the farmout package.

The farmout terms may also be affected by timing relative to the stipulations of the lease. The farmee can generally demand and obtain a much higher residual interest in a lease with only six months remaining in its primary term compared to one with several years remaining.

Division of Interest between Production Streams

In broad terms crude oil is the most valuable yield from the drilling and producing operation (referring back to item 4 in the first paragraph of this discussion). The gas associated with the crude oil production, generally known as casinghead gas, is its byproduct. A good deal of casinghead gas may be consumed on the lease as fuel for the operation. Sometimes the excess is flared. In most jurisdictions it must be compressed and either reinjected into the subsurface or sold. The expense of compressing the casinghead gas and providing the necessary flowline connections to a gas trunkline may be substantial, making the economics of the operation rather lean at best. ORRI which may be quite acceptable for the crude oil production from a lease may constitute too great a financial burden for the sales of casinghead gas to carry. For this reason it is not unusual to see one ORRI for crude oil and a lesser ORRI value assigned to casinghead gas sales.

Natural gas represents a somewhat comparable situation. The value of natural gas as a commodity is less than that of crude oil. Gas wells may also produce condensate as a valuable byproduct. Accordingly there are sometimes different ORRI levels assigned to natural gas and condensate as distinct from crude oil and casinghead gas. Still other ORRI levels are possible with regard to the products from a gasoline plant in the field which may recover ethane, propane, normal and isobutane, natural gas liquids and sulfur if these are assigned back to the producing lease.

Division of Interest in Deeper Rights

Many geological trapping configurations serve to hold hydrocarbons in a series of separate reservoirs at varying depth. The natural development of an oil or gas discovery is to explore in the most shallow zone first. That is the first one the driller encounters and his funds may be running short. Exploration of deeper possibilities takes on additional risk which may prompt a whole new set of Divisions of Interest for these new deeper prospects. Oftentimes these may be negotiated several years after the shallower zone has been on production.

Reversionary Divisions of Interest

The importance of Casing Point as a pivotal stage in Divisions of Interest with respect to exploratory risk has already been discussed. A second important phase in many Divisions of Interest involves a change, known as a "Reversion," in the financial relationships between the parties at the time of Payout. Figure 7B-7 shows a chronology, patterned after Figures 7B-5 and 6, in which the Working Interest relationship reverts to a simple split with no carried or overriding interests after there has been enough production revenue from the override or carry to reimburse the driller and other operating parties (e.g., Messrs W, or X, Y and Z) for their agreed costs in fulfilling their obligations in the deal. Not every ORRI or Carried Interest has a reversion provision, in fact, most do not.

Reversions can take place when Production Payments or certain other conditions are satisfied. Reversions may also be triggered by the calendar, i.e., a period of time, or by the attainment of a specified cumulative production volume, in addition to the Payout of dollar amounts, or the physical achievement of a Casing Point drilling depth as already described.

FIGURE 7B-7
DECISION POINTS - CASING POINT AND PAYOUT

Figure 7B-7
DECISION POINTS - CASING POINT AND PAYOUT

TIME FRAME

Before Casing Point
•Generate Project
•Lease Acreage
•Spud well
•Drill to Csg. Pt.
 (Core?) & Log

After Casing Point
•Set Production String
•Complete & Test Well
•Install Lease Tanks
 & Flowline Hookup

After Payout
•Continue Production
•Ongoing Operating
 Costs, Well Repairs,
 etc.

Casing Point

Payout

COSTS

•Entirely by WI
Owners as defined
"before Csg. Pt."

•Entirely by WI Owners
 as defined prior to
 payment including
 those carried only to
 Csg. Pt.

•Entirely by WI
Owners

REVENUES

•Not Applicable

•Shared by Owners of
 - Royalties
 - Working Interests
 - ORRI
 - Production Payments

•Shared by Owners of
 - Royalties
 - Working Interests
 - ORRI
 - Production Payments

*

*

* WI ownership may change at these points as provided by the contract controlling their relationships;
i.e., lease, farmout, PSC, service agreement, etc.

299

NET INCOME, "THE BOTTOM LINE"

CHAPTER

VIII

"Accounting is the art of recording, classifying, and summarizing in a significant manner and in terms of money, transactions and events which are, in part at least, of a financial character, and interpreting the results thereof."

—AIA

In the world of business, the computation of net income or "profit" is the means by which the accountant keeps track of how well the firm is doing financially. Profit, in its simplest concept is the excess of revenue over all of the implicit costs of conducting the business within a specified time period. In the oil and gas producing business profit is directly determined by the three factors: price, costs, and volume. These are the same basic concerns of a host government whose NOC is charged with providing the basic domestic petroleum requirements and, if possible, exporting crude oil to augment the country's balance of payments.

The first seven chapters of this text have dealt with the various economic factors which focus more or less on individual project undertakings. There are, however, several highly important factors which must be considered at the corporate level. These include:

● corporate financial structure, and its cost of capital, which continues from the last chapter

● corporate income tax

● capacity for risk, as introduced in chapters V and VI, which is a function of the size of the enterprise

● the opportunities and benefits of vertical integration, that will be dealt with briefly in the final chapter

● the determination of the company's profit or loss and the payment of dividends to its shareholders, which is the thrust of this chapter

An oil company, like any commercial enterprise in the private sector, is in business to make a profit for its shareholders. If a company fails to provide a return on investment, to the satisfaction of its

301

owners, the firm will go out of business. This liquidation can occur either through bankruptcy, or more likely, by merger of the assets into a more successful company. Either of these situations is a relatively sudden and dramatic occurrence even though bankruptcy is always the culmination of a long series of poor decisions and business results.

Unfortunately, there currently is not a universally accepted set of accounting rules. Various countries or areas of the world have slightly different rules that the accountants must follow when preparing financial data. These rules, commonly referred to Generally Accepted Accounting Principles (GAAP), are significantly different and require companies to keep several sets of records that conform to GAAP in the different areas in which they operate. A number of countries have adopted International Financial-Reporting Standards (IFRS) to simplify international accounting, but this is not complete. The U.S. hopes to adopt IFRS by 2009.

PETROLEUM INDUSTRY ACCOUNTING

The process of looking at an oil company from a financial and/or accounting viewpoint involves a different perspective than the one to which most operating people in the industry may have become accustomed. One objective of this text is to demonstrate more clearly how, when, and why an oil company makes a profit, or fails to do so.

There are many factors in the categories of finding costs, oil and gas prices and volumes, and the large investments required to develop and produce petroleum, all of which are exceedingly important to oil industry's profit or loss. Most of these are within the direct control of the firm. A significant number, however, such as price and demand, are beyond the basic influence and control of an individual oil company, or operator.

The Accountant starts with the same concept of profit as does the Manager of Operations, (i.e. the excess of revenues over costs of doing business). The accountant immediately becomes concerned, however, with the equity capital employed by the firm, its capitalized assets, taxes, and their related costs, as primary business expense.

Operating personnel, as previously explained, tend to concentrate their efforts on maintaining the maximum possible Cash Flow from their segment of the operation, with the understanding that with a healthy cash flow, a suitable profit will follow. Operating people generally deal with the ongoing operations of producing properties which may span a number of years, and with the evaluation of new projects.

In studying the profitability of the oil and gas producing industry the nature and limitations of the world's established accounting systems have to be recognized. These are the procedures by which profit and loss are computed. These accounting procedures have been designed traditionally to handle either:

1. Retail firms whose books will look good if in their activities they can buy low and sell high, and at the same time keep their inventories in balance; or

2. Manufacturing enterprises which seek to purchase their raw materials cheaply and in a timely a manner, hold their production costs to a competitive level, and market the product in a way that will show a profit.

Unfortunately, the exploration and production segment of the oil business doesn't fit either of these more common patterns of commercial enterprise. One of the most troublesome hurdles toward improving profit in the producing sector of the petroleum industry relates to this mismatch with the traditional business accounting system.

These basic difficulties in adapting modern accounting practice to the oil business is profound to say the least. The three basic principles of traditional accounting are:

1. **The Cost Principle** which provides that assets are reflected on the balance sheet at actual original cost unless there is a decline in utility value.

2. **The Matching Principle** which provides that current revenues should be matched against the costs incurred to produce those revenues in the same accounting period.

3. **The Realization Principle** which holds that revenue is recognized at the time that a transaction is completed with a third party.

Under the cost principle, the primary asset of an oil producing company, its major oil and gas reserves, is largely ignored on the firm's balance sheet. The premise that a company's financial statements accurately evaluate the enterprise as a "going concern," obviously fails when the system lacks any objective means of pin-pointing the all important asset value of the petroleum reserves in the ground. This effectively compromises the Matching Principle.

Further, since the industry's cost of finding new petroleum reserves is constantly changing with the ongoing exploration effort, the true economic cost of a company's current oil and gas production is not reflected in its income, or "profit and loss" statement. All of this makes the reading of an oil company's annual report, as discussed in Appendix B of this Chapter, an interesting and challenging exercise.

The Realization Principle creates only minor problems when the oil and gas are sold and delivered to third parties. When the volumes are "sold" from one subsidiary to another in a vertically integrated company "transfer" prices sometimes create distortions. These often are the result of the integrated company's efforts to reduce overall taxes which may differ in the various industrial segments.

Accounting Terms

Several basic accounting terms (see 'Glossary of Petroleum and Financial Terms' at the end of this book) are key to understanding the accountant's financial results, however. These include:

1. Capitalize or Expense

2. Tangible or Intangible

3. DD&A (Depreciation, Depletion and Amortization)

According to basic accounting theory expenditures for long-term capital assets of the firm are capitalized and recorded on the balance sheet. Expenditures that offset revenues and have a short term effect on profit are normally expensed, and listed on the income statement. The detailed format of the financial statements including the balance sheet and income statement are discussed later in this chapter.

Tangible assets are those of a physical nature such as lease tanks, wellheads, buildings, etc. which have a useful life of more than one year and a recognized salvage value. Intangible assets lack physical substance. These include such items as a portion of well drilling costs, patents, oil and gas leases, and other rights of various kinds. The process of methodically expensing the cost of a physical asset over time, thereby progressively reducing its value on the balance sheet is called depreciation. The similar process as applied to intangible assets is called amortization. Depletion applies to the progressive removal, or depletion, of natural resources. Collectively these write-offs are frequently called DD&A.

Cash Flow and Profit

There is frequent confusion between cash flow and profit.

Cash flow is an operating concept which quite simply totals up all of the revenue for a project, or even an entire company, and then deducts all of the operating expenditures. It is easily understood by all organization levels. Cash flows can be computed for any group of operations and for any convenient period of time. Cash flow, for instance, might be calculated for the entire producing life of a project; or for a calendar period; or perhaps only until payout of the particular investment.

The accounting profession continually modifies the actual cash flow by including the accruals of amounts owed to the business, or owed by the business to others. They also isolate capital expenditures, or investments, and charge off an annual depreciation amount in order to reach what they term, "profit."

Net Earnings, or profit, are always reported for a unique time period, usually one year. This is the "bottom line" that everyone likes to talk about. It is the most important item in the company's Annual Report. However, it is possible to have a good looking profit reported for one year and terrible looking one for the next. Maximizing profit for a single year is a dubious corporate objective largely for this reason. The firm's management should look upon maximizing shareholder's wealth over a period of years as the fundamental corporate goal.

The two ideas of maximizing profit, as well as the growth of the firm, in terms of shareholder's wealth need not be in conflict with one another. A consistent growth in a company's net earnings or profit over a span of years should increase the value of its common shares. Accordingly, there is nothing sinister about management stressing the desirability of increased profit—seemingly for profit's sake. Let's first review how profit is computed in order to gain a better feel for how to improve it.

Profit is what remains for the common shareholders out of the gross revenue after deducting all costs and taxes. If a way can be found to increase gross revenue in relation to costs and taxes, then profit goes up. Correspondingly, if costs and/or taxes can be reduced relative to gross income or revenue, profit will be increased.

Reducing costs is a particularly fertile field for improving operating profit. So much emphasis has been placed on this aspect in the past, however, that some line supervisors have carried their field operating cost cutting to extremes, sacrificing longer term profit in order to improve the looks of their company's current financial reports. All of this emphasizes the importance of keeping cost cutting programs in balance with the company's objectives, in both the long term as well as the short term.

TABLE 8-1
SIMPLE COMPARISON OF FINANCIAL AND CASH BOOKS

CASH BOOKS–Cash Flow Computation
- Cash Flow Concept -

Period	Investment	Revenue	Depletion, Depreciation & Amortization	Other Expense	Net Cash Flow	Cumulative Cash Flow
1	1,000	500	n/a	100	-600	-600
2		500	n/a	100	400	-200
3		500	n/a	100	400	200
4		500	n/a	100	400	600
5		500	n/a	100	400	1,000
Totals	$1,000	$2,500		$500	$1,000	$1,000

(handwritten annotations in right margin: "discont ½ yr", "-572", "346.7", "315.19", "286.5", "260.5", "PV at 10% discont rate", "-639")

FINANCIAL BOOKS–Accounting Approach
- Accounting Concept -

Period	Investment	Revenue	Depletion, Depreciation & Amortization	Other Expense	Profit
1	1,000	500	200	100	200
2		500	200	100	200
3		500	200	100	200
4		500	200	100	200
5		500	200	100	200
Totals	$1,000	$2,500	$1,000	$500	$1,000

(handwritten annotation below table: "PV at 10% = 79J.")

One further reference to cash flow: *One can go bankrupt with insufficient cash flow—no matter how profitable the enterprise is reported to be.*

If an investment were evaluated using both the cash flow and accounting methods, the undiscounted profit at the end of the lifetime of the investment would be the same for each. Because of the difference in handling of annual items, the annual net cash flow differs from annual profit. This difference then will cause a difference in the present value of profit between the two methods. Table 8-1 presents an example of an investment evaluated both with the cash flow and accounting methods. The difference in annual values and the present value of profit are evident. It is noteworthy that the Total Net Cash Flow and Total Profit are identical at the end of the project life.

The difference in the two approaches lies in the fact that the accountants in preparing the Financial Statement choose not to charge off the entire investment in the year incurred. Rather they expense one fifth of it each year through the DD&A account. The workings of the DD&A account are discussed later in this chapter.

HOW PROFIT IS COMPUTED

Profit is such a familiar word, that it is worth while to pause and review what actually goes into its calculation. The simple starting equation has already been discussed:

$$\text{Profit} = \text{Gross Income} - (\text{Expenses} + \text{DD\&A} + \text{Taxes})$$

Gross Income

The meaning of Gross Income was discussed under Cash Flow in Chapter IV. In the usual petroleum exploration and production situation it is the sum of the income from the sale of crude oil, natural gas, natural gas liquids and sulfur. However, the accountants consider the sale to occur when ownership is transferred, not when payment is made.

Costs

Keeping track of the costs of operating a business occupies much of the activity of the firm's accounting department. Accounting theory is not static, and is occasionally subject to controversy within the accounting profession. Our discussion of costs will be confined to some of the more basic concepts. Specific and detailed accounting terminology will differ between oil companies. It is well to familiarize one's self with the specifics used by his/her own company. If there are differences, a little discussion between the participants should produce a common understanding of the terms and techniques in use.

Operating Taxes

Operating taxes are just another expense that is incurred by a project and in principle no different than other direct operating costs. These taxes are generally called 'excise taxes', which include all taxes other than income taxes. Such taxes are based upon a variety of parameters; e.g. production rate, gross revenue, purchase price, property value (ad valorem), oil price, etc., but are not profit based. These taxes will persist as long as the operation upon which it is based continues, irregardless of whether or not it is generating a profit. On the other hand, income tax is profit based and will be in proportion to the profit earned.

The petroleum industry carries one of the heaviest, if not the heaviest, tax burdens of any type of commercial enterprise. Other lines of business pay taxes, but not to this extent. The importance of taxes in computing profit has been mentioned earlier. Each level of government, (federal, state and local) has the power to tax—and all of them do so. Following the dramatic rise in crude oil prices during the 1970's several oil producing countries levied some form of excise, or "Windfall Profits" tax, on oil companies operating within their borders . These taxes skimmed off a part of the gross income that is deemed to have come from world crude oil price increases which the host government concluded did not result from the local oil company's efforts, and therefore should not accrue to them, but rather to the government. Essentially all of these excise taxes were eliminated, or substantially reduced as oil prices declined.

State and provincial governments may also impose direct taxes on oil and gas production per se. These may be in the form of severance taxes on the theory that a portion of the state's natural resources is being severed from the state's bounty and thus should be taxed. On the Crown, e.g., Provincial Lands, of the western provinces of Canada the principal tax, or "government take," is through the Crown Royalty which may run as high as forty-six percent of production. Technically this is not a tax, but a royalty. Royalties do result in a very substantial increase in the Outside Share and corresponding decrease in cash flow to the net working interest owner, who must pay all the operating costs out of his share.

Another important group of taxes, which may be charged directly to the operating accounts at various lower levels within the company, includes the broad range of local taxes. These are most frequently in the form of 'ad valorem' taxes assessed on the value of individual properties and their surface facilities.

Income Taxation

The federal governments in nearly every country, except the Cayman Islands (and a few other tax haven countries), levy a tax on net income, or profit, reported by any person or corporation operating within their boundaries. Since most oil companies are able to make a profit, they are subject to income taxes. The maximum tax rates are normally a bit less than half of the net income of most corporations. State and Provincial Governments also impose income taxes, although normally not as high as the federal rates.

Income Tax Accounting

The accounting rules and procedures previously described generally are followed in determining the corporate profit for purposes of reporting the financial condition of the company to its shareholders and the financial community. This includes the calculation of income tax liability shown on the income statement. That value is not the actual amount of income tax paid, but following the matching principle it is the amount of tax due on the profit earned that year, if the tax laws followed the accounting rules. Tax laws rarely coincide with accounting procedures, so the actual amount of income tax paid is frequently less than is shown on the income statement, with the difference carried to the balance sheet as a future liability. The appropriate accounting rules are standardized within a country and must be followed by all companies under its jurisdiction. These accounting rules and procedures vary from country to country. When the rules differ between tax and financial accounting it becomes necessary to maintain more than one set of accounting records, each individually tailored to the appropriate set of rules and procedures of that country. Thus, profit for paying income tax may not be the same as the profit reported to the shareholders and financial community. It is important to apply the applicable rules and procedures to the appropriate analytical computation.

The current U.S. federal income tax rate for corporations in the U.S. is 35% but a flat fifty percent is often used as a reasonable approximation of the total government tax take within the accuracy of most of the other factors employed in the analysis. At this rate of taxation the company nets 50 cents on each dollar of net income. This reduction in net income after tax highlights the fact that reducing the firm's tax obligations may be one of the most effective ways of improving its profit. Tax rules differ in each country, and are often complex. Tax interpretations also change continually due to new court rulings, new legislation, or tax service practice.

A variety of novel approaches to figuring the taxes emerged during the early 1980's. In some jurisdictions the effective rate was sometimes so high as to make the development and production of many fields uneconomical. The fact that investment and operating costs per barrel tend to be higher for smaller oilfields compared to those in giant fields has proven to be a problem for the lawmakers and taxing authorities. This has led to a number of changes in the taxation of the industry around the world.

Many host governments seeking to attract foreign capital for their fledgling petroleum industry now advertise for oil companies to come and bring money and expertise. Some of these host government approaches to taxation have given rise to quite serious problems in figuring the allowable foreign tax credits of the multinational oil firms in their home countries.

Timing of Deductions

The timing of expense deductions for income tax purposes mainly has to do with DD&A charges as permitted by the tax regulations. Here the principal concern deals with the depreciation rates which the country's taxing authorities allow. Companies generally prefer accelerated depreciation in one of the several forms which may be permitted. The taxing authorities, on the other hand, desire just the reverse, with the resulting higher tax collections. Tax rates and regulations are constantly subject to change. Thus, the business axiom: "Take the largest possible deduction at the earliest possible date."

Intangible Drilling Costs

In many jurisdictions around the world the operator has the option regarding the capitalizing or expensing of Intangible Drilling Expenses. In practice, the maximum permissible intangible drilling costs are generally expensed since this reduces taxable income and thus the tax obligation, leaving more cash available for other revenue generating opportunities.

Investment Tax Credits

Many countries provide investment tax credits (ITC's) as a means of stimulating new industrial development. ITCs are directly subtracted from the tax liability, whereas deductions are subtracted from revenues for tax computations. Thus at a 50 percent tax rate a tax credit has essentially twice the impact of a deduction. ITCs are usually offered as a specified percentage of qualified investments made by the taxpayer, normally within a stipulated period of time.

Capitalization of Expenditures

Most companies with taxable income will choose to expense for tax purposes everything permitted by the tax authorities. This practice has been the source of a great deal of contest and litigation through the years. At the present time, for example, the IRS in the U.S. insists that all successful G&G expenditures be capitalized, with very few exceptions. These costs are then recovered through the units of production method of depletion over many years. The North American oil business is notorious for the great variety and complexity of property trades, assignments of interests, full or partial, and contributions of acreage and/or money to get a test drilled. The tax consequences of these types of dealings can be important, often in surprising ways, to one or more of the parties involved.

All of this discussion of taxation reminds you of what a minister to King Louis XIV of France once remarked---"The art of taxation consists in so plucking the goose as to obtain the largest amount of feathers with the least amount of hissing."

EXPLORATION COSTS

Exploration costs are borne entirely by the working interest owners. Outside of North America this is generally the contracting company, or companies if it is a consortium. The NOC of that host country may be one of the participants depending upon the terms of the concession or production-sharing agreement. Major exploration activities include G&G (Geological and Geophysical) studies, acquisitions of exploration rights, and the drilling of exploratory wells. Geological studies are performed by the company's geological staff with occasional assistance from outside consultants. Geological costs include salaries and benefits of company geologists and their assistants, fees for outside consultants, costs of preparing, reproducing and distributing geological maps and reports, purchasing geological data, well logs, and reports from outside sources. Under the "successful efforts" accounting method all of these costs are expensed up to the point that a discovery is made, reducing each years profits accordingly. When a discovery is made, the cost of the successful wildcat well and all subsequent geological costs associated with the discovery are capitalized. Thus a high percentage of all geological and geophysical costs are expensed in the year the money is spent because there is no evidence of a successful event.

Geophysical studies are generally prepared by the company's geophysists, with occasional assistance from outside consultants. Seismic surveys may be conducted by company crews and equipment, but more often by outside geophysical contractors. The processing, reprocessing, interpretation, preparation of seismic sections and maps may be done either in-house, or outside the company facilities, or both. All of these costs are posted to the books for the property involved.

The cost of acquiring exploration rights (signature bonuses) are capitalized when incurred. The costs of lease rentals, property taxes, etc. are expensed. Undrilled exploratory properties are assessed periodically to ascertain whether their asset value has been impaired, i.e., reduced. A property might be impaired, for instance, by the drilling of a dry hole on, or adjacent to the area in question. When such impairment occurs the asset value is reduced on the company's balance sheet and the amount of the reduction is shown as amortization expense on the income statement. The relinquishment, or dropping, of an undrilled exploratory lease, perhaps merely by omitting the payment of rentals, requires that all remaining capitalized acquisition costs be expensed immediately in that year.

Properties are reclassified from unproven to proved when oil or gas reserves are discovered or otherwise attributed to it. When this occurs, the full acquisition costs are recapitalized on the company's books, and are rescheduled to be written off over the estimated life of the production. This same method is employed for purchased producing properties.

Exploratory well drilling costs are capitalized upon completion pending determination (which may be in a subsequent fiscal year) as to whether the well has found commercial reserves. If the well is commercial the drilling, completion, and facility costs remain capitalized and are depreciated over time as the reserves are produced. If the well is abandoned as a dry hole then one of two methods is used to account for the well costs. Under the "full cost" method, the well costs remain capitalized and are depreciated against production, along with other company assets, if the reserves are judged sufficient to do so. Under the "successful efforts" approach the costs of the dry hole are written off, i.e., expensed, immediately when the well is reported as completed and abandoned on the company's records. "Full Cost" and "Successful Efforts" accounting are discussed more fully in a subsequent section.

DEVELOPMENT COSTS

Development costs, include the drilling and equipping of producing wells, installing production facilities and eventually providing secondary and EOR programs. Under a concession type of operation all costs of development are paid by the operator. In a production sharing type of arrangement all or a specified part of the development costs may be recoverable out of the cost oil share. However, this usually requires that the operator fund the program initially.

The costs of drilling development wells are capitalized, regardless of whether they produce, or are found dry. This includes drillsite preparation onshore and platform costs offshore. All of the production facilities such as flowlines, separators, treaters, dehydration units, lease storage tanks, etc., are capitalized.

In secondary recovery projects all of the injection wells and equipment are capitalized. The costs of injection fluids such as water, steam, carbon dioxide, or LPG, are generally capitalized on the company's financial books and expensed on its tax books.

All of these capitalized costs are depreciated as the related reserves are produced.

COST OF FINDING AND DEVELOPING NEW RESERVES

An oil company produces itself out of business unless it replaces production by new discoveries or purchase of new reserves found by someone else. The cost of replacing these reserves is a basic part of the oil business. The expenses involved in purchasing proven reserves is fairly straight forward. The lack of precise definition of the ultimate recovery which will be realized is the principal risk or uncertainty associated with this approach. The initial expenditure is, of course, definite.

Unfortunately, there is no recognized, or established methodology for ascertaining the *cost of finding* new reserves resulting from the company's exploration effort. Several approaches will be discussed. Each has its advocates; and its limitations.

There are several quite fundamental problems involved:

1. The long period of time which elapses between the initial exploration concept and the eventual drilling, if ever. In many cases there is preliminary conceptual discussion long before any concerted exploratory effort is mounted. The exploration project may wait several years for a place on the exploration budget. How much of this general overhead should be charged to the specific project.

2. In the oil business the successful exploratory efforts must also pay for the much more frequent, but nevertheless costly unsuccessful effort. Not every unsuccessful drilling venture has a direct bearing on an individual discovery, but many do.How much of the cost of unsuccessful exploratory effort should be charged to the new success?

3. Forecasting the magnitude of reserve additions following the estimate attributed to the original exploratory discovery, creates a major obstacle to the cost of finding computation.

Finding costs are oftentimes figured as of some intermediate point in time after the year of discovery but before final abandonment. The established guidelines for posting recovery estimates to a new discovery provide for a minimal initial value which is revised upward after additional stepout drilling, and later with the actual initiation of secondary or enhanced recovery operations.

The absence of any standard by which to figure the cost of finding and the lack of any common understanding as to what factors should and should not be included means that nearly every company and consultant does it differently. This does not create a problem when results are compared within that company and everyone understands how the numbers were derived. It does mean, however, that comparing one company's numbers with others becomes essentially meaningless.

Probably the most common approach to cost of finding would be to take an arbitrary period, say, three years of exploratory costs and divide those dollars by the number of barrels the engineers attribute to new discoveries drilled and completed during the succeeding year times plus an arbitrary scale-up experience ratio of perhaps 25 percent. The scale-up ratio might be derived from company, or industry experience in relating the ultimate recovery estimates to those posted to the company's books immediately after completion of the discovery well.

McGill in studies reported in the AAPG, tabulated reserve changes for Texas RRC District 8A over a period of 31 years. This showed very significant increases in estimates of proved reserves from the same discoveries in each of the first four years of new field life, on the average. This period generally sees the field fully developed, after which for the next six years reported remaining reserve changes are negative reflecting production, until year ten when additional reserves are again recognized and posted to the books.

There can be great divergence of these figures between various geological basins. For example, there would be little similarity between the scale-up ratios of discoveries in the pinnacle reef fields of Northern Michigan and Southeastern Ontario, many of which are one well fields, and say, the Permian of West Texas as reflected in RRC District 8A. Some companies figure their *cost of finding* on a regional basis.

Another common approach to the *cost of finding* question is simply to divide one year's exploration expenditure by that year's figure for total reserves added. The technical deficiencies of this simplistic one-year approach are obvious. Nevertheless, the input numbers, although arbitrary, are readily available. Comparing such figures over a period of years may give Management some idea as to whether finding costs have been increasing or decreasing. There may be a somewhat better basis for comparing a series of one-year based cost of finding figures from one company to another, than with any of the more involved, but less comparable methods.

PRODUCTION COSTS

Direct Production Costs

Production expenses comprise (1) the labor costs, including personnel benefits, to operate the wells and surface equipment, (2) well and equipment repairs and maintenance, (3) materials and supplies, (4) fuel, water and services, and (5) property taxes. Depreciation and amortization of all previously capitalized expenditures for production per se must be accounted for on the company's financial books and may be included as part of production expense. Costs of well repairs, or workovers, including labor, outside service companies, cement, acidizing and frac materials, etc. are all expensed. The costs of recompleting to another producing horizon, on the other hand, are all capitalized and depreciated out of the ensuing production.

311

Allocated Production Costs

This is a far ranging category. Most NOC's, and private sector oil companies operating abroad, i.e., outside the jurisdictional boundaries of their home country, prefer to handle production costs on an country-by-country basis. In some countries a "ring fence" rule is imposed which requires that all expenses be allocated and charged on a prospect-by-prospect basis. Hopefully this means on a field-by-field basis if exploration is successful. In the U.S. the government requires that accounts be kept on an individual lease, or property-by-property basis.

The direct charges already described pertain specifically to wells, lease roads, tank batteries, flow lines, and the like. Allocated costs are those incurred by higher offices which benefit a broad range of the firm's activities. A portion of those costs is properly allocated to each of the several producing properties which benefit from the cost or activity. One must consider the costs of the office that looks after a large number of properties scattered over a geographical region, with a number of field vehicles, a pipeyard, etc. These and a host of other charges at the field level have to be prorated, or allocated to the various properties which they serve. The salary of the superintendent who is responsible for the whole group of leases is allocated on a per well, or per property basis, along with the corresponding expenses of his staff. Then the cost of the employee benefits for all of the superintendent and his staff must be added. This includes such items as the prorated portion of the company's contribution to the employees' pension fund, group hospitalization insurance, and the like. Charging employee benefits to outside owners of a jointly owned property is sometimes limited to a maximum percent of payroll under the terms of the Operating Agreement.

It should also be recognized that every succeedingly higher level in the company organization does work which directly benefits each individual property, or lease. Each company has its own procedures for furnishing operating cost figures, (i.e., the totals of the direct and allocated expenses), back to the specific lease on a periodic basis. Summary totals are also provided for each supervisory grouping of leases and operating level. These figures are used by management to appraise the performance of supervisors at subordinate levels. In this regard, timing of posting charges can sometimes result in misleading indications of cost control performance. Not all fixed charges are allocated costs.

The "cost oil" provisions of production sharing contracts should include all of the expense of the producing operations, although there frequently are limitations. Under concession agreements production costs are the responsibility of the working interest owners.

Work in Progress at End of Accounting Period

Another troublesome aspect of non-cash charges in the matter of accounting reserve for various types of work that are incomplete at year-end. One of these may be the incomplete well account. As an example it is assumed that there is a major gas well development program in which it is important to get a number of wells drilled to hold a group of leases that are about to expire. However, there is no immediate opportunity to connect the production to a gas market. In this scenario, the possibility exists of a well in which the drilling has been accomplished, and at year-end the well is waiting on a completion rig.

The account handling this situation in most companies will total up the work done to that point, and charge the costs associated with this incomplete well to expense, with a corresponding reduction in profit for financial and tax purposes. The costs related to an identical offset well tested, and completed as a producer, before year-end would be largely capitalized, and only a part of the total expenditure expensed, with a resulting increase in reported profit for the year.

ACCOUNTING FOR CONSUMPTION OF ASSETS

NON-CASH CHARGES: Depreciation, Depletion and Amortization (DD&A)

Next along the road to determining the company's profit figure for the year is the matter of non-cash charges. These are the ones which many operating people find a bit difficult to understand and appreciate. Here we include a big dollar item, theDD&A account (Depreciation, Depletion and Amortization). Although these non-cash charges are frequently lumped together, for purposes of discussion and review, they involve different accounting techniques.

In theory, the owner employs DD&A charge-offs against income to match a proportionate part of the cost of investments with the revenue generated by their utilization over their useful life. Strictly speaking, tangible assets (other than natural resources) are depreciated, intangible assets such as lease acquisition costs are amortized, and natural resources are depleted.

One might be led to expect that there is a cash "kitty" being accumulated somewhere in the company's vaults to buy replacements for worn out machinery, or to handle theDD&A account. This is not the way it works. Actually there is no cash involved: thus the term "non-cash charges." Depreciation is not really a source of funds, nor is it truly a negative cash flow. It is a bookkeeping deduction from revenue, applied in calculating taxable and book income, to account for the using up of assets over the long term.

DD&A as non-cash expense reduces profit, and thus income tax for the year. It also reduces the retained earnings shown on the balance sheet as will be discussed later in this chapter. Depreciation also provides the mechanism whereby a major investment (which of itself does not affect profit and retained earnings) is charged progressively into items of expense, which do impact on the firm's profit.

A drilling rig or other piece of oilfield machinery is a tangible asset. Its value declines over the period of time in which it is employed. This reduction in the asset's value is conceived as piecemeal consumption or expenditure of capital. For instance, a truck is a unit of capital. The tires that wear out with use of the truck are actually small units of capital consumed in the intended service of the vehicle. Similarly, the wear of machine parts and the deterioration of elements such as a drilling rig's drilling line comprise physical consumption of capital. Expenditures of capital in this way are often difficult to observe in a physical sense, and are correspondingly difficult to evaluate precisely in monetary terms. They are real, nonetheless.

This concept of depreciating the cost of an asset, in such a manner that the profit and loss statement is a more accurate reflection of current asset value, is basic to modern day financial and income tax reporting. The accountant seeks through the technique of depreciation to keep a current reading of the value of the firm's unexpended physical capital. Accountants employ the term "book value" to represent the original value of an asset less its accumulated depreciation at any point in time. Depreciation thus serves as the means of getting the used-up portion of the initial asset investment off the balance sheet as the asset is used up or worn out.

The physical loss in capital should to be accounted for in arriving at the true cost of production. Theusen (loc.cit.) explains the accounting concept of depreciation in terms of the cost of an asset using a prepaid operating expense to be charged against profits over the life of an asset. Rather than charging the entire cost as an expense at the time the asset is purchased, the accountant attempts to spread the anticipated reduction in value over the life of the asset.

Depreciation, or "Capital Cost Allowance" as it is known in the Canadian regulations, may be complicated further by the fact that the lessening of value of an asset both with use, and the passage of time, should both be recognized. Estimating the useful life of an asset for depreciation purposes may be a matter of engineering judgement. However, current techniques and requirements in the engineering and accounting professions for making this assessment are not always in sync with one another.

In computing depreciation, the original cost of the asset is reduced yearly according to one of several arbitrary mathematical procedures. Usually the firm's accountants employ a duplicity of depreciation methods, one for computing depreciation (which will appear on the company's financial statements), and a second technique that will be used for tax purposes.

Common depreciation methods based on the expected life of the asset include the *straight line* and *declining balance methods.* Those based on a measure of use include units of production which ties each year's depreciation charge to the actual volume of crude oil produced, and *service performed,* based on some other measure of use.

The general preference among the major companies in the oil producing business is to employ straight line, or units of production, depreciation on the firm's financial books. This shows conservative, and hopefully, improving profit figures year to year. For tax book purposes the accelerated depreciation methods (as may be permitted by the taxing authorities) are preferred. These tend to reduce taxable income and therefore taxes. Equations for calculating depreciation and depletion are presented in Appendix VIII-A.

What is the significance of having fully depreciated an asset to the point that the remaining undepreciated balance is zero? This can occur before the assets is physically taken out of service, so this does not mean that the asset is of no value. It only means that you have fully charged the cost of the asset against income and will no longer reduce profit by depreciation on that asset. The value of an asset is not the undepreciated balance in that asset's account, but its fair market value; i.e., what it could be sold for on the open market. Thus at best, the undepreciated balance in an asset account is a very poor approximation of its true value at that time.

Straight Line Depreciation

Straight line depreciation is the most common method employed in the industry. This technique assumes that the value of an asset decreases at a constant rate over time. Thus if an asset cost $10,000 and has an estimated useful life of 10 years, the depreciation will be:

$10,000/10 = $1,000 per year, or 1/10 of the purchase price per year.

Straight line depreciation is the most conservative of the several methods in use.

Units of Production Depreciation

Oil producing properties typically produce less and less income over time once the field starts to decline in productivity. The straight line method burdens the early years' revenues relatively lightly, when the production and profit are highest, and the later years' income unduly, when production is down and operating costs are still significant. The units of production (UOP) method tends to minimize this problem. UOP also has the advantage of reducing the asset value to zero only when the reserves are fully depleted. It is, however, subject to revisions of the engineer's estimates of remaining reserves which may go through considerable refinement as depletion proceeds. The units of production method does assure that the asset will be written off over its actual economic life rather than according to an estimated life which in practice may either not be achieved or may be exceeded.

■ EXAMPLE 8-1
■ UNITS OF PRODUCTION DEPRECIATION

A lease and its producing well have a year end undepreciated capitalized value of $1,000,000 for well tangibles and lease facilities. It is expected to produce an estimated 500,000 barrels of net revenue interest oil including the 500 bbls. of test oil produced and sold in year 1. The following producing rates are projected:

Year	Production
0	0 barrels
1	500
2	75,000
3	60,000
4	60,000
5	50,000
6	45,000
-	—
-	—
n	5,000
Total	500,000

DETERMINE:

A UOP depreciation schedule for the situation described.

SOLUTION:

The UOP method charges off annual depreciation in the same proportion that annual production reduces the remaining reserves. Since the total value to be depreciated is $1,000,000, and ultimate recovery is 500,000 barrels, every barrel of oil produced should bear $2.00 of depreciation ($1,000,000/500,000 bbl. = $2.00/bbl.). Therefore the schedule of depreciation is determined by multiplying $2.00 by each year's annual production, as is shown in Table 8-2.

Alternatively, UOP depreciation can be considered as writing-off the remaining undepreciated balance in the proportion that annual production bears to the remaining reserve. Using this approach it is necessary to determine the fraction that each year's production comprises of the remaining reserves at the beginning of that year, and multiplying that changing fraction by the undepreciated balance at the beginning of that year. This method of computation allows for reserve revisions and additional capital expenditure during the life of the producing property. It is the procedure most frequently used by accountants. This can also be observed in Table 8-2. If the estimated ultimate recovery remains unchanged and no additional capital investments are made, both methods will yield the same result.

TABLE 8-2
UNITS OF PRODUCTION METHOD OF DEPRECIATION EXAMPle

Year	Reserves at Beginning of Year (bbl)	Production During Year (bbl)	Depreciation Amount for Year ($)	Year-end Undepreciated Balance ($)
0	0	0	0	1,000,000
1	500,000	500	1,000	999,000
2	499,500	75,000	150,000	849,000
3	424,500	60,000	120,000	729,000
4	364,500	60,000	120,000	609,000
5	304,500	50,000	100,000	509,000
6	254,500	45,000	90,000	419,000
⋮	⋮	⋮	⋮	⋮
⋮	⋮	⋮	⋮	⋮
n	5,000	5,000	10,000	0
Total		500,000	1,000,000	

Accelerated Methods of Depreciation

Accelerated depreciation methods generate a greater depreciation charge in the earlier years of the life of a tangible capital item than would straight line depreciation. The annual write-offs under these methods decline with time.

Various provisions for accelerated depreciation write-offs may be available for tax purposes. These methods may be granted by governments as a means of encouraging economic activity in a particular area, or time frame. Faster write-offs are generally looked upon with favor by business as a means of reducing the "tax bite" from a revenue stream.

Declining Balance Methods

In declining balance depreciation the first year's depreciation is a multiple of straight line depreciation and then each subsequent year's depreciation is reduced by a fraction related to the total depreciation life. The most commonly used version of this method is the "double declining balance" method in which the first year's depreciation is double the value which would be calculated for straight line. Each following year's depreciation is reduced by the fraction of two over the depreciation life. The declining balance methods approach depreciation in the following manner. The book value of the asset is multiplied by a constant fixed percentage at the beginning of each fiscal year to determine the depreciation charge against income for that year. Since the book value is reduced each year by the amount of the previous year's depreciation, the absolute size of the depreciation charge also declines with time each year.

The declining balance "factor" may be any constant value between, for example, 125 and 200 percent of each year's straight line depreciation, all of which is deducted from the first of that year's remaining book value. By definition the Double Declining Balance method uses a 200 percent factor. In application of the 200 percent declining balance method, the first year's depreciation for a ten year project life would be twice the straight line rate of ten percent, i.e., twenty percent for year one.

The declining balance method is the only one of the depreciation methods in this discussion that does not produce a depreciation schedule whose sum over the life of the capital asset will equal the initial depreciable base. (Figure 8-1 and Table 8-3). This situation occurs because the factor is applied each year to the remaining undepreciated base rather than to the initial investment base.

Summary

Figure 8-1 depicts the remaining undepreciated balance as a percent of the initial book value, depreciated according to the methods discussed, for an assumed productive life of 10 years. It is common practice to start with double declining balance depreciation for several years and then switch to straight line, to fully depreciate the asset. This type of switching can be economically advantageous, but may be regulated for tax purposes by local authority.

FIGURE 8-1
COMPARISON OF DEPRECIATION METHODS

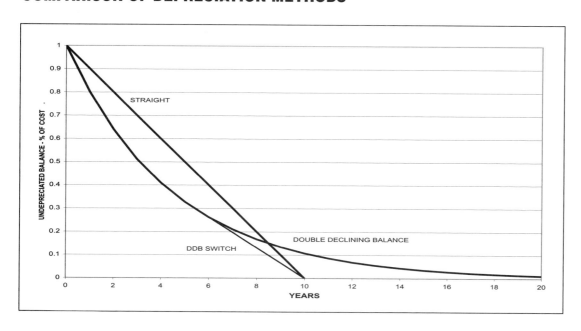

Table 8-3 illustrates the impact of depreciation on a firm's profit for a ten year project, using several methods. It can be seen that the declining balance methods do not fully depreciate the cost of the asset. To fully depreciate an asset using the declining balance method it is necessary to switch to straight line during the depreciation period, as indicated by the line identified as DDB Switch on Figure 8-1. The optimum time to switch to straight line can be determined from Equation 8.6 presented in Appendix VIII-A at the end of this chapter. Table 8-3 shows the annual depreciation amounts for a $3,000 investment having a life of 10 years for the depreciation methods presented in Figure 8-1 and for units of production depreciation.

TABLE 8-3
SUMMARY OF DEPRECIATION METHODS

Year	Straight Line	100% Declining Balance	Double Declining Balance	Double Decl. Bal. St. Line	Units of Production Vol.	Units of Production Depn.
1	$300	$300	$600	$600	5	$119
2	300	270	480	480	20	476
3	300	243	384	384	18	429
4	300	219	307	307	16	381
5	300	197	246	246	15	357
6	300	177	197	197	13	310
7	300	159	157	196	12	286
8	300	143	126	197	10	238
9	300	129	101	196	9	214
10	300	116	81	197	8	190
Totals	$3,000	$1,953	$2,679	$3,000	126	$3,000

Service Performed

In some cases it may be inappropriate to assume that capital is recovered according to one of the aforementioned value-time models of depreciation. An important alternative procedure is to assume that depreciation occurs on the basis of actual service performed, without reference to time. Thus an airplane tire might be depreciated on the basis of actual aircraft landings. If the cost of the tire were $500 and it was estimated to be good for 5,000 landings, then the depreciation charge per landing will be:

$$(\$500) / (5,000) = \$0.10/landing$$

The annual depreciation would depend upon the number of landings that occurred during the year.

Salvage

Accounting authorities and tax authorities agree that depreciation should be based upon the original actual cost of an asset with no reduction for the value (salvage) that the asset may have when it is taken out of service. This recognizes that it is almost impossible to estimate the value of a worn out piece of equipment a long time in the future, especially in an inflationary economic climate.

Discont factor should
only be applied to cash in/cash
out actuals.

CHAPTER VIII / NET INCOME, "THE BOTTOM LINE"

If an asset does have a recognizable value when it is taken out of service, the difference between its then "fair market value" and the undepreciated balance in that asset's account is treated as a profit or a loss against ordinary income. The asset will be removed from the company's books at that point as an abnormal retirement by taking irregular depreciation equal to the undepreciated balance remaining in that account.

Depletion

Depletion differs in theory from depreciation due to the fact that the latter is the result of use over time while the former is the result of the intentional, systematic recovery or removal of certain types of assets. Depletion refers to operations that deplete, or use up the supply or deposit. The word depletion literally means to empty. When natural resources are produced and sold depletion indicates the lessening in value of the reserve as the process proceeds. Examples of depletion include the mining and removal of coal, cutting and removal of timber from a forest, the removal of stone from a quarry or the recovery of crude oil from a reservoir.

The manner in which the capital recovery accounting is handled under depletion differs significantly from the alternative methods of calculating depreciation. In the case of depreciation it is reasonable to assume that the used up asset will be replaced with a like asset. In the case of the depletion, however, specific replacement is not possible. The amounts charged for depreciation, for example in a manufacturing activity, would theoretically be reinvested in new equipment to continue the operation.

In the exploitation of a natural resource, however, the amounts charged to depletion cannot be used directly to replace the ore deposit or oil reserve and the venture could eventually sell itself out of business. The revenue in such cases must include both the profit earned and the return of an appropriate portion of the capital which was invested. The long term continuation of operations involving the exploitation of non-renewable natural resources normally calls for the acquisition of new properties to replace those being depleted so that the business remains a going concern.

Expenditures which are subject to cost depletion include geological, geophysical and land expenditures incurred in making a discovery. It also includes the portion of any purchase price which was allocated to the value of the reserves in the ground. Using the UOP method for calculating depletion will allocate the same unit cost to the first barrel of oil produced from the reservoir as it will the last barrel produced from the reservoir. This of course assumes that there are no additional capital expenditures for discovery or purchase of those reserves and that the estimated ultimate recovery turned out to be correct.

Computing depletion by the cost method is similar to the units of production methods of figuring depreciation which were discussed in a previous section. Under the cost depletion method the depletion charge is figured on the amount of the resource that is actually produced and sold as related to the initial investment in the total amount of the resource available.

If it is assumed, for example, that a reservoir which contains an estimated 1,000,000 barrels of recoverable oil required an initial outlay of $5,000,000 to find and develop. The units of production depletion rate for this property is $5,000,000/1,000,000, or $5/bbl for each barrel produced and sold. This depletion charge is accumulated on an annual basis. Thus, if 50,000 barrels are produced during

The faster the depreciation, the better
economic results in NPV calc

the twelve month period the non-cash charge for cost depletion will be:

$$(50,000 \ bbls) * (\$5/bbl) = \$250,000$$

The number of units of remaining reserve will change each year according to the previous year's withdrawals coupled with any reevaluation of the remaining recoveries as the result of a new reservoir engineering study. Thus, cost depletion must be recalculated annually by dividing the unrecovered cost of the reserve by the currently estimated beginning of the year barrels of reserves (see Equation 8.8, Appendix VIII-A).

Certain extractive industries in the U.S., and some other jurisdictions, may qualify for an optional method of calculating tax depletion allowance. This alternate method is known as percentage depletion. Under percentage depletion a fixed percentage of gross income from production at the wellhead, limited to the net taxable income for the period (including expensed IDC's), is permitted as the depletion write-off. Some of the stipulated percentages for depletion include: sulfur and uranium, 22%; coal and salt, 10%; and gravel 5%. Oil and gas produced by non-integrated companies, i.e., those which are not engaged in refining and marketing, and whose net production is less than 1000 equivalent BOPD, can qualify for a 15% percentage depletion allowance in the U.S. The cost depletion basis is reduced by the accumulated amount of any percentage depletion charged against the property, but percentage depletion is not limited to the actual costs.

EXAMPLE 8-2
COMPARING PERCENTAGE AND COST DEPLETION

To illustrate the working of percentage depletion let us return to the production example mentioned above. Further, let us assume the price of the produced crude oil is $20/bbl at the pipeline connection, pipeline gauging and transportation charges to the pipeline connection are $1.80/bbl and the total direct and allocated operating expense of the property for the year is $500,000. The allowable percentage depletion for the property is then computed as follows:

Year's Production 50,000 bbl at $20/bbl	$1,000,000
Plus delivery charges (50,000 bbls at $1.80/bbl)	$90,000
Gross depletable wellhead income	$1,090,000
Percentage depletion, @15%	$163,500

A check is also necessary to determine whether the figure $163,500 exceeds the maximum depletion allowance permitted as follows:

Gross depletable income	$1,090,000
Less operating expenses	500,000
Maximum allowable percentage depletion allowance	$590,000

Since $163,500 is less than $590,000, the full percentage depletion of $163,500 will be permitted. Otherwise, only the limitation of 100% of net income would have been allowed for that tax year. Comparing the $163,500 figure with the $250,000 cost depletion value computed earlier demonstrates that the cost depletion method is to be preferred for the year under study. All of these computations must be made for each property each year to determine the depletion method which is the most advantageous for the period for tax purposes.

EXAMPLE 8-3
EFFECT OF CHANGE IN RESERVE ON DD&A

A producing property was purchased during the past fiscal year for $30 million, which was comprised of $20 million for the reserves and $10 million for tangible equipment. After the purchase the operator had an additional $500 thousand tangible drilling expenditure. The proved developed reserves after royalty for the property at the end of the fiscal year was 10 million barrels. The property produced 2,000 BOPD after royalty over the entire year for which the operator received $20 per barrel.

DETERMINE:

1. What are the operator's cost depletion and UOP depreciation figures for the year?

2. What are the corresponding values if the estimated reserve is increased to 20 million barrels as the result of a new engineering study?

SOLUTION:

(1) Depletion expense:

$$\$20,000,000 * \frac{2,000 * 365}{10,000,000 + (2,000 * 365)} = \$1,360,671$$

Depreciation expense:

$$\$10,500,000 * \frac{2,000 * 365}{10,000,000 + (2,000 * 365)} = \$714,352$$

$$\text{Total DD\&A} = \$2,075,023 (\$2.84/\text{bbl})$$

(2) Depletion expense

$$\$20,000,000 * \frac{2,000 * 365}{20,000,000 + (2,000 * 365)} = \$704,293$$

Depreciation expense:

$$\$10,500,000 * \frac{2,000 * 365}{20,000,000 + (2,000 * 365)} = \$369,754$$

$$\text{Total DD\&A} = \$1,074,047 (\$1.47/\text{bbl})$$

This represents a difference in DD&A resulting from the engineering reevaluation of reserves, of $2,075,023 - $1,074,047 = +$1,000,976 which is reflected as a direct increase in the year's before-tax profit.

The result of reserve recalculation can work the other way, too. A cut in the estimated life or recoveries from a project will mean a boost in the DD&A charges, a sudden reduction in the year's profit, and a potentially very disturbed and unhappy management and ownership group.

Amortization

Intangible capital can be amortized using the same methods as has just been described for depreciating tangible capital. All of the depreciation philosophy and concepts apply equally to intangible capital. However, in the oil business the most commonly used method of amortization for accounting purposes is the unit of production method. The most frequently amortized cost is intangible drilling expense. Just as with the tangible portion, the intangible capital expenditure will benefit the well throughout its entire life, in proportion to production.

As was discussed earlier, methods for distributing capital costs over time for income tax purposes or for other government calculations often differ from those recommended by the accounting community. In some cases intangible capital is completely amortized in the year of expenditure, just like an expense, or spread on a straight line basis over a only a few years, or a combination of the two methods.

Lease bonus is another capital expenditure that is frequently amortized. This type of expenditure is made prior to making an oil or gas discovery and there is a significant probability that exploration on the tract will be unsuccessful. Therefore, for accounting purposes, it is assumed that the venture will be a failure and the bonus is amortized over the primary term of the lease or exploratory period of the venture. Such amortization is usually on a straight line basis. If the lease bonus is spent on a tract where a discovery is actually made, then all of the amortization previously taken will be restored and the bonus will then be recovered by depletion.

DIVIDEND PAYOUT

Once the after-tax profit has been figured the investor immediately looks to his dividend. The dividend is the portion of the profit that is paid out on a per share basis to equity owners of record (on the date established by the Board of Directors) at the time the dividend is declared. If you are not the holder of record of the equity shares on that date you don't get the dividend: payment will be made to the previous owner. The dividend represents cash in hand and income to the share owner. Dividends are a part of the shareholder's return on investment (equity ownership), just as the interest which would have been paid on a bond or debenture comprises the return on a debt security.

Dividend payouts vary among companies, and industries. Probably a common mean of payout would be forty to fifty percent of net earnings, with the rest of the retained earnings plowed back into the business to help the firm grow in the expectation that earnings and dividends in subsequent years will be even larger.

COMPANY FINANCIAL ORGANIZATION

The financial affairs of the company are directed by and are the responsibility of an individual designated as its Chief Financial Officer. This is usually not the title of a company officer, per se. He or she may carry the title of a company office anywhere from Vice President to Vice Chairman of the company's Board of Directors. The initials "CFO" are frequently shown as a supplemental title, so to speak, just as CEO may be used along side the title of the Chief Executive Officer, who may be President, or Chairman of the Board.

The organization which the CFO directs includes the company's treasury functions, accounting, and investor relations. Although the Treasury Department, led by the company's Treasurer, is normally one of the smaller units of the company, personnel-wise, it handles all of the monies actually received by and disbursed by the firm.

322

A good part of the Treasurer's duties, including the prompt collection of funds due, and paying the accounts due, involves timing of the cash management function. An oil company's cash income stream is normally fairly steady over time as oil and gas production is paid for month by month. Major projects, and the payment of dividends, require large disbursements at specific points in time. This necessitates accumulating funds in anticipation of these requirements, usually in the form of commercial paper.

Commercial paper comprises short term obligations with maturities of 2 to 270 days issued by banks, major corporations and other borrowers to investors with temporary idle cash. Company Treasurers like the flexibility and security of this type of instrument which is issued only by top-rated concerns and is nearly always backed by a bank line of credit. Both Moody's and Standard and Poor's assign rating to commercial paper as discussed in Appendix VII-A.

OIL AND GAS ACCOUNTING

Early accounting for tax assessment, as well as commercial accounts, in England before the availability of paper was accomplished by physically notching small-diameter tree stumps known as "tallies." Thus, the expression "tallying up the account." More modern accounting techniques are reportedly an early German invention which provides a detailed standardized arrangement of numerical account codes for assets, liabilities, revenues and expenses. France has a mandatory code of accounts which greatly simplifies the taxman's and the auditor's work but presents some difficulties in moving from industry to industry.

The Accounting Organization

The principal task of the firm's accountants is the periodic preparation of the two most basic financial statements for the enterprise, the income statement and the balance sheet. A highly important accompanying role is their service to management in budget preparation and internal control reporting to the operating departments. Figure 8-2 shows a typical accounting organization for a small oil and gas producing company.

There are five basic categories of accounts of a business. These are assets, liabilities, equity, revenue and expense. The first three of these are evaluated as of a fixed date and comprise the balance sheet. Revenues and expenses must be accumulated over a stated period of time, usually one year, ending with the effective date of the balance sheet.

Each of these financial statements is prepared from a number of accounts which are kept and totaled up at specific times for the purpose of preparing the basic accounting statements. The individual accounts are designed to provide summaries of operating data which are useful to management in monitoring the activities of the enterprise. This is a fairly labor-intensive activity compared to the rest of the oil industry. Examples of commonly employed accounts include cash, accounts payable, capital stock, operating costs, pipeline runs, accounts receivable and a number of others.

FIGURE 8-2
ORGANIZATION CHART

For an Oil and Gas Producing Company Accounting Department

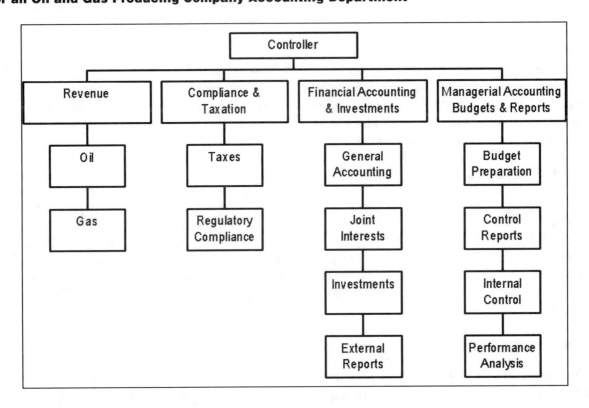

Successful Efforts and Full Cost Accounting Methods

The oil and gas accounting rules provide a choice between the industry's two distinct types of accounting systems: full cost and successful efforts. The larger oil companies employ the successful efforts accounting method while independents and those companies which rely heavily on outside sources of capital, such as drilling funds, may generally prefer the full cost approach. The accounting firm, Ernst&Young, in their brochure on the full cost method states:

"The full cost method is based on the concept that oil and gas exploration is a single activity, not a series of individual activities. Full costing regards the costs of unsuccessful acquisition and exploration activities as necessary for the discovery of reserves. Because all acquisition and exploration costs, whether productive or not, are incurred in anticipation of discovering oil and gas reserves and are essential to the ultimate discovery of reserves, the full cost method calls for these costs to be capitalized in their entirety. Costs capitalized are amortized on the units-of-production (UOP) basis. General and administrative expenses applicable to current production and general corporate matters and other costs related to current production are charged to expense as incurred."

The philosophy of the method is appealing, and it is widely followed, particularly among the smaller companies in the oil business. Full cost accounting assumes that all exploration expenditures will have a long term effect upon the company, so no matter if a discovery is made or the exploration effort ends in failure, accordingly, all of the costs are capitalized and amortized over a number of years. Under this accounting procedure "dry holes" are listed as assets of the company, just like successful wells, and are amortized along with all of the exploration costs incurred for successful and unsuccessful ventures. Full cost accounting has the effect of increasing profits for a new company or one that significantly increases its exploration program during the first few years of its use, but will reduce profits in later years, at times when big discoveries are made, or in years when exploratory expenditures are reduced.

Successful effort accounting capitalizes only those exploration and acquisition costs which are related to actual oil and/or gas discoveries. All other exploration costs are expensed in the year in which the money is spent. Normally, exploration funds are spent not knowing whether they will lead to a successful effort or not. In these circumstances, it is customary to expense all drilling costs until the well is completed, and at year end to reverse those entries and capitalize the funds that were spent on the wells that were completed as producers. Most large major oil companies use the successful efforts accounting method. These, and other comparisons of the two methods are summarized in Table 8-4.

The petroleum industry in the U.S. is subjected to an interesting conflict of federal bureaucracy relative to the two methods. The IRS, even though they set their own rules for keeping the company's books for tax purposes, rather favors the full cost method simply because it results in higher indicated profits and thus higher tax collections. The SEC, another branch of government, on the other hand, has a strong dislike for the full cost method. The SEC feels that the full cost method can be, and all too frequently is, misleading to the investing public. Current profit figures for an exploration company clearly are higher under the full cost method since so little of the cash flow of the firm is deducted from income as exploration expense for the inevitable dry holes. The higher beginning book profit figures under the full cost approach may look unduly appealing to the potential investor.

The real difficulty arises in the case of the less successful small exploration companies. A company that has nothing but dry holes to show for its recent exploration efforts can still show a handsome profit from its established cash flow, and a large figure on the asset side of its balance sheet for unamortized exploration expenditures. "Buyer beware!"

The IRS imposes a degree of control of the situation by requiring a write-off of assets against current profit if there are insufficient projected future revenues from the company's existing proven oil and gas reserves to eventually offset all costs which were associated with the finding, developing and producing those reserves. The full cost method requires a quarterly evaluation of reserves to determine whether such write-offs are necessary. During periods of declining crude oil prices this can create quite a burden. In the U.S. one is permitted a one-time switch from full cost to successful efforts—never the reverse.

The successful efforts accounting method takes a much more conservative approach with regard to reporting current earnings, since approximately 70 percent of exploration expenditures are subtracted from revenue as incurred, and only the successful wildcat wells are capitalized. The assets shown under this accounting method thus come closer to reflecting the current exploration results. This is not to say that under the full cost method there are no reserves behind the asset figure, but merely that there is no way to tell.

Expensing dry holes under the successful efforts method has the effect of reducing current profits by the amount of dry hole writeoffs and also understates the total amounts capitalized which are matched with the future revenues as the reserves are produced. One of the accounting problems in keeping track of the upstream oil business is that even though the well may have been drilled and completed within a certain fiscal year, the effort actually culminates activities which span several time periods. Since expensing is confined to a single accounting period the capitalization process can conveniently be linked to several time periods. All booked assets are eventually charged to non-cash expense, but over a prolonged period of time.

Over time, the full cost method capitalizes a greater amount than the successful efforts method, and will probably result in larger cumulative earnings, which are reflected in the company's retained earnings statement. The fundamental difference between the two methods really has to do with the relative size of the cost center employed. Full cost uses the largest possible cost center, i.e. the whole company, whereas successful efforts actually applies the charges to an individual property, or lease, as the cost center.

Full cost may come closer to reflecting the results of cumulative drilling efforts, over time, by allowing a comparison of unamortized costs with the present values of future income. An illustration of this type of situation is included as the Elmworth example in Chapter X.

It should also be recognized that the full cost and successful efforts methods are not diametrically opposed methods of accounting. Neither method allows capitalized costs to exceed the present value of future net revenues. The advocates of the successful efforts method feel that expenditures (dry holes) which yield no economic benefits should be written off at the time incurred and that only those expenditures that lead directly to producible reserves, and thus future revenues, should be capitalized. The successful efforts devotees argue that other types of business do not capitalize failures.

Since successful efforts can yield overly conservative financial indications, and full cost fails to highlight poor short run exploration results, neither method by itself provides adequate managerial performance information. This again demonstrates the inadequacy of conventional balance sheet accounting to the oil business. Nevertheless, petroleum is part of the world's commercial community which has evolved financial reporting procedures dating back to the days of Charles II in England. Such a capital intensive industry as petroleum has to look to outside funding for a significant portion of its needs. Accordingly, it must report back to the investor in as conventional a set of financial statements as possible. Both full cost and successful efforts are widely employed to fill this need.

INTERNAL COMPANY ECONOMIC REPORTS

Lease Operating Statements

One of the most basic financial statements in the petroleum production industry is generally referred to as the Lifting Cost Statement. This seeks to collect, usually on a monthly basis, all of the direct operating costs for an individual producing lease or group of wells. It is a particularly effective device for monitoring finite costs of field operations. If lifting costs are up, the production supervisor knows straight away that he will have to explain to his management the reasons which led to the charges.

The principal difficulty with lease operating statements as a management tool has always been that of current posting. If a particular charge gets lost in the shuffle and does not get posted for several months the chances are that the operating personnel will have forgotten the details and that portion of the statement only serves to generate frustration and actually fails as a management tool. Computer handling of these direct charges has greatly alleviated this shortcoming.

TABLE 8-4
SUMMARY—TAX TREATMENT VS. FINANCIAL BOOK TREATMENT

	Successful Efforts	Full Cost	Tax Treatment
1. Dry holes–Exploratory	Expense	Capitalize	Expense*
2. G & G Costs:			
Core drilling	Expense	Capitalize	Expense
Seismic crews	Expense	Capitalize	Expense
Data processing	Expense	Capitalize	Expense
Interpretation	Expense	Capitalize	Expense
Test well contributions	Expense	Capitalize	Expense
Portion of overhead	Expense	Capitalize	Expense
3. Acquisition of Exploration Rights	Capitalize	Capitalize	Capitalize
4. Non-producing tracts	Amortize	Expense	Inventory
When surrendered	Expense Balance	Capitalize	Expense
5. Discovery & Development wells			
Intangible well costs	Capitalize	Capitalize	Exp./Amort.
Tangible well costs	Capitalize	Capitalize	Capitalize
Delineation G&G costs	Capitalize	Capitalize	Capitalize
6. Taxes and lease rentals	Expense	Expense	Expense

Methods of Handling Depletion, Depreciation, and Amortization (DD&A)

		Successful Efforts	Full Cost	Tax Treatment
7. Depreciation:				
Lease and well equipment				
	–New	UOP	UOP	Tax Schedule
	–Used	UOP	UOP	Tax Schedule
8. Depletion		Cost depletion	Cost depletion	Cost depletion**
9. Amortization:				
Intangible drilling costs		UOP	UOP	Tax Schedule

*Unsuccessful G & G may be capitalized for tax purposes at the taxpayers election.
**Royalty owners and small independent producers may be permitted percentage depletion of 15% of gross income from production not to exceed 100% of net income.

Financial Statements Above the Field Operating Level

As one moves up the organization in designing the periodic statements of costs the matter of overhead appears, and becomes more and more prominent. Overheads are expenditures that are incurred on behalf of the entire operating entity. There is no straightforward relationship between overhead costs and operating or development expenditures. The allocation of overheads is generally accomplished in a rather arbitrary manner, but can be very far reaching in their economic impact. The most simple methods of allocation rely on ratios to some easily determined physical count. The number of producing wells in the operation, or the number of daily barrels of oil production are the most common bases of allocation.

Examples abound in the industry of situations in which the application of simple ratios to the allocation of overhead has resulted in the untimely, or premature abandonment of otherwise viable operations. A classical example from the marketing side of the oil business involved a small one pump gasoline sales situation outside a rural grocery in a remote area. Costs were low, volume was steady, and the company supervisor came past only once every several months. The company allocated marketing overhead on a per service station count. Charging the same dollar overhead expenses each month to the small rural outlet, as to a 500,000 gallon a month station in a major city, eventually resulted in the demise of the small, otherwise worthwhile, rural station.

Items of allocated operating expense at the field level were reviewed in Table 4-3. Overheads may apply to all levels of the company organization, however, depending on the company's choosing. Some companies, for instance, charge the costs of research and development only to the upper levels of the corporation. The costs of administration and top management are often allocated down the organization, but some attention should be paid as to how much of the CEO's salary should be charged against each barrel of production.

The major breakdown of overhead allocation is generally shown in the series of Control Reports discussed in Chapter IX.

Operating Statements at Higher Levels in the Company Organization

Good management practice calls for the periodic production of operating cost statements for each organizational unit. The difficulty with many of these statements stems from the inclusion of such a large proportion of allocated costs which are outside the control of the individual supervisor of the unit. The allocated costs at these levels may include such things as accrued, but untaken, vacation time and other accounting refinements which are of little help in the day to day management of the activity.

FINANCIAL STATEMENTS AT THE CORPORATE LEVEL

Accounting for the financial performance of the entire firm follows an infinite maze of rules and regulations. Many of these procedures are traditional, and many are legal requirements, in society's effort to protect the innocent investor in the company. Accounting is the method of keeping score in business.

The scores depicted on the company's financial reports do not represent the availability of real spendable dollars. In contrast to the company's cash books, or records, the financial books are based on accruals which keep track of transactions that create assets or liabilities.

At the corporate level a series of several scoreboards, known as financial statements, are used. These always include the balance sheet and the income statement as a minimum. Other additional financial statements are also provided depending upon the requirements of the host government and the preference and persuasion of the Chief Financial Officer of the company.

The balance sheet shows the relationship of the company's assets, liabilities and owners equity at a specific point in time made to coincide with the closing date of the income statement. The term "balance sheet" derives from the accounting requirement that the total of the firm's assets exactly balance, i.e., equal, the sum of the liabilities plus the owner's equity. Thus:

Assets = Liabilities + Owner's Equity

The company's financial activities which don't have direct, or immediate impact on the year's profits (long term financing, accrued tax liabilities, etc.) which do not show on the Income Statement are normally reflected on the balance sheet.

The income statement and the balance sheet are linked through the Owner's Equity section of the balance sheet. If company's earnings (Net Income) in excess of dividends paid out are positive over a period of time then owner's equity increases. The converse is also true, of course. Thus activities which affect one financial statement will eventually affect the others as well.

The balance sheet represents the financial picture as it existed at the close of business on a particular day. It is normal practice to show the balance sheet as of the last day of the current fiscal year along side a set of similar data for the previous year which is included for comparison. Showing the data for more than the single year being reported isn't mandatory, however.

Figures 8-3 and 8-4 comprise simple descriptions of the balance sheet and income statement formats. A general explanation of the various items follows the balance sheet format. Figure 8-5 then shows how these two statements relate to one another through the owner's equity section.

THE BALANCE SHEET

The balance sheet format for an oil and gas producing company shown in Figure 8-3 sets out the assets, liabilities, and equity of the firm as of a certain date.

The balance sheet is divided into two parts. The assets of the firm are listed in the upper part and are subdivided into several major groupings. The company's liabilities are listed on the lower part which shows what the firm is obligated or will be obligated to pay. Below the liabilities section is a statement of the stockholders equity, i.e., paid up common shares plus retained earnings and surplus. The equity figure, in effect, lists the amount the stockholders would share, or "split up," if the company were liquidated at full market value on the balance sheet date. The top and bottom parts of the balance sheet always total to exactly the same figure, i.e., the Assets must equal the sum of the Liabilities plus the Shareholders Equity, the basic accounting equation.

There are two important facts that should be recognized with regard to a company's balance. First, the asset values shown are what was actually paid for the listed assets, when they were acquired less accumulated DD&A, unless the DD&A is listed separately. If the assets were acquired a long time in the past they probably bear little relationship to current or replacement value due to inflation. Second, any change that is made to the balance sheet must flow through the income statement. This means that any accounting changes that affect the balance sheet will also affect profit on the income statement. Any revaluation of assets, such as inventory or marketable securities will also affect profit.

FIGURE 8-3
BALANCE SHEET FORMAT

Financial condition at a point in time, as December 31, 200Z

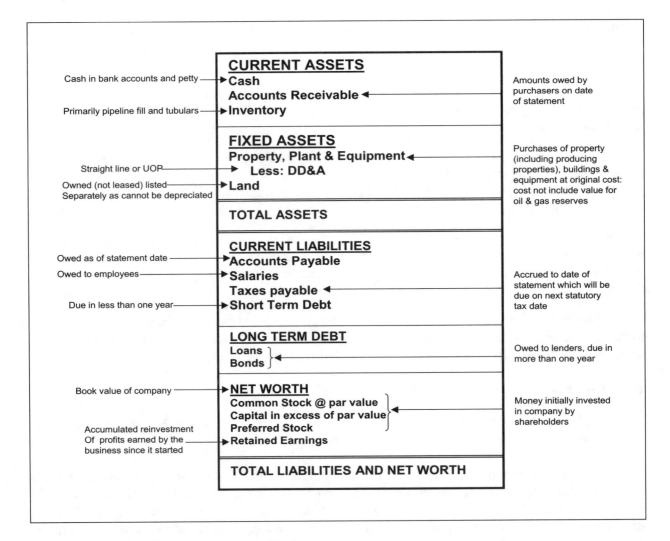

THE TOP OF THE BALANCE SHEET

ASSETS

A company's list of assets reflects how the enterprise has used monies made available to it.

The assets section of a company's balance sheet is normally divided into current assets and fixed assets. Sub-categories are listed in order of their presumed liquidity, i.e., the ease with which the asset can be converted to cash.In North America they are listed in descending order. In most of Europe the listing is reversed, i.e., the most liquid asset is shown last.

Current Assets

In general, current assets include cash in the bank and other liquid assets that will be turned into cash in the normal course of business within one year of the balance sheet date. The items in this category are similarly listed in the order of facility in their conversion to cash.

Accounts Receivable

This is the amount still to be collected from pipeline purchasers of the firm's oil and gas production and from other customers as of the date of the balance sheet.

Inventories

Unlike most commercial establishments oil and gas producers don't create and maintain a significant inventory of production, as would a manufacturing enterprise. There are exceptions, however, such as the great blocks of stored sulfur recovered from the sour gas production in Alberta. Producing companies often do maintain significant inventories of tubular goods and other oilfield equipment. The largest dollar amount of a producing company's inventory may involve a share of pipeline fill of crude oil in certain pipeline systems which may serve the operation. These are carried as inventory for purposes of the financial statements. Neither depreciation or other accounting or tax write-off are calculated for any item classified as inventory. Items in inventory are only depreciated after they are removed from inventory put to their intended use.

Fixed Assets

Fixed assets include the major assets of the company which are not intended for sale and which are used in the conduct of the company's business. In the U.K. fixed assets are listed at the top of the asset listings.

Property, Plant and Equipment

This is the largest sub-category of the fixed asset account in most businesses. The usual approach to valuing these assets for balance sheet purposes is to post the original cost minus depreciation accumulated to the date of the balance sheet. The resulting figure is not a reflection of its current market value, nor its future replacement cost. This is particularly true of a resource industry where its largest single asset is its reserves, which may have been discovered by the company itself with little of that cost having been capitalized. The cost of any reserves which a company purchases is included in this account, but that may have been a long time in the past and represent only a small fraction of its current value.

THE BOTTOM OF THE BALANCE SHEET

LIABILITIES

The liabilities indicate what money has been made available to the enterprise and its source. This area of the balance sheet is also divided into current and long term categories.

Current Liabilities

Current liabilities are those financial obligations which are due within one year. Current assets and current liabilities are related to some degree since the current assets represent the source from which payments are made to satisfy the current debts.

Accounts Payable

Accounts payable represents all amounts owed by the firm in the regular conduct of its business with creditors from whom it has purchased materials and services on open account and which have not yet been settled up on the balance sheet date.

The balance sheet records actual accounting liabilities not financial commitments. Thus, if the company has entered into a contract for purchases over ten years this financial fact will not appear on the balance sheet. The commitment may be revealed in one of the notes to the financial statements, however.

Wages and Salaries Payable

The amount of wages and salaries accrued, but not as yet paid out on the date of the balance sheet, appear on the statement under this heading.

Taxes Payable

Accrued taxes owed but not yet paid at the closing date of the Balance Sheet.

Short Term Loans

Short term money owed to a bank or other financial house is shown under the notes payable balance sheet heading.

Long Term Debt

Debts of a business which are due more than one year from the date of the financial report are classed as long term obligations and appear under this heading.

NET WORTH

Shareholder's equity, which is the net worth of the company, represents the total equity interest that all of the shareholders as a group own in the corporation. The category is also variously known as owner's equity, proprietorship, capital, and net worth. The company's net worth is determined after subtracting all liabilities, although it must be remembered that the assets are recorded at their initial, or original cost rather than at their current replacement value. Also, in the case of oil and gas exploration and producing companies there is no direct recognition on the balance sheet of the often very substantial value of the firm's underground reserves which have been found through its own exploratory effort.

Common Stock

Every publicly owned company is permitted to issue a certain amount of equity in shares. This is known as its authorized capital. A company may not need its entire authorization initially. The amount actually issued to stockholders is known as the issued capital. This may equal but not exceed the total authorized by the shareholders.

Basically the category of the firm's common stock, or contributed capital represents the shareholder's ownership of the company. These shares are represented by the stock certificates issued by the corporation to its shareholders. This section of the statement has two parts: 1) the face or par value of the issued stock, and 2) the amounts contributed in excess of the par value per share. If the par value for the common shares of the company represented in Figure 8-3 is $10 per share, and the company sold 10,000 shares at $15, then there would be $50,000 included under the second category of "contributed capital in excess of par."

The number of shares of stock in the company may be increased with the approval of the shareholders. Frequently this is done in the form of a stock split or stock dividend. The purpose of a stock split or stock dividend is to lower the price of the stock, in hopes that it will be more attractive to purchasers. The value of the company is unaffected by such "paper" transfers, as is the value to the individual shareholders. The shareholder will just have more shares of lesser value to represent his same proportionate share of the company. However, the earnings per share (EPS) after the split will be substantially reduced which could make the shares appear less attractive to potential new shareholders.

Preferred Stock

A secondary form of oil company financing is known as "Preferred Stock" . These , usually non-voting, certificates represent contributions of capital to the company with a commitment that dividends will first be paid to the owners of the preferred shares before any dividends are paid to the company's common share owners. The magnitude of preferred share dividends is often specified on the certificate. Convertible preferred shares carry the option of converting each preferred share into a specified number of common shares.

Retained Earnings

Retained earnings, which is sometimes designated as "Earned Surplus," is the accumulation of the firm's earnings that have been reinvested in the company and not paid out as dividends. It is quite properly included as part of the shareholder's equity, or "Net Worth" of the enterprise.

The U.K. Companies Act of 1981 permits an alternative form of balance sheet on a single page. This deducts Current Liabilities from Total Assets and balances this net value against Long-term Liabilities and Capital.

The simulated balance sheet for Able Oil Company, a North American company, is shown in Table 8B-2 of Appendix to this chapter.

THE INCOME STATEMENT

The income statement is the second of the principal financial statements. While the balance sheet shows the assets, and how they are financed as of a particular date, the income statement indicates the amount of profit the company made during the year; how much was paid out to shareholders as dividends; and how much retained for the growth of the business. Figure 8-4 outlines the contents of the income statement showing the revenues and expenses of the business for a stated period of time. It should be noted that interest is shown as a financial cost and is included before the calculation of income tax since it is a tax deductible item. However, dividends are listed after Net Income because they are paid out of after tax net income. Normally dividends are less than the net income for a given year, with the balance being added to the company's retained earning and becoming part of the shareholders equity as shown on the balance sheet.

This figure shows the revenues, costs and expenses and net income for the reporting entity for the period covered. Usually historical data for one or more previous periods is also shown for comparison purposes. In an annual report the figures shown will be for the latest fiscal year and one or more previous fiscal years. The fiscal year is commonly a calendar year, but this is not necessarily the case. It may be some other 12 month period designed to coincide with the seasonal nature of some businesses.

The income statement shows four different types of profit. These are: gross profit, operating profit, profit before tax, and net profit after tax. It is this latter figure that is often referred to as "the bottom line." Net Income is the company's "profit" for the reporting period. All of the adjustments previously described are included in arriving at the Net Income. This figure represents the funds that are available for dividends to the shareholders or for reinvestment in the business. In the example income statement you will see in the section labeled "Retained Earnings," that part of the net income was paid out as dividends and the remainder reinvested in the company. The retained earnings figure is the cumulative of all previous years' reinvestment in the company. The end of year figure is greater than the beginning year balance by the amount of Net Income which was not paid out as dividends. If "Cash dividends" exceed Net Income, then the end of year balance will be less than that at the beginning of the year.

FIGURE 8-4
INCOME STATEMENT FORMAT

Covers a period of time, such as January 1- December 31, 200Z

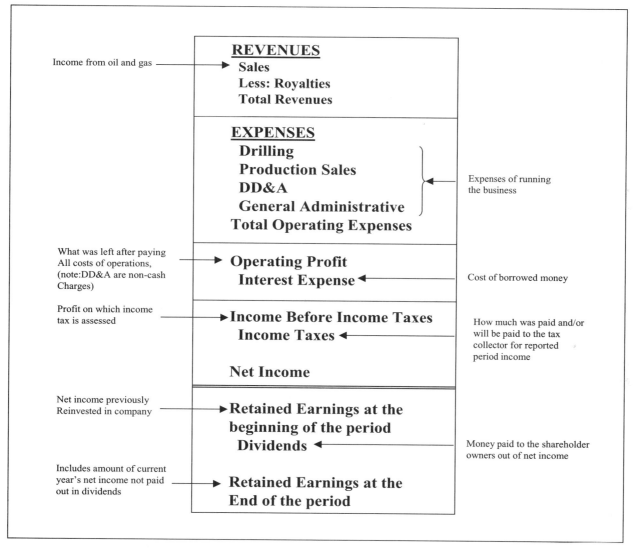

The after-tax profit for the period, usually one fiscal year ending with the date of the balance sheet, is comprised of the revenues earned, expenses incurred, the gains, losses and net income or net loss for the period. All of this information is detailed in the Income Statement which is sometimes referred to as the Profit and Loss statement, or merely as the Earnings Report.

FIGURE 8-5
SERIES OF THREE FINANCIAL STATEMENTS

CURRENT ASSETS
Cash
Accounts Receivable
Inventory
	REVENUES
	  Sales
	  Less: Royalties
	  Total Revenues
		CURRENT ASSETS
		Cash
		Accounts Receivable
		Inventory

FIXED ASSETS
Property, Plant & Equipment
    Less: DD&A
Land
	EXPENSES
	  Drilling
	  Production Sales
	  DD&A
	  General Administrative
	  Total Operating Expenses
		FIXED ASSETS
		Property, Plant & Equipment
		    Less: DD&A
		Land

TOTAL ASSETS		**TOTAL ASSETS**
CURRENT LIABILITIES Accounts Payable Salaries Taxes payable Short Term Debt	**Operating Profit** Interest Expense	**CURRENT LIABILITIES** Accounts Payable Salaries Taxes payable Short Term Debt
LONG TERM DEBT Loans Bonds	**Income Before Income Taxes** Income Taxes **Net Income**	**LONG TERM DEBT** Loans Bonds
NET WORTH Common Stock @ par value Capital in excess of par value Preferred Stock Retained Earnings	**Retained Earnings at the beginning of the period** Dividends	**NET WORTH** Common Stock @ par value Capital in excess of par value Preferred Stock Retained Earnings
TOTAL LIABILITIES AND NET WORTH	**Retained Earnings at the End of the period**	**TOTAL LIABILITIES AND NET WORTH**

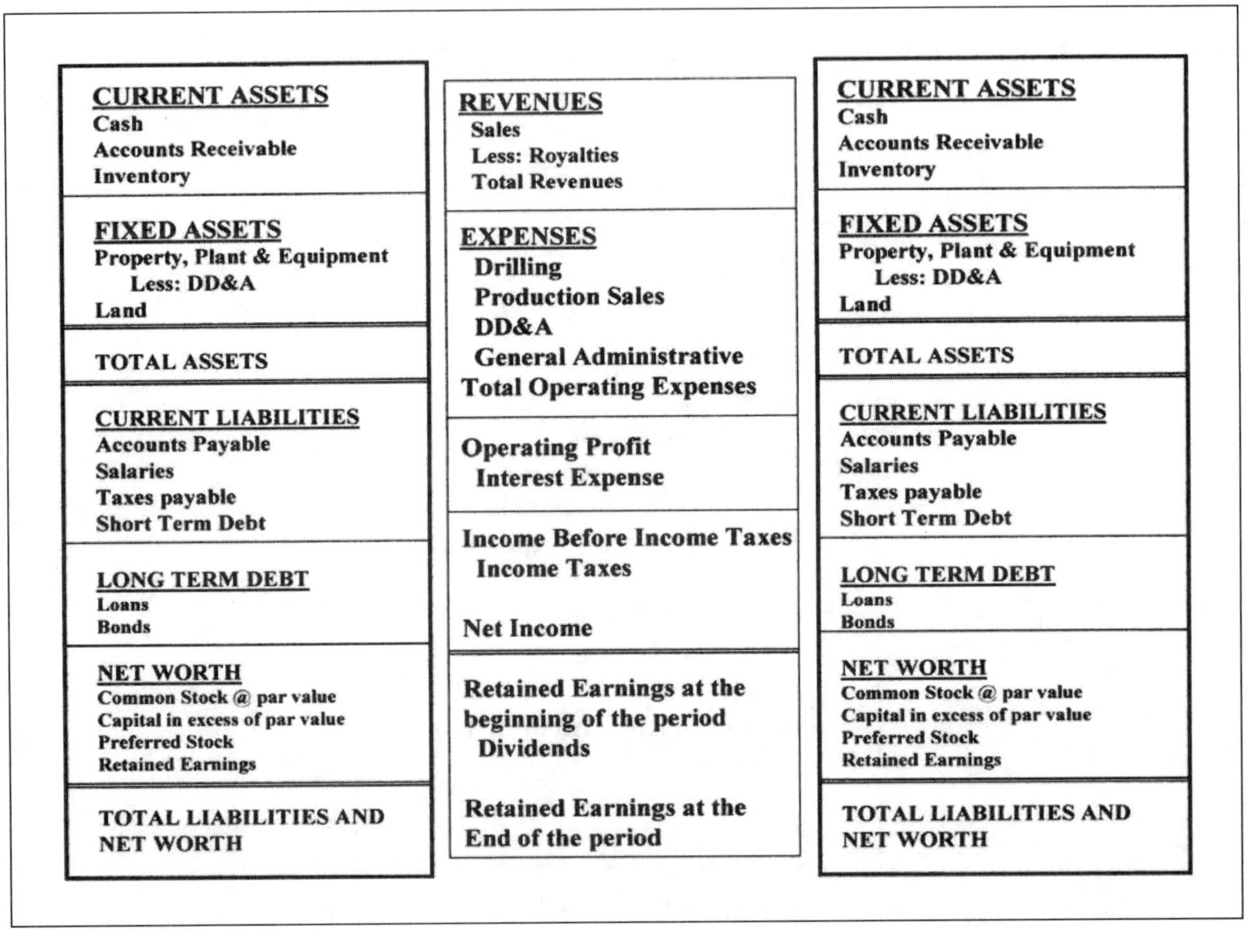

Figure 8-5 depicts the relationship of the income statement covering one fiscal year to the balance sheets for the beginning and end of that year.

The monetary amounts shown in the financial statements must conform to certain accounting practices. The International Accounting Standards Committee with representation from over 70 countries seeks to coordinate and unify these practices. In the U.S., the procedures must be acceptable to such authorities as the Financial Accounting Standards Board (FASB), the Securities and Exchange Commission (SEC) and the Internal Revenue Service (IRS). In the U.K. the Accounting Standards Committee (ASC) with its Statements of Standard Accounting Practice (SSAP) serves a similar function. There are counterpart organizations in the other countries. Each of these bodies has its own rules. These do not generally vary a great deal from one another as to the basics, but they can certainly differ in the specifics—and accounting has to be specific.

A frequent source of these differences stems from our previous observation that accounting theory is not static. This means that changes in one country, while they may be fully acceptable to the next country, are not always implemented at the same time. On the other hand, there is also the frequent case that each country likes to be a little different, if for no other reason than to draw attention to its independence.

STATEMENT OF CASH FLOWS

The Statement of Cash Flows is also an important table found in the corporate annual report. The data on this table is similar to cash flow discussed in Chapter IV and subsequent chapters, because it does present the cash-out and cash-in for the period of time indicated. The normal time interval used for the Statement of Cash Flows is the same as the Income Statement, ending at the same date used for the Balance Sheet. This statement shows how much net cash was provided by operating activities and financing activities, as well as how much cash was spent for capital expenditures, other financing activities, or repayment of debt. This statement also shows a reconciliation of the company's cash position by listing the cash balance at the beginning of the year, increase or decrease in cash for the year, and the cash balance at the end of the year. The year end cash balance must be greater than or equal to zero.

The starting point for the Statement of Cash Flows is the net income for the year as presented on the Income Statement. This number is then modified by all of the accounting adjustments made to arrive at the net income number; i.e. DD&A, payables, accruals, prepayments, deferred charges, changes in working capital, noncurrent items, and other accounting adjustments made to reconcile net income.

The following is the general format followed in preparing the Statement of Cash Flows.

FIGURE 8-6
STATEMENT OF CASH FLOWS FORMAT

Covers a period of time, such as January1- December 31, 200Z

FROM OPERATING ACTIVITIES	
Net Income ←	— From income statement
(+/-) Accounting adjustments ←	— DD&A,payables,accruables,‹
Net cash provided by operating activities	
INVESTMENT ACTIVITIES	
Capital Expenditures ←	— Capital budget
Other Investments ←	— Subsidiaries,other cos, etc.
Net cash used for investment activities	
FINANCING ACTIVITIES	
Dividends paid ←	— Paid to shareholders
Change in Long-Term Debt	
Change in Short-Term Debt	— New debt or debt retirement
Net cash used for financing activities	
INCREASE (DECREASE) IN CASH	
CASH POSITION	
Balance at beginning of year	
Increase (decrease) cash for year	
Balance at end of year ←	— Must be positive

FINANCIAL DISCLOSURES REGARDING OIL AND GAS PRODUCING ACTIVITIES

Normal financial accounting gives little asset value for oil and gas reserves found by the company's own exploratory efforts. The only value which will appear on the balance sheet will be capitalized exploration costs, bonuses, and the amount paid for purchase of producing properties. Even though petroleum reserves may be the major asset of an oil company, the balance sheet may indicate a value which is much less than their true worth.

The Securities and Exchange Commission (SEC) in the U.S. a number of years ago developed concern that the investing public might be misled by this discrepancy between the long established industrial accounting methods which failed to recognize the full asset value on the balance sheet, of oil and gas reserves as a major segment of the true total value of oil and gas producing companies. The SEC initiated the requirement that all oil and gas producing companies whose equities are traded on the public stock exchanges include in their financial statements certain information to supplement the traditional corporate financials.

The early efforts, known as Reserve Recognition Accounting (RRA), represented an attempt by the SEC to include in an oil company's published annual financial report specific acknowledgement of its proved reserves. There were a number of serious shortcomings in the RRA procedures. Subsequently, at the prompting of the SEC, the Financial Accounting Standards Board (FASB) published FASB Standard 69, which acknowledges:"The underlying causes of the problem economic characteristics of oil and gas producing activities." The statement further concludes that:"Mineral interests in proved oil and gas reserves may have significantly different economic values because of such features as location, qualitative properties, development status, and tax status. Reserve quantity information does not give a comparable base for comparison over time, or among companies."

The SEC now requires the inclusion in the oil company's financial statements of publicly traded oil companies a supplement to their annual financial statements, incorporating the following, with regional breakdown, which normally separates the totals for the U.S. and other countries:

- Proved oil and gas reserve quantities
- Capitalized costs relating to oil and gas producing activities
- Costs incurred in oil and gas property acquisition, exploration, and development activities
- Results of operations for oil and gas producing activities
- A standardized measure of discounted future net cash flows relating to proved oil and gas reserve quantities

"Standardized Measure" of Discounted Future Net Cash Flows

The most useful disclosure about oil and gas producing activities required by FAS 69 is the value of future production discounted at 10% per year using current prices and costs. This is known within financial circles merely as the "Standardized Measure." This provides a somewhat limited basis for comparing one company with another, although these figures have only minor relationship to the real asset value of an oil company.

This requirement constitutes a marked departure from the traditional accounting Matching Principle, discussed earlier, which would specify that cash flow, from which profit is derived in the Income Statement, must relate directly to actual investment expenditures within that financial period, in this case, oil and gas exploration.

The specific requirements for preparation of the Standardized Measure table as presented in FAS 69 by the Financial Accounting Standards Board (1995) are;

a. *Future cash inflows.* These shall be computed by applying year-end prices of oil and gas relating to the enterprise's proved reserves to the year-end quantities of those reserves. Future price changes shall be considered only to the extent provided by contractual arrangements in existence at year-end.

b. *Future development and production costs.* These costs shall be computed by estimating the expenditures to be incurred in developing and producing the proved oil and gas reserves at the end of the year, based on year-end costs and assuming continuation of existing economic conditions. If estimated development expenditures are significant, they shall be presented separately from estimated production costs.

c. *Future income tax expenses.* These expenses shall be computed by applying the appropriate year-end statutory tax rates, with consideration of future tax rates already legislated to the future pretax net cash flows relating to the enterprise's proved oil and gas reserves, less the tax basis of the property involved. The future income tax expenses shall give effect to permanent differences and tax credits and allowances relating to the enterprise's proved oil and gas reserves.

d. *Future net cash flows.* These amounts are the result of subtracting future development and production costs and future income tax expenses from future cash inflows.

e. *Discount.* This amount shall be derived from using a discount rate of 10 percent a year to reflect the timing of future net cash flows relating to proved oil and gas reserves.

f. *Standardized measure of discounted future net cash flows.* This amount is the future net cash flows less the computed discount.

Table 8B-4 in Appendix VIII-B of this chapter shows the format for the Standardized Measure. These requirements, and several other required supplemental tabulations, adds substantially to the burden and complexity of preparing the oil company's annual reports and extends their financial statements by several pages each year. There is still a good bit of controversy within the industry as to the reliability of the standardized measure as a gauge of the value of the oil and gas assets of the company, and as a means of comparing one company with another. The required tabulations in most Annual Reports are accompanied by disclaimers to the effect that... 'other assumptions than those prescribed by the regulations could be used in these calculations producing substantially different results. The standardized measure information is not representative of the fair market value of the company's proved oil and gas reserves, and the company cautions readers that actual future net cash flow may vary dramatically from these estimates.' 'Because of unpredictable variances in expenses and capital forecasts, crude oil and natural gas price changes,—management believes the usefulness of these projections is limited' 'Both reserve estimates and production forecasts are subject to revision as additional technical information becomes available and economic conditions change.' 'The volatility of crude oil prices reinforces the possibility that significant revisions in the computation may occur in the future.' 'Management cautions against relying on the information presented because of the highly arbitrary nature of assumptions on which it is based.' 'It should be recognized that applying current costs and prices and a 10 percent standard discount rate does not convey absolute value. The discounted amounts arrived at are only one measure of the value of proved reserves.'

FINANCIAL RATIOS

Companies are judged by the financial community and the investing public on the basis of the net income, or profit figures they report. The potential investor is always interested in knowing how the reported profit figures compare with those of a year earlier. Many analysts carry the comparisons back as far as ten years. Comparisons of the most recent quarter with the corresponding period of a year ago is also common information usually carried in the business press as news of interest to the financial community.

Financial analysts normally undertake an examination of a firm's financial statements in terms of a series of ratios. Christopher Nobes comments (loc.cit.) that "ratio analysis is a popular sport of investment analysts, financial journalists, textbook writers and examination setters. It involves the comparison of a company with its past or with other companies, by setting one piece of financial data in the context of another. "In effect, ratio analysis serves to raise questions; i.e., points out areas in which further study may be warranted.

The more common of these so called "Financial Ratios" which are generally employed in analyzing oil company's performance are:

- Price-earnings, or earnings per share (EPS)
- Cash flow/share
- Book value/share
- Debt to Equity
- Net reserve replacement after production
- Number of years of production at current rate (R/P)
- Current Assets to Current Liabilities
- Net Income to Interest + Principal
- Conversion of debt to preferred, or common equity
- Dividend payout
- Ratio of Net Income to Investment

There are also many other financial ratios that individual analysts have their own favorite way of calculating. Ratios are specific for companies in different industries. Comparisons between a company involved in automobile manufacture and another in the petroleum industry are apt to be quite misleading.

Earnings per Common Share (EPS)

Earnings per share represents the net income, or profit divided by the average number of shares outstanding during the period. Owners and prospective purchasers of common stocks are frequently are more interested in the earnings per share than they are in current dividend payouts. If the capital structure of the firm includes securities, e.g., preferred shares, which at the owner's option can be converted into common shares, one may see reference to "fully diluted" earnings per share. This involves the common stock equivalents such as convertible preferred stock, convertible bonds, and stock options. These securities are deemed to be but one step removed from common stock. Their trading price quite obviously relates closely to the value of the common stock to which they can be converted. The fully diluted earnings per share adds these common stock equivalents to the average number of outstanding shares of common as the denominator of the ratio. When and if the convertible securities are exchanged for common stock the EPS will be reduced as a result.

The proportions of each kind of security issued by the company is important to know in the oil business, as it is in all business. Many small producers are able to finance their new ventures internally from retained earnings. Other producers have availed themselves of senior securities (i.e. bonds and debentures) as means of funding new undertakings. A high proportion of bonds may reduce the attractiveness of both the preferred and common stock, while a large amount of preferred may reduce the attractiveness of the common. This comes about because bond interest must be paid before preferred stock dividends, which, in turn, must be paid before any dividends on the common shares.

The following are some of the official and unofficial terms that are commonly being used to describe earnings - - - or lack thereof.

OPERATING EARNINGS: This is another name for 'pro forma' earnings, in which companies present their financial results 'as if' certain ordinary items (usually expenses) did not exist. Also sometimes called 'core income,' 'economic earnings,' 'ongoing earnings' or 'earnings excluding special items.' None of these terms have any particular meaning under GAAP.

OPERATING INCOME: Sounds like operating earnings, but has a strict definition under GAAP. Revenue less cost of goods sold and related operating expenses stemming from a company's normal business activities. It excludes, for example interest income and expenses, dividend income, taxes and extraordinary items.

INCOME FROM CONTINUING OPERATIONS: Also sounds like operating earnings, but this, like operating income, has a real meaning under GAAP. It is revenues and expenses stemming from a company's ongoing operations, after taxes and includes interest income and expenses and other nonoperating gains and losses. It excludes only three things: discontinued operations, cumulative effects of changes in accounting principles and extraordinary items.

EXTRAORDINARY ITEMS: This is a real term under GAAP, meaning items that are both unusual in nature and infrequent in occurrence. Such as earthquake-related losses in an area where quakes are rare. Extraordinary items count when calculating net income, though not when calculating 'income from continuing operations.'

SPECIAL CHARGES: A term companies use for expenses that are ordinary costs of doing business, but that companies want investors to exclude when valuing their stocks. Also sometimes called 'one time,' 'unusual' or 'exceptional' charges. These terms have no standard definition under GAAP and the items don't meet the GAAP test of an extraordinary item.

CASH FLOW: a GAAP term meaning cash receipts minus cash disbursements for a given reporting period. It is separated into three categories on a financial statement: cash flow from operating activities, cash flow from financing activities and cash flow from investing activities.

EBITDA: Earnings before interest, taxes, depreciation and amortization. But some use the term to refer to earnings figures that exclude not only these expenses, but others as well, such as start-up costs for new ventures and other cash charges. Though often described as cash flow, Ebitda is not the same. For starters, it doesn't necessarily reflect changes in companies' liquidity.

Debt to Equity

The measure of the amount of "leverage," or "gearing," in a company's capital structure is its debt-equity ratio. Some companies prefer to employ a debt to debt plus equity ratio. In either case, the higher the proportion of loans, the higher the firm is said to be leveraged. In a highly leveraged company any increase in profit is magnified by the time it reaches the stockholders. This is because the cost of the borrowed capital is normally at a fixed rate of interest and the shareholders portion varies with the earnings. The debt-equity ratio is of prime concern to the bond rating agencies described in the Appendix to Chapter VII.

Liquidity Ratios

A key question is always whether a company will be able to pay its obligations when they come due. Failure to pay its debts on time damages the firm's credit rating. More serious, however, is the prospect that failure to pay on time may jeopardize the company's very existence if the unpaid creditors choose to take legal action to enforce collection. The sharp rise in business failures within the oil and gas industry during the early 1980's underscores the importance of keeping a close watch on the debt paying ability of a business. Current liabilities comprise debts that come due within one year. Current assets represent the ability to satisfy those obligations.

Other financial ratios which are used extensively by the financial community in analyzing other types of business are of only limited interest to the appraisal of oil and gas producing enterprises. They are, of course, quite applicable to the drilling and oilfield service industries. These factors would include such ratios as the Acid Test Ratio which basically omits inventory from the current asset numerator. Since the oil and gas producer seldom has very much of an inventory burden (produced oil and gas go down the pipeline to someone else's account) this ratio is not generally of great concern.

Similarly, financial ratios dealing with Profit Margin, Return on Assets, and Net Profit on Sales are normally not critical to the oil and gas producer. They are, however, an ever present worry in other industries, to bankers and other members of the financial community who may not be overly familiar with the oil business per se.

Working Capital

In addition to the usual ratios, investors are increasingly looking to the figures of "Working Capital Provided by Continuing Operations," or sometimes merely "Changes in Working Capital," which is to be found in the company's financial statements as an important index of the firm's fiscal health. This figure is a good indicator of whether the firm is generating enough cash to meet its payroll, conduct its operations, and service its debt. Another good indicator of fiscal health is the company's record of continuity in paying dividends to its shareholders.

One measure of corporate fiscal efficiency that is immediately available from the balance sheet is the net working capital. This is merely the figure for current liabilities subtracted from current assets and represent the company funds retained for current operating expenses. Any marked change in net working capital from one year to the next would certainly warrant further inquiry.

Net Book Value per Share

The net book value per share ratio is interesting as a check when comparing one oil producer with another similar firm. Here again, these values differ quite markedly in level from those of most other modern commercial enterprise, largely due to the method of dealing with oil and gas reserves as a booked asset. In the normal course of events, net book value per share of common stock is intended to mean the amount of money each share would receive if the company were liquidated, based on the balance sheet values. However, in the case of an oil company a basic abnormality persists in the fact that the financial books do not fully recognize the asset value of oil and gas reserves found by company exploration. Similarly, initial booked values are not adjusted for changes in the price of oil, changes in recovery estimates of purchased reserves, or for a number of other such factors.

Summary of Financial Ratios

There is not a unique value for any of the previously described financial ratios. Each one will differ from one company to another, depending upon the amount of risk investors associate with a particular company and its expected future potential. The ratios for all companies will also vary from time to time depending upon general economic conditions and current investment climate. Table 8-5 shows the usual range and median value for three of the ratios for stocks of a group of companies (30) that comprise the Dow Jones Industrial Average (NYSE & Nasdaq). The range and median values were collected on a monthly basis since February, 1905, and include 80% of the data during that period. In 10% of the cases the average traded higher than the range and 10% lower. In recent years values for the three ratios shown in Table 8-5 have been greater than the usual range indicated.

TABLE 8-5
FINANCIAL RATIOS

	USUAL RANGE	MEDIAN
PRICE/DIVIDENDS	15.4 - 32.9	23.9
PRICE/BOOK	1.0 - 2.5	1.5
PRICE/EARNINGS	7.4 - 23.0	13.8

TECHNICAL NOTE ON INFLATION ACCOUNTING

Conventional net cash flow analyses are prepared in terms of monetary funds. Unfortunately, the buying power of money does not remain constant. In an inflationary environment, the value of a given amount of money gradually declines. Each month it takes more Argentine Pesos, or Brazilian Cruizeros, to buy a loaf of bread, or a barrel of crude oil in the international market, which is universally priced in U.S. dollars. To add further to the complexity, the price of crude oil in U.S. dollars also inflates as was shown in Figure 3-4.

In the oil business long payout projects with low percentage returns may not earn enough to repay the initial investment, when measured in terms of actual purchasing power. Inflation can cause great distortion of a company's financial performance. This is particularly true of its profit history, the importance of long-term debt and the DD&A account (discussed previously in this chapter). For example, the company's profit and loss account might show a profit of $200 which may encourage the management to pay a dividend of $100 for the year. However, when adjusted for inflation, the profit of $200 might actually be a loss of $83 in real terms. Thus, the dividend of $100 reduces the equity base (shareholder's funds) and the financial working capital of the company.

Unfortunately inflation is not a one-time situation which can be contained in one accounting period. Rather it continues, building upon itself, much like compound interest which was discussed in Chapter V. Since the production of crude oil and natural gas spans a number of years any recognition of inflationary effects should incorporate this type of compounding. The best approach to handling the problem of inflation in a cash flow projection is to convert the spreadsheet to one of buying power estimates (both revenues and expenses) expressed in "current dollars," or "Money of the day" as the Europeans refer to it.

Net cash flow projections can be indexed to provide forecasts in terms of actual buying power rather than current dollars. The purpose of this type of indexing is to reduce the economic risk due to inflation. This indexing procedure involves multiplying each year's forecast of prices, and costs by pre-determined escalation factors so as to represent the monetary values of those costs and prices in the years ahead. The exercise is rendered more complex by the fact that crude oil prices and the costs of drilling, development and production characteristically do not all inflate at the same rate. It is also noteworthy that the higher the inflation rates, the greater the uncertainty in their reliability. Nevertheless, inflation is a serious matter to be dealt with in economic analyses, and indexing is the best approach toward coping with this business risk.

In some countries contract terms are indexed according to some official government report of price levels. Brazil and Israel are examples. Contracts may be fully indexed, or partially indexed. For example, a 10 percent increase in general inflation as shown by the government figures might call for a 7 percent increase in the partial indexing of a contract.

Depreciation is an exception to the current dollars handling since it is a portion of the fixed historical investment cost. This yields a somewhat larger taxable income projection than would be the situation in the "constant dollar" case, which ignores the effects of inflation. This income tax increase is one of the many unfortunate results of inflation. The tax collector doesn't seem to complain, however.

A cash flow spreadsheet provides a year by year projection of after-tax income. To convert this to a forecast of future buying power it is necessary to apply stepwise deflation indexing. This set of discount factors could be the same as those employed for adjusting the revenue and expense cash flows. In the petroleum industry, however, history has shown that each of the three sets of indexing factors is likely to differ somewhat one from the other. In other words changes in the price of crude oil are not immediately reflected in the costs of drilling, or offshore construction. Simultaneously, staff salaries and administrative costs may project more realistically according to some generally recognized "cost of living" forecast. The resulting annual cash flow stream after all this year by year indexing can then be measured by the various economic "yardsticks," described in Chapter V, employing discount factors that have not otherwise been modified for inflation. If future buying power is not considered, the yardsticks which involve present value can give quite misleading results according to Davidson (loc. cit.).

For example, capital intensive projects may look unduly favorable because the tax liability is understated.

This procedure, which is rigorous in concept, suffers from the reliability, or in some cases the extreme unreliability, of the several indexing forecasts required. In many analytical situations it is expedient to simply employ a hurdle rate which recognizes the cost of money and the real return desired, say 10 percent. If costs are rising 5 percent per year, a discount factor of 15 percent (i.e., 10% + 5%) applied to the net cash flow, without indexing either the revenue or expense streams, may come close. This over-simplification obviously equates all of the indexing streams.

Inflation affects the profitability and financing for both external published financial statements and internal financial reporting. While most accountants agree that inflation must be acknowledged, there is no real agreement as to how it should be done or as to how valid and useful are the results. The two basic methods for accounting for inflation are: GPPA (General Purchasing Power Accounting) which uses one general government index of retail prices and CCA (Current Cost Accounting) which uses multiple price indices.

The sources and cost of capital to the petroleum industry are ones in which petroleum engineering input is vital. The engineers provide the oil and gas production forecasts upon which loan paybacks, and indeed the credit worthiness of the enterprise are based. The engineers also furnish the reserve estimates that comprise the security for the majority of the borrowings. Yet, it is doubtful that many engineers working in field and district offices are aware of this employment of their efforts by the financial community.

CONCLUSION

While accounting, as has been discussed in this chapter, is important to the functioning of a company and to the financial community, cash flow is still the basis of company management. However, many contract terms and the income tax laws of most countries closely follow many of the accounting principles discussed in this chapter. Therefore, it is important that anyone doing economic evaluations at least understand the basic principles of accounting, so that they can be applied when required. Such knowledge will also enable the evaluator to distinguish between cash flow items and non-cash flow items, to prevent misuse of accounting values. The true value of an asset is based upon a cash flow analysis not accounting principles. Warren Buffet was once asked how he determined what he should pay when buying a company. He replied, "You just want to estimate a company's cash flow over time, discount them back and buy for less than that."

REFERENCES

American Institute of Certified Public Accountants, "What else can Financial Statements tell you?"

Arthur Young, "The Full Cost Method of Accounting for Oil and Gas Producing Activities"

Blecke, C. J., Financial Analysis for Decision Making, 1966, Prentice-Hall, Inc., Englewood Cliffs, N.J.

Brealey,R. and Myers,S., Principles of Corporate Finance, 2nd Ed., 1984, McGraw-Hill, New York

Brock,H.R., Klingstedt,J.P.,and Jones,D.M., Accounting for Oil & Gas Producing Companies, Professional Development Institute, North Texas State University, Denton, TX 76203

Financial Accounting Standards Board, "Statement of Financial Accounting Standards No. 69," November 1982

Follett,R., How to Keep Score in Business, 1978, Fawlett Publ., New York 10019

Gallun,R.A., and Stevenson,J.W., Fundamentals of Oil&Gas Accounting, 1983, PennWell, Tulsa,OK

International Labour Office (ILO), How to Read a Balance Sheet, 2nd (Revised) Ed., 1985, International Labour Organisation, Geneva 22, Switzerland

Johnson, Daniel, Oil Company Financial Analysis in Nontechnical Language, 1992, PennWell, Tulsa,OK 74101.

Johnston, B.&Purser, J.,"Measuring Performance of Oil and Gas Firms," O&GJ, p.130, Dec. 17, 1984

Merrill Lynch Pierce Fenner & Smith Inc., How to Read a Financial Report, Fourth Edition, Merrill Lynch, New York

Nobes, Christopher, The Economist Pocket Accountant, 1983, Basil Blackwell Publisher Ltd., Oxford OX4 1JF, and The Economist, London SW1A 1HG

Theusen, H.G., Fabrycky, W.J., and Theusen, G.J., Engineering Economy, Fifth Edition, 1977, Prentice-Hall, Inc., Englewood Cliffs, N.J. 07632

Tracy, J.A., How to Read a Financial Report, 2nd Ed., John Wiley & Sons, New York

APPENDIX VIII-A

DD&A FORMULAS

Straight Line Depreciation (D_{SL})

$$D_{SL} = (C / t_L)$$ (8.1)

where:

C = cost of tangible capital item

t_L = depreciation life

Declining Balance Depreciation (DB_n)

For first year (n = 1)

$$DB_1 = C(M / t_L)$$ (8.2)

and for any year "n" the depreciation is the first year depreciation rate determined from Eq. 8.2 times the remaining undepreciated balance;

$$DB_n = [C(M/tL)] * [1 - (M/tL)]^{(n-1)}$$ (8.3)

or as a shortcut method;

$$DB_n = [DB_{(n-1)}] * [1 - (M/t_L)]$$ (8.4)

The cumulative declining balance depreciation at the end of any year "n" is:

$$\sum_{i=1}^{n}(DB_i) = C\left\{1 - [1 - (M / t_L)]^n\right\}$$ (8.5)

where:

C = cost of tangible capital item

t_L = depreciation life

M = multiplier (for double declining balance = 2)

n = number of years since start of depreciation

If it is desired to fully depreciate a capital cost by the end of the depreciation life, using the declining balance method, then it is necessary to use a modified method. This involves switching from declining balance to straight line at the point that the declining balance depreciation is equal to straight line depreciation of the remaining undepreciated balance. The year at which this switch is made can be determined with the following equation.

$$N = t_L - (t_L / M) + 1$$ (8.6)

where:

N = year to switch from declining balance to straight line

t_L = depreciation life

M = multiplier (for double declining balance = 2

Unit-of-production Depreciation or Cost Depletion (UOP_n)

$$UOP_n = C(Q_n / U)$$ (8.7)

or:

$$UOP_n = \left[C - \sum_{i=1}^{n} UOP_{(i-1)} \right] (Q_n / R_n)$$ (8.8)

where:

C = cost of tangible capital item or leasehold cost

Q_n = annual production (BOE) of year n

U = Ultimate recovery = $\sum_{n=1}^{t_L} (Q_n)$

R_n = reserves at beginning of year n

n = depreciation year

t_L = Producing life

APPENDIX VIII-B

HOW TO READ AN OIL COMPANY'S ANNUAL REPORT

The security and exchange authorities in essentially every country require that each corporation, in which the shares are publicly traded, publish an annual report for the firm's present and prospective shareholders. These reports normally consist of a number of pages of text, pictures of the company's operations, and a standardized set of financial statements. The reports are widely distributed and contain a great deal of useful factual information. In fact, the penalties for false, or misleading disclosure are so severe that one can rely almost completely on any information gleaned from an annual report.

Reading annual reports is somewhat of an art. The following is an outline of a few hints on how to go about such an exercise.

Annual reports as a class of publications are admittedly difficult to "read." Actually they are not intended to be read like a magazine, even though an annual report may be somewhat similar in appearance. The bare legal requirements which govern a company's publication could usually be met with the filing of a few pages of data and footnotes. The fancy covers, photos in color, and glossy text have become the norm, however, in the modern world business scene. More realistically an annual report presents the firm's financial statements for the year. The reports contain a great deal of valid, useful and interesting information if one has the patience for careful study.

Before one starts into an annual report, it is well to review what is already known about the company.

An annual report generally contains several prominent sections. The financial statements are the most important, and probably the most difficult to understand. These consist of:

(1) the firm's Balance Sheet, which tells how strong the company's finances are by showing what the company owns,—-and what it owes—-as of a certain date;

(2) an Income Statement to show the net profit for the year reported, and usually the previous year as well for comparison. This is where we look to see how the company did this year compared to last year, whether the firm had a profit or a loss, and the extent of either;

(3) the Statement of Cash Flows, which tells us a great deal more about how the company conducted its business during the year; and

(4) a Standardized Measure Statement which provides the SEC required projection of the discounted value of the company's proven oil and gas reserves.

There may be some combining of these later statements for some of the smaller producing companies. The report also includes the highly important statement from the firm's independent auditors.

Annual reports are usually prefaced, perhaps on the inside front cover, with a letter to shareholders signed by the Chairman or chief executive (CEO) of the company. The largest number of pages in the report will usually be taken up by a review of the company's activities during the year, department by department.

349

The first section of this chapter discussed the problems of fitting normal accounting practice tailored to the majority of Western world business to the unique nature and business practice of the oil industry. All of this holds true with regard to reading and deciphering the annual reports of the exploration and producing sector of the oil business.

Where to begin? — At the back.

Probably a good first step in reading the annual report of a producing oil company should be an automatic one. That is, to determine whether the firm employs the "Full Cost," or "Successful Efforts" method of accounting. This information may be found in the text relating to the company's financial matters or in the footnotes to the financial statements. The two principal methods of accounting employed in the oil and gas producing industry were discussed earlier in this chapter.

Secondly, toward the back of the report one finds the Auditor's Report. This tersely worded statement certifies either that the independent outside auditor finds the financial statements are fairly presented in conformity with "generally accepted accounting principles" and affixes the external auditing firm's name to the report, or that the auditor is issuing a report that is something other than the one-paragraph report signifying a "clean" or unqualified opinion. There are two categories which auditors use when dissatisfied with the financial data presented. An auditor may qualify an opinion by stating what specifically prevented him from expressing a "clean opinion."

In the words of the American Institute of Certified Public Accountants (AICPA) (loc.cit.) "The second alternative is a more drastic response. Theoretically, the auditor may disclaim an opinion—which means he can't express any conclusions—or express an adverse opinion—which means the financial statements are misleading. In practice, these reports are seldom issued. Such reports are not acceptable to the SEC and major stock exchanges and could lead to de-listing of a company's securities."

The reader will want to make sure that the auditor's opinion is a 'clean' one rather than a qualified, or 'dirty' one. A dirty opinion, in the jargon of the accounting profession, is one which contains disclaimer words like "subject to" and "except for," indicating that the auditors have reservations about giving a clean bill of health to the enterprise. Generally, this happens only if the company is in rather serious, or potentially serious financial difficulty. Normally, for a clean report the auditor's report is not more than two paragraphs long, such as the one shown in Figure 8B-1. Anything longer can usually be considered a 'red flag' which should alert the reader to look further. A 'clean bill of health' does not necessarily mean that the company is doing well financially. All that a clean opinion signifies is that the financial records have been properly maintained, and fairly presented.

Additionally, the accountants may include footnotes to the financial statements explaining the accounting methods used to prepare the statements. Major items, of an unusual nature, in the financial statements are explained in detail in the footnotes. In order to compare one company's financial performance to another, with reasonable accuracy, the footnote content for both companies must be given serious consideration.

FIGURE 8B-1
TYPICAL AUDITOR'S STATEMENT

To the Shareholders of
Able Oil Company

In our opinion, the accompanying consolidated balance sheet and the related consolidated statements of income, of cash flow, and of shareholder's equity present fairly, in all material respects, the financial position of Able Oil Company and its subsidiaries at December 31, 200Z and 200Y, and the results of their operations and their cash flow for each of the three years in the period ended December 31, 200Z, in conformity with generally accepted accounting principles. These financial statements are the responsibility of the Company's management; our responsibility is to express an opinion on these financial statements based on our audits. We conducted our audits of these statements in accordance with generally accepted auditing standards which require that we plan and perform the audit to obtain reasonable assurance about whether the financial statements are free of material misstatement. An audit includes examining, on a test basis, evidence supporting the amounts and disclosures in the financial statements, assessing the accounting principles used and significant estimates made by management, and evaluating the overall financial statement presentation. We believe that our audits provide a reasonable basis for the opinion expressed above.

Price Waterhouse
Denver, Colorado
January 28, 200Z+1

Now, back to the front of the report:

Next we should hear from the chief executive of the firm in his letter "To Our Stockholders." The CEO's letter frequently makes the most interesting reading of the whole document. If anything negative is noted in these remarks, it is reasonable to assume that the problem is serious enough that the company management felt that the problem had to be put `up front'. Again, such wording as, `except for', `in spite of', and `notwithstanding' should be considered as additional red flags signaling trouble spots.

It is almost universal that the CEO's letter closes with a word of appreciation for the diligent efforts of the company's employees.

Director's Report

In the U.K. annual reports include a Director's Report section which contains detailed disclosures of director's shareholdings, charitable and political donations, a review of the past year and discussion of future plans, and other company matters. The Director's Report is examined by the company's outside auditors for consistency with the firm's financial statements.

There is no direct equivalent to the Director's Report in the U.S. and Canada although the Form 10-K which is filed with the Securities and Exchange Commissions contains similar information.

Now the Financials:

A typical annual report worldwide will contain a series of financial statements, as already mentioned. The format, title and detailed content of these may vary somewhat between countries as well as individual companies. The two statements which are certain to be included are the firm's Income Statement and Balance Sheet. In the U.S. the SEC also requires a Statement of Cash Flows (similar to the traditional S&A, or Source and Application of Funds Statement), and the Standardized Measure of Discounted Net Cash Flows. In the U.S. the SEC also requires other disclosures covering a three year period on such matters as proved reserves, exploration and producing activities, and shareholder's equity. Tables 8B-1 through 8B-4 represent the four typical formal financial statements for a ficticious producing oil company in the U.S., adapted from a number of independent sources.

TABLE 8B-1
Able Oil Company

STATEMENT OF INCOME AND RETAINED EARNINGS
for the years ending December 31st, 200Z, 200Y and 200X

(millions of dollars)				
	Years ended December 31	200Z	200Y	200X
Revenues	Sales and other operating revenue	26,489	23,695	23,066
	Less: Consumer excise and sales taxes	2,066	1,992	1,996
		24,423	21,703	21,070
	Equity earnings, interest and other income	367	245	329
	Total revenues	24,790	21,948	21,399
Costs and Expenses	Purchases and operating expenses	18,733	15,711	15,286
	Selling, general and administrative expenses	806	747	740
	Exploration, including dry holes	461	400	423
	Research expenses	204	201	194
	Depreciation, depletion, & amortization	1,991	1,824	1,947
	Interest and discount amortization	330	337	350
	Income and operating taxes			
	Operating Taxes	721	659	617
	Federal and other income taxes	508	664	638
	Total costs and expenses	23,754	20,543	20,195
Income from Operations		1,036	1,405	1,204
	Cumulative effect of accounting changes	—	—	35
Net Income		1,036	1,405	1,239
Retained Earnings	Balance at beginning of year	14,004	13,336	12,797
	Net income	1,036	1,405	1,239
	Cash dividends	(750)	(737)	(700)
	Balance at end of year	14,290	14,004	13,336

While the balance sheet indicates the financial soundness of the company as of a given date, the other statements are designed to present a record of its operations throughout the year. Two consecutive years are shown for Able Oil in the Balance Sheet, Table 8B-2, and three years in the other statements.

TABLE 8B-2

Able Oil Company

BALANCE SHEET

December 31, 200Z, and 200Y

	(millions of dollars) As of December 31	200Z	200Y
ASSETS	**Current Assets**		
	Cash	337	139
	Short-term securities at cost	527	387
	Receivables & payables, less allowance for doubtful accounts	3,298	2,952
	Owing by related parties	149	97
	Inventories of oils and chemicals	770	733
	Inventories of materials & supplies	345	324
	Total Current Assets	5,426	4,632
	Investments, long-term receivables and deferred charges	1,878	2,101
	Property, plant and equipment at cost, less accumulated depreciation, depletion and amortization	21,192	20,866
	Total Assets	28,496	27,599
LIABILITIES & SHAREHOLDER'S EQUITY	**Current Liabilities**		
	Accounts payable—trade	2,166	1,768
	Other payables and accruals	565	569
	Income, operating and consumer tax	1,046	463
	Owing to related parties	184	163
	Short-term debt	475	660
	Total Current Liabilities	4,436	3,623
	Long-term debt	3,014	2,840
	Owing to related parties	356	436
	Deferred income taxes	4,355	4,651
	Total Long-term Liabilities	7,725	7,927
	Shareholder's Equity		
	Common stock—300,000,000 shares $1 per share par value authorized and outstanding	300	300
	Capital in excess of par value	1,745	1,745
	Earnings reinvested	14,290	14,004
	Total Shareholder's Equity	16,335	16,049
	Total Liabilities and Shareholder's Equity	28,496	27,599

*The notes to the Able Oil's simulated financial statements have been omitted for simplicity. Traditionally, Note 1, refers to the accounting procedures employed.

The third in the list of financial statements is the Statement of Cash Flows shown in Table 8B-3. This statement provides additional information on how Able Oil actually conducted its business, during the report period by recording the sources and uses of its funds.

TABLE 8B-3

Able Oil Company

STATEMENT OF CASH FLOWS

	(millions of dollars) Years ended December 31	200Z	200Y	200X
Cash Flows from Operating Activities	Net income	1,036	1,405	1,239
	Adjustments to reconcile net income to net cash provided by operating activities	—	—	(35)
	Depreciation, depletion, amortization and retirements	1,991	1,824	1,947
	Dividends in excess of (less than) equity income (increases) decreases in working capital	29	6	17
	Receivables and prepayments	(398)	(236)	51
	Inventories	(58)	(93)	11
	Payables and accruals	998	413	(21)
	Deferred income taxes	(5)	219	207
	Other noncurrent items	(460)	(132)	363
	Net Cash Provided by Operating Activities	3,133	3,406	3,779
Cash Flows Provided by (used for) Investing Activities	Capital expenditures	(2,344)	(2,033)	(2,895)
	Proceeds from issuance of long-term debt	2	280	139
	Other investments and advances	308	(54)	133
	Net Cash Used for Investing Activities	(2,034)	(1,807)	(2,623)
Cash Flows Provided by (used for) Financing Activities	Proceeds from issuance of long-term debt	—	—	500
	Principal payments on long-term debt	(534)	(513)	(680)
	Dividends	(750)	(737)	(700)
	Increase (decrease) in short-term obligations	523	(295)	(346)
	Net Cash Used for Financing Activities	(761)	(1,545)	(1,226)
Net Cash Flows	Increase (Decrease) in Cash and Short-term securities	338	54	(70)
Cash and Short-term Securities	Balance at beginning of year	526	472	542
	Increase (decrease) in cash and short-term securities	338	54	(70)
	Balance at end of year	864	526	472

STATEMENT OF CASH FLOWS

This statement records the changes that have taken place in the period between the two balance sheets reflecting the new funds which have come available to the company and how they have been used. The sources of new funds may include:

- net profit (after DD&A and Taxes)
- DD&A
- new share capital issued
- sale of fixed assets
- new borrowings

DD&Ado not exactly represent a source of new funds but since no money actually leaves the company the amount by which gross profit was subjected to these non-cash charges to obtain the net income, or profit figure, is now added back in determining the true flow of funds. DD&A charges can be very large under certain circumstances and will have a corresponding impact on the profit figures computed for a given year as well as the impact they have on the Cash Flow Statement.

The negative cash flow, or where the cash went, occupies the lower half of the statement. This part of the statement shows that Able had captial expenditures of $2,344 million last year, which was 15% greater than the year before, but this was 24% less than two years before.

The application, or use, of the funds typically includes the following items:

- payment of dividends (Able paid $750 million this year)
- purchase of fixed assets (properties)
- repayment of loans or capital
- increase in net working capital

The fourth major financial statement, Standardized Measure of Discounted Future Net Cash Flows is shown in Table 8B-4. The history, advantages and disadvantages of this statement were discussed earlier in this chapter.

TABLE 8B-4

Able Oil Company

STANDARDIZED MEASURE OF DISCOUNTED FUTURE NET CASH FLOWS

(millions of dollars)

	200Z			200Y		
	U.S.	Foreign	Total	U.S.	Foreign	Total
Future cash inflows	73,429	6,951	80,380	59,206	4,306	63,512
Future production and development costs	35,412	2,590	38,002	31,886	2,212	34,098
Future income tax expenses	11,536	1,887	13,423	8,110	992	9,102
Future net cash flows	26,481	2,474	28,955	19,210	1,102	20,312
10% annual discount for estimated timing of cash flows	14,785	1,061	15,846	10,506	489	10,995
Standardized measure of discounted future net cash flows	11,696	1,413	13,109	8,704	613	9,317
Aggregate change from previous year			3,792			1,992
Weighted average year end crude oil prices ($/bbl)	21.13	25.26		16.93	17.14	

Principal Changes in Standardized Measure of Discounted Future Net Cash Flow
(millions of dollars)

	200Z	200Y	200X
Sales and transfers of oil and gas produced, net of production costs	(2,988)	(2,498)	(2,233)
Net changes in prices and costs	5,148	3,679	(1,979)
Extensions, discoveries, additions and improved recovery, less related costs	1,288	1,233	367
Net purchases and sales of reserves	55	(63)	199
Development costs incurred during the period	1,081	861	929
Revisions of previous reserve estimates	285	106	(116)
Accretion of discount	1,371	1,015	1,195
Net change in income taxes	(1,871)	(1,567)	779

Watch those Footnotes

Some of the most interesting and important reading in an annual report is contained in the Footnotes to the Financial Statements. These footnotes will contain information on any unusual, one-time items such as income from the sale of a major asset. Other examples of appropriate footnotes would be:

- Contingent liabilities representing claims or pending lawsuits.

- Changes in the method of depreciating fixed assets or the handling of G & G expenditures.

- Changes in the value of company stock issued and outstanding due to splits or stock dividends.

- Details of stock options granted and exercised by officers and employees.

- Employment contracts, profit sharing, and retirement plans and the state of funding of the latter.

FINANCIAL ANALYSIS

The analysis of the four principal financial statements is the next step which one should undertake. This is usually accomplished by means of some simple but perceptive arithmetic.

Able Oil's net Book Value for the year reported is arrived at from the balance sheet as follows:

Total Assets		$28,496 mm
less: Current Liabilities	($ 4,436 mm)	
Long-term Liabilities	($ 7,725 mm)	
Preferred Stock	0	
	($12,161 mm)	
Net Book Value		**$16,335 mm**

Net Book Value per Common Share:

$$\frac{\$16,335mm}{300,000,000sh.} = \$54.45$$

Financial Ratios

Capital structure is normally derived from the balance sheet in terms of the company's debt/equity ratio at year-end. In our example Able Oil's figure is:

$$\frac{\text{Total Long - Term Debt}}{\text{Total Stockholder's Equity}} = \frac{\$3,014 \text{ mm}}{\$16,335 \text{ mm}} = 18.45\%$$

which is quite healthy for a company such as Able.

Similarly, the ratios of earnings per share, and cash flow per share, and how they vary fro year to year are of particular interest to potential investors.

200Z	eps $3.45	Net Cash Flow/share $1.13	Debt/Equity Ratio 18.45%
200Y	eps $4.68	Net Cash Flow/share $0.18	Debt/Equity Ratio 17.70%

Current Ratio: The Basic Test of Short-Term Solvency

Just what a comfortable level of working capital should be varies somewhat from industry to industry. The most widely used ratio to test the short term capability to pay the firm's debts is the Current Ratio. This is simply the current asset figure from the balance sheet divided by the current liability figure.

The Current Ratio of Able Oil for 200Z is:

$$\frac{\text{Current Assets}}{\text{Current Liabilities}} = \frac{\$5,426 \text{ mm}}{\$4,436 \text{ mm}} = 1.22$$

A current ratio of 1.22 is healthy for the oil and gas industry, and means that for every $1 of current liabilities there is more than a dollar's worth of current assets to back it up on a current basis.

OTHER PARTS OF THE REPORT

Officers and Directors

It is often interesting before putting down the annual report to check through the list of company officers to see who has the recognition, and presumably the authority in running the company. The list of directors and their affiliations outside of the firm under review may also shed interesting light and perspective. This is particularly true when officials of outside financial institutions are listed as directors.

Other Reports to Governmental Authorities

Public companies in the U.S. are required to file three quarterly reports (10-Q's) and an annual report (10-K) with the SEC. These are public documents available for a minor fee from the SEC headquarters in Washington, D.C. In Canada the comparable report is the AIF (Annual Information Form) which is filed with each of the ten provinces. Copies of these filings can generally be obtained from the company merely for the asking. These reports contain a great deal of additional information presented in a very terse legal manner without pictures or fancy covers.

The Long View

The annual report and the financial information which it presents relates to a single corporate entity during a specific time frame. It is important to study and compare the company's performance over a period of years. This may be accomplished with a series of annual reports supplemented with such other pertinent information as becomes available to the reviewer.

| CHAPTER IX | # BUDGETING, SCHEDULING AND CORPORATE PLANNING |

The budget is a central cog in the business machine that must mesh with many other business activities

Budgeting is a very general term representing a type of activity which is universally practiced in its various forms by the private sector oil companies, the NOC's, and their governments the world over. The reason that budgeting does not have a standardized definition is due mainly to its all encompassing nature. It includes industrial and corporate concepts, an annual routine, techniques and format, all laced together by the philosophy of the firm's management under the general heading of a budget. Since a budget is a time-related plan it goes a step beyond the basic decision-making criteria that have been discussed in previous chapters.

A company's budgets are the most important documents that are prepared during the year. The capital budget is the means for the company to achieve its goals and objectives through implementation of its strategies. The success or failure of a company's investment program will determine the success or failure of the company. The decision with regard to each investment opportunity must support the company's strategy and be governed by the criteria that it establishes to achieve its short term and long term goals. The cash budget determines the survival or downfall of a company, as a company must have sufficient cash at all times to pay its bills, repay debts, pay interest on its debts, and provide a return to its investors. Since these budgets are just short term forecasts, they must be continually monitored to immediately detect variations as they occur and to implement changes which are necessary to achieve the company's goals.

Budgets usually cover a period of twelve months which normally coincides with the company's fiscal year. Once approved by the firm's management, the budget becomes their plan of action for the designated period of time. It also becomes management's principal method of monitoring the organization's overall performance, as well as that of each of the firm's component units.

The budget, like any other plan or forecast, should not be "cast in stone," and followed no matter what develops. It should be reviewed periodically throughout the year to assure that the premises upon which it was prepared are still valid. If it is determined that changes have occurred in the economic climate, or in execution of the plan, it may be necessary to revise the budget one or more times during the year. A sharp drop in the price of hydrocarbons, or the drilling of an abnormal number of dry holes, are examples of situations that might dictate revisions of a current year's budget. Thus, it is important to review budget performance periodically so that timely revisions can be made. Periodic reviews also serve as a surveillance tool to assure management that the plan is being followed.

Budgets of all the various types—and detailing the firm's program at all organizational levels— constitute the principal management vehicle for the formal expression of the company's plans and objectives for a specified period of time, the budget period. No other management tool is capable of providing the operational direction that can be achieved through a well-planned and developed budget. A budget can be used in many ways and covers numerous functions.

These may include a:

- Capital Budget
- Cash Budget
- Expense Budget
- Exploration Budget
- Personnel Budget

Each has a somewhat different objective as a planning tool, although they must all tie together to provide a consistent program for the budget period.

In order that the overall budget cover all of these activities in a comprehensive manner it is best that they be prepared with input from all corporate levels from the field level to top management. This requires considerable time and talent on the part of all who participate in the budgeting function. This often translates into a considerable item of overhead cost. Depending upon the size of the organization and the degree of detail which management requires to be included in the budget plan, this effort easily may require a sizeable commitment of time from three to nine months in advance of the period specifically covered by the budget.

Although there are advantages in having most of the budgeting process start at the field and district level, it should be kept in mind that recommendations from that level will, by their very nature, tend to be related to that unit's existing activities and very little deviation from the status quo should be expected. Another potential problem with initiating budget submittals from the field operating level is the danger that some urgent, or particularly profitable, projects may be deferred until annual budget submittal time which may be months in the future.

Once the final budget is prepared, and approved by all departments, it goes to top management of the organization for approval. In many organizations, approval of the final budget also includes authority for expenditure of certain specifically budgeted items by the appropriate management level. It is fairly common, however, for top management to reserve that authority for themselves in cases of very large expenditures, for high risk projects, or for projects that could have a significant impact upon the organization. Any project proposal that may arise during the budget year, which was not anticipated in the budget, requires separate approval to either be added to the budget or replace an approved budget item.

Thus, it can be seen that a budget cannot be a static plan, but must reflect those necessary changes which occur throughout the budget period. The revised budget is also a useful tool in preparing the following year's budgets.

All budgeting relies on forecasting the continuity of a series of basic factors operating through the budget period. The budget variables, e.g., capital investments to be undertaken during the period, are then superimposed on this forecast. The basic continuity forecast includes such assumptions as a continuation of current production from existing facilities, a continuation, or a predetermined projection of crude oil and natural gas prices, a continuation of existing tax regimes, and similar postulates regarding the ongoing business.

Essentially all of the concepts and techniques discussed earlier in this text are employed in budget preparation. This includes the production rate and price forecasting discussions from Chapters II and III and cash flow concepts from Chapter IV. Risk and uncertainty, the economic yardsticks used to choose between investment alternatives, DD&A, and the means of computing profit, as well as the means of presenting the finances of the company, all feature in the budget plan. What is new at this stage, as all of these components are brought together in our discussion of budgeting, is the concept of the timing as a strategic element.

Primary consideration in the budgeting process is developing a projection of the cash flow which the corporation can rely upon in achieving the profit target while at the same time doing all the other things that the company hopes to accomplish within the budget year. Optimizing the use of a company's available cash flow is a critically important activity in any organization. Budgeting, scheduling and planning are all involved in seeking this best approach both short term (one year), and long term (five-plus years). The former is generally known as budgeting and the latter as long range planning.

It should not be implied that every project listed on the approved budget is to be fully completed within the budget period. In the petroleum industry many projects require more than a year to plan, implement and complete. This "carry over" from one budget period to the next has to be recognized in each budget. The limitations of manpower and equipment, as well as the necessity of seasonal operation, for example, in northern Canada and the North Sea dictate the continual necessity of including "carry-in" and "carry-out" items for the budget period.

Scheduling of projects and activities throughout the budget year is also important. If the budget or fiscal year is the same as the calendar year, there may be a tendency within the organization to try to start every budget approved project in January. Unless some effort is made to spread the work on budget approved projects throughout the budget period, a preponderance of January starts can create a severe distortion of the cash flow. Revenues from the majority of oil producing operations run fairly uniformly throughout the budget period. Peak gas sales may be the exception, but actual payments under the terms of the sales contract may still not match up with too many January starts. Optimum allocation of company personnel may also present a problem.

CAPITAL BUDGETING

The capital budget, or fixed asset spending plan, tends to be on a project by project basis. This is in contrast to the other types of budgeting listed in the introductory paragraphs of this chapter. Capital budgeting, in particular, has important financial implications for the firm, both on the balance sheet, as well as the more immediate effects on cash flow. The capital budgeting process which will be described here is typical although each individual step may vary considerably from one organization to another.

361

Capital budgeting activity is initiated by the Board of Directors or the top level of management in most companies. This will occur early in the year when the budget is being prepared for the following year. The "Board" will establish the basis for the capital budget by reviewing the corporate goals and objectives, ratifying the corporate hurdle rate, approving price and inflation forecasts, affirming the level of capital expenditures for the following year, and allocating those funds among the various divisions of the company. Once the "Board" has completed these activities, budget preparation can begin using the guidance that has been provided.

A question frequently asked is; "Where does the money come from for the capital budget?" The answer is; "From the sale of oil and gas." As was discussed in Chapter VII, the oil industry is essentially self funded, with only a small percentage of investment funds coming from outside of the industry. This is why the size of oil companies' budgets is so dependent upon the price of oil. In years of high oil prices oil companies announce budget increases and in years of declining oil prices companies may announce budget cuts.

The following graph, which shows the correlation between the price of oil and the increase and decrease in capital budget funds over the past 25 years, clearly shows the dependency of a company's capital budget on the price of oil and gas. The oil prices have been shifted one year to the right to recognize the fact that there is usually a one year lag before changes in the price of oil fully affect the budgets of most companies. However, abrupt changes may necessitate increases or decreases as the budget is being implemented.

FIGURE 9-1
E&P SPENDING vs. OIL PRICE

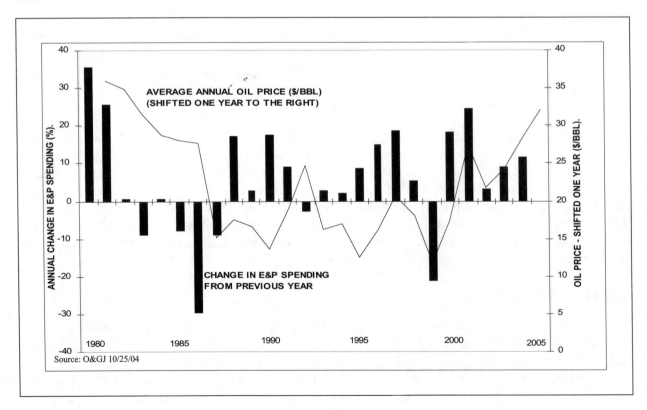

Source: O&GJ 10/25/04

In the context of accounting principles and practices discussed in Chapter VIII, the money for capital budgets (CAPEX) is mainly provided by net income (profit) and DD&A. Remember that DD&A is not a cash flow item, so even though income was reduced by DD&A to determine profit, this money never left the company, so that cash is still there to be used for capital investments. The following outline shows the way in which the amount of cash available for capital expenditures is estimated each year;

PREVIOUS YEAR'S CASH BALANCE

+ Net income

- Dividends paid to shareholders

+ Depreciation, Depletion, & Amortization

+/- Change in outstanding stock

+/- Change in company debt

+ Deferred income tax

_____

= Cash available for capital expenditures (CAPEX)

It may be necessary to make this calculation several times during the year as new information becomes available. It isn't uncommon for companies to change their estimates of yearly capital expenditures as those expenditures are being made to reflect changes in oil prices.

Non-discretionary, or mandatory, items which must be undertaken for various reasons are listed first for the allocation of available funds. These may reflect the impact of government regulations or the result of new corporate policy. Occasionally funds are budgeted for projects previously justified or under development. If funds were included in a previous justification, but not budgeted at that time, it may not be necessary to rejustify their expenditure when they are included in a subsequent budget. They might be carried in a "non-profitability" category along with the non-discretionary or mandatory items. Most, but not all, capital budget items are discretionary in terms of financial outlay, or time of undertaking. A healthy organization will generally have more capital improvement proposals than it has funds or personnel to handle. This calls for a selection, or culling, process in order to identify the economically most desirable projects. The projects are then ranked accordingly, and the available funds allocated to the top ranking projects. The rest are either rejected or deferred for reconsideration in a subsequent budget period. Some companies also include a "Contingent Not in Budget" category for these projects. So if additional funds become available during the budget year they may be undertaken or they may be substituted for approved projects that do not materialize.

The amount budgeted for each project is the total capital required to complete the project and is considered a commitment amount. Since not all projects will be completed during the 12 month budget period, it is also necessary to estimate the amount of cash that is expected to be spent on each project during the 12 month budget period, commonly referred to as CAPEX. CAPEX may be less than the amount budgeted for each project, with any committed budget funds not spent during the budget period carried forward to subsequent budget period. Such "Carry-outs" must be funded by cash available in future budget periods. However, once a project is budgeted and initiated it need not be re-budgeted to secure the needed cash. The CAPEX forecast for the budget period is needed for the Cash Budget which will be described later in this chapter.

Final approval of the capital budget is the responsibility of the Board of Directors and is normally done at their last meeting of the year. The approved budget usually grants authority to spend the funds to lower level managers consistent with the company's established "Delegation of Authority", as described later in this chapter.

Ranking and Selection of Budget Submittals

Not all organizations employ the same technique in ranking investment opportunities. Ranking will reflect the organization's goals and strategy. Large organizations tend to rank investments with different emphasis than smaller companies. The immediate cash position of the organization also materially affects the way it looks at new investment opportunity, with the cash short organization placing a greater emphasis on a rapid return. The volume and quality of investment opportunities available to an organization obviously also has an impact upon how investments are perceived and ranked.

The optimum ranking criteria for any organization should lead to the selection of investments which are commensurate with the cost of capital from all sources, utilize all of the capital available, and support the continued operation of the organization. An organization will generally find that one, or more likely a combination, of commonly used criteria discussed in Chapter V meets its particular needs in ranking investment opportunities. This may include: net present value (NPV), discounted return on investment (DROI), internal rate of return (IRR), and profit on investment (POI).

Once the investment opportunities have been graded according to the criteria which best reflect the goals and strategies of the organization, the selection process is undertaken.

Outline of the Budgeting Process

1. Identify all necessary non-discretionary capital expenditures. These may include investments required to meet safety and/or environmental standards. The alternative to these investments may be curtailment or termination of certain activities. Continued funding of previously initiated projects may also be non-discretionary.

2. Establish the level of funds available for discretionary expenditures.

3. Select discretionary investments in descending order of rank until either the total available funds are exhausted or the minimum acceptable yardstick value is reached.

   a. If there are more acceptable investment opportunities than funds available, the selected investments should be reviewed to determine if any of them could be deferred to a subsequent budget period without a significant loss in profitability. If such projects are identified, they may be deferred to enable undertaking a lower ranking opportunity that cannot be deferred.

   b. If available funds exceed acceptable investment opportunities, this may indicate that the investment criteria is not compatible with the investment climate, or that some investments have either been overlooked or improperly rejected. In any event, this situation may indicate that the organization's goals and strategies should be reviewed.

The problem of arranging projects in the order of decreasing economic desirability often leads to some interesting and challenging paradoxes. Example 9-1 will be used to illustrate some of these problems.

EXAMPLE 9-1
RANKING AND SELECTION OF INVESTMENT OPPORTUNITIES

Ten investment opportunities (A through J) totalling $128 million have been identified which must be evaluated to determine if they meet the corporate investment criteria and if so, which ones would be selected for various amounts of investment funds. The investment, cash flow, and ultimate profit for each investment opportunity is listed in the accompanying table. Net present value (NPV) and discounted return on investment (DROI) have been calculated for each using a hurdle rate of 10%. The internal rate of return (IRR), payout time, and profit on investment (PIR) have also been determined for each investment. All of the projects have a common order of necessity and there are no subjective factors which might influence their ordering. It is assumed that either all projects have an identical degree of risk, or that each investment has been risk discounted for the uncertainty of each project. It will be necessary to rank these projects only on the basis of the economic yardstick data presented.

ANALYSIS:

A. If there is at least $128 million available for investment, all projects would be selected. All investment opportunities have a positive NPV discounted at the corporate hurdle rate of 10% and the IRR of all projects is greater than the 10% minimum required rate of return.

B. If less than $128 million is available for investment, the projects must be ranked according to some criteria in order to select the projects which will benefit the company most, for the limited amount of investment funds available. Corporate goals and objectives will determine the choice of ranking criteria. At the bottom of the accompanying table you can see investments which would be chosen for various levels of funding according to each of the ranking criteria.

EXAMPLE 9-1
BUDGET PROJECT RANKING

		CASH FLOW OF INVESTMENT OPPORTUNITIES – MM$									
	YEAR	A	B	C	D	E	F	G	H	I	J
INVESTMENT	0	-10	-12	-8	-5	-15	-13	-11	-16	-18	-20
	1	5	7	7	7	7	6	6	6	7	5
	2	4	6	6	7	5	5	5	5	6	5
	3	4	5	5	6	3	4	4	4	5	4
	4	3	3	5	5	2	3	4	3	4	4
	5	3	2	3	4	1	2	3	3	4	4
	6						2	3	2	3	4
	7						1	2	2	3	3
	8						1	1	2	2	3
	9								1	1	3
	10								1	1	3
	11										2
	12										2
	13										2
	14										2
	15										2
	16										1
	17										1
	18										1
	19										1
	20										1
NET INCOME		19	23	26	29	18	24	28	29	36	53
NET CASH FLOW		9	11	18	24	3	11	17	13	18	33
HURDLE RATE = 10%											
NPV		5.49	7.27	13.35	18.66	0.46	5.87	10.37	5.40	8.50	9.74
DROI		54.9%	60.6%	166.9%	373.1%	3.0%	45.2%	94.2%	33.7%	47.2%	48.7%
IRR		39.6%	48.6%	111.5%	262.5%	12.2%	31.7%	50.3%	23.0%	27.8%	20.6%
PAYOUT		3	2	2	1	3	3	2	4	4	5
ROI		190%	192%	325%	580%	120%	185%	255%	181%	200%	265%
PIR		90%	92%	225%	480%	20%	85%	155%	81%	100%	165%

(handwritten above NPV row: "highest" over D, "lowest" over E)

Budget Funds available

```
|———————————————————+ $128
|————————————————$100
|——— $70
```

RANKING BY

NPV	D, C, G, J, I	B, F, A	H, E
DROI*	D, C, G, B, A, J	I, F	H, E
IRR*	D, C, G, B, A, F	I, H	J, E
PAYOUT	D, B, C, G, A, E	F, I	H, J
PIR	D, C, J, G, I	B, A, F	H, E

* most commonly used ranking criteria

(handwritten margin notes, not part of printed text: "see prev. page", "But all positive", "descending order highest first", "DROI only thing goes at portfolio with highest value", "only be carried out if we have $100", "only be carried out if we have $128", "Favors more value per $ invested but lower NPV")

Figure 9-2 depicts the ordering of a set of budget submittals according to Internal Rate of Return (IRR). If this set of investment opportunities represents the entire spectrum of opportunities available to an organization, then the IRR of the last project selected within the constraint of the total funds available to that organization would be an indication of the "opportunity cost of capital" for that organization. As shown in Figure 9-2, the opportunity cost of capital is somewhere between the IRR of the last project selected and the first one not selected.

FIGURE 9-2
RANKING OF CAPITAL BUDGET SUBMITTALS BY IRR

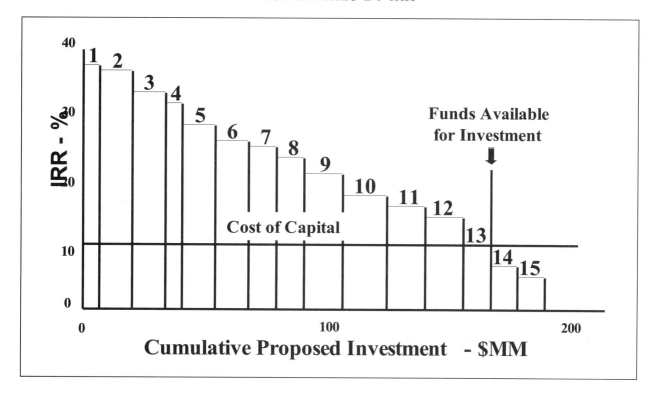

Figure 9-3 shows a ranking of investment opportunities according to the Discounted Return on Investment (DROI). Again this represents all investment opportunities available to an organization, with the level of funding also indicated. It can be seen that the DROI of the lowest ranking project to be selected within the constraint of funds available establishes a cut-off value for this ranking criterion. Such a cut-off is more severe than a cost of capital cut-off, as all 13 projects have a rate of return which is greater than the cost of capital. If an acceptable project cannot be undertaken within the budget constraints, there are several alternatives. If the project can be deferred, it might be considered for a later budget. It might be undertaken as a joint venture, with some other organization supplying most or all of the capital, or an outright sale might permit the realization of some value from the project.

If a project should fail to meet the cut-off before an organization reaches its total allocation of budget funds, there are several alternatives. The project might be dropped from further consideration; an exception to the cut-off might be made for special circumstances; or the project might be referred back to the originating location for further study. It might be possible to redesign the project with a lower investment, which would increase its profitability above the cut-off value.

367

FIGURE 9-3
RANKING OF CAPITAL BUDGET SUBMITTALS BY DROI

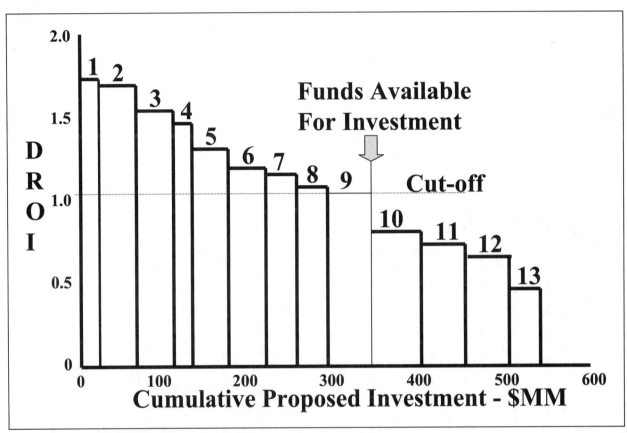

Portfolio Effects and Diversification

The objective of the budget process is to select a group of projects, or a portfolio of investments, which in total will meet the company's goals and objectives. This may mean that in the final selection process, some investment opportunities with poorer investment criteria values may be selected over better projects, because of some other desirable characteristic, or because there is some synergism created by several projects that may not be apparent as each is evaluated. For this reason it may be necessary to use several criteria for choosing projects and determine which combination of projects will yield the best results for the company. (See portfolio analysis: Chapter VI)

Markowitz (loc.cit.) pointed out a number of years ago that projects that may not be acceptable when considered individually may sometimes merit consideration when analyzed in combination with other projects. The risk-return features of a proposed undertaking examined in isolation may indicate that it does not warrant investment, but the way the opportunity interacts with other prospective projects (the covariance, or portfolio effects) could result in its acceptance. As Clark, et al (loc.cit.) remarks, "The business firm, after all, is merely an amalgam of previously accepted projects."

Regional emphasis, or de-emphasis, of exploration activity is a frequent portfolio problem in which the application of the normal economic parameters may be modified temporarily to accommodate the changing management strategy.

An important aspect of portfolio effect deals with the corporate desirability of diversifying its new discretionary investment. This can be particularly important in the company's exploration program. As has been mentioned previously, the larger, or giant oilfields tend to be more profitable than the smaller ones. Unfortunately, the smaller ones are much more numerous, and statistically, easier to find. Everyone hopes to find the big new giant oil field with its inherent high, long-life profitability. These successes occur only rarely, however. Thus, it is poor strategy to concentrate all of the company's effort on finding the big field in a new area, where the probabilities are against finding anything. It is much more prudent to allocate a portion of the company's exploratory funds to "elephant hunting," but to diversify, or spread the effort, to include a goodly portion of the available funds to work in known areas in which the company has good experience at finding oilfields of acceptable size.

Some oil companies have chosen to diversify into other lines of industrial activity in the hope of improving the return on the shareholder's investment.

EXPENSE AND CASH BUDGETING

These budgets are largely non-discretionary since the overall operation must continue if the firm is to stay in business. In reality they become forecasts. The expense portion is the precursor to the cash forecast which may be key to many items during the capital budget preparation.

Expense forecasting must include all items of cash outlay incorporating both direct and administrative costs for the budget period. The direct operating cost forecast, broken down by operational units, can also serve as an important supervisory control instrument. This amplifies the desirability of having the field operating people participate in its preparation.

Budgeting Expense

Most expense involves the maintenance of the organization and its ongoing activities. Thus, a logical place to start in formulating the projections of these costs for the forthcoming budget year is from the accounting records of the past several years.

Fixed and variable costs should be differentiated. Office rents and the maintenance of lease roads, or docking facilities, represent fairly fixed or reasonably constant types of ongoing expenditures. On the other hand, there are a number of types of costs which vary directly with the number of wells being operated, the amount of oil being produced, or the number of personnel employed in a given organization unit. These are designated as variable costs.

Time spent analyzing those costs which will simply be repeats of the previous year, and those which will change due to some accomplished or predicted change during the budget period, is always a worthwhile exercise. Both types of costs must then be predicted for the budget period since this takes up one of the very first allocations of budgeted cash.

A check of the current year's budget against the actual expenditures for the year to date is also worth undertaking in order to determine how closely the two sets of data correlate. If there is a significant discrepancy a study should be undertaken to ascertain the reason, so that the same deficiency is not repeated.

Cash Budgeting

A cash budget is a short term forecast that coincides with the capital and operating budget period and includes all cash flow items for the entire organization, as discussed in Chapter IV. It recognizes income from all existing projects plus the projected income from those projects in the proposed budget and any projects carried forward from previous budgets. This budget should also include cash-in from financing or other non-operating sources. On the other side of the ledger, it includes projected expenses for all ongoing operations, any anticipated non-discretionary expenses, expenses associated with proposed and carry-forward projects and indirect or overhead expenditures. It must also include the portion of capital expenditures for all carry-forward projects and proposed projects to be spent during the budget period. Thus, this budget or forecast must include all cash-in and cash-out items anticipated for the budget period. A negative annual cash flow would indicate the need for outside financing or call for a reduction of outlays. A positive cash flow may indicate that funds are available for payment of a dividend or for additional investments.

The following outline indicates the steps that must be followed in preparing a cash budget or short term cash flow forecast;

> Cash available at the beginning of the year
> + Income for the year
>   From existing projects
>   From new projects
> - Expenses
>   For ongoing operations
>   For indirect (overhead)
>   For non-discretionary
>   For new projects
>   Income Tax
> - Capital Expenditures (CAPEX)
>   For uncompleted previous year's budget projects
>   For new projects
> - Dividends paid to shareholders
> +/- Changes in corporate debt or stock
> ―――――――――――――――――――――――――――――――
> = Cash available at the end of the year $\geq O$

Most of the income will come from existing projects and most of the expenses will go to ongoing operations and overhead, but the other items listed should not be ignored. If the cash available at the end of the year is a positive value then there should be sufficient cash to conduct the year's business. If not, some plans will have to be changed, some projects delayed, or a source of outside funding established. The company could cut the dividends paid to the shareholders, but this is probably the least desirable solution as that usually leads to a drop in the price of the company's stock. So it is obvious that cash is the key item to a successful company. The Cash Flow Statement of a firms annual report, as described in Chapter VIII, is the actual achievement of this budget.

Economic Modeling

The generic term which is applied to the mathematical development of net cash flow streams on a cash basis, financial basis, or on an after tax approach is called "Economic Modeling" (Chapters IV and VI). A complete array of modeling techniques exists varying from simple arithmetic spreadsheets to complex systems of mathematical equations, such as linear programming or dynamic programming, involving relationships between several economic variables and constants. Some variables are explicit; some must be derived. An example might be a model designed to forecast the price of crude oil by seeking to forecast worldwide gasoline demand from projections of the number of motor vehicles on the road in future years. These more complex approaches are beyond the scope of this text, and are mentioned merely to identify the concepts. In reality, cash budgeting involves the creation of an economic model of the entire operation, however simple or complex it may be.

Personnel Budgeting

The availability of qualified personnel, or lack there of, can be just as limiting a factor in proper budgeting as the availability of funds. The addition of new operations to an existing work unit does not necessarily require additional personnel, but often it does. This needs careful study both by the company's operating people as well as by its personnel people.

The personnel budget should recognize the need for continued training of both technical and operating employees and provide sufficient corporate personnel and financial resources for this function. There is a normal turnover of employees in every company due to retirement, job hopping, and for a myriad of other reasons which should be anticipated on a statistical basis. An engineer, or geologist, who has just graduated from university certainly is not equal to a seasoned engineer with many years of experience. The personnel budget must forecast needs by level of experience, either through a job/experience title system, years of service, years of experience, level of training, or other appropriate statistics. For these reasons, a personnel budget is an important part of corporate planning.

Additional personnel add immediately to the expense budget forecasts. This must recognize not only salaries but also the employee benefits, office space and furnishings, company vehicles, and all of the other related costs.

Limits of Monetary Authority

The efficient operation of an enterprise is dependent upon a logical and effective delegation of authority to each level of organization. This has to be recognized in the budgeting process. Many of these authority delegations can be expressed in monetary units.Atypical, but obviously over simplified, delegation chart is shown in Figure 9-4.

The design of a corporation's delegation pattern is an interesting exercise in itself. Most originate from the top down depending upon each subordinate level's confidence in the people below. This is effective from the standpoint of management control. To be workable, however, each level has to have sufficient local authority to allow it to get its job done without having to go to the next higher level every time it needs to buy another drum of demulsifier or paint the lease "doghouses." This suggests the desirability of building the delegation chart from the bottom up. Obviously both approaches have to be accommodated.

The budgeting process in most companies, in addition to allocating adequate funds to each organization level to permit it to do its assigned operations throughout the budget period, also provides that certain types of projects which technically are beyond the local authority can be undertaken if they appear as detailed items on the final approved budget.

371

FIGURE 9-4
TYPICAL OIL COMPANY DELEGATION OF AUTHORITY

	Authorization Level
Board of Directors	Unlimited
President	$25 Million
Vice Presidents	$10 Million
	$7.5 Million
	$1.0 Million
	$750,000
	$250,000
Field Level	$50,000

Authorizations for Expenditure (AFE)

The final "go ahead" to conduct a specific operation or capital expenditure is usually handled ,with a sheet of paper known throughout the industry as an AFE. This may have a number of signatures denoting the final approval of the various staff groups which may be involved. If the project is a joint venture of several working interest owners, as so many are in the petroleum industry, then the AFE originated by the operating company is also "signed off" by each of the other owners before the job commences. This document also authorizes payment of invoices as they are submitted by vendors who supply materials and labor for budgeted projects.

The smoothest handling of AFE's for outside operations is accomplished by having the non-operating company's local office execute the AFE based on its delegated or budgeted authority. This greatly facilitates the local office's ability to monitor the progress and competency of the operation.

An important function of the annual budget is its planned timing of the undertaking of the larger projects so as to balance them with the manpower available to the operating unit. A company's budget must provide not only for its own operations but also for the outside operated projects which it is obligated to fund during the budget year. One of the frequently claimed advantages of being the operator is the control that this provides with regard to the actual timing of the undertaking within the budget period.

Timing and Cost of Budget Preparation and Updates

In the larger companies, due to time constraints if the fiscal year is the same as the calendar year, it frequently becomes necessary to start the budget planning process during the previous summer, a full half year ahead of the actual budget period. In this world of fast moving events this advanced planning

372

consideration presents a serious challenge to the field and district staff. These operating personnel who are occupied with day to day operational problems naturally find it difficult to envision income and expenses, and mandatory and discretionary investments that far into the future.

Capital budgeting has grown to become such an important activity among the large major companies that many of these firms have whole departments devoted solely to budgeting and planning on a year-around basis. The administrative cost of such an entire department obviously makes budgeting a very costly exercise. A great deal of the cost of budgeting may also be hidden in engineering and other overhead expense categories. The time of these professionals is diverted from other work in order to prepare the basic input data for the annual budget and any interim revisions.

PLANNING AND SCHEDULING — TWO SEPARATE ACTIVITIES

Planning is generally considered as defining and establishing the sequence and duration of activities that must be followed so that the authorized projects will be completed in an economical and timely manner. There is a tendency in planning to recognize only the more costly of the activities required in a major project. To achieve realistic planning for economic analyses, it is important to recognize that activities which of themselves may be trivial in cost can be very significant in time, which becomes money when the scheduling is applied. This is particularly true when Present Value factors may be applied later in the analysis. Scheduling then details when these activities are to occur within a very specific time frame.

Scheduling, per se, can only be undertaken when all of the budget-approved activities have been defined and interrelated from a fiscal standpoint. Scheduling is the process of determining the elapsed time relative to a starting time, zero, from which the series of events unfolds. These times derive from the times required in completing the preceding activities. The scheduling of company-owned or long leased offshore drilling rigs, and purchasing of long lead time materials, may require special consideration by their respective departments once the budget is approved.

BUDGETING EXPLORATION ACTIVITIES

Exploration programs, due to their nature, are on-going over a period of years. They normally involve a phase of office study and environmental research, a leasing or concession acquisition, geophysical and perhaps geochemical work in the field followed by extensive interpretation and correlation. Each of these phases calls for budget allocation of time and experience in addition to the eventual exploratory drilling effort.

A well designed exploration program will attempt to spread out these activities as part of the budgeting process so that one year's budget doesn't have a disproportionate amount of geophysical field work with, say, little or no interpretative load this year with a revisal of the professional personnel requirements the following year.

Exploratory budgets tend to fluctuate widely in the initial stages of the budgeting process. Then, as the months go by there may be a sudden expansion, or sometimes curtailment of exploratory funds as the company gains a better feel of how the firm's earnings for the year are progressing. Operating people sometimes find difficulty in learning to accept this annual phenomenon. Drilling operations generally fall within the producing department's domain, and the engineers are called upon to handle the planning and execution of the exploratory drilling as well as their own development wells.

A wildcat well may have been designed, the pipe ordered and the contractor preparing to spud, only to see the whole venture suddenly canceled, or deferred until 'sometime next year'. The inverse of the situation also occurs in which a whole series of exploratory wells are suddenly authorized for drilling and hopefully completion just before year end. This phenomenon makes it difficult for the engineer to comprehend why the exploration department's operations has many ups and downs, while the producing department appears much more able to set a series of budget goals and predictions, and to stick with them.

The major part of the exploration budget allocation in North America may be for lease acquisitions. Oil and gas leases are capitalized as tangible assets and amortized over their primary term. Undrilled leases normally also require an annual rental fee. Part of the cash flow of the firm must be earmarked for these lease acquisition, amortization and maintenance costs. A typical exploration budget also includes general exploration expense, which is comprised of overhead, staff, and support expenditures.

An exploration budget for a single twelve month period will represent expenditures for many individual programs in various stages of evolution. Some exploration programs will be in the initial "idea" stage, some in the follow-up phase, and others have advanced to the drilling stage.

Exploration activities move through part or all of the seven phases listed below:

PHASE I - GEOLOGICAL CONCEPT
PHASE II - SEISMIC/GEOLOGICAL INVESTIGATION
PHASE III - ACQUISITION OF EXPLORATION RIGHTS
PHASE IV - DETAILED SEISMIC/GEOLOGICAL STUDIES
PHASE V - DRILLING
PHASE VI - DEVELOPMENT
PHASE VII - TERMINATE PROGRAM IF UNSUCCESSFUL

The budget for any year will include numerous projects in various phases from I to VI. As an exploration prospect moves from one phase to the next, capital requirements expand accompanied by increasingly more sophisticated economic evaluation and technical effort.

Phase VII, termination of unsuccessful projects, can occur at any point in the evaluation process, as failure can occur during any of the other phases. For example, a geologic concept may turn out erroneous, or seismic data may condemn a prospect, or exploration rights may not be acquired, or detailed geologic studies nullify the prospect, or a dry hole is drilled, or development cannot be economically justified.

Movement from one phase to the next higher phase may be delayed for many reasons, such as budgetary constraints, need to develop new technology, or in the case of Phase III waiting for exploration rights to become available. For these reasons a particular prospect may not appear in each successive year's budget as it moves from Phase I to Phase VI or VII. The typical exploration prospect will be funded over many years before it is either recognized as a success or failure. A particular prospect may change drastically between Phase I and Phase VI as additional information is acquired and the outcome of each expenditure is evaluated.

Impact Of Accounting Methods On The Budgeting Process

Under the basic oil and gas accounting rules it is necessary for a company to choose between the industry's two distinct types of accounting systems: full cost and successful efforts. The very significant difference between the two accounting systems shows up on the assets side of the balance sheet as was discussed in Chapter VIII.

Under the successful efforts method all exploration dry holes are expensed. The exploration cash budget allocation of these companies is approached with high regard for the company's profit target for the year on the assumption that all of the exploratory wells will be dry holes. The exploration allocation thus becomes the 'swing' item in the overall budget. Most major oil companies, i.e., those employing the successful efforts method, find that the number and cost of the occasional development dry holes very nearly offsets the expenditures for the successful wildcat discovery wells. G & G, (geological and geophysical expenditures) as already mentioned, are mostly expensed under the successful efforts method, but capitalized in the alternative full cost method.

Example 9-2 shows how all of the various expenditures involved in a typical exploration program finally appear on a typical budget under both the successful efforts and the alternative full cost accounting methods.

■ EXAMPLE 9-2
■ ACCOUNTING FOR THE EXPLORATION BUDGET

By comparison with the successful efforts group, budgeting under the full cost accounting method becomes more of a business planning, or forecasting, exercise rather than an allocation procedure. Cash flow projections are equally important in both the full cost and the successful efforts accounting methods.

The full cost method requires the budgeting of the organization's overhead costs consisting principally of salaries, office rent, utilities, etc., as previously discussed. These costs must be forecast and provided for out of available funds. The cash demands of operating the firm's production have also to be anticipated. Since the expenditures for exploratory drilling, both successful and unsuccessful, are to be capitalized the cash demand for this part of the firm's activity does not detract in a very substantial way from the company's profit objectives. The budget year's expenditure for this activity is thus determined only by the availability of cash from other activities, plus any sources outside of the company. The companies which employ the full cost method of accounting are generally the ones most interested in the promotion of drilling funds and other means of raising cash to put into their exploration programs.

Proponents of both accounting methods are very interested in showing the best possible bottom line figures at year-end. In general, companies employing full cost accounting seem to be most concerned about immediate earnings per share, perhaps because these companies tend to be the smaller and their EPS ratio may vary significantly from month to month.

EXAMPLE 9-2
EXPLORATION BUDGET - 200X Year

		Successful Efforts Accounting		Full Cost Accounting	
Budget Category	**Proposed Budget**	**Capitalized**	**Expensed**	**Capitalized**	**Expensed**
Exploratory drilling	9,500	2,800	6,700	9,500	—
Lease purchases	4,500	4,500	—	4,500	—
Stratigraphic core tests	300	300	—	300	—
Geophysical crews	3,000	—	3,000	3,000	—
Lease Rentals and Taxes	1,100	—	1,100	1,100	—
Geological expense	1,600	—	1,600	1,600	—
Geophysical interpretation	2,000	—	2,000	2,000	—
Land, scouting, EDP	1,100	—	1,100	1,100	—
Dry hole contributions	400	—	400	400	—
Overhead	2,500	—	2,500	—	2,500
Tape & core storage facility	1,050	1,000	50	1,000	50
Total Budget	27,050	8,600	18,450	24,500	2,550

Thousands of Dollars

These differences in accounting methods are reflected in Example 9-2. Data show that under successful efforts accounting approximately 70% of the total exploration budget is expensed while less than 10% is expensed when the full cost method is used.

This accounting representation involves the corporate financial books. The way these same expenditures are treated for the purposes of federal income tax in the U.S. was discussed in Chapter VIII.

Handling Uncertainty

Uncertainty and risk are just as much a problem in budgeting as they are in the analysis of individual projects and programs. There may be a great tendency to ignore the matters of risk and uncertainty when composing a large budget with many entries. Sensitivity analysis, as discussed in Chapter VI, of major parts of the total budget is a good technique for dealing with the uncertainty problem. Sensitivity analysis is hardly practical for each individual item, but some such analysis (as for price forecasts) is generally warranted.

Other Considerations

A common practice in most large companies is to employ the budget for categorizing expenditures into major groupings, for example, so many dollars to be spent in Basin A; or, alternatively, the proportions of exploratory drilling to development drilling. A certain amount of cutting of the recommended projects to conform to specified economic yardstick guidelines is also done.

EVALUATING OPERATING PERFORMANCE AT THE CORPORATE LEVEL

Most well managed companies review their annual budget several times as the money is being spent. This is an excellent device for quickly determining the impact of the budget on the firm's financial condition. There are three facets to performance measurement:

1. The need to monitor major projects under construction to ensure that there are no acute delays or cost overruns

2. The conduct of post-investment appraisals on selected major projects shortly after they have started up to determine what, if any, problems need fixing or adjustment. These are generally referred to as "post audits."

3. The ongoing performance measurement of the entire firm, and its major departments, through a system of periodic "control reports."

Control Reports and Periodic Updates

This is done through a series of monthly or quarterly Control Reports which tabulate the actual physical and financial performance for the period of each division of the company against both the approved budget figures for the current year as well as the corresponding figures for the previous year.

Table 9-1 shows the format of such a control report prepared monthly for the top management of a major oil company. Comparable reports are also prepared and issued for each of the major geographical operating divisions of the company. Other pages of a typical control report break out the total company's activities by function, such as Exploration, Production etc., without reference to geographical boundaries. These control reports frequently include graphical presentations comparing the performance of the various functional activities within each operating division, and its major levels of management.

TABLE 9-1
CONTROL REPORT (M$)

200Y			FOUR MONTHS		
			200X	200Y	
APR	MAR		Actual	Budget	Actual
		GROSS REVENUES			
11,260	11,775	Crude Oil	43,767	48,373	46,070
560	590	NGL	2,185	2,415	2,300
5,100	5,200	Gas	19,570	21,630	20,600
50	75	Other	238	263	250
16,970	17,640	TOTAL REVENUES	65,759	72,681	69,220
850	975	Less: Purchases for Resale	3,468	3,832	3,650
		Royalty	9,828	10,863	10,346
13,582	14,030	NET REVENUES	52,463	57,986	55,225
		COSTS & EXPENSES			
3,800	4,200	Production — Direct	15,200	16,800	16,000
1,150	725	Well Expenses	3,563	3,937	3,750
951	982	Operating Taxes	3,672	4,059	3,866
975	975	Capital Recovery, DD&A	3,705	4,095	3,900
0	100	Dry Holes	190	210	200
750	775	Net Plant Operations	2,898	3,202	3,050
875	900	Administrative & Other	3,373	3,720	3,550
8,501	8,657	Total	32,600	36,032	34,316
750	750	Directly Incurred Exploration	2,850	3,150	3,000
200	300	Geophysical Expense	950	1,050	1,000
150	150	Amortization of Unproved Properties	570	630	600
75	50	Rental Expense	238	263	250
25	0	Cash Contributions Dry Holes	48	52	50
100	115	Administrative & Other	409	451	430
1,300	1,365	Total	5,064	5,596	5,330
55	58	Alloc.Gen.Office Exp.	215	237	226
97	89	Direct Charges - Parent	353	391	372
30	30	Regional Office - Staff	114	126	120
15	17	Research & Development	61	67	64
5	7	All Other	23	25	24
202	201	Total	766	846	806
10,003	10,223	TOTAL COSTS & EXPENSES	38,429	42,474	40,452
3,579	3,807	INCOME BEFORE INCOME TAXES	14,034	15,511	14,773
1,074	1,142	Federal Income Taxes - Current	4,210	4,653	4,432
143	152	- Deferred	561	620	591
215	228	State Income Taxes	842	931	886
1,432	1,523	TOTAL INCOME TAXES	5,614	6,205	5,909
2,149	2,284	NET INC.BEFORE INDIRECT CHARGES			
			8,420	9,307	8,864
350	350	Capital Employed Adjustment	1,330	1,470	1,400
1,798	1,934	NET INCOME	7,090	7,837	7,464

LONG RANGE PLANS

Corporate long range planning and the development of company strategy became popular about forty years ago. Long range planning is now firmly established as an important staff function in most large corporations including those in the oil industry. Most of these long range plans under the successful efforts accounting process are updated annually, with the first year matching up with the current year's budget, and cover the next 5 to 20 years or more.

A certain amount of long range forecasting is required by the U.S. Government Securities and Exchange Commission (SEC) in an effort to overcome the deficiencies of both accounting methods. All U.S. oil and gas producing companies are required to file a report "Standardized Measure of Discounted Future Cash Flows Related to Proved Oil and Gas Reserves," (see Chapter VIII) which is an indication of the value of their proven reserves. Preparation of this report requires a year-by-year forecast of future cash flow from currently proven hydrocarbon reserves, based upon current oil and gas prices and the current cost of producing those reserves. Any future capital expenditures which may be required to produce the reserves are also included in the cost of production at current price levels. For example, secondary recovery investments and anticipated production increases can not be included until injection operations actually commence.

The forecast prepared to satisfy the SEC "Standardized Measure" requirement can easily be modified to satisfy internal needs for a long range forecast. This is accomplished by substituting internal forecasts of inflation and oil and gas pricing for the strict SEC values and applying the company's hurdle rate. It would also be necessary to add any non-hydrocarbon businesses conducted by the company.

Planning should not be looked upon as a means of reducing or eliminating risk, but rather as a process of selecting the risks to be taken. The key to long range planning in the oil business lies in the reliability of the cash flow projection. This requires a good production forecast for the period coupled with a realistic price projection. It is also necessary to forecast new oil and gas discoveries during the period, their producing rates and the impact of any "discovery bonuses" which may apply. The decline in productivity of older oilfields during the prolonged planning period should be recognized. This is seldom required in the normal one-year budgeting process. The matter of predicting prices to apply to future production is not a simple task, yet the whole plan is dependent upon the realism with which this can be accomplished (Chapters II and III).

The outlays for development drilling by years (including confirmation wells to the new discoveries) have to be forecast for the planning period. The major capital expenditures each year for such projects as natural gasoline plants, secondary recovery projects, etc., as well as non-discretionary expenditures also must be recognized. Operating costs have to be projected as part of the planning process with due recognition of anticipated increases in wages and salaries, along with possible staff reductions, or enlargements, etc.

The next step involves the trade-off between discretionary major capital expenditures and the exploratory program projected for the period. Various timing scenarios are usually run through analytical procedures to derive an after-tax earnings forecast for each trial case. Finally a balance is struck between exploration and capital improvements to provide the base case.

Most companies then apply a further series of sensitivity cases to the long range plan. These normally involve such matters as computing the effects of ranges of price forecasts in order to determine the impact of price variations on the plan. Other sensitivity cases might include variations in the levels of operating expense, and/or natural gas sales and price over the forecast period.

When the long range plan is finished it then goes forward through channels to the company's top management for their information and review. If the plan projects a large cash surplus during the period, management may wish to pursue a major corporate acquisition of some sort, or alternatively a change in the dividend policy. If on the other hand the plan forecasts a serious deficiency of funds, management may find a curtailment of the projected program to be in order, with discretionary projects being dropped first. A secondary possibility is that a cash deficiency will be handled by means of borrowing additional funds to permit the enlarged program to go forward and thereby enhance the firm's rate of growth. The long range plan, with its integrated staff preparation, enables top management to reach intelligent decisions regarding the company's goals and strategies.

REFERENCES

Anthony, R. N. and Reece, J. S., Accounting Text and Cases, 7th Ed., 1983, Irwin, Homewood, IL 60430

Clark, J. J., Hindelang, T. J., and Pritchard, R. E., Capital Budgeting, 1979, Prentice-Hall, Inglewood Cliffs, NJ 07632

Harrison, F. L., Advanced Project Management, 1981, John Wiley & Sons, New York

Markowitz, H., Portfolio Selection: Efficient Diversification of Investment, 1959, John Wiley & Sons, New York

Steiner, G. A., Strategic Planning, 1979, Macmillan, New York 10022

Sweeny, H. W. A. and Rachlin, R., Editors, Handbook of Budgeting, 1981, John Wiley&Sons, New York

Willis, R. J., Computer Models for Business Decisions, 1987, John Wiley & Sons, New York,

CHAPTER	ECONOMIC ANALYSIS

ECONOMIC ANALYSIS OF OPERATIONS

CHAPTER

X

A profitable company is the combination of many profitable projects.

Oilfield operations may be categorized as being either ongoing (repetitive in nature) or non-repetitive project type of activities. The economic analysis and monitoring of the two types of operation differ significantly from one another.

ANALYSIS OF ONGOING OPERATIONS

Managerial analysis and control is the process of assuring that the operations are being performed or accomplished in line with the desired results in terms of both volumes of production and gross well expense. Operations may or may not be going according to plan. It is oftentimes necessary to modify the plan depending upon the results of actual observed performance. A good and timely reporting system is essential to effective management control of petroleum producing operations.

Four business elements are required for the continuous production of oil and gas and delivery of a saleable product to the pipeline: 1) a predetermined goal or plan; 2) an adequate and timely system of operating reports; 3) comparison of actual operations to the goal; and 4) a procedure and history of successful corrective action. The corrective action may be to bring the operation into line with the plan, or to modify the plan more in line with the actual performance. In the upstream segment of the oil business, control involves surveillance of reserves, production rates, timing of operations, and costs.

Field operating costs are generally analyzed according to several different categories. There are a number of direct costs such as labor, power, fuel, repairs, and treating costs. These were described in Chapter IV as factors directly affecting the cash flow of the operation. Some of these costs are primarily dependent upon the volumes of daily production. Others are more directly related to the number of wells being operated on the particular property being analyzed.

Well dependent costs are mainly the result of well workovers and the repair and maintenance of production equipment. Volume-dependent costs relate proportionately to the quantities of fluid, oil and/or water that is lifted, transported, treated, and injected. Volume-related costs are fairly easy to analyze and control. Well-dependent costs may be more difficult, particularly if they include allocated costs for supervision, engineering, accounting, timekeeping, warehousing, and general transportation.

The costs of operations are usually monitored and assessed on both a dollars per barrel of oil produced, and dollars per producing well per month basis. Either, or both, of these indices should serve promptly to identify changes in the economics of operation. These may come from any of a number of causes, or a combination of several.

Different wells and different fields may vary widely in terms of lifting costs. Max Ball (loc. cit.) writing in 1939 observed:

> "A shallow flush well putting thousands of barrels a day into the tanks produces oil at a few cents a barrel; oil from a small deep pumper may cost two or three dollars a barrel. The most worrisome fact in an oil producer's life is that as production goes down the cost per barrel goes up, for it costs nearly as much to pump a well when it is producing ten barrels per day as when it is producing fifty, and there are only one-fifth as many barrels to which to charge it."

The same fundamentals persist today although the dollar magnitudes require updating.

Various Factors Which Affect Ongoing Operating Costs

The efficiency of the operating labor in the oilfield is probably the prime factor in determining the ultimate lifting cost. It is far from the only factor, however.

The General Economic Conditions existing in the oil producing region are the prime determinant in the level of wages. To a lesser degree, as described in Chapter III, the prices received for crude oil and natural gas are reflective of the general economic conditions, either currently prevailing, or recently past. This, in turn, clearly affects the costs of oilfield services, as well as utility costs.

Depletion Effects are another source of increasing lifting costs with time. Petroleum is a depleting resource, which manifests itself in a gradual decline in the daily productivity of the individual wells. This effect, although gradual, is not persistently observable each twenty-four hour period. It is rather like watching the day to day transition in the weather with the change of seasons. Nevertheless the productivity does decline, usually a good bit faster than the ongoing costs of the operation.

These declines in productivity are generally coincident with changes (declines) in reservoir pressure which affects the physical chemistry of the reservoir fluids. More treating chemicals are normally required later in the life of the typical oilfield. Paraffin removal, and sulfur removal (depending on the nature of the reservoir fluids) tends to be a greater problem in fields which have been on production for longer periods of time.

Equipment wear is unavoidable in the oilfield, as anywhere else. In older fields equipment repair becomes more frequent, and costs rise. The costs of upkeep of lease operating areas including fences & cattle guards, firewalls, etc. go on even with declines in the oil and gas production.

Corrosion is another unfortunate oilfield problem. Any time one puts metals, of even slightly differing composition, in contact with dissimilar electrically conductive fluids there will be expenses for corrosion prevention or protection, or costs of equipment replacement. These can be substantial, and oilfields are made to order environments for corrosion. These costs also tend to accumulate with the age of production.

In most oilfields Secondary Recovery by waterflooding, or gas injection, comes a bit later in the life of the field after the true nature of the reservoir and its fluids have been determined. Eventual breakthrough of the injected fluid is to be expected and the costs of handling the increasing volumes of these by-products adds to the operating costs as time goes by. Another secondary recovery cost, which is oftentimes encountered in older waterfloods, is the development of hydrogen sulfide production from bacterial action made possible from the introduction of foreign injection waters into the reservoir. This H2S creates an additional Corrosion cost.

Well completions do not stand the test of time forever, so workover expenses increase as reservoirs deplete. All of these factors, and their sometimes not too easy to decipher effects on month to month operating costs are, of course, reflected in the periodic operating data. It may take a good analyst to recognize and isolate the various problems for appropriate corrective action.

Technical Efficiency vs. Financial Efficiency

In the oilfield technical and financial efficiency have always gone "hand-in-hand." Technical efficiency is usually measured in terms of daily production volumes. Increasing production generally, but not always, results in improved financial efficiency, i.e., profit. Reduction in lifting costs should also increase operating profit even with no increase in producing rate. The various means of reducing lifting costs are therefore a prime center of managerial attention.

INCREMENTAL ECONOMICS

One of the most common types of economic evaluations involves the difference between two cash flows. Incremental or differential economic evaluations are made to determine if the benefits of additional expenditures can be justified. The base case cash flow, frequently called the "do nothing" cash flow, is a forecast under existing conditions. The second cash flow reflects the future including the results of an additional investment or significant one time expense item. Well workovers, supplemental recovery, delay decommissioning, acceleration projects, additional drilling, debottlenecking, and project expansion are typical examples of incremental or differential economics. The incremental or differential economics is the difference between the two cash flows that can be analyzed as any other cash flow.

EXAMPLE 10-1
INCREMENTAL ECONOMICS

This is a situation where the incremental investment is small compared to the base investment in the project and will completely mask the value of the incremental investment if only the combined cash flow is analyzed. The base investment is $1,000,000 while the additional investment is only $100,000. Both the base investment and the base investment plus the additional investment are very profitable when evaluated individually, but the contribution of the added investment is very poor by comparison. This can only be determined by evaluating the incremental change in cash flow. The following table shows that the contribution of the additional $100,000 investment is very poor and might not be justified as a stand-alone investment.

| DISCOUNT | NET CASH FLOW | | |
RATE	BASE CASE	BASE CASE+	INCREMENTAL
0%	$1,635,270	$1,649,345	$ 14,075
3%	$1,249,535	$1,263,237	$ 13,701
5%	$ 955,485	$ 966,835	$ 11,350
10%	$ 544,783	$ 548,919	$ 4,136
15%	$ 277,297	$ 273,513	($ 3,785)
20%	$ 91,971	$ 80,785	($ 11,186)
25%	($ 43,028)	($ 60,774)	($ 17,745)
IRR	23.24%	22.68%	12.60%
DROI (@10%)	54.48%	49.90%	4.14%
INVESTMENT	$1,000,000	$1,100,000	$100,000

From above table and Figure 10-1 it is apparent that it would be very difficult to distinguish between the two cash flows, but the incremental cash flow tells a completely different story.

FIGURE 10-1
INCREMENTAL ECONOMIC ANALYSIS

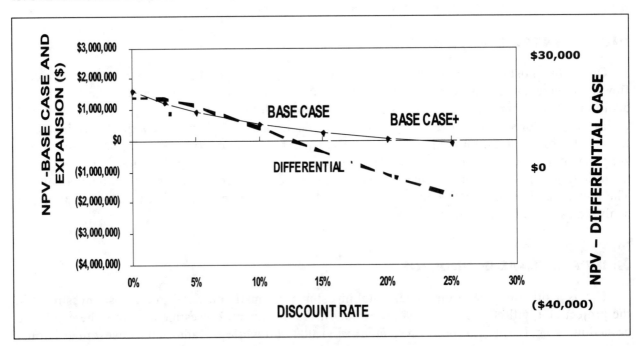

Well Workovers

Money spent on well workovers is money well spent because most of them are successful and they usually payout quickly. Failure to keep wells in good operating condition will lead to rapid decline in production and premature abandonment causing loss of reserves and lower profits.

Example 10-2 is an economic analysis of a well workover as an illustration of spending money to accelerate the cash flow, and increase the ultimate income. The cost of this particular type of operation is "expensed" for accounting purposes, rather than "capitalized" as an investment.

■ EXAMPLE 10-2
ECONOMIC ANALYSIS OF A WELL WORKOVER

A well was completed and went on production at an initial rate of 90 BOPD. Five years later the well had established a hyperbolic decline in production with an initial decline rate of 30 percent and an h-factor of 0.7. The following year the well's productivity began to drop off even more sharply. Two years later the operator decides the well should be reperforated, at a cost of $75,000, to restore its productivity. This is expected to result in a producing rate of 56 BOPD compared to the current 44 BOPD.

Assume a constant oil price of $23 per barrel, a net revenue interest of 80%, operating costs of $5,000 per well per month. The economic limit is reached at about 10 BOPD.

DETERMINE:

The economics of the well workover.

SOLUTION:

The incremental economics of the well repair are determined by assuming that the production rate after reperforating will be greater, but that it will continue to follow a hyperbolic decline with a somewhat higher initial decline rate (40%) and an h-factor of, say 0.8. This is the basis for the following incremental cash flow.

YEAR	Before Workover BOPD	Before Workover $/mo	After Workover BOPD	After Workover $/mo	Incremental $/yr	Incremental Cum.	Incremental Cum. NCF
-2	64	$30,524	N.A.				
-1	57	$26,620	N.A.				
0							($75,000)
1	44	$19,159	56	$25,912	$81,031	$81,031	$6,031
2	35	$14,269	44	$19,118	$58,192	$139,223	$64,223
3	29	$10,854	35	$14,334	$41,758	$180,981	$105,981
4	24	$8,356	29	$11,023	$32,014	$212,995	$137,995
5	21	$6,460	25	$8,612	$25,834	$238,828	$163,828
6	18	$4,980	21	$6,785	$21,660	$260,489	$185,489
7	15	$3,280	18	$4,936	$19,872	$280,361	$205,361
8	13	$2,176	15	$3,280	$13,248	$293,609	$218,609
9	11	$1,072	13	$2,176	$13,248	$306,857	$231,857
10	10	$520	11	$1,072	$6,624	$313,481	$238,481
11	9	($32)[1]	10	$520[2]	$6,240	$319,721	$244,721
TOTAL	78,995[3]	$853,120	99,632	$1,173,225	$319,721		

[1]Abandoned year 10 / [2]Abandoned year 11 / [3]Years 1-10

IRR calculated on the incremental cash flow is 140%, net present value at 15 percent is $140,525, DROI at 15% is 187%, payout occurs in a little over 11 months, and reserves are increased by 20,637 barrels. The workover appears to be a very profitable project and is acceptable.

Delay Platform Decommissioning

When abandonment was discussed in Chapter II it was mentioned that it might be advantageous to delay it as long as possible, if the operator was confronted with a significant expenditure to decommission an offshore platform, as many operators face in the North Sea.   An analysis to decide if such a delay is warranted is best performed using an incremental analysis. The base case would represent the existing situation, abandonment occurring when the current operation reaches its economic limit. It would be necessary to create a second cash flow representing the results of an additional expenditure that would result in an extended life for the platform and thus delay the date that the platform would be decommissioned. Justification of the additional expenditure would be determined by analyzing the difference between these two cash flows to determine if the increased benefits are sufficient to warrant the expenditure. All of these analyses should be done on an after income tax basis, as there may also be some tax advantage to postponement of decommissioning.

Activities that might delay decommissioning could be the drilling of an additional well(s) to improve reservoir drainage or access new reserves or reserves that had been prematurely abandoned in the past when lower prices for oil and gas prevailed. It may be possible to side-track existing wells to cut the cost of drilling. Or existing producing reservoirs might be redeveloped by re-acidizing, re-fracing, or re-perforating currently producing wells to increase their productivity. Or inject water or gas to enhance recovery or maintain the pressure of currently producing reservoirs. Or it may be desirable to modernize or debottleneck the processing facilities to achieve a higher production rate. These activities when analyzed individually might have only marginal economic value, but when coupled with the value of delaying a major expenditure and prolonging the producing life of a field might be quite attractive investment opportunities. The value of delayed decommissioning comes from several sources; (1) it reduces the present value of the decommissioning cost, (2) adds reserves by extending the producing life of the reservoir, and (3) it adds reserves as a result of the expenditure. When all of the benefits are added together, what might initially be thought to be a marginal investment opportunity may turn into a good investment opportunity.   Even if you can only postpone decommissioning for a few years, it may prove worthwhile. The only method to determine if this is the case is to analyze the entire project with an incremental analysis as was described.

Apache's 2003 purchase of the Forties Field from BP is a good example of what can be done to an old field and thus delay its abandonment. They have done most of the things listed above and turned this field from one that was rapidly approaching its economic limit to one that may continue to produce for the next twenty-five or so years, by more than doubling its reserves at the time of purchase.

EQUIPMENT REPLACEMENT

After equipment has been used for a number of years, consideration may be given to replacing it. The need to replace equipment can arise for any of a number of reasons. The existing equipment may have become obsolete due to advances in technology or it may no longer provide the desired service. Due to its age or condition, it may require excessive maintenance and repair or operate inefficiently. Conditions may have changed from those that existed at the time the equipment was originally purchased such that the existing equipment is either oversized or undersized for the job to be performed. There are many reasons why consideration should be given to replacing existing equipment, but the final decision should be justified on economic values. This decision is the same as deciding between alternative pieces of equipment to provide the same service, with the exception that the existing equipment must be considered a sunk cost. So, it is necessary to show that the total cost of the new equipment to provide the desired service, including both operating and capital costs, is less than the operating and maintenance cost of the existing equipment.

If the justification for the new equipment includes the need for a greater volume, then the incremental profit increase should be included in the valuation of the new equipment. If changed conditions would require modification of existing equipment to obtain satisfactory service, this is an added cost of maintaining the existing equipment. Thus, it is necessary to quantify all factors which would be unique to every alternative before comparing the total cost of each. Just because a new piece of equipment may be bright and shiny and include the "latest" technology, it may not necessarily represent a lower total cost than the greasy rusted model of several years ago.

There are several additional items that the analyst should be aware of when evaluating the replacement of equipment. Be sure to deduct the "fair market" salvage value of the existing equipment from the new equipment cost. The existing and replacement equipment should be evaluated over the same time interval. If the replacement equipment is expected to last longer than the existing equipment, adjustments must be made to compensate for this difference. One way to do this is to terminate the evaluation of the replacement equipment at the terminal point for the existing equipment and credit it with its fair market value at that point in time. Another way would be to compare replacement to continued use of the current equipment on the basis of total annual cost. When replacing equipment with a similar one, the cost is usually expensed rather than capitalized for both tax and accounting purposes. However, this can be one of those grey areas, so expert advice should be obtained in those cases where the difference between expensing and capitalizing an item is significant.

Two examples are presented to explain these ideas more fully.

■ EXAMPLE 10-3
ECONOMIC EVALUATION OF REPLACEMENT

An automated diatomaceous earth silo and mixer package which would cost $250,000 is being considered to replace a manually operated one currently utilized in a waterflood. It is estimated that the new equipment would save approximately $50,000 per year over the next eight years. The existing equipment has no salvage value now and the new equipment is also assumed to have no salvage value eight years from now. If the operator uses a hurdle rate of 10%, is the new equipment justified?

Year	Cost Savings	Replacement Equipment	Net Cash Flow	PVNCF* @ 10%
0		$250,000	($250,000)	($250,000)
1	$50,000		$50,000	$47,673
2	$50,000		$50,000	$43,339
3	$50,000		$50,000	$39,399
4	$50,000		$50,000	$35,818
5	$50,000		$50,000	$32,561
6	$50,000		$50,000	$29,601
7	$50,000		$50,000	$26,911
8	$50,000		$50,000	$24,464
Total	$400,000	$250,000	$150,000	$29,766

*Mid-year discounting

Replacement of a filter system for a waterflood can be evaluated as shown in the above table. In this case, cost savings were treated like income and the new equipment expensed as it was assumed that the replacement was with a similar piece of equipment. Therefore, a before tax evaluation will lead to the same decision as an after tax analysis would. In this case the new equipment would be justified because the total present value net cash flow is $29,766 in favor of the new equipment.

This could also be solved using the annuity concept with mid-year discounting:

$$PVNCF_{10} = \$50,000 \, (1.1)^{0.5} \left[\frac{1-(1.1)^{-8}}{0.1} \right] - 250,000 = \$29,766$$

$$DROI_{10} = \$29,766 \, / \, \$250,000 \; 11.9\%$$

EXAMPLE 10-4
REPLACEMENT OF BEAM PUMPING UNITS

Consider producing operations in the Cerro Dragon field of Chabut Province, Argentina. The wells are equipped with gas powered API 228 beam pumping units that were installed four years ago at a cost of $25,000 per well. Consideration is now being given to replacing these units with electrically powered API 640 units at a cost of $35,000 each. The API 228 units, operating on purchased gas, cost $700 per month to operate. The projected operating expense for the new larger electric units is forecast to cost only $400 per month. The best bid per unit for the API 228 units with their engines is $15,000. The salvage value for these after another ten years of service is $10,000, the new API 640 units are assumed to have a service life of twenty years and a salvage value $25,000. These values are diagrammed in the following illustration:

ANALYSIS OF PUMPING UNIT REPLACEMENT

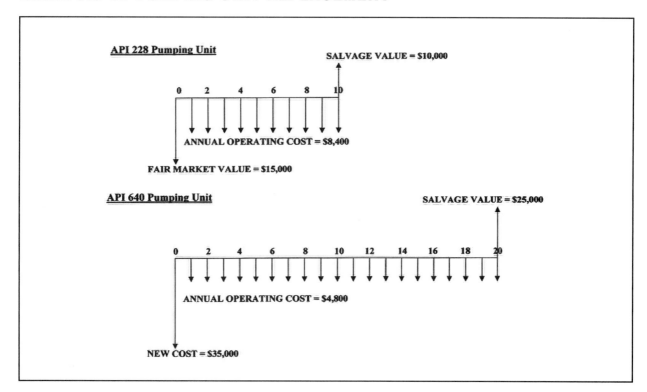

DETERMINE:

Calculate the total annual cost of continued use of an API 228 pumping unit and a replacement API 640 unit, using a hurdle rate of 15%. Is the replacement justified?

SOLUTION:

Annual capital cost can be determined by calculating the annual annuity which would be equivalent to the cost minus the salvage value when the unit is retired from service. The formula for an annuity was developed in Chapter V and is presented as Equation 5.8. Annuity values are presented in Tables 5B-3 and 5B-4. The total annual cost is calculated as follows:

API 228 UNIT

From Tables 5B-3 and 5B-4 of Appendix V-B:

$$1/PVIFA_{15\%,10\text{ yr}} = 0.1858$$

$$1/FVIFA_{15\%,10\text{ yr}} = 0.0459$$

Annual capital cost = \$15,000 * 0.1858 - \$10,000 * 0.0459	= \$2,328
Annual operating cost = \$700 * 12	= \$8,400
Equivalent annual total cost	= \$10,728

API 640 UNIT

From Tables 5B-3 and 5B-4 of Appendix V-B:

$$1/PVIFA_{15\%,20\ yr} = 0.1490$$
$$1/FVIFA_{15\%,20\ yr} = 0.0091$$

Annual Capital Cost = \$35,000 * 0.1490 - \$25,000 * 0.0091	= \$4,987
Annual operating cost = \$400 * 12	= \$4,800
Equivalent annual total cost	= \$9,787

The above calculations indicate that the replacement API 640 unit has a lower annual cost than the existing API 228 unit, so that the replacement should be justified. Another question that may arise is, what is the continued use value of the existing unit compared to the replacement unit? This can be determined by setting the annual cost of the API 228 unit in the above calculation equal to the annual cost of the API 640 unit, with the current value of the existing unit unknown. This leads to the following computation, if X is the unknown current value of the existing unit.

$$(X * 0.1858) - (\$10,000 * 0.0459) + (\$700 * 12) = \$9,787$$
$$X = \$9,935$$

This further supports the decision that the unit should be replaced with the more efficient one, because the continued use value of \$9,935 is less than the \$15,000 which has been offered for the unit. A number of other factors may also have to be considered before the final decision is reached. For example, no change in oil productivity has been assumed for the larger pumping unit.

The economic analysis of replacement of used, but still operable, equipment by new and more efficient equipment generally encompasses sunk cost considerations. These decisions often involve a defender (who purchased the old equipment) and a challenger (who proposes the new equipment). Objective analysis employs what is called the "outsider viewpoint" as illustrated in Example 10-4. This is a means of avoiding sunk costs by employing the current sales value, or "fair market value," of the old equipment rather than its original investment or purchase price.

PURCHASE / SALE OF PRODUCING PROPERTIES

Evaluation of the purchase and sale of producing properties differs from other economic evaluations previously described in a number of ways. Producing properties may include three different types of assets, each of which are treated differently for tax and accounting purposes. The value of a producing property may not be the same for the buyer and seller and there are a number of different ways to structure the transaction which may also affect the value of the exchange.

In parts of North America where the surface owner may also be the owner of subsurface minerals, a transaction involving subsurface minerals may also include the surface land overlying the minerals. Ownership of the subsurface and/or surface (land) is referred to as ownership in "fee" or "fee ownership." When purchase/sale of producing properties include fee interests, the surface (land) and mineral interest must be evaluated separately at their "fair market value." Fair market value is the cash price that a willing buyer would pay to a willing seller for the property in its existing state, under prevailing market conditions at the time that the evaluation is made.

The operating interest, frequently called the working interest or leasehold, in a property must also be evaluated separately. Its value is dependent upon the profit which may be realized in producing the hydrocarbon reserves existing at the time of purchase. This value is some function of the net present value to be realized from the property, after deducting all future expenditures required to produce the reserves, including the cost to acquire the property. It may be the net present value discounted at the hurdle rate, at some higher or lower discount rate, or any other relationship determined to be appropriate by the buyer and seller. Producing properties also include tangible capital in the form of equipment and facilities necessary to produce and process the hydrocarbons for sale. These too must be evaluated separately at their fair market value at the time of the valuation. The fair market value of tangibles is not their value to the continued operation of the property, but their value if dismantled and removed from the property. This value will be much less than the value for continued operations because it will reflect used equipment values further reduced by the cost of removal.

Each of these three types of assets are treated separately for tax and accounting purposes. Surface (land) is considered perpetual and even though it can be changed and transformed, it cannot be used up, so its value is not depreciated over time. For both tax and accounting purposes the value of land is kept on the company's books at acquisition cost until it is subsequently sold and then deducted from the sale price to determine a profit or loss on that asset.

The portion of the sale price attributed to either a mineral interest or a leasehold interest is reduced over time in proportion to the volume of hydrocarbons produced from the property. In other words, the cost is recovered through "cost depletion" for both tax and accounting purposes. This procedure allocates the same fraction of the acquisition cost to each barrel of oil and/or cubic foot of gas produced.

Tangible capital acquired through purchase of a producing property is treated for both tax and accounting purposes just like any other equipment which is purchased. Its cost is spread over its remaining operating life through depreciation. While depreciation for tax purposes may differ from that for accounting purposes, depreciation of used equipment is usually very similar to that for new equipment for both tax and accounting.

The value of a producing property may not be the same for the buyer and the seller for a number of reasons. First, they may have different estimates of reserves, future production rates, and/or future expenditure requirements. Or, one party may feel that there are additional drilling or completion prospects on the property. These differences could lead to widely different values for the leasehold and mineral interest. The tax status of the two parties may also be vastly different. The purchaser must recover the purchase price through depreciation and depletion over the prescribed number of years. The seller may have a gain on the sale subject to taxation or a loss which would give rise to a tax credit. A gain or loss by the seller will depend upon his cost and the amount of depreciation and depletion already taken. The actual tax that each party may have to pay as a result of the transaction is also a function of the rest of their business. Other losses might shelter a gain on this transaction, or losses on this transaction might shelter other gains. Thus, it can be seen that the actual tax and accounting effects of a producing property purchase/sale can differ widely among organizations and must be calculated explicitly for each party.

The way that the transaction is structured can also affect the value to either party. The simplest transaction is a cash sale with payment made at the time that ownership is transferred. However, it is not uncommon for the seller to partially finance the sale and receive the payments, including interest, over a specified period of time. The purchaser might also go to a third party for financing. In this case it would be a cash sale as far as the seller is concerned, but the purchaser would have to repay the loan. A popular variation of this is financing a producing property purchase through a production payment on that property, with a portion of the production dedicated to repay the loan. There also may be some advantage to exchange properties, where each party has a property of similar value it wishes to sell. In such a case both properties are evaluated in a similar manner and any difference in value is paid in cash. In what is commonly referred to as a "like-kind exchange" each party retains the remaining cost in the property disposed of for the new property acquired for tax and accounting purposes. Usually no gain or loss is recorded for this type of transaction.

If no purchaser can be found for a given property, or if the price is unacceptably low to the seller, there are two options which should be considered. The seller can continue to operate the property to depletion or until an acceptable sale can be made. Or the seller could abandon the property at any point in time. At abandonment the operator would plug and abandon all wells, remove all equipment and sell or reuse whatever can be salvaged from the operation. When abandonment is complete the operating rights are then surrendered back to the mineral owner.

The value of abandonment is the amount realized from sale of salvageable equipment minus the cost of abandonment and salvage plus tax credits that may arise by taking any remaining depreciation, depletion or amortization left for the property. The value of continued operation or abandonment, which ever is larger, establishes the minimum acceptable sale price for a producing property.

The following is an example of a producing property purchase and sale evaluation. Since the value of a producing property can be highly dependent upon tax implications, a tax status has been assumed for the seller. This is only a single example and may not be appropriate to other evaluations.

EXAMPLE 10-5
EVALUATION OF A PRODUCING PROPERTY PURCHASE / SALE

Gusher Oil Company wishes to sell the R. U. Lucky property, which it owns in fee. The Bullet Oil Co. has made an offer to buy the property for $2.0 million cash. You are to determine if this is an acceptable price. The following information was obtained from various company sources to make the evaluation.

Original purchase price of surface (land)	$100,000
Original cost of equipment	$1,750,000
Original cost of leasehold (minerals)	$500,000
Current "fair market value" of surface (land)	$275,000
Current "fair market value" of equipment	$500,000
Cumulative depreciation of equipment	$1,382,500
Cumulative cost depletion of leasehold	$400,000

NPV of cash flow of continued operations, after tax (at hurdle rate) = $1,120,000

Income tax rate	36.6%
Hurdle rate	10.0%

The first thing that must be done is to separate the offering price into the amount attributable to each type of asset.

Sale Price (cash)	$2,000,000
less: Surface (land) Value	$275,000
less: Value of Equipment	$500,000
Leasehold Value	$1,225,000

Next it is necessary to determine the tax consequences of the sale to the seller. The alternative to selling the property would be to continue to operate it, which has a net present value, after tax of $1,120,000. It is assumed that this includes the cost to plug and abandon the property and the salvage value of any equipment, but excludes the value of the surface (land).

Type of Asset	Tax Calculation		Tax on Gain
Land	0.366 [275,000 - 100,000]	=	$64,050
Leasehold	0.366 [1,225,000 - (500,000-400,000)]	=	$411,750
Equipment	0.366 [500,000 - (1,750,000 - 1,382,500)]	=	$48,495

For the sale price to be acceptable, the net amount received, after tax, for leasehold plus equipment must be greater than the net present value which would be obtained from continued operation of the property.

Cash sale price	$2,000,000
less: Value of surface (land)	($275,000)
Sale price of leasehold & equipment	$1,725,000
less: Tax on leasehold gain	($411,750)
less: Tax on equipment	($48,495)
Net cash from leasehold & equipment	$1,264,755
plus: Value of surface (land)	$275,000
less: Tax on surface (land)	($64,050)
Total Net Cash Realized from Sale	$1,475,705

The offered $2.0 million for the producing property is above the minimum acceptable sale price since the amount realized for leasehold and equipment of $1,264,755 is greater than $1,120,000, the net present value of continued operations. This sale will provide the seller net cash of $1,475,705 to be used as the seller wishes.

Frequently the result of a sale of producing properties are reported or compared on the basis of dollars per barrel of equivalent hydrocarbons reserves purchased. Over the past few years this index has been in the range of $3/BEQ - $10/BEQ (barrels of energy equivalent) for large purchases. While such an index is interesting and may give an indication of market conditions and trends, properties should not be evaluated by using some convenient index. Even under similar economic conditions the index can vary greatly between properties because of differences in capital requirements, operating cost, stage of development or depletion, production rate, value of the oil, etc. When evaluating a producing property for purchase or sale it must be analyzed using data for that property which reflects its operating cost, reserves, additional development, hydrocarbon value and any other factors unique to that property.

SUNK COSTS

When making economic analyses the decision maker's objective is to determine the course of action which should result in the most favorable future benefits. Since it is only the future consequences of investment alternatives that can be affected by present decisions, it is appropriate to the analytical process that costs already incurred in the past which cannot be recovered, be disregarded. A past cost, or "sunk cost," is one which cannot be altered by present or future action.

Suppose one buys and installs an automated pump shutdown system for $3,000 and after two months of operation it either is found unsatisfactory, or more probably is rendered obsolete by new developments, and one finds that he can sell the already installed equipment for no more than $2,000. The $1,000 loss is a sunk cost that resulted from a poor, or unfortunate, decision that cannot be altered by present or future action. Economic decisions related to future action should not be adversely affected by sunk costs. In practice, emotional involvement sometimes makes this difficult.

In the above example, the current $2,000 value of the automated pump shutdown represents the "fair market value" of that piece of equipment. Fair market value and sunk cost are concepts that must be considered together in many economic evaluations. Neither of these concepts excludes the other, but the fair market value of an asset reduces the amount of sunk cost that is unrecoverable. The fair market value is often treated as a salvage value as is seen in Example 10-4. So if a previously acquired asset is being used in a new project or replaced in a continuing project, its fair market or alternate use value should be included in the evaluation.

The most common type of costs that are treated as sunk costs when making development or production decisions are the cost of discovering or purchasing the reserves. Exploration costs or the cost of acquiring producing properties should not be considered when making decisions after these expenditures have been made, however these are not sunk costs for any evaluation made prior to their payment.

While the principle of sunk cost dictates that unrecoverable (sunk) costs should be ignored when evaluating a subsequent investment opportunity, frequently there are tax credits which are generated by the sunk cost which should not be ignored. Tax credits in the form of depreciation, depletion, or other write-offs yet to be taken should be included, where appropriate, in economic evaluations of investment opportunities. Such credits are treated as cash-in and may have a significant impact on decision alternatives. For example, the tax credit which can be taken immediately upon the surrender of a non-producing property, as the result of the write-off of its cost (sunk), may be greater than the value of developing a marginal producing property. If a property is developed the tax credits for the sunk costs are spread over the entire life of the producing property and will have a lower present value than if taken as a lump sum immediately.

There also may be accounting considerations which have an impact upon strictly applying the principle of sunk cost. When a decision is made which results in the write-off of a sunk cost, such as the abandonment and surrender of operating rights (lease), the accounting effect is to reduce the annual net profit by the portion of the original cost remaining on the company's books. If this is significant it may influence management decisions.

Occasionally it may be desirable to look at past expenditures to review past performance. Such reviews may be helpful in evaluating forecasting ability and in establishing realistic goals. The principle of sunk cost is not involved in these type of analysis because the analysis is not being performed for the purpose of making an investment decision.

LOOK BACK OR FULL CYCLE INVESTMENT ANALYSES

It is good management practice to periodically look back over past investments to ascertain that the company's objectives are being achieved. The question that management would like answered is: "What is the rate of return of the investments that were made in 'ABC' field during the period between year 'X' and year 'Y'?" Management knows what was forecast for the entire program, and were assured that each individual investment was justified as it was made, but actual performance may not match predicted performance, and economic conditions change. So a "Look Back" is appropriate periodically.

A look back economic evaluation differs significantly from the more common justification economic evaluations discussed previously in this text. Look back evaluations generally do not lead to an investment decision related to the project, or projects, being analyzed. That investment has already been made, and the cost is a "sunk cost." If the evaluation indicates that the investment does not meet the corporate objectives, it cannot be "unspent"! However, if a project is performing so poorly that future economics, ignoring the sunk costs, are unsatisfactory, then the look back analysis may indicate that the project should be discounted at that point. The purpose of a look back economic evaluation is to compare actual economic performance with that which was predicted, in an effort to improve prediction techniques and to assist management in selection of future investments. Frequently this type of analysis is referred to as a "full cycle" analysis because the entire life cycle of the investment is included in the evaluation. The performance of the project from its start to the date of the evaluation will be actual historical data, while from that date forward it will be forecast as is done in a normal evaluation. However, the historical performance may improve the forecast over that which was originally made.

Decisions in the oil business, such as whether or not to drill a well, are normally undertaken with the aid of the Economic Decision Tools described in Chapter V. Frequently large investment programs are comprised of many individual investments, with each justified separately and initiated over a period of a number of years. Evaluation of such programs may be further complicated when it is part of a much larger ongoing operation such that they cannot be isolated to permit a standard economic analysis from first investment to final abandonment. A good example of this situation would be the expansion of an existing oil field.

Many companies find it useful to review the economics of all projects on a regular basis. It is a good idea to review each budget project several years after it was budgeted. A delay of several years is desirable so that all budgeted wells and projects have been completed before undertaking the look back analysis. With this delay the success or failure of all projects will be known and there will be some performance upon which to base the analysis. It may also be desirable to look back at all major projects periodically, not only to improve forecasting techniques but to assure management that the projects are still profitable or to determine if some changes should be made. Both of these look back programs will assist in forecasting new proposed projects and help management identify types of projects that should or should not be pursued.

The Buy/Sell technique can be used to evaluate a complex series of investments. The name comes from the basic procedure followed by this method. It is assumed that you buy the assets which are in place at the start of the evaluation period at the "fair market value" of those assets at that time. The initial value of the assets must not only include the value of surface and subsurface facilities, but also the value of all oil and gas reserves considered to have been developed at that time. Then, theoretically, you Sell the investment at the end of the evaluation period, again for the "fair market value" of the assets which are in place at that time. Once these values are established, the evaluation is just like the analysis of any other cash flow. The theoretical purchase price is a "cash-out" item at the start of the evaluation period and the theoretical selling price is a "cash-in" item at the end of the evaluation period, with all "cash-out" and "cash-in" items that occur within the evaluation period applied when they occur.

Determination of the purchase and sale prices should be made in the same way as would be done for an actual purchase or sale, i.e., the present value of all future cash flows at the date of purchase or sale. This can be done by completing a full evaluation at each point in time or by analogy with contemporary sales of similar properties. It is important to recognize the actual economic conditions prevalent at each evaluation date, e.g., hydrocarbon prices, equipment cost, abandonment cost, etc.

Example 10-6 shows how the Buy/Sell technique can be used to determine the rate of return of a series of investments in an oil field, which was partially developed prior to the first investment in the evaluation period.

▮ EXAMPLE 10-6
▮ ELMWORTH BASIN PLAY OF WESTERN ALBERTA

Figure 10-2 depicts one operator's drilling history over a thirteen year period, beginning in 1977, for the Elmworth Basin Play of Western Alberta. The economic history of this program is shown in Figure 10-3 and in Table 10-1. The program's first positive net cash flow occurred in 1984, after which another $64 million was spent on expansion of the productive area over the next four years. After the gathering system was installed and a market was secured for the gas, positive cash flows were achieved in each of the last two years of the study. Separate studies determined that the fair market value of the project at the beginning of 1979 was $19.1 million and had increased to $136.8 million at the end of 1988. During the 10-year evaluation period, the sale of hydrocarbons generated $115.7 million, but this was $11 million less than the $126.7 million investment that was made during that period.

TABLE 10-1
ECONOMIC HISTORY OF THE ELMWORTH AREA 1979-1989
(Millions Canadian $)

	1979	1980	1981	1982	1983	1984	1985	1986	1987	1988	Total
Buy/Sell	-19.1									136.8	117.7
Op. Incm.	0.1	3.5	3.6	5.0	9.4	16.8	16.0	12.9	25.2	23.2	115.7
Cap.Exp.	-11.3	-5.6	-11.2	-9.3	-14.2	-10.7	-22.8	-25.5	-7.9	-8.2	-126.7
CashFlow	-11.2	-2.1	-7.6	-4.3	-4.8	6.1	-6.8	-12.6	17.3	15.0	-11.0
Nct	-30.3	-2.1	-7.6	-4.3	-4.8	6.1	-6.8	-12.6	17.3	151.8	106.7
Compounded @ 10% to 1-1-89											
CashFlow	-27.7	-4.7	-15.5	-8.0	-8.1	9.4	-9.5	-16.0	20.0	15.7	-44.5
Net	-77.2	-4.7	-15.5	-8.0	-8.1	9.4	-9.5	-16.0	20.0	152.5	42.8

TABLE 10-2
PRESENT VALUE NET CASH FLOW OF THE ELMWORTH AREA

	Undiscounted	Present Value at December 31, 1988	
Interest Rate	0.0%	10.0%	14.0%
Beginning Value at 1/1/79	19.1	49.5	71.0
Added Value During Period	117.7	87.3	65.8
Ending Value at 12/31/88	136.8	136.8	136.8
Net Cash Generated 1979-89	-11.0	-44.5	-65.8
Net Increase During Period	106.7	42.8	0.0

FIGURE 10-2
DRILLING HISTORY / Elmworth Area,Canada

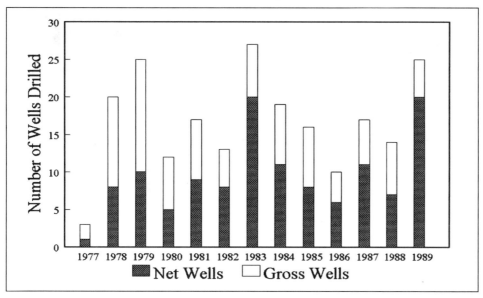

In addition to the economic history of the Elmworth Area, the theoretical purchase (buy) and sales (sell) prices are also included in the table. The purchase is assumed to be on January 1, 1979 and the sale on December 31, 1988. A present value date of 1/1/89, the end of the analysis period, was selected for this analysis. To obtain present values, it was necessary to compound all annual values to this date, as is also shown above. If the operator's cost of capital is 10%, then the net present value of the cash flow, including buy and sell values, increased by $42.8 million during the 10 year period. Additionally, by trial and error, it was determined that an interest rate of about 14% would reduce the present value of the cash flow to zero, which by definition is the rate of return (IRR) of all of the investments made during the 10-year period of analysis.

As can be seen from the data in Table 10-2, even though the net cash position on both a discounted and undiscounted basis is still negative at the end of the 10-year analysis period, the large increase in value of the reserves more than offsets this deficit. This has been a successful venture because the 14% rate of return is greater than the company's cost of capital (10%).

FIGURE 10-3
ECONOMIC RESULTS / Elmworth Area, Canada

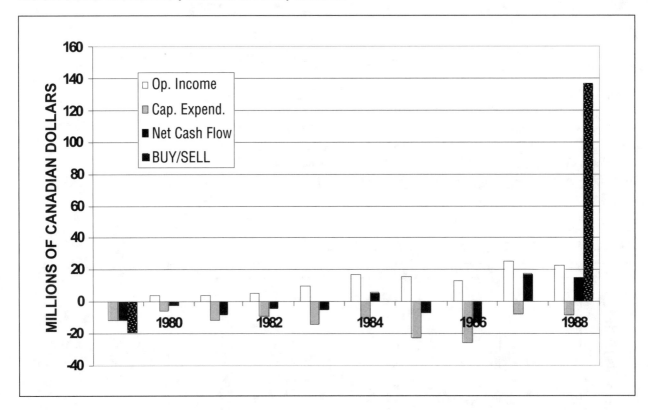

REFERENCES

Albanese, R., *Managing Toward Accountability for Peformance*, 1978, Irwin, Homewood, IL 60430

Campbells, *Analysis of Management of Petroleum Investments Risk, Taxes and Time*, 1987 CPS, Norman, OK 73072.

Capen, E. C., "Rethinking Sunk Costs—A Value Approach," *JPT*, December 1991, pp 1418-1423.

Ball, Max, *This Fascinating Oil Business,* 1940, Bobbs-Merrill, Indianapolis, IN

Harrison, F. L., *Advanced Project Management,* 1981, John Wiley & Sons, New York

Hatley, Allen G., Editor, *The Oil Finders: A Collection of Stories About Exploration*, 1995, Centex Press, Utopia, Texas 78884.

Masters, J. A., Editor, Elmsworth—"Case Study of a Deep Basin Gas Field," *Memoir 38*, 1984, AAPG, Tulsa, OK 74101.

Park, W. R., *Cost Engineering Analysis*, 1973, John Wiley & Sons, New York

Villarreal, J. V., *Evaluacion de Proyectos Con Applicacion a Instalaciones Petroleras,* 1984, Inst, Politech. Nac., Mexico D.F.

OIL AND GAS PRODUCTION AS A WORLDWIDE BUSINESS OPERATION

CHAPTER
XI

*Petroleum knows no boundaries,
but man has created many.*

The history of the development of the international oil business is captivating to say the least. There had been long-time knowledge and limited special uses for petroleum since the dawn of history. However, it was the hand-in-hand technical development of oilwell drilling techniques and the mass production of the automobile in the U.S. that fostered the modern petroleum industry. At the present time, oil as a commodity accounts for about five percent of the gross national economy of most of the industrialized world.

PETROLEUM GEOGRAPHY AND HISTORY

The centers of oil exploration activity and success around the world have varied through history. The modern petroleum industry is generally considered to have started with James Williams' hand dug oilwell in Southwestern Ontario in 1858 followed by the drilling of the Drake well in Pennsylvania the following year.

The Standard Oil (Exxon) empire soon after got its start in the Pennsylvania oilfields. Some of the early work in the Dutch East Indies, now Indonesia, led to the birth of Shell's great worldwide petroleum empire. By 1938, just before the outbreak of World War II, Venezuela had become the largest producer of crude oil outside of the U.S. and Russia. Saudi Arabia and Russia now lead the list of producing countries. The world's largest concentration of crude oil reserves lies in the Middle East. The U.S. relinquished its role as the world's largest producer of crude oil in 1974 and has been a net importer since 1973. Now the six largest oil producers in order of daily production are: Saudi Arabia, Russia, U.S., Iran, China, and Norway.

401

E&P IN THE U.S. DIFFERS FROM THE REST OF THE WORLD

The principal dissimilarity between the petroleum exploration and producing business in the U.S. (and Eastern Canada) and the rest of the world stems from the definition of mineral ownership. Onshore in the U.S. any minerals and everything else that lies under the surface of the ground, to the center of the earth, belongs with the land. The rights to petroleum and other minerals can be legally "severed" and sold off to another owner. The holder of the mineral rights has the authority to execute an "oil and gas lease" with an oil company. The lessee agrees to explore for oil and gas, entirely at the oil company's expense. If hydrocarbons are found in commercial quantities the oil company pays all of the costs of drilling, development and production. The mineral owner receives an agreed fraction of the oil and gas produced and actually sold. This royalty fraction was traditionally one-eight, but an astute landowner may negotiate any royalty fraction that "the traffic will bear." This kind of mineral leasing arrangement derives from the early days of iron ore and coal ownership and production in England.

Crude oil and natural gas flow in response to pressure differentials and can easily migrate across lease boundaries, particularly if there are wells producing from the adjacent lease which are not offset on the lease in question. Land ownership areas in the U.S. tend to be fairly small, averaging only about sixty acres. This has meant that once oil is discovered on a tract of land there is an immediate rush to drill on the adjoining tracts to prevent one's oil from being stolen out from under his property. This has certainly had a profound effect on the pace of oilwell drilling in the U.S. compared with other parts of the world where tracts are large, and the minerals are owned by the government, rather than the surface owner of the land. More than three million oil and gas wells have been drilled in the U.S.—more than in all the rest of the world.

THE MULTI-NATIONALS

The decade following World War II has sometimes been referred to as the 'ten golden years' of the petroleum industry. The oil companies prospered, the host governments seemed reasonably content with the concession agreements, demand for oil was growing rapidly and supplies were ample. The only major exception was the crisis in Iran in 1950.

This dispute culminated in the nationalization of the Anglo- Persian (now British Petroleum) interests in Iran. The controversy dragged on for two years, eventually involving the Inter national Court, the United Nations and intervention by the U.S. The fall of the Mossadegh government in 1953 permitted the opening of negotiations under a new government led by the Shah. This finally led to the establishment of an International Consortium of oil companies to operate the oilfields. Title to the oil remained in the National Iranian Oil Company (NIOC). An important development of the controversy was the establishment of a fifty-fifty division of the profits between the government and the companies. This seemed to stabilize the whole international industry which settled back happily. Just a few years later Amoco, which was not a participant in the consortium, made a 50/50 deal with NIOC to explore and produce on a joint venture basis new offshore areas in the Persian Gulf. This, in effect, gave the Iranian government through their 100 percent ownership in NIOC a 75/25 take.

Through all of this golden era the international oil business was pretty much dominated by the 'Seven Sister' companies discussed in much detail by Sampson (loc.cit.) and listed as follows:

THE "SEVEN SISTERS"

British Petroleum (now BP PLC)

Chevron

Exxon (now ExxonMobile Corp.)

Gulf (acquired by Chevron)

Mobil (merged with Exxon)

Royal Dutch/Shell Group

Texaco (merged with Chevron)

These companies exerted a strong stabilizing influence on the oil industry throughout the world during this era. Each of the surviving four companies remains as a fully integrated corporation with extensive operations in exploration, production, transportation, refining, and marketing of petroleum and its products.

Throughout this same period the domestic petroleum industry in the U.S. was also prospering, but with a markedly different composition. The industry was, and still is, composed of literally thousands of oil companies varying in size and capitalization rivaling that of the Seven Sister companies down to individual independent oilmen. Production "allowables" from individual oilwells were set by government regulatory bodies in the various states. The regulatory process was dominated by the "allowable" rates of production set each month by the Texas Railroad Commission.

Vertical integration was considered to be the key to maintaining profitability in the oil industry prior to the 1970's. Refining, marketing and transportation were generally considered necessary, if marginally profitable, adjuncts to maintaining a "home for the crude." The industry's profitability centered in the producing segment of the business. In large measure this situation persists today for those firms which can handle the combined problems of geological and financial risk.

THE NATIONAL OIL COMPANIES

Following World War I with the establishment of the state-owned Compagnie Francaise des Petroles (CFP Total) in France, and more prominently during the decade of the 70's, most countries in the world established national, or government oil companies (NOCs). NOCs generally espouse three principal objectives on behalf of their respective governments:

1. to reduce dependence on the multinationals for their oil supplies

2. to provide the government with an "inside window"on the petroleum industry to enable its bureaucracy to judge the performance of the multinationals within the specific country,

3. to assure continuity of supply both at the crude oil producing, and refining and marketing stages at home.

It is fundamental that a government needs a corporate entity if it is to participate in the oil industry. All national oil companies have, in varying degrees, broader constraints and obligations than private corporations. Many of these obligations are contrary to strictly commercial endeavor. The national oil companies, as public enterprises, are expected to help the country in capital formation by generating a profit and to generate foreign exchange from their operations. In this respect it should be possible to gauge this portion of their performance in terms of the usual commercial criteria. A principal exception involves the cost of capital which is generally contributed "gratis" by the NOC's government. On the other hand they have their more general aims such as promoting the working conditions and pay scales of workers; providing social benefits to the employee's family through medical, health and schooling programs; and many other political goals.

Since the fall of the Berlin Wall in 1989 there has been a strong trend throughout the world to privatize companies that were previously partially or wholly owned by governments. This has been a worldwide phenomenon, not limited to formerly communist countries alone. The first significant privatization in the oil industry actually occurred prior to 1989, when the British government sold their 51% share of British Petroleum, which they had held since 1914, when it was known as Anglo-Persian. Other companies that have recently been privatized include Argentina's YPF, France's CFP Total and Elf Aquitane, Spain's Repsol, Hungary's MOL and Italy's Ente Nazionale Idrocarburi (ENI). Privatization of ENI is particularly noteworthy. It was organized in 1953 by gathering various state hydrocarbon companies together in competition with the major private oil companies of the world to ensure Italy of its own supply of international oil. Its founder was the controversial Enrico Mattei, who first popularized the phrase 'The Seven Sisters.'

A number of national oil companies, such as Turkiye Petrolleri AO (TPAO), and Statoil, are exploring for oil outside of their own borders. In a trend counter to this, a number of countries that had previously limited exploration and production activities within their borders to the national oil company of that country have opened up these activities to outside private interests. Venezuela and Qatar, two members of OPEC, and Brazil are notable examples of countries that have recently opened upE&P activities to outsiders. Many other OPEC countries have continued to have a mixture of NOC and outside operation of their oil fields.

Another trend in national oil companies is movement into the downstream sector of the petroleum industry outside of their country. Conspicuous companies in this area are Kuwait Oil Company's marketing of gasoline in Western Europe under the "Q-8" brand, Norway with Statoil stations, Venezuela's gasoline sales in the U.S. using the CITGO name, and Pemex's joint ownership of Shell's refinery in Houston. This is in addition to refinery construction and product sales in the home country of many national oil companies.

Many of the countries of the world still have their own NOC's although the list is shrinking. Table 11-1 is a list of some of the larger of the national oil companies and their country of ownership.

TABLE 11-1
NATIONAL OIL COMPANIES

COMPANY	COUNTRY
ADNOC	Abu Dhabi
SONATRACH	Algeria
Y.P.F.B.	Bolivia
Petro-Canada	Canada
Ecopetrol (to be privatized)	Colombia
ENAP	Chile
CNOOC	China
CNPC	China
Ecopetrol	Colombia
CEPE	Ecuador
EGPC	Egypt
Gazprom	Russia
GNPC	Ghana
Pertamina (to be privatized)	Indonesia
ONGC	India
NIOC	Iran
INOC	Iraq
JOGMNC	Japan
KOC	Kuwait
LNOC	Libya
Lukoil (Entrepreneurial)	Russia
Petro China	China
Petrobras (Entrepreneurial)	Brazil
Petronas (Entrepreneurial)	Malaysia
Pemex	Mexico
NNPC	Nigeria
Statoil (partially privatized)	Norway
OGDC	Pakistan
Petroperu	Peru
QGPC	Qatar
Saudi ARAMCO	Saudi Arabia
Sonangol	Angola
Petrotrin	Trinidad & Tabago
TPAO	Turkey
ENOC	United Arab Emirates
ANCAP	Uruguay
PDVSA	Venezuela

The most notable exceptions, i.e., countries without a government oil company are the United States, the United Kingdom and more recently Argentina, France and Italy.

ORGANIZATION OF PETROLEUM EXPORTING COUNTRIES (OPEC)

OPEC was organized in 1960 in protest of the Seven Sisters move to reduce posted prices for a number of exporting countries. This was due in part to new competition from independent newcomers to the international oil business, primarily in Libya. The OPEC charter states:

"Its Objective is to co-ordinate and unify petroleum policies among Member Countries, in order to secure fair and stable prices for petroleum producers; an efficient, economic and regular supply of petroleum to consuming nations; and a fair return on capital to those investing in the industry."

OPEC's current membership and year of association, along with their crude oil production for recent years, are listed in Table 11-2. Saudi Arabia is the largest producer of crude oil in the world. Two other OPEC countries, i.e., Iran and Venezuela are among the top ten producers. Other countries in the top ten are: Russia, U.S., China, Norway, Mexico, U.K., and Canada. These 10 countries account for about 58% of the world's crude oil production.

TABLE 11-2
OPEC PRODUCTION / Millions of Barrels per Day

Country	1990	1995	2000	2005	2007	Quota 7/07
Saudi Arabia (1960)	6.38	8.00	8.26	9.30	8.70	9.10
Iran (1960)	3.12	3.61	3.68	3.89	3.92	4.11
Venezuela (1960)	2.12	2.61	3.03	2.11	2.39	3.22
Iraq (1960)	2.08	0.60	2.57	1.81	2.08	- - -
UAE (1967)	2.06	2.20	2.23	2.46	2.47	2.44
Nigeria (1971)	1.81	1.89	2.03	2.41	2.17	2.31
Kuwait (1960)	1.24	2.00	2.1	2.43	2.44	2.25
Libya (1962)	1.37	1.38	1.41	1.64	1.70	1.50
Indonesia (1962)	1.27	1.33	1.27	0.95	0.84	1.45
Algeria (1969)	0.80	0.75	0.81	1.35	1.36	0.89
Qatar (1961)	0.39	0.44	0.66	.80	0.80	0.73
Angola (2007)[1]	- - -	- - -	- - -	- - -	1.70	- - -
Ecuador (2007)[2]	- - -	- - -	- - -	- - -	0.50	- - -
Total	22.64	24.81	28.06	29.15	31.07	28.00

[1] Joined OPEC 1/1/07
[2] Rejoined OPEC 11/1/07

OPEC's export revenues reached a high point in 1979-80 of about $280 billion. These fell by one half to some $135 billion in 1985, and halved again to about $75 billion following the price collapse in 1986. Since 1982 OPEC has set quotas for its members' production in an effort to support a minimum world crude price. The July, 2007 quotas are included in Table 11-2. Although the financial situation of the individual members of OPEC varies widely, most have run current account deficits since 1982. Their uniform desire to maximize income forces each member to a delicate balance between over pricing in the world market and losing share, and increasing export volumes (perhaps in excess of their OPEC quota).

Saudi Arabia is seen to have taken the brunt of the "swing" in world demand. Their production has varied from 8.5 million BOPD (barrels of crude oil per day) to as little as 2.5 million BOPD, at one point, in OPEC's effort to stabilize the world price. The Saudi's current deficit would seem to restrict such strong swing producer action on their part in the future. The cutbacks in other producing countries and reduction in crude oil price inflicted severe impairment on the economies of many exporting countries.

Venezuela, whose economy has been particularly hard hit by the price and volume reductions of her crude oil exports, has announced the termination of its development efforts in the Faja heavy oil belt. This is similar to what has happened in North America with the oil shale in Colorado and in Canada's Athabasca tar sands, which is probably the world's largest single accumulation of hydrocarbons. Canada recently added 175 billion barrels of tar sand bitumen to its proved reserves as currently economically recoverable and a major tar sand project is scheduled to start-up soon.

CHANGES IN THE PETROLEUM INDUSTRY DURING THE 1970'S

The first major nationalization of foreign controlled petroleum exploration and producing interests occurred in Russia in 1917. Later takeovers took place in Mexico during the 1930's, and in Iran in the 1950's. More widespread takeovers began in the early 1970's when Libya, Kuwait and other middle eastern countries undertook to force the multinationals from their traditional concession ownership roles.

As Ed Morse (loc.cit.) relates: The 1970's were clearly a period of petroleum nationalism. Oil and mining companies were expropriated, and the concession system was replaced by production-sharing agreements and a movement toward service or fixed-fee contracts. In most cases, a government was assumed to own its country's oil and mineral resources, which became regarded as precious depletable elements of national sovereignty. Competition among companies for access to crude oil allowed governments to impose increasingly stringent terms on those wishing to explore for and produce oil, as well as to feel able, in many cases, to change contract terms unilaterally in order to increase their take.

"The situation has been rapidly reversed. For the rest of this decade governments will be competing with each other for oil company capital and, in particular, competing against the standard set by the investment climate in the United States."

This turn around in the attitude of host governments, since the crude oil price declines of late 1985 and early 1986, toward inviting private sector foreign companies to bring money, and know-how, and explore for oil has been quite remarkable. With the price decline the traditional multinational oil companies severely curtailed their exploration programs in an effort to maintain profitability and their negotiating teams stayed home. In their place a number of North American independents became intrigued with the possibilities of working overseas. By 1988 the majors were back talking with host governments about new contracts but the independents were also there.

In 1970 private sector oil companies produced and controlled 90 percent of the crude oil production in the non-communist world. Ten years later this had dropped to around 30 percent. As privately owned petroleum companies found their ownership reduced they progressively lost the financial incentive to continue their effective worldwide stabilization of crude oil prices.

Some of the private sector multinational companies, which had been relieved of their ownership of large overseas concessions, initially undertook to continue their crude oil supply from those host countries by arranging long-term commitments for crude oil liftings at prices unilaterally set by the host governments. This made the oil companies vulnerable to the price fluctuations in the retail marketplace to a degree which these firms had not experienced before. At the same time world demand for petroleum products was declining in response to the much higher prices. The first reaction to this situation was an industry-wide program of refinery upgrading designed to process the heavier, sour, and traditionally cheaper crude oils. At the same time the throughput of existing refineries was enlarged and smaller, less efficient plants were sold off or shut down in order to achieve a greater economy of scale.

After the Iranian revolution of 1979-1980, marked differences in prices of crude oil from various producing countries developed. Production volumes from non-OPEC countries were increasing rapidly. The producers in the non-OPEC countries were operating with no restrictions on crude oil pricing and were soon looking for external markets. Oil, in excess of the long term contractual liftings from the OPEC countries, now produced by the government oil companies of the same host countries, was also searching for a market. Increasing quantities of OPEC oil began to appear in the European spot markets for sale at reduced prices. A number of OPEC countries also embarked on some large countertrade or barter dealings in which the discounted price for the crude oil could be neatly disguised. These discrepancies in raw material costs were soon reflected in the retail market place.

The intense competition for crude oil markets which soon resulted, and the simultaneous actions of some consumer country governments to artificially control product prices, made the long term contracts for the high officially priced OPEC oil untenable. The whole world crude oil pricing situation at the time led to the rapid development of the "spot," or free, market for crude oils. Term coverage of crude oil supply dropped from 90 percent to less than 60 percent between 1979 and 1983. Ironically, during this same period, several of the largest traditional exporting countries launched massive export refinery construction programs. These new refining projects have no agreed OPEC pricing or export volume restrictions. They do run counter to the traditional economic pattern of minimizing the transportation element in product costs by transporting the crude raw material to refining centers closer to the final consumer.

The rapid increase in crude oil prices brought on a degree of conservation throughout the world which most economists never thought possible. World markets for petroleum actually reversed their traditional pattern of constantly increasing demand. At the same time new exploration efforts in non-OPEC countries brought forth impressive new discoveries of crude oil deposits and resulting producing capacity. This growth of non-OPEC production, principally in the North Sea countries and Mexico, as indicated in Table 11-3, finally forced OPEC in late 1985 to abandon their efforts to sustain high prices by restricting OPEC production.

Loss of control by the private sector of crude oil pricing overseas has led the major multinationals to become much more selective in their crude oil purchases, to rely increasingly on the spot market, to close a number of marginal refineries, and often to substantially curtail their marketing territories. Another result has been a reduced emphasis on total integration of the industry. Ever since the days of John D. Rockefeller the industry has adhered to the fundamental premise that integration of all of the company's activities from exploration through to the consumer into a single company was the best way to insure and improve profitability. Each of the multinational oil companies' industrial segments, e.g.exploration or marketing, is now more than ever expected to demonstrate its individual profit contribution to the consolidated firm, with an acceptable return on employed capital.

TABLE 11-3
MAJOR CRUDE OIL EXPORTING COUNTRIES

Pre-1960	1960-1974	Post 1974
*UAE (Abu Dhabi)	Plus:	Plus:
*Algeria	Brunei	Angola
*Indonesia	Colombia	Cameroon
*Iran	*UAE (Dubai)	China
*Iraq	***Gabon	Congo
*Kuwait	*Nigeria	**Ecuador
*Libya	Oman	Egypt
Neutral Zone		Malaysia
*Qatar		Mexico
Russia		Norway
*Saudi Arabia		Syria
Trinidad		United Kingdom
*Venezuela		Yemen

*OPEC Members
**Withdrew from OPEC 1992, rejoined 2007
***Withdrew from OPEC in 1995

Within the last couple of years, however, we have seen several of the principal members of OPEC moving downstream. It is still too early to tell how extensive this reintegration will be. Nevertheless we have seen the Kuwaiti NOC purchase the former Gulf Oil's refining and marketing operation in Europe and more recently to take substantial equity positions (20%) in British Petroleum and CFP Total. PDV the Venezuelan NOC has purchased half interests in refineries in Corpus Christi, Texas and Lake Charles, Louisiana. Other downstream OPEC purchases are constantly being rumored.

The petroleum industry's preoccupation with the present and future price of crude oil is a relatively new problem to its exploration and producing segments. No longer can the oil companies count on a stable price for new production as their exploration teams ponder the geologic risk of new expenditures.

INTERNATIONAL PETROLEUM AGREEMENTS

The earliest international petroleum concessions were granted at the turn of the century and included provisions such as:

1. Very large contract areas

2. Long concession periods of up to 99 years

3. No relinquishment requirement

4. No participation by the host country

5. Royalty was paid in cash or gold on the exported tonnage of crude oil

6. The contractual provisions were guaranteed by the host government and its sovereign for the duration of the concession.

The old concession system resulted in the discovery of some very large reserves of petroleum. Nevertheless changes in the nature of the world oil business in the mid-1950's after World War II led to changes in a number of the principal host country-industry agreements. The concession format has evolved into total or partial ownership by the host government on a joint venture basis in most of the older oil producing countries in the Middle East. In 1966 Indonesia negotiated the first production sharing agreement and a number of other countries, such as Egypt, followed soon after.

During the period from mid-1960 through 1980 the industry was aggressively seeking new places to explore outside of OPEC's domain. It became increasingly popular among host governments to turn to Risk Service contracts in which the foreign contractor takes on all of the risk and expense of exploring and developing production from a contract area. In return the foreign contractor is paid a stipulated fee per barrel produced for the account of the NOC. In the 1980's with the industry's move to more frontier areas with higher risk, and higher upside rewards for success, the host country-contractor agreements tended more toward what is known as the rate of return based profit sharing type of contract.

The economics of crude oil production are greatly affected by specific contract terms. This fact alone dictates the necessity for a thorough knowledge and understanding of: 1) the other party or parties to the agreement; 2) the structure and provisions of the various types of contracts in vogue; and 3) the respective probabilities of the geological, political and fiscal risks involved. All of these factors must then be negotiated into a binding contract which will accomplish the aims of both parties, to the benefit of each.

EXPLORATION AND PRODUCTION CONTRACTS

Every country has laws pertaining to the exploration for and extraction of petroleum or minerals. These tend to be fairly general, providing that the host government has complete sovereignty over these resources. Certain labor and tax provisions may be laid down, particularly as they apply to foreign investors. By and large the responsibility for administering the country's interest in the resources is vested in a designated ministry of the government. Most generally these interests are directed through the country's national oil company.

The petroleum legislation will normally allow the government through the ministry or its national oil company to negotiate and commit the country to binding agreements with individual oil companies. The contracting companies are frequently foreign owned. They undertake to explore for and, if successful, to produce and export crude oil for their own account. These agreements can involve very large amounts of money and may be politically very sensitive in the host country.

The basic provisions of these international petroleum agreements deal with:

Risk and Financing

It is common practice in all types of exploration and development agreements to place the risk, both geological and fiscal, on the contracting company. The contractor is expected to provide the financial and technical resources for exploration of a designated area within the country, and to assume the risk of failure. In the event of success, the contracting company expects to receive compensation in the form of recovery of its initial investment, interest on the funds invested in the venture before payout, plus a reasonable profit. This may be either in cash or in oil. Both should be freely exportable from the country involved.

Economic Return (Profit)

In the event of success, as stated above, the contracting company expects to earn, and take from the country, a profit commensurate with the risk. Essentially the same economic return can be achieved by any of several types of agreements including: concession, royalty, production sharing, profit sharing, service, risk, and toll contracts. These, in turn, can all be modified through fiscal devices such as cash bonuses, varying taxation rates, varying the production or profit split between the contractor and host, depreciation/amortization rules, crude oil purchase and settlement prices, rates of return, and fees. None of the basic types of agreement is inherently superior to the others. By applying a sufficient number of such parameters in varying degrees it should be possible to achieve an equitable arrangement between contractor and host government through any of these basic types of contract.

Management

The actual direction of the operation, and the degree of influence which the contracting company is allowed to exercise in the planning and conduct of its exploration and development investment schedules, is closely defined in the contract. Usually the contracting company is designated as the operator of the venture with some recognition or provision for the national oil company to assume operation at some, perhaps indefinite, date in the future. In practice, the contracting company, as operator, is subject to supervision by government inspectors and, oftentimes, a management committee chaired by the host government. The amount of control can be expected to vary with the experience of the ministry and its national oil company, and the degree of confidence they place in the contracting company.

HOST COUNTRY OBJECTIVES

The primary aim of the host government is to ensure the economic benefits to the country. This is generally interpreted to mean control over terms and conditions, price, and levels of production so as to capture the maximum economic rent at an acceptable level of risk. The energy ministries of the respective countries strive for efficient resource development and insist on control over the way oil production is developed (cost, drilling programs, conservation, maximum producing rates, etc.) and

411

over its timing. Manpower development and thus the training and participation of nationals, is also an important consideration.

The outcome of worldwide petroleum exploration and production can never be explained entirely on the basis of geology. Politics and economics have often precluded, fostered or inhibited such E & P activity. In the main, however, in the world's current economic situation the amount of petroleum activity in individual developing countries varies with that country's need or demand for new supplies of crude oil within its borders. New supplies are desired to satisfy local requirements and reduce the economic drain of imports and/or permit the export of oil as a means of gaining foreign exchange.

At the present time there is a growing competition among many developing countries desiring technical and financial assistance in connection with their petroleum exploration and development programs. With the world industry in a current state of over supply the major multinational oil companies who only a few years ago were the aggressors in seeking to acquire new places to explore are now the reluctant suitors. Host countries accordingly are making overtures to smaller companies in the hope of achieving their nation's petroleum objectives.

The critical needs of developing countries in being able to find oil and carry out these objectives are:

KNOW-HOW - i.e., access to the technology, organization, skilled manpower, entrepreneurship (handling risk), and specialized hardware of the petroleum industry.

CAPITAL - access to investment funds for development.

MARKETS - access to external outlets for oil and gas exports.

To varying degrees each of these needs carries a price tag, ordinarily in terms of some sacrifice, in maximizing the government's economic return.

CONTRACTING COMPANY OBJECTIVES

The evolution of the petroleum industry in recent years has modified the contracting company's objectives in regard to new overseas exploration ventures. Previously an integrated multinational oil company had to concern itself with "finding a home," i.e. a place to refine and market its production from its worldwide activities. The recent advance in spot marketing of crude oil has all but alleviated this concern. With the spot market, however, has come a growing problem of price volatility to the producing company which increases the fiscal risk borne by the foreign investor.

Thus, the financial objective of the contracting company has become that of balancing geological, political and fiscal risk against the economic parameters of prospective return on out-of-pocket investment achieved in a quick and orderly manner.

Foreign investors have three prime concerns regarding the fiscal regime of the host country. First, the foreign investor is concerned that the fiscal arrangements may be changed ex post facto. The contracting company will strive to protect its contractual rights, including fiscal understandings against such eventualities as cancelling or curtailing rights of foreign exchange of income from the project, or a unilateral change in the fraction of exportable crude oil from the operation.

Second, the foreign investor is concerned with possible changes in the petroleum tax regime which may prove discriminatory to some companies due to:

412

1. size and financial capabilities,

2. degree of foreign ownership, or

3. tenure and past degree of success in the host country.

Third, the foreign company will worry about the country's income tax structure and whether payments to the host government currently are, and will remain creditable in the contracting company's home country.

There are a number of differences in priorities between the two parties regarding contractual agreements. The following list delineates the principal objectives:

Ranking of Priorities:

HOST GOVERNMENT	CONTRACTING OIL COMPANY
1. Don't Disrupt the Political "Ship of State"	1. Maximize, and Expedite Economic Returns
2. Earn Foreign Exchange and build Financial Capital Base of the Country	2. Receive Reasonable Return for Degree of Risk Undertaken
3. Maximize Economic Returns and Build Local Industry Based on Cheap Fuel	3. Minimize period during which Investment Capital is at Risk, i.e., Payback Period
4. Further Domestic "Social Progress" Policy Goals	4. Ensure Repatriation of Funds and Export of Crude Oil Entitlement
5. Maintain and Increase Control over Country's Natural Resources	5. Retain Ownership of the Project and Consequent Claim on its Profits
6. Further Foreign Policy Goals and Reduce Imports and Efficiency	6. Retain Operating Control to Assure Production Economics
7. Promote Local Ownership	7. Avoid Creating Precedent in Contract Terms that the Company may wish to Avoid in other Countries
8. Develop Local Industry to Produce Oilfield Equipment	8. Maintain Global Standards, Efficiency and Reputation
9. Encourage Education Grants and maximum Transfer of Technical R&D to the Host Country	9. Develop Overseas Managers
10. Develop a National Pool of Oil Industry Talent	10. Balance Worldwide Crude Oil Supplies and Increase Oil Reserves

Political and Cultural Considerations

Wherever oil and gas are found in exportable quantities, the regional importance of the fortunate country is inevitably enhanced. Thus Venezuela, Nigeria and Indonesia all play more formidable roles in their respective regions than might be expected from other economic indicators. Mexico's oil activity in the 1970's greatly increased that country's importance in Central America and Latin American politics.

Corporate activities, even though their motives are apolitical, nonetheless affect, and are affected by the political conditions in the host countries. Political events and decisions often have a direct impact on even the most routine operations.

The multinational oil companies have traditionally negotiated in the expectation that Western legal and political traditions will hold and prevail, in every part of the world. Agreements between a private company and a sovereign government differ significantly from agreements between two private sector companies. Sovereign governments have the right to violate or terminate agreements under their domain—and they frequently do so. There can also be substantial cultural differences between the parties which may affect the continuity of contracts overseas.

TYPES OF AGREEMENTS BETWEEN OPERATING COMPANIES AND HOST GOVERNMENTS

The initial arrangement with the host government may involve permits for geological, geochemical or geophysical surveys.

These may be in the form of a "spec-shoot" in which the contractor receives the exclusive right to conduct seismic work in pre-established areas and then to sell the information under pre-agreed terms for a specified time period. The host government normally receives a complete copy of the information free of cost.

Another variation of such agreements is a "group-shoot" of pre-established areas in which a number of companies join with the national oil company in sharing the cost of the seismic surveys. This affords the participating companies an edge in bidding for the geologically most desirable tracts when offered. Geophysical rights to a wide area may also be the initial phase of an exploration agreement in which the contracting company may then be in a better position to pursue further exploration and drilling in a reduced "contract area," or to withdraw if the company finds nothing of interest. In either event the host government, or its national oil company, must be provided a copy of the survey results.

Concession Contracts

The most common form of agreement for ongoing exploration and development rights is the concession, sometimes referred to as a license or royalty/tax agreement. Concessions were the principal instrument for acquiring the right to explore and produce through the mid-1950's. In the concession type of agreement the host government assigns the right to the contracting company to explore and develop surface defined areas for mineral (petroleum) resources in return for a share of the proceeds (royalty) and taxes. The contracting companies compete for concession rights in a number of ways, including, bonus (front-end payment), royalties, and tax arrangements. The reliance placed on a given approach to granting the award varies among countries.

Originally a concession area might include a whole country or province. Since the latter part of the 1950's host countries reduced these rights from the original very large areas to blocks of much smaller size. Agreements of this type are currently in effect in some 120 countries. Most countries in Europe

414

and America still utilize the concession method. They include the United States, Canada, Norway, the Netherlands, West Germany, Trinidad, and the United Kingdom. The individual concession areas are limited in size, but companies are permitted as many areas as they wish to obligate themselves to undertake.

In the concession-type agreement the foreign company has a simple equity interest in the project. Such agreement stipulates that the foreign company will pay all of the exploration, development and operating expenses. The company also pays royalties on the value of production, and income taxes on net earnings. The royalty is taken from gross revenue (or production) with the rate commonly in the range of zero to twenty percent, although Canada's Province of Alberta has extracted royalties as high as 43 percent. The royalty arrangement forms a floor under the government's take. Royalty can usually be taken in cash, or in kind (oil), at the host government's option. Determining the value of the produced petroleum (settlement oil price) becomes important if there is a cash payment. Since the production may leave the host country only on intra-company transfer to the contracting company's own refinery with no third party or spot market sales to establish the oil's true market value, it is important to work out a pricing clause in the agreement.

In the early concessions, the operators posted (determined) the price of their crude oils and paid royalties on those prices. During the 1950's and 1960's world surpluses forced down the market prices for crude oil. In 1963 the newly formed OPEC rallied its member countries to refuse to allow the lowering of oil prices for royalty and tax purposes. Royalties and taxes were then set by the host governments. These were almost universally higher than market prices. This situation was the forerunner of the decade of OPEC pricing. It afforded the host country the opportunity at any time to increase its take without changing the country's tax system or the agreement into which it entered assuming, as was the case with the early agreements, that there was no provision dealing with pricing arrangements. The operator took care of all that in the early days.

Joint Venture Agreements

During the late 1950's joint ventures between the contracting companies and the host governments began to appear in the Middle East. The first such agreement was reached by Amoco in Iran. Under these arrangements the host government formed an operating company jointly with the contracting oil company for the exploration and production of petroleum within a specified area.

Host governments usually held 50 percent, or slightly less, of the venture. After the Libyan agreement with Occidental in 1973 government shares of over 51 percent became common. It is normal in the joint venture agreements to have the contracting company assume all of the exploration risk by carrying the host government's participation until discovery. If there is a commercial discovery, the host government contributes its proportional share of the development costs, although frequently the foreign company must also carry the government through the development phase as well. Reimbursement for development costs and any reimbursable exploration expenditures are usually made from an agreed fraction of the host government's share of production.

Most royalties are deductible when computing income taxes within a host country. Other deductions normally include operating costs, some specified allowance for overhead, depreciation, and interest payments. The resulting taxable income under the concession agreements is normally taxed at the country's basic corporate tax rate. Some countries employ special tax rates for the petroleum industry operating within their boundaries.

The oil and gas industry employs two basic types of agreements for the development and production phase. These are concession-type agreements and contracts which come in several forms. Contractual agreements between foreign investing companies and host governments, as distinct from concession agreements, include production or profit sharing with the state, service contracts, risk contracts, and toll contracts.

Production Sharing Contracts (PSC)

Production sharing arrangements have become more prevalent in the older producing areas of the world such as Indonesia, where they were first introduced in 1966, Egypt, and Libya. Such agreements normally provide that the costs and risks through development for production are to be borne by the contractor and recovered from a negotiated fraction of production. The remaining production is shared, or divided according to an agreed formula, and income taxes are levied on the contracting company's profits.

The first significant E&P production sharing agreement was used in Indonesia in 1966 by Independent Indonesian American Petroleum Company (IIAPCO) for exploration of 14,000,000 acres offshore Northwest Java. A discovery was made in August 1970 and production began in 1971. In excess of two billion equivalent barrels of oil, LPG, and gas have been produced from that tract. Significant features of that contract are presented on the following page. Amoco in Egypt and Mobil in Indonesia followed shortly after with PSC's related to already established production.

As in a concession contract, the company is granted the right to explore for oil and gas and develop any commercial discoveries. Unlike the basic type of concession agreement, the production is split between the host government, usually through its national oil company (NOC), and the contractor (see Figure 4-2). The shared fraction of production from the project is designated as "profit oil." Profit oil is the residual of production after first reimbursing the contractor for capital and operating expenditures. Typical splits of profit oil between parties to production sharing agreements are: Egypt - 85 (government)/15 (contractor), Indonesia - 85/15 for oil and 65/35 for gas, and Libya - 81/19.

IIAPCO (Indonesia) PSC - 8/19/66

No bonuses
6 year exploration period extendable to 10 years
Exploration Program: $7,500,000 (6 years)
                             $2,100,000 (firm-3 years)
30 year life from date of approval
Relinquishment: 25% at end of year 3
                     25% at end of year 6
                     Size and shape approved by Government
                     Acreage drop delayed by bonus payment
Provision for transfer of ownership with government
                approval (not to be reasonably withheld)
Right to lift Contractor's share of oil
NOC responsible for payment of government fees, charges
                and taxes on imports
Annual cost recovery limit 40%
Government/Contractor profit oil split 65%/35%

There are a number of contracts which include a provision that host country income taxes owed by the foreign contracting company will be paid by the national oil company (NOC). More commonly the contractor pays its own income tax on its share of the profit oil.

Risk Contracts

Risk contracts basically provide that the contracting company assumes the entire risk of exploration and development. If a commercial discovery is made and brought into production, the contractor is reimbursed in cash, or in crude oil, for all costs which often include some interest provision for the capital outlay. Sometimes there may be a provision for an ongoing joint venture between the two parties.

Service Contracts

Service agreements vary widely. The contracting company may undertake to explore and develop production for a payment, usually in cash, or possibly in oil, for services rendered. In this type of agreement the contracting company is working for the host government as a supplier of services and know how. The contracting oil company has no equity position. Neither does the contractor have any ownership in the oil he finds and produces. The contractor is compensated for the production delivered to the government by means of a per barrel fee which is specified in the contract. In some cases he may have the privilege to buy back a fraction of the production at discounted market prices. The NOC may pay the contractor's taxes so that all the fees are after-tax.

This type of agreement has great popular appeal in countries where the government finds it highly advantageous to be able to say that there is no taint of ownership, or claim on the host country's oil resources by a foreign entity. The subset of service contracts now referred to as toll contracts include the type of arrangement pioneered in Argentina during the late 1950's. In tolling contracts the operating company takes on the entire economic burden of exploring, developing and pipelining and is paid only for each barrel, or tonne, of crude oil eventually delivered to the NOC at the designated point of delivery.

GENERAL PROVISIONS

There are a number of contract terms or provisions that may be applied to any of the types of agreements listed above. These include:

Signature and Production Bonuses

These bonuses involve cash payments in lump sum to the host government at certain trigger points (prearranged levels of production, or cumulative recovery). Signature bonuses are payable within a specified time after the initial signing of a contract for exploration. They are usually of the order of one million, to as much as several million U.S. dollars. They are often looked upon as similar to an earnest money payment and are generally publicly acclaimed and favorably received by the host government. From the standpoint of the contracting company such agreements detract from the economic attractiveness of the venture by loading the front end of the project in year zero, thereby reducing the net present value of the project. If the bonus requirement is too large it may raise the barriers to entry of foreign investment, particularly for the smaller independent oil companies who may not be able to afford the increased startup costs of overseas exploration programs.

Discovery and production bonuses are less of a problem from this standpoint because there at least is, or soon will be, production income to cover the cash requirements.Discovery bonuses are usually called for within thirty days of a commercial oil discovery. Production bonuses are usually triggered and payable when the rate of production from the property exceeds a specified figure over a period of thirty days. These normally do not affect the overall economics of the project but might cause the operating company to defer production-stimulating work in the field.

Work Commitments

Certain minimum amounts of geophysical work, normally specified in kilometers of field seismic coverage, and exploratory drilling are usually agreed to and spelled out in the agreement. Drilling minimums may be defined in terms of the number of wells; total footage of all exploratory wells; and oftentimes the penetration of a specified geological formation or objective. There may also have to be a means of identifying which provision will prevail in the event of a conflict of requirements. The location of the seismic effort and subsequent drilling is generally left to the contractor. There may be a provision regarding the time frame within which this minimum exploratory effort must be completed.

Royalties

The required payment of a royalty share to the owner of the mineral being extracted from the natural deposit has been a tradition since antiquity. Petroleum royalties either in cash or in oil are well established and understood throughout the petroleum industry. Royalties may be applied either as a fixed, or sliding scale fraction of the production before applying any of the various contract terms discussed previously.

Sliding scale (or variable rate) royalties may be specified as a function of levels of production prices, crude oil gravity,location of production (on or offshore, water depth, etc.). Sliding scale arrangements, as a class, are referred to as "progressive taxes" because they tend to adjust a portion of the profit between the contractor and the government.

Royalty payments may be looked upon as a floor, or guarantee, of a minimum flow of revenue to the government once production is established, even before the operation becomes taxable. Host governments generally collect both a royalty on production as well as taxes on the company's net income.

Taxes on Profits

Most governments of the world employ taxes on business profits as a fundamental part of their tax regimes. Income taxes and surtaxes are looked upon with favor by the contracting companies in contrast to some of the other forms of sharing the production benefits with the host. These are payable only when, and if, a profit is realized. A period of some years after the start of successful exploration may be required before profit is achieved. This type of tax can be structured so that the contractor's initial expenditures for exploration and development can be recovered early in the life of the project, which is highly desirable from the foreign investor's standpoint. Surtaxes are generally imposed as a progressive tax when the company's income exceeds a stipulated rate of return or other gauge of profitability.

Income taxes increase the host government's exposure to revenue risk as a result of crude oil price fluctuations and declining production levels due to depletion. Marginal fields may never be developed, and secondary recovery projects may not be undertaken, if the host government's tax rates are too high, and as a result, make such projects unprofitable.

Central Bank Guarantees

Most private sector companies will insist that their contract with the host government include a central bank guarantee of:

1. the national oil company's various payment obligations, and

2. the availability of foreign exchange for payments and obligations under the contract.

This is considered a necessity since the NOC's obligations are normally in local currency which might not be freely convertible or exportable. Some multinational oil companies have been embarrassed to find that their contracts did not provide both guarantees.

Local Content, or "Buy National" Policies

Most countries will insist that the contract include a provision that the contracting company buy locally all goods and services which the host country can provide. This is an entirely reasonable means of encouraging local industry and promoting the training and utilization of local labor. The other side of the coin is that the host government may feel that their people are more qualified than they really are, considering the complexity of the industry. Many lives have been lost due to inexperienced human error in the international oil business, particularly in hostile climates offshore.

For the mutual benefit of both parties to the agreement it is best to require strict adherence to internationally recognized standards, such as the API and ASTM, as part of the buy national provisions of the contract. This also makes it easier for the national company to convince local suppliers that construction rebar won't substitute for API sucker rods in a seven thousand foot well, or that the local building cement may quickset during well completion operations causing the loss of hundreds of thousands of dollars of drilling effort.

Customs Import and Export Duties

The international oil industry has hundreds of stories of delays in exploration and production activity due to holdups in dealing with the local customs officials regarding the import or export of specialized equipment. Since this branch of government is almost invariably responsible to some other than the petroleum ministry there is advantage to obtaining prior contract approval and commitment from the authority concerned.

Structure and Format of the Agreement

Each country has a law or laws which bear on agreements with multinational firms. These are revised from time to time. Some countries will enact what is known as a "protocol" or supplement to their petroleum law to accommodate a special situation pending the next complete rewrite of the legislation. Such a protocol has the same force and effect as a law.

Although there is no standard form for international petroleum contracts, a degree of uniformity in such agreements has evolved with time. Outlines of typical contracts for petroleum exploration, development and production are presented in the appendix to this chapter.

THE PETROLEUM INDUSTRY BUSINESS

The oil business is unique among the world's commercial activities. Crude oil is the largest item of world trade even surpassing foodstuffs.

It is important to understanding the economic fundamentals of the oil and gas exploration and producing industry to have an appreciation of how and why the entire petroleum industry, from well to service station, differs from every other type of commercial enterprise.

Petroleum exploration and production involve a number of distinctive features as a commercial enterprise. Note: The first several items which pertain to the E&P segment of the industry are repeated from Appendix VII-B where they relate directly to the discussion of "Oilfield Deal Structure."

1. Petroleum is an extractive industry.

Deposits of petroleum are difficult to locate. They are generally found far from the country's centers of population, which makes transportation an industry challenge.

There is a financial relationship between the owner of the resource, (who may be the landowner in the U.S. and Eastern Canada, or the government over the rest of the world), and the lessor, who undertakes at his entire expense, to explore and produce the resource while providing the minerals owner his proportionate share of any proceeds.

It is a fundamental concept of most commercial enterprise that any eventual profit is shared on the same basis as the original investment in the project. In the extractive industries, where substantial royalties may exist, the royalty owner shares in any revenue while the owners of working interest pay all costs and investments and share only in the net income after royalty, costs and investments. Producing properties are normally plugged and abandoned when the working interest owner's net income from the declining production no longer pays the cost of the operation. Thus, an abnormally high royalty will force an earlier abandonment of the producing stream.

2. Petroleum is a risk industry.

Participants in the business may wish to reduce their interest in one venture in order to participate in others as a means of spreading their risk. This is very similar to the insurance business where the carrier of a substantial policy will often "lay off a portion of the risk" by disposing of part of his ownership to others.

3. Petroleum producing wells may yield a variety of commercial streams.

These include crude oil, casinghead gas, natural gas, condensate, natural gas liquids, and sulfur. All of these must be disposed of currently, and commercially.

4. Petroleum may be produced from a number of separate sources, or strata, underlying the same tract of mineral ownership.

The ownership and Division of Interest may vary with the different sources.

5. Petroleum is an extremely capital intensive industry.

Looking across the whole industry from exploration to marketing, the single most important factor which distinguishes the oil business from all others is its capital intensity.

Some industries, e.g., clothing manufacture, are very labor intensive with minimum fixed cost. The owner of such a business can hire his help directly as needed. His investments are minimal and his costs vary merely with the level of orders in process. The petroleum industry, by contrast, requires the investment of tremendous amounts of money in any of its several activities. This provides exceptional economy of scale, but more importantly, it means that the industry's variable operating costs are minor in comparison to the magnitude of its required investment.

This capital intensity, and the resulting high fixed costs and low variable, or per barrel operating costs, involves all phases of the petroleum industry from exploration through marketing. It will reappear as we consider the various segments.

There is "ease of entry," where an individual entrepreneur can establish himself as an independent businessman, at only two or three points of the Integrated system.

6. Operating costs do not diminish appreciably with reduced throughput.

One of the first effects of this capital intensive refinery investment is that you do not save any money by running it at reduced capacity—so, one is prompted to operate at full volume. A similar situation exists throughout the other segments. This phenomenon is a direct result of the capital intensive nature of the business.

Operating any of the phases at 50 percent capacity costs nearly the same as 95 percent of full capacity. The tendency, then, is to use the capacity—sometimes even with a negative return in one or two individual segments of a fully integrated enterprise.

Refinery startups and shut downs are very costly in terms of heat efficiency and maintenance of product quality, so one runs continuously except for annual turnaround for cleaning and inspection.

There is not much direct labor involved in refining. Most of the giant operations are automated, or controlled from a central point in the refinery complex. The labor force tends to be very specialized and difficult to replace in the short range so the full payroll is normally kept on even during the infrequent shutdowns and turnarounds.

Generally speaking refineries are built near the consumer whereas crude oil seems to be found in nature in more remote areas worldwide. Basically, it is always cheaper to transport the raw material (crude oil) than to move a complex of finished products.

TRANSPORTATION of liquid and gaseous petroleum is best done by pipeline. Here again, the amount of capital required to construct the line is tremendous compared to the cost of operating it. This also leads to a "keep the line full" philosophy.

The other major mode of transportation in the worldwide oil industry is by ocean going tanker. Here is another capital intensive example of economy of scale. It takes the same size crew for a 500,000 DWT vessel as for a World War II T-2 of 16,000 DWT.

The same capital intensity pertains to the PRODUCTION of crude oil and natural gas. Most of the money that will ever be spent on a development well occurs before the first barrel of oil is ever produced. Offshore development drilling, particularly in hostile environments, such as offshore Newfoundland or the Beaufort Sea, can be more costly than the initial discovery wildcat. This is quite a contrast to the traditional land-based exploration and development situation. It points up the capital intensive nature of the production segment, however.

7. The oil business is comprised of a series of distinct activities, or segments.

The petroleum business includes four principal components:

Exploration and Production
Refining
Transport
Marketing

EXPLORATION and PRODUCTION is often referred to as the "Upstream" part of the business, and Refining and Marketing as "Downstream." In most integrated oil companies, which include these several components, each segment is expected to stand alone as a separate profit center.

Taking these in reverse order, MARKETING is the segment which has the most day to day contact with the ultimate consumers. It seems remarkable that the marketing of automotive motor fuels started with a gasoline pump located outside the local grocery store. By the 1930's we had competing brands of service stations on each of the four corners of most major intersections. Now we have come full circle and the service stations sell groceries. Actually, at many locations, the cash flow from the canned pop and junk foods may approach that realized from the gasoline sales.

In strictly physical terms, no other business deals with such volumes of liquid commodities, nor sells directly to the consumer through a hose and liquid flow meter.

In REFINING we differentiate between fixed and variable costs. The fixed costs are primarily the investment, or capital, costs which get very large.

It would be nice to be able to avoid the investment, and contract only for what the market requires. Then there would be only variable costs. Unfortunately, someone has to put up the large amount of money to build the refinery, which becomes the high fixed capital costs. One industry observer has stated that "he who builds a refinery becomes its economic slave."

8. Exploration is the most economically and physically risk-prone of the segments.

All business has commercial risk of various kinds. That is fundamental to the free enterprise system. In the upstream segment of the oil business we usually think first of geological risk—most wildcats, and some development wells are dry holes. Geological risk, along with the industry's capital intensity, represent the principal reasons for the surviving major oil companies being so large.

Frankel (loc. cit.) characterizes this phase of the industry as a "mix of technology and Monte Carlo." Most operating companies, if they are large enough, play the game of "averaging out":

Economic Consequence

- 1 well drilling program, gusher	= owner is well off
- 1 well drilling program, dry	= economic disaster
- 100 + well drilling program	= indifferent, should average out and the company at least remain afloat.

Even with substantial "averaging out" essentially all of the successful major companies have found it essential to also have a big, or giant oil field to lean back upon.

Some companies attempt to side-step the exploration risk by purchasing proved reserves in the ground for less than their company average cost of finding new crude oil production. Another recent corporate approach has been "going international" where the alleged finding costs may be less.

Another important fundamental stemming, from this big company, high investment background is:

9. The profitability of the several individual segments of the industry varies substantially and periodically, one from another.

One rationalization of the variation in profitability among the segments is the time required to establish a successful operation in each individual activity, e.g.,:

10-15 years to find and develop new oil production. (Refer to Figure 1-1.)
10 years to develop a substantial product marketing base.
And, only 2 years to build a refinery.

There is also an ebb and flow of profitability with time in each of the segments. All are capital intensive. For many years production was by far the most profitable segment. Finding costs were low, and the depletion allowance in the U.S. gave favorable tax treatment. At present, in distinction with just a few years ago, Petrochemicals, Refining and Marketing are where the money is being made.

The financial performance of the industry, and its oil company participants, continually confounds the financial analysts. They keep tripping over the fact that it takes, perhaps ten years from the first geological concept to the report of a new commercial discovery. The analysts attempt to convert everything into forecasts of quarterly earnings. They also are forever trying to tell us that a ten cent increase in the price of WTI will mean "X" cents increase in the price of gasoline at the pump. The industry is far too complex for any such simple, immediate cause and effect.

10. The industry has a strong tendency to bring these several segments of activity under a single coordinated management, referred to as VERTICAL INTEGRATION.

Although efforts are made to maintain the profitability of each individual segment, the principal reason that vertical integration has been such a pervading industry objective stems from the desire to stabilize the total company's profit position. A temporary, or even permanent, deficiency in the bottom line of one segment may be compensated from another. Stability of earnings can be very important in raising and maintaining the company's ever demanding capital structure.

The degree of success—or lack of it—in finding new crude oil supplies is the major uncertainty to any operating oil company. Historically, if several operators found major new supplies, at or near the same time, it was nice to have one's own refinery to serve as a "home for the oil." This management intrigue with vertical integration has seen some recent decline among the smaller major and larger independent operators who are learning to live with the spot and futures markets. In contrast, however, the major National Oil Company, (NOC), exporters and even the "Wall Street Refiners" and traders are now actively seeking to add, i.e., integrate, "downstream" refining capacity.

11. The petroleum industry is readily adaptable to physical exchanges of both raw materials (crude oils) and most finished products (e.g., motor fuels and lubricants).

Although no two crude oils are the same, modern petroleum refining technology permits the manufacture of products of standard quality from a variety of source materials. The fact that it costs more to transport finished product than raw material makes it desirable to locate the refining centers near the ultimate customer. This, in turn, suggests the exchange of finished product among the local distribution and retail outlets.

12. The principal assets of an oil company are usually its reserves in the ground of crude oil and natural gas. If these were acquired as the result of the company's own exploration effort they do not show as an asset on the firm's balance sheet.

Standard business accounting methods, as were discussed in this text, are unable to cope with this all-important aspect.

13. The Oil Business is the enigma of Wall Street.

Crude oil prices vary in a cyclic pattern which doesn't match the general economic cycle. Variations in corporate financial performance from one calendar quarter to the next, which are such a mainstay of security analysis in most industries have little, if any, meaning in the petroleum industry. Income from new capital expenditures in the upstream sector of the business will take many years to achieve, if at all. Cash flow per share replaces earnings per share (EPS) as the best gauge of current financial performance of an oil company. This is primarily due to the complexities of non-cash charges employed in the industry, and whether the company employs the optional "full cost" or "successful efforts" method of financial accounting.

Pundits persist in attempts to directly relate minor current changes in crude oil price to prospective changes in motor fuel price at the pump. Although motor gasoline is the industry's principal product, it is far from the only source of revenue in a multifaceted and multi-functional chain of profit centers. The industry is far too complex for any such simple, immediate cause and effect. And all of this is before the tax man takes his series of cuts of the cash flow through the many steps from the well to the service station.

14. The Oil Business is not self-regulating. Throughout its more than 140-year history it has been susceptible to significant outside political and mercantile control.

For its first 125 years the growth of the modern petroleum industry advanced and receded in lock-step with the economy of the Western world. The number of competitors in the business had apparently little effect on the overall supply/demand of petroleum. There has always been a potential oversupply of both crude oil and natural gas. Up until the mid-1980's it had been possible for one person, or group, to control crude oil supply:

- John D. Rockefeller through his Standard Oil Trust effectively controlled the supply until its dissolution in 1911.

- Then followed a period of uncertainty. This permitted the establishment of the "Seven Sisters" and the great worldwide competition between Exxon and Shell, which culminated in the Achnacarry agreement in 1928. About the same time the Texas Railroad Commission was directed to set each month the allowable crude oil producing rate for each well in the state, which effectively determined the producing rate for the whole country.

The relinquishment of the Seven Sister's control to OPEC in the mid-1970's continued the industry's economic stability until the early 1980's when the growth of non-OPEC production, stimulated by the high prices, created such a capacity surplus that OPEC lost control.

The industry is again in a period of uncertainty learning to live with the volatility of spot and futures market pricing for the production of its high risk and high priced exploration efforts.

15. There are two ways of optimizing the Operations of the combined industrial segments which one company may control.

These alternate methods of optimizing cash flow involve:

1. Volume, or
2. Margin, but not both, as OPEC keeps learning.

In other words, OPEC, or other major producers, can adjust the volumes going to market in an effort to stabilize prices; or alternatively they can target a margin of profit in the expectation that if it is too high, or too low, sales volumes will adjust accordingly.

The single exception could be Saudi Arabia whose reserves are so large and operating costs so low that it may do either, and effectively (barring political intervention) wipe out all of its crude oil producing competitors, taking over to supply the world from its own sources for a long time.

16. The Petroleum Industry is probably the Most Heavily Taxed of All Commercial Enterprise.

The petroleum, natural gas and petrochemical industries have been a prime objective of the taxing authorities around the world almost since the industry began. There are several basic reasons for this. First, the industry is very large, and has some of the world's largest corporations. The evolution of corporate size as a natural defense against risk was discussed in Chapter VI. The financial figures of the industry, and even of the major companies, tend to be so tremendous in the eyes of the general public that they obviously can afford to pay a lion's share of the cost of government.

Secondly, the industry since the days of John D. Rockefeller has been rather "tight lipped" in making public its many problems and accomplishments. Secrecy has bred suspicion in the minds of the public.

Thirdly, the fact that the modern petroleum industry impacts on every man, woman and child in the modern world makes it an ideal tax collector, who is forced to treat everyone in a reasonably equitable manner.

REFERENCES

Barrows, G. H., *Worldwide Concession Contracts and Petroleum Legislation,* 1983, PennWell, Tulsa,OK 74101

Ellis Jones, P., Oil—A practical guide to the economics of world petroleum, (1988), Woodhead-Faulkner Ltd., Ca+mbridge CB2 1QY

Frankel, P. H., "The Structure of the Oil Industry," lecture given at the Institute of Petroleum, London July 4-6, 1988

Frankel, P. H., Common Carrier of Common Sense—A Selection of His Writings, 1989, Oxford Institute for Energy Studies, Oxford University Press, Oxford OX2 6DP.

Grossling, B. F., and Nielsen, D. T., *The Search for Oil and Its Impediments,* Financial Times Business Information, London EC1M 5SA (1985)

International Petroleum Encyclopedia, published annually by PennWell, Tulsa, OK 74101

Johnston, Daniel, *International Petroleum Fiscal Systems and Production Sharing Contracts*, 1994, PennWell, Tulsa, OK 74101.

McKechnie, G., et al, Energy Finance, (1983), Euromoney Publications, London EC4

Morse, Edward L., et al, Joan Pearce, Ed., The Third Oil Shock, (1983), Routledge & Kegan Paul Ltd., London WCD1E 7DD

Morse, Edward L., Contract Negotiations for Petroleum Exploration and Production, The Petroleum Finance Co. Ltd., Washington, D.C.(1985)

Petroconsultants, Inc., Houston, TX, 1987, Private communication.

Sampson, Anthony, The Seven Sisters (New Post-Gulf War Edition), 1993, Coronet Books, Hodder and,Stoughton Limited, London.

Simon, D. A. G., "Oil Trading—A Changing World," paper presented at the 1984 Offshore Northern Seas Conference held in Stavanger, Norway

The Economist Intelligence Unit, "OPEC and the World Oil Outlook"1983, Special Report No. 140, The Economist, London

Treat, J. E., et al, Energy Futures - Trading Opportunities for the 1980's, PennWell, Tulsa, OK 74101

Turner, Louis, Oil Companies in the International System, George Allen & Unwin (Publishers) Ltd., London WC1A ILU, UK

Yergin,D. The Prize, 1991, Simon & Schuster, New York 10020

Zartman, I. W., The 50% Solution, Anchor Books, Garden City, N.Y. (1976)

APPENDIX XI-A

OUTLINE OF A TYPICAL CONTRACT FOR PETROLEUM EXPLORATION, DEVELOPMENT AND PRODUCTION

Preamble **GOVERNING LAW**

This introductory section indicates that the contracting parties agree to abide by the country's petroleum legislation and recites the specific date and legal designation of that law.

Article I. **DEFINITIONS**

This article may contain five or ten pages of fairly basic definitions of words found in contracts varying all the way from "exploration" and "production" to "geological basement."

Article II. **OBJECTIVE OF THE CONTRACT**

The second article typically sets out several of the most basic objectives of the agreement, such as to explore for, develop and produce petroleum; to train local personnel; and who shall pay exploration costs.

Article III. **CONTRACT AREA**

The contract area is precisely defined, employing an appendix to the contract for further specification, if necessary. This article may also clarify that the contractor has no claim under the agreement to the surface rights, nor claim on any other minerals.

Article IV. **CONTRACT TERM**

Three or four pages may be required to carefully define the period of time in which the contract will remain in full force and effect. The extent of the exploration term at the start of the contract is always specified in this part of the document.

Article V. **WORK AND EXPENDITURE OBLIGATIONS**

The time by which the contractor must commence Petroleum Operations is specified along with provisions for extending the Exploration Periods. Provision is made for possible failure to comply with the minimum obligations.

Article VI.   RELINQUISHMENTS

This article is particularly important to the contractor. It specifies at which time or times the contractor must give up a portion(s) of the original contract area. The portion(s) relinquished exclude(s) areas where discoveries have been made, but the final relinquishment may include all of the original contract area deemed to be non productive.

Article VII.   MANAGEMENT ORGANIZATION AND ITS FUNCTION

The organizational relationship between the contractor and the host government through its ministry and its NOC is spelled out along with the authorization limits and decision making procedures for the operator. This section may run eight or ten pages depending on the complexity of the arrangement.

Article VIII.   OPERATOR

This designates the contractor as operator of the exploration and development project. The article will generally provide for transfer of operatorship from the contractor to the host government's NOC at some time (usually not specified) in the future.

Article IX.   WORK PROGRAMS AND ANNUAL BUDGETS

The schedule for submitting the next year's work program and budget is outlined in this article. The possible over or under expenditure of the approved budget on the part of the contractor will probably also be dealt with.

Article X.   ASSISTANCE PROVIDED BY NATIONAL OIL COMPANY

In this article, a number of specific items, which the NOC will undertake on behalf of the contractor, most of which involve necessary dealings with other branches of government, is outlined.

Article XI.   DETERMINATION OF COMMERCIALITY

This important section defines the procedure for notification, testing and agreement between the contractor and the host government confirming that a commercial discovery has been achieved. This is critical to the agreement because it switches the program from exploration to development. As a result, the host government may become involved in sharing the subsequent expenses and will trigger a whole new array of fiscal provisions in the contract.

Article XII.   CONFIDENTIALITY AND DATA RELEASE

Part of the agreement between the contractor and the host government may involve title to geologic, geophysical, drilling and production data. The contract should recite this understanding for the benefit of all concerned.

The NOC (not necessarily the ministry unless so specified) usually wishes to maintain confidentiality of data and reports pertaining the contract area in accordance with the laws of the host country and with general industry practice. On the other hand, this kind of information cannot be released by the contractor, even beyond the contract period, without the prior approval of the NOC. Specific exceptions may be provided so that certain types of data may be released to lending institutions which are financing parties to the agreement. News releases to home country Securities and Exchange Commissions may be granted blanket approval with the host government being provided a copy in advance.

Article XIII.   **OWNERSHIP OF ASSETS**

Generally speaking, all of the assets that are purchased for and permanently installed on the project belong either to the contractor, during the exploration phase, or to the project itself, and eventually revert to the NOC. An important exception involves equipment owned by a sub contractor such as a drilling rig or seismic equipment which is brought into the host country to perform a service for the project and then exported from the country for work elsewhere.

Article XIV.   **SOLE RISK**

This section provides that if, following determination of commerciality, either the contractor or the NOC chooses not to participate in the ongoing operation, or a portion thereof, the other party may take over and continue at its "sole risk." If the sole risk operation eventually proves profitable the declining party can rejoin after payout plus a specified large financial penalty out of production.

Article XV.   **HOST GOVERNMENT PARTICIPATION**

Defines the government's options to have their NOC join in the operation and share the development and production expenditures after giving proper notice and reimbursing the ongoing participants for the proportionate share of the expenditures to date.

Article XVI.   **BONUSES AND ROYALTIES**

Specifies what cash bonuses are to be paid the host government (usually not directly to its NOC) and when. Any royalty due to the government is also specified.

Article XVII.   **FINANCING, COST RECOVERY AND PARTICIPATION**

This article recites again that the contractor pays all expenses during the exploration phase. It will probably recite the basic changes following the first commercial discovery within the contract area. These issues will be treated in more detail in subsequent sections.

Article XVIII. **CRUDE OIL PRODUCTION AND ALLOCATION**

This critical section might be entitled "To whom go the spoils?" This article spells out the specific splits in production revenue; to whom they are paid; and the priority of the payments. For example, royalty is paid first and the disbursement may be specified as going directly to the federal treasury. Cost recovery oil may be paid second and may be either a fraction of the gross barrel or some other specifically defined split. Cost recovery oil normally goes to the operator to be shared according to the participation, if it is a joint venture. The co venturers may be other foreign investors, and can include the NOC if it has also taken on a role as a joint venturer participant in the operation of the project.

There may be development costs to be recovered, which usually come next until that account is satisfied. Next, there is the so called profit oil to divide and appropriate payments to be made. The tax collector will then want his part of the contractor's "profit" and disbursement must be made to that authority.

Article XIX. **QUALITY, QUANTITY, AND PRICE OF CRUDE OIL**

Crude oil gravity, sulfur content and BS&W affect the price received for the production. Therefore, provision for testing samples of the production in a locally approved laboratory is usually included in the agreement. There also needs to be provision for all parties to the contract to verify the measurements of crude oil production and sales. This section will also detail the procedure for setting the per barrel price upon which the various disbursements will be made.

Article XX. **TRANSPORT AND FINAL DESTINATION OF CRUDE OIL**

The importance of the transport provisions depends upon the specific project. For natural gas development transportation provisions are vital. The contractor may be required to transport crude oil to a specified location at his expense, or it may be necessary for the contractor to share a common pipeline with other producers whose throughput of natural gas may compete. The foreign contractor may be vulnerable to tariff increases if he doesn't own the line. It is also important to work out ahead of time, and to note in the contract, which party is responsible for the transportation costs of royalty production, and other splits of the production.

In some jurisdictions there may be a requirement that crude oil produced offshore be landed before export. This may incur very large undersea pipeline investment. The landing and export can also involve the host government's customs officials. The final destination of petroleum exports may be of concern to the host government, and may require its prior approval.

Article XXI.   **PREFERENCE FOR LOCAL GOODS AND SERVICES, AND EMPLOYMENT OF LOCAL PERSONNEL**

These two requirements are often made the subject of separate individual articles. Their intent is obvious. The contractor always desires that there be reasonably specific procedures set out in this section so that there will not be undue delays in the processing of requests for exceptions. In the case of goods and services there frequently is a stipulation that the local supplier cannot charge more than a specified percentage above the imported international price, say, fifteen percent.

A great deal of variation exists among different countries in procedures for granting entry visas and work permits for expatriate personnel. Generally speaking, the more of the detailed requirements and specific procedure that can be worked out in the contract ahead of time is preferable for all parties concerned.

Article XXII.   **IMPORT AND EXPORT**

The contractor is granted the right to import all equipment and materials necessary for Petroleum Exploration free of all duties and taxes. Imports related to Petroleum Development and Production are subject to duty, not to exceed 10% of the CIF value.

Contractor may export produced Petroleum and equipment no longer needed for Petroleum Operations free of export duty and taxes.

Article XXIII.   **PERSONNEL TRAINING AND TECHNOLOGY TRANSFER**

As with the previous article, these two items sometimes warrant separate sections or articles in the contract. Training of local nationals is becoming an important part of any new overseas exploration and development contract. The training may be of several types, such as: First, on the job understudying of professionals in specific positions in the enterprise within the host country, or sometimes even in another country; second, attendance at public, or in house training programs conducted by an international petroleum training contractor; and thirdly, enrollment at the foreign oil company's expense in a university academic program to prepare the participant for a professional position in his or her home, i.e., the host country.

The issue of "transfer of technology" is a matter of great concern in the developing countries. It has never been fully understood by the multinational companies. Any provision which isn't completely understood by all parties to the agreement is an invitation to difficulty in the future. In the exploration and production segment of the petroleum industry where patent rights may play a more minor role in the fast moving and complex nature of the operations, "transfer of technology" is generally understood by the contractor to mean "know how."

431

The whole question of transfer of technology unfortunately, is vague at best. The multinational companies may agree to the wording, and commit to trying to convey their know how (acquired usually over many years of international operating experience) to the host country nationals, and hope for the best.

Technical data acquired during the exploration phase generally belongs to the Contractor with a copy being furnished to the host government Ministry, or in some jurisdictions, to the NOC. Sale of the data may require host government approval and they may require a share of the proceeds.

Article XXIV.   **NATURAL GAS**

A section of the contract must speak to the subject of natural gas since this type of petroleum is becoming easier to find, increasingly prevalent, and harder to sell. The matter of natural gas is exceedingly complex physically, fiscally and operationally. Separate provisions are necessary for associated, or "casinghead," gas as distinct from non associated natural gas.

Article XXV.   **CURRENCY AND EXCHANGE CONTROL**

Specifies the currency (usually U.S. Dollars) in which payments will be made to the Contractor. Also provides for freedom of export and import of foreign currencies and domestic exchange for the conduct of the operations and payments to subcontractors and expatriate personnel associated with the project.

Article XXVI.   **TAXES**

Host government generally likes to have a specific article acknowledging that the contracting foreign company and all of its sub contractors are taxable in the host country.

Article XXVII.   **AUDITING**

The contract will usually rely on a separate appendix dealing with accounting procedures. These agreements are frequently very complex in order to provide both contractor and host government, as well as all of the respective government departments in both the host country and the home country of the foreign co venturers all of the required data.

All non operating parties to the agreement deserve the right to audit accounting statements. This should be clearly stated in the contract.

Personnel costs represent a major part of operating expense. Such costs should be spelled out as to what parts of the expatriate employees' salaries and benefits are chargeable to the joint project. For example, items such as furlough expense and the employee's group insurance coverage with his foreign oil company employer are expenses that should be addressed.

Article XXVIII. **SUBCONTRACTORS**

This article confirms the right of the operator to employ subcontractors in connection with the project but specifies that they are subject to the same types of regulations and restrictions as the Contractor's own organization and personnel.

Article XXIX. **GOODS AND SERVICES**

Contractor is required to provide necessary goods and services giving preference to local suppliers. Competitive bidding is called for and there is usually a percentage limit on any excessive price for local goods over imports. At the termination of the Contract any inventories revert to the NOC.

Article XXX. **LAND SURFACE AND FACILITIES**

This article indicates that the NOC will arrange for the assignment of public lands to the project and assist the Contractor in acquiring privately held lands which may be necessary. It also requires the Contractor to construct the necessary facilities and to share them with the NOC.

Article XXXI. **SUPPLY OF NATIONAL PETROLEUM NEEDS**

The local refining and utility requirements have priority in the acquisition of any production of crude oil and natural gas from the Contract Area before any possible export.

Article XXXII. **ENVIRONMENTAL PROTECTION AND SAFETY**

A recent feature of petroleum contracts deals with protection of the natural surface and subsurface environment. The principal concern is pollution from oil spills or releasing salt water into lakes and streams.

Conservation of the petroleum resource is also generally recognized as a contractual item and requirement. The Libyans in the early 1970's recognized that "conservation" was a "motherhood issue" which had OPEC's blessing and which was so subjective that it could be argued interminably. Contract terminology in this area usually is still vague.

Adherence to international safety standards is another rather new provision. Drilling, and other safety standards are still not completely worked out in the North American exploration and development industry. There is, as yet, no clear cut set of standards for worldwide application.

Article XXXIII.   **UNITIZATION**

Requires unitization of operation of producing oil and gas reservoirs which are found to extend beyond the boundary of the Contract Area, if so ordered by the NOC.

Article XXXIV.   **INDEMNITY AND INSURANCE**

Insurance coverage is generally of somewhat lesser concern in the developing countries than in the home country of most foreign companies. Nevertheless, the contract should speak to the types and extent of coverage which the contracting oil company will carry for the benefit of the joint account since the costs are properly chargeable to it.

Article XXXV.   **ASSIGNMENT AND TRANSFER**

Assignment rights have been a fundamental problem in international petroleum agreements through the years. Normal practice involves the granting of full rights of assignment to the contractor's subsidiary or affiliated companies with prior notification to the NOC. There may be stipulations regarding financial and performance guarantees.

All other assignments must be subjected to full screening and prior approval of the host government coupled with a performance guarantee and perhaps a right of first refusal by the NOC. The desire of the host government to protect itself under the terms of the agreement is understandable. The problem from the standpoint of the developing countries who are striving to get into the oil business is that these approval procedures greatly inhibit the drilling of prospects considered marginal by the original contracting company. Since the tremendous geological uncertainties can only be clarified by the actual drilling of a well on a prospect, the more wells drilled the better the chances of discovery.

In North America a large number of oil and gas discoveries are made on farmouts to operating companies who were not parties to the original leasing. The normal procedure for approval of assignment has the unfortunate effect of greatly reducing the number of farmout drilling ventures. This essentially deprives the host country of this additional multiple exposure to discovery with the drill bit.

Article XXXVI.   **TERMINATION OF THE CONTRACT**

This article provides the necessary provision outlines the means and procedures for winding up the agreement.

Article XXXVII.   ABANDONMENT OF OPERATIONS

Termination of production of oil and gas from the contract requires the plugging of wells, removal of surface facilities, equipment, structures and pipelines, and restoration of the surface to its original condition. This article further provides for funding of these operations from revenues generated by sale of oil and gas from the contract area.

Article XXXVIII.   APPLICABLE LAWS AND REGULATIONS

Provides the Contractor and all subcontractors comply with all of the host country's laws and regulations. Exception to regulations may be granted if they interfere with operations under the Contract.

Terms and conditions of the Agreement and Petroleum Law as of the date of the Agreement shall prevail over any subsequent law if it diminishes the rights currently granted.

Article XXXIX.   FORCE MAJEURE

A force majeure clause is always necessary. The article will probably also provide for certain notifications to various authorities in the event that such a claim is made by the operator.

Article XXXX.   DISPUTE SETTLEMENT

An arbitration clause is always included, but essentially never invoked. Some countries have refused on the basis of sovereignty to agree to arbitration clauses which yield any authority for the settlement of disputes to any judicial body outside of the host country. Nonetheless it is the norm to provide for final and binding arbitration by one of the several international bodies of arbitration.

Article XXXXI.   LANGUAGE OF THE CONTRACT AND WORKING LANGUAGE

This article provides that although the contract may be written in both the language of the host country and the language of the contracting company's home country, in case of any difference or dispute the language of the host country shall prevail.

Article XXXXII.   NOTICES

Here the official addresses of the parties for the delivery of official notices are spelled out.

Article XXXXIII.   **EFFECTIVE DATE**

Specifies when the Agreement actually becomes effective, which in most cases is its ratification by the host government.

Article XXXXIV.   **MISCELLANEOUS PROVISIONS**

This may include the legal nicety that this Agreement comprises the entire understanding between the Parties, which in reality is much too restrictive for an operation of this type. It may also list the signatories and their dates of signing.

Appendix I.   **MAP OF CONTRACT AREA**

This complements Article III of the Agreement.

Appendix II.   **ACCOUNTING PROCEDURE**

Specifies specific procedures for accounting personnel.

Appendix III.   **PURCHASING PROCEDURES**

Specifies specific procedures when purchasing for the joint account

Appendix IV.   **CRUDE OIL LIFTING PROCEDURE**

Presents detailed procedures for measuring, loading, paying, and accounting for crude oil produced and moved from the contract area. These matters were ignored in many earlier contracts.

GLOSSARY OF PETROLEUM
AND FINANCIAL TERMS

*The difference between the right word
and the almost right word is the difference
between lightning and the lightning bug.*

THE BIGGEST PROBLEM MOST PEOPLE HAVE WITH A NEW OR UNFAMILIAR ACTIVITY IS THAT OF NOMENCLATURE. THEY ARE APT TO FEEL UNEASY BECAUSE THEY DON'T KNOW THE LANGUAGE. THE PETROLEUM INDUSTRY DOES HAVE A UNIQUE SET OF TERMS AND EXPRESSIONS. THE FOLLOWING IS AN ATTEMPT TO EXPLAIN A GREAT MANY OF THESE.

A

ABANDONMENT — The act of terminating all operations, plugging wells, removing all surface equipment, and restoring the surface to its original condition. Offshore it would involve platform decommissioning and/or removal.

ABSOLUTE OPEN FLOW POTENTIAL — of a gas well is a theoretical value employed in some property evaluations, and by some states in their regulatory processes. It is the rate at which a gas well could flow into the wellbore against zero pressure. The AOFP is a function of reservoir pressure, the product of reservoir permeability times thickness (Kh), the degree of formation damage and other physical factors. Many statewide rules require that theAOFPbe determined at initial completion of the well as part of the regulatory procedure.

ABSTRACT OF TITLE —A collection in summary form of all of the recorded instruments, usually from the County Records, affecting the title to a tract of land.

ACCELERATED DEPRECIATION —These are computation techniques that allow the owner of the asset to take greater amounts of depreciation in the early years of its life, thereby deferring some of the taxes until later years. (See Chapter VIII).

ACCEPTING HOUSE —A British financial institution that buys accounts receivable, designated as Bills of Exchange, from companies who prefer to accept a discounted amount of cash now rather than the full cash amount in due course. A bill accepted by the bank carries that institution's guarantee.

ACCOUNTING METHODS

FULL COST — Accounting method in which all exploration costs, e.g., Geological and Geophysical and drilling, are charged against income on a units-of-production basis. This system is based on the view that in order to find promising acreage it is necessary to incur losses in unrelated areas. Full cost accounting involves capitalization of all costs. Expenditures for exploration of properties which are not acquired, or which are released, are treated as indirect costs of acquired or retained acreage.

SUCCESSFUL EFFORTS — An accounting system which allows a firm to expense the exploration failures and capitalize its successes. FASB-19 states, "Under this method, the test-year exploration costs are allocated to basic types of oil or gas reservoirs on the basis of the intangible drilling costs of successful exploratory wells, as identified with those two types of reservoirs, or by other measures of the successful exploratory effort as identified with such reservoirs, such as footage drilled or the cost of acreage transferred from non-producing to producing. Subsequent allocations of the costs thus allocated to reservoirs are made on a common unit-of-measure (such as BTU's) or other basis, with the resultant cost thereby allocated to the lease liquids and gas then being attributed to test-year production volumes."

ACRES (OWNERSHIP)

FEE ACRES —Acreage on which the landowner has complete ownership, including both surface and mineral rights.

GROSS ACRES —Total surface acres of a property.

LEASEHOLD ACRES — Acreage on which a party (leasee) has acquired, by lease, the right to explore for and produce oil, gas, and/or other minerals in return for a stated royalty, and possibly other considerations.

MINERAL ACRES —Acreage on which ownership of the minerals in place has not been divided into royalty and working interests. It may be an undivided interest in a larger tract.

NET ACRES —Gross acres times fractional economic interest owned (working interest). Term may apply to leasehold, fee, mineral, royalty acres, etc.

ROYALTY ACRES — Acres on which a royalty interest is owned. A party owning such a royalty interest does not have the right to produce the minerals or reserves underlying such acreage.

ACRES(MEASUREMENT) —May be expressed on either a gross or net basis. Gross refers to total acres, while net refers to company-owned acres.

438

SURFACE ACRE — 43,560 square feet of surface area.

PROVED ACRES — Relates to acres believed to contain economically producible reserves. May also be used in the context of the maximum limits of the productively proven portions of all producing formations.

ACREAGE-BASED ROYALTY — A term used to describe a royalty payable on a per acre basis, e.g., a royalty expressed as a fixed sum per acre on a shut-in oil or gas well.

ADMINISTRATIVE COSTS —The costs associated with managing an enterprise as distinct from the provision of goods and services.

ADP — Alternate Delivery Procedure. A futures contract provision which allows buyers and sellers to make and take delivery under terms or conditions which differ from those prescribed in the futures contract. An ADP may occur following termination of futures trading in the spot month, and after long and short positions have been matched for the purpose of delivery.

AD VALOREM TAX — An annual tax assessed as a percentage of the value of the item being taxed. This tax is usually applied to real property, such as land, buildings, equipment and oil and gas reserves.

AFE — See AUTHORIZATION FOR EXPENDITURE

AFIT—After Federal Income Taxes. Usually used in connection with profit, as in "AFIT net income."

AFRA —International crude oil tanker rates established monthly by the London Tanker Brokers Panel (LTBP) based on term charters in existence during the previous month (mid-month to mid-month). The "Award Panel" meets monthly and declares the Average Freight Rate Assessment (AFRA) of Worldscale rates for the previous month. See: WORLDSCALE

ALKYLATION — A catalytic refining process that combines lighter hydrocarbons, such as butanes, into hydrocarbons of the gasoline range.

ALLOCATION FORMULA —The formula employed in fixing the allowable production for a well, pool, or field.

ALLOWABLE —The common meaning in the industry is that volume of oil or gas production a well or leasehold is permitted to produce per day under proration orders of a state, federal or provincial regulatory body.

ALTERNATIVE FUNDING — A source of financing for a project other than internally generated funds from the company.

AMORTIZE —The procedure of allocating the cost of long-lived assets to the periods in which their usefulness applies. Amortization is applied to intangible capital. When it is applied to fixed assets it is called depreciation expense, or capital recovery allowance, whereas for wasting assets (natural resources), it is referred to as depletion expense. (See also DEPRECIATION, and DEPLETION.) The term may also refer to the liquidation of a debt on an installment basis.

ANNUAL PERCENTAGE RATE (APR) — An interest rate that reflects the actual cost of borrowing as a yearly rate when compounded more frequently than once a year.

ANNUAL REPORT—Popular term for the yearly report made by a company to holders of its securities. U.S. law requires all registered corporations to prepare and distribute such reports. They typically contain the firm's Balance Sheet, a statement of the Source and Application of funds, and a list of changes in retained earnings.

The form of the report filed with the SEC in the U.S. is called Form 10-K, which expands on items in the annual report. Shareholders may request Form 10-K from the corporate secretary of the company for additional information.

ANNUITY —A method of spreading a cost or scheduling the repayment of a debt with interest evenly over a specified period of time. Each payment is the same and includes both interest and debt repayment in varying amounts.

API GRAVITY —The density (weight per unit volume) of crude oil on a scale adopted many years ago by the American Petroleum Institute. On the API scale, the higher the gravity the lighter the oil. Most crude oils range from approximately 27 degrees to 40 degrees API gravity; lower than that are termed heavy, lighter than 40 degree, light oils.

$$\text{API Gravity} = \frac{141.5}{\text{specific gravity at } 60° \text{ F}} - 131.5$$

API (specific) Gravity bears a physical relationship to true specific gravity, or density, but is more convenient to work with than the decimal fractions.

APPRAISAL DRILLING — Wells drilled in the vicinity of a discovery well in order to evaluate the extent and importance of the find.

ARA RANGE —A term used to indicate the Amsterdam/Rotterdam/ Antwerp area for delivery and trading of oil products on the spot market.

ARBITRAGE —1) As verb: the act of simultaneously buying and selling the same security in different market places to profit from a disparity in market prices. 2). As verb: the act of buying one security coupled to a short sale of the same security to profit from a disparity of prices. 3) As noun: an offsetting security position that has a built-in profit. 4) As adjective or as past participle: to describe a security position that establishes a profit. For example, "His long position was arbitraged by a short sale."

In more recent times, the term has come to be applied to buying stock in companies that are, or are rumored to be takeover targets, and selling short stock in the acquiring company to lock in their profit and limit their risk.

AREA OF MUTUAL INTEREST — Any area which, by prior agreement, is the subject of mutual sharing of ownership of any leasing rights acquired.

AROMATICS —A class of hydrocarbons characterized by the six-carbon ring of the benzene series and related compounds.

ASPHALTENES — A brownish-black semi-solid mixture of very complex heavy hydrocarbons plus other chemical constituents obtained in the distillation residue of many crude oils.

ASSETS —

CAPITAL ASSETS — These consist of the dollar value of assets that are tangible and which can be appraised by physical inspection. This include buildings, rigs and machinery, and equipment of all types which depreciate in value with age and usage. This category of assets also includes negotiable financial instruments such as stocks and bonds, and any unencumbered cash.

A second category of Capital Asset is known as INTANGIBLE Assets which includes all types of minerals, including oil and gas.Novalue may be established by direct inspection (all reservoir engineering reserve assessments comprise indirect evaluations). Intangible Assets have a further characteristic in that they do not necessarily deteriorate, or depreciate with time. This class of asset loses value only when produced, and cannot be replaced in its initial undiscovered state. Intangible assets, among many other things, include the G&G expenditures incurred in connection with the acquisition or retention of acreage. The exact differentiation between tangible and intangible assets is frequently somewhat arbitrary, but is highly important from an income tax standpoint, since they are treated quite differently in the tax computations.

CURRENT ASSETS —Assets which are, or may be, converted into cash in a short period of time, generally one year. In general, the term includes the following resources:

1. Cash, and equivalent items

2. Inventories

3. Receivables such as trade accounts, notes, and acceptances

4. Marketable securities representing the short term investment of cash for current operations.

5. Certain prepaid expenses, which may be paid on an annual basis which does not coincide with the date of the financial statement.

ASSOCIATED GAS — Is gas that occurs together with oil, either as free gas or in solution. Gas which exists only as a physically separate phase in a reservoir is termed non-associated gas.

AUTHORIZATION OF EXPENDITURE (AFE) — A request for approval of funds for a given project. The term may also refer to the paper form used during the planning process for a well, or other project, soon to be undertaken. It includes the estimated costs in both the intangible drilling costs, (IDC's), and the tangible equipment categories. A breakdown of costs is normally included.

AUTHORIZED UNISSUED STOCK (See Stock)

AVERAGE CAPITAL EMPLOYED —End of year capital employed plus beginning of year capital employed, divided by two.

441

B

BACK-IN INTEREST —Form of carried interest (see below) in which the latter converts to a regular working interest after payout, i.e., after the carrying parties have recouped their costs, the carried party converts to a regular fractional working interest, thereafter paying its share of costs and receiving its share of revenues.

BACKWARDATION — A relationship between prices of a commodity in which the cost of immediately available supplies exceeds the price of comparable volumes and quality available in the future. Oil prices were in backwardation during much of the late 1980's when prices were sliding from their earlier peaks.

BALANCE SHEET — The basic financial statement showing the nature and amount of a company's assets, liabilities, and its net worth (shareholders' or stockholders' equity) as of a given date. BARNBURNER—An unusually productive well, i.e., every investor's dream a super producer that will assist one toward early retirement.

BARREL (bbl) —A unit of volume measurement of petroleum and its products. 1 barrel = 42 U.S. gallons, 158.97 liters, or 35 Imperial gallons. A barrel of crude oil may weigh between 280 and 389 lbs. depending upon its API gravity. The first b in its abbreviation (bbl) comes from the early days of the oil industry when Standard Oil Company painted their uniform size barrels blue.

BARRELS PER DAY —The customary term of measurement of the rates of production from a well or oilfield, throughput of pipelines, refineries and other petroleum facilities. Generally the reference is to the average number of barrels per calendar day over a specified time, usually a year.

BARRELS OF OIL EQUIVALENT (boe) —The quantity of NGL or natural gas necessary to equate on the basis of relative heating value, or on a relative price, with a barrel of crude oil. For natural gas, approximately 6 million BTU equals one barrel of oil equivalent. For NGL, 1.455 barrels of NGL is equal to one barrel of oil equivalent.

BASIS —The differential that exists at any time between the futures price of a given commodity and its comparable current cash or spot price.

BASIS RISK —The price risk that derives from the physical differences in quality and location between a hedging instrument, such as a futures contract, and the actual commodity itself.

BASEMENT — Hard, igneous or metamorphic rock lying below sedimentary formations. Seldom contains petroleum.

BASIS POINT — By common agreement, 0.01% of yield on a fixed-income security. For example, if a bond's yield to maturity changed from 9.05% to 10.35%, there was a change of 130 basis points. The term is also used with yield changes for securities sold at a discount from face value.

BATCHING SEQUENCE — The order in which petroleum or product shipments are sent through a pipeline.

BATTERY — A group of at least two lease storage tanks in the oilfield.

BFIT — Before Federal Income Taxes.

BLIND POOL — Where the drilling program prospectus does not specify the drilling prospects.

BLOCK — A group of leases, usually contiguous, held in one ownership.

BLOWOUT —A sudden, violent escape of oil and/or gas from a well caused by uncontrolled high pressure. Usually occurs when the well is being drilled. A wild well.

BLOWOUT PREVENTER — Specially designed assembly of heavy-duty valves and equipment installed on the wellhead while drilling or working over a well to control well pressure and help prevent blowouts.

BOND —Along-term debt instrument which obligates a borrower to pay a given rate of interest and to repay a fixed amount of money on a specific future date.

BONUS —The money paid by the lessee for the execution of an oil and gas lease by the owner of the mineral rights. In the international industry this is referred to as a signature bonus. Other overseas bonuses paid to the host government include discovery bonuses paid at the time of the first commercial oil discovery, and production bonuses paid when certain specified rates of oil production are attained. Another form is called an oil or royalty bonus. This may be in the form of an overriding royalty reserved for the owner of the mineral rights in addition to the usual one-eighth royalty.

BOOK PROFIT — A financial book (accounting) assessment of earnings.

BOOK RATE OF RETURN (BROR) — A measure of corporate performance.

BROR = (Financial Book Earnings) + (After Tax Interest Expense)÷(Average capital employed)

BOOK TRANSFER — Transfer of title of a cash commodity to the buyer without a corresponding physical movement.

BOOK VALUE —Net amount at which an asset is valued on financial books. This may vary from its market or intrinsic value. Book value is unrecovered cost, i.e., original cost less accumulated depreciation, depletion, amortization, abandonments, etc. Although book value can be a deceptive measurement, it is used by many to make a gross selection of common shares that may be underpriced. Also called the net tangible value or the liquidating value per common share.

BOTTOMHOLEMONEY —Monetary contribution made to the drilling party for reaching a specified well depth or stratigraphic equivalent, irrespective of productivity, in exchange for well information.

BREAK-EVEN ANALYSIS — A quantitative technique to ascertain the point where total undiscounted revenues equal total undiscounted costs. Although break-even analysis may be before or after tax, the after-tax calculation usually is of more use to the firm since tax is a very real portion of deductions from revenue.

BREAKUP VALUE — The monies that would be realized if all of a company's operating entities were sold and applicable debt repaid.

BRIDGING LOAN —Short-term financing provided while the borrower seeks a longer-term loan.

BRENT BLEND —Production from the Brent and other fields in the East Shetland Basin of the U.K.. sector of the North Sea brought ashore by pipeline to Sullom Voe in the Shetlands. Brent (approx. 38°API) is the principal marker crude for trading other North Sea crudes to which differentials for quality and location are applied. Brent is quoted on an f.o.b. basis at Sullom Voe.

BRENT MARKET —Awidely traded forward market in North Sea Brent crude which evolved in the early 1980's. Brent has become a key benchmark crude in international oil price risk management in both term and spot sales. It is the designated crude in the IPE's London futures market.

BRINGING IN A WELL — Completing a well for production.

BS&W — The traditional abbreviation for basic sediment and water; the produced formation water and extraneous material present in crude oil.

BTU —British Thermal Unit; the amount of heat required to raise one pound of water one degree Fahrenheit.

BULK PLANTS —The storage and distribution plants for petroleum products usually centrally located to a number of service stations or retail outlets.

BUNKER FUEL — Residual fuel oil used for bunkering, i.e., fueling, ships.

C

CALENDAR — Industry term for list of upcoming securities offerings. Usually preceded by type of security. For example, municipal calendar, corporate bond calendar.

CALL ON PRODUCTION —The right to purchase hydrocarbons produced from a particular property.

CALL —A type of option which gives the right to buy a commodity at a fixed price in the future. In oil trading it is an adjunct to the futures market activity on both the NYMEX and IPE. (See "options" and "put").

CAP (or Ceiling) —A swap transaction in which the buyer seeks a maximum price position for his supplies. The antithesis of a Floor Agreement. A comparable price hedge may sometimes be accomplished using futures and options.

CAPITAL

BORROWED CAPITAL — Amounts owed with an original maturity of more than one year. Also includes amounts owed on long-term debt maturing within the current year.

INVESTED CAPITAL —Total assets less total liabilities. This equals the shareholder's investment in a firm as determined on the financial books (shareholder net worth).

SOURCES OF CAPITAL — Capital may flow into the firm from long-term debt (repayment in more than one year), short-term debt (repayment in less than one year), equity, and retained earnings from operations.

TOTAL CAPITAL — Sum of borrowed and invested capital.

CAPITAL ASSET — Has a useful life greater than one year and a recognized salvage value.

CAPITAL EMPLOYED—Capital Expenditures less noncash charges (i.e., capital recovery, deferred tax, and amortization of non-producing properties.).

CAPITAL EXPENDITURES (COSTS)—Those expenditures that are recorded as assets on the financial books and are subtracted from revenues over an extended period of time through some form of capital recovery. In the E&P sector these include the total cost of lease acquisition, geological and geophysical (seismic and well log surveys) exploration, and others including tangible property acquisition.

CAPITAL GAIN (OR LOSS)—The excess (or deficit) of proceeds realized from the sale or exchange of a capital asset over (or under) its book value.

CAPITAL RECOVERY — The method by which a capitalized investment is deducted over time from revenues in the determination of earnings or taxable income, i.e., depreciation, depletion, and amortization.

CAPITALIZE — (1) In accounting terms, the periodic expensing of capital costs through DD&A. (2) Recording investment outlays as additions to asset value rather than as expenses.

CAPITAL SURPLUS—Dollar amount by which price paid by original purchaser exceeds the par value of securities. For example, if the issuer sells stock at $15 that has a par value of $5. The $5 is entered in common stock account on balance sheet, and $10 is entered in capital surplus account. The same principle is used if a company sells stock reacquired and held in the treasury above its par value. Also called paid-in capital.

CARRIED INTEREST — Where a party or parties have their expenses paid (in effect, are carried) by other parties up to a specified limit. A fairly standard industry deal is known as third for a quarter, in which an investor pays 1/3 of the expenses for 1/4 of the revenue. The promoter is thus "carried" for a quarter. The industry phrase "carried to casing point" means the party or parties being carried have all costs paid by other parties through drilling and up to the point of setting production casing, i.e., the commencement of completion operations. "Carried to the tanks" means carried up to the point of sale of the product. In this latter case the carrying parties pay the total cost of the completed well and the surface production facilities. Depending upon the terms of the agreement, the carrying parties may then have the right to recover such costs from the carried party's share of the reserves if, as, and when produced from the property.

CARRIED PARTY—The party, or interest owner, for whom funds are advanced in a carried interest arrangement.

CARTEL — A group of businesses or nations that agree to control prices by regulating production and marketing of a product.

CARVED OUT OIL (OR GAS) PAYMENT—A payment in oil or gas assigned by the owner of a working interest. The payment may be denoted in dollars, in barrels, in MCF, or in a fraction of the production over a specified period of time. The payment is made from a fractional part of the working interest or fee interest which is to be satisfied in a period shorter than the life of the interest from which it is carved.

CASH BOOKS — Refers exclusively to the company's cash transactions irrespective of accounting treatment. Non-cash charges are naturally excluded. Companies normally do not maintain separate 'cash books' per se.

CASH FLOW—The movement of money into or out of the firm's treasury, or from a specific project under study. Expenditures are recorded as negative and receipts or revenues are shown as positive cash flows. Both the outflow, e.g. expenses, capital expenditures and taxes, and inflows, or revenues and reimbursements, are added algebraically and the net summation is referred to as Net Cash Flow.

CASH OPERATING INCOME DEDICATION LOAN (COI) — A type of loan which is repaid from an agreed portion of the Cash Operating Income.

CASING — Steel pipe run in the wellbore and cemented in place to prevent cave-ins during subsequent drilling, and to allow production after the well is completed. There may be several strings of casing in an individual well, one inside the other. The first string to be run in a new well is called surface pipe which is cemented in place to shut out shallow formation water, and to serve as an anchor or foundation for all subsequent drilling activity.

CASING POINT — The point in time during the drilling of a well when it has reached a predetermined depth, and usually a suite of logs has been run, that the major decision must be made as to whether to abandon it as a dry hole, or complete it as a producing well.

CASINGHEAD GAS — Natural gas that flows out of a producing well together (associated) with the crude oil and is separated from the oil in the oilfield separator.

CATALYTIC ("Cat") CRACKING—A refining method utilizing a catalyst, and moderate heat and pressure relative to Thermal Cracking, to break apart larger hydrocarbon molecules into smaller ones of gasoline range.

CEILING — See Cap

CEMENTING — The process as part of the drilling operation of pumping a thin slurry of a special portland cement at high pressure down the casing pipe. This comes out of the bottom of the casing underground, and as pumping is continued, the cement is forced up outside of the casing to fill the small annular space between the pipe and the wall of the hole. When enough cement slurry has been pumped through, a plastic plug is inserted in the casing and drilling mud is pumped in to force the plug to bottom. Pressure is then maintained for a sufficient period for the cement to set. Finally, the plug is drilled out and operations are resumed.

CERTAINTY — The case of a single outcome of an event for which the probability of occurrence is 1.0.

CHARTER, TANKER — A contractual arrangement between a charterer who requires a vessel to meet his transportation needs and an owner who places his vessel at the disposal of the charterer.

CHRISTMAS TREE—An oilfield term for the intricate assortment of control valves, gauges and chokes assembled at wellhead to control production. Its shape, complexity and number of pressure gauges (ornaments) led to its name.

CIF—Cost, Insurance, freight: The term refers to a sale in which the buyer agrees to pay a unit price that includes the FOB (free on board) value at the port of origin plus all costs of insurance and transportation. It differs from an FOB transaction in that the seller, for a fee, arranges for the transportation and insurance. It also differs from a "delivered" type of agreement since it is generally ex-customs duty and the buyer accepts the quantity and quality at the loading port rather than to pay on quantity/quality as determined at the unloading port. Risk and title in a CIF deal are transferred from seller to buyer at the loading port, although seller is obligated to provide insurance in a transferable policy at the time of loading.

CLEARING HOUSE — An exchange associated organization charged with the function of insuring the financial integrity of each trade. Orders are legally cleared so that the clearing house becomes the buyer to all sellers and the seller to all buyers.

CLEAN CARGO — Refined products: gasoline, kerosene, distillate, and, jet fuel carried by tanker, barges and tank cars. Bunker fuels and residuals are not included.

COLLAR — A type of swap transaction which establishes a price range between a floor and ceiling price which is hedged against a single fixed price. This reduces the price risk to a limited amount of market volatility.

COMMERCIAL PAPER — Unsecured promissory notes of corporations, issued to provide short-term financing, which are sold at a discount and redeemed at face value. This type of security is exempt from registration if maturity is 270 days or less. Highly competitive with other money market instruments. Unlike most other money market instruments, commercial paper has no developed secondary market.

COMMERCIAL WELL — A well which produces crude oil and/or natural gas in sufficient quantities to be economical for commercial operation; i.e., a well in which the revenues exceed the costs of operating and maintaining the well.

COMMINGLING — The intentional mixing of crude oil, petroleum fractions or products having similar characteristics.

COMMISSIONED AGENT—Agasoline distributor who works for a petroleum refiner and is paid on a commission basis.

COMMISSION HOUSE — An financial organization which trades commodities and/or futures and options contracts for customer accounts for a fee.

COMMON CARRIER —A person or company (usually a pipeline) who has state, provincial or federal authority to perform public transportation for hire.

COMMON STOCK — The stock, or share interest of the owners of the firm. These owners have claim to earnings and assets remaining after all claims of bondholders, other creditors, and preferred stockholders are settled. The charter of incorporation defines the rights of the common stockholders. The common stockholders control the company's management through its Board of Directors which they elect, and company policy through voting rights.

COMMON STOCK EQUIVALENT —Name given to securities that may, with or without the addition of money, be exchanged for common stock. Convertibles, rights, warrants, and long calls are typical common stock equivalents, but only under the precise conditions specified.

COMPLETION —That part of the operation in the drilling of a well that takes place between the time the producing formation is penetrated and recognized, usually by means of wire line logging or coring, and the time when the well is ready to produce. The term also is used to refer to each separate physical setup within a well bore which provides for the production of hydrocarbons, generally restricted to a single zone, or "common source of supply" as defined by a government regulatory agency.

COMPENSATING BALANCE —A bank balance of a given amount that the firm maintains in its demand deposit account. It may be required by either a formal or informal agreement with the firm's commercial bank. Such balances are usually required by the bank (1) on the unused portion of a loan commitment, (2) on the unpaid portion of an outstanding loan, or (3) in exchange for certain services provided by the bank, such as check-clearing or credit information. These balances raise the effective rate of interest paid on borrowed funds.

COMPENSATORY ROYALTY —A payment from the lessee to the lessor in lieu of drilling an offset well. It is a negotiated payment representing a compromise where the operator does not feel that an offset well, even though required by the lease agreement, would be economically justified.

COMPOUNDING —The computation process of determining the future value of a payment, or series of payments when applying the concept of compound interest to the current value. Thus the term refers to the growth whereby the interest and principal from a previous period become the principal for the next period and grow at the original, or same rate of interest.

CONCESSION —An arrangement between the host government and the contracting company in which the petroleum production accrues to the concessionaire while the government collects commensurate taxes and royalty.

CONDENSATE —Hydrocarbons which are in the gaseous state under reservoir conditions of temperature and pressure, but which become liquid at the surface and may be recovered by conventional separators. Synonym: distillate (older usage).

CONFIRMATION WELL —A well drilled to confirm, or "prove" the productive zone found in an exploratory discovery well.

CONNATE WATER —Fossil sea water trapped in the pores of a geologic formation while it was formed.

CONSTANT PERCENT DECLINE —Aproduction decline behavior in which the producing rate decreases by the same percentage every time step, also known as "Exponential," or "Semi-log Straight Line" decline.

CONTANGO —A time relationship between prices under which immediately available crude oil sells at a discount to future supplies. The reverse of backwardation. Oil markets were in contango over most of the first half of 1990.

CONTRACT FOR DIFFERENCES —Afinancial agreement in a swaps or other commercial arrangement in which the arithmetic difference between two similar but opposite transaction are sold or exchanged rather than the total amounts involved.

CONVERSION CLAUSE —A clause in an assignment or other legal document providing for the conversion, either automatically or at the option of one of the parties, of one type of interest or ownership into another type of interest (e.g. conversion of an overriding royalty or net profits interest into a fixed share of the working interest). Conversion clauses, or rights are also included in some types of securities (e.g. a convertible debenture which under certain stipulated conditions can be converted, or exchanged for a fixed number of shares of common stock in the firm).

CONVERTIBLE — Class of corporation securities that is convertible into a fixed number of shares of other securities of the same corporation. Convertibles usually are debentures or preferred shares that may be exchanged for a fixed number of common shares.

COPAS (Council of Petroleum Accounting Societies)—Manual of accounting procedures for handling joint operations, particularly for allocating costs as either direct operating expenses or indirect overhead charges incurred by the operator of the joint property.

CORE —A cylindrical sample of earth taken vertically at various depths. Cores are utilized to determine porosity, permeability, lithology, fluid content, and geologic age. This information is used by petroleum engineers and geologists to determine the probable productivity of a formation.

CORPORATION —An association, usually of many persons, chartered by a state, provincial or federal government for the conduct of business within its jurisdiction.Acorporation is a "legal person" and thus functions independently of its owners. It provides limited personal liability to its owners for any debts of the corporation.

COST DEPLETION — A depletion allowance method in which the annual allowance is a fraction of the Leasehold investment. The fraction is calculated by dividing the production in the year by the estimated remaining oil and gas reserve at the beginning of the year.

COST OF CAPITAL —The payment (expressed as a percentage) which must be made to the sources of the firm's capital, e.g. equity, bonds, bank loans, etc., for the use of those funds.

COST TO CONDEMN —The total cost of exploratory drilling and G&G necessary to decide that a lease or area is not commercially productive of oil or gas.

COST OF FINDING —The exploratory expenditures per barrel of oil equivalent consumed in locating reserves. (See Finding Costs)

COST OIL — That part of the production revenue which is reserved to pay operating and a specified portion of the development costs under a production sharing agreement. (See Profit Oil)

C.O.S.T. WELL (Continental Offshore Stratigraphic Test Well) — A well funded by a number of firms for the purpose of obtaining subsurface geologic data in frontier areas offshore. The well is drilled without the intention of accomplishing a commercial completion.

COUPON – 1) The interest rate, expressed as a percentage of par, or face value, that the issuer promises to pay over lifetime of a debt security. Coupon rates are annual. Normal practice is to pay half the amount semiannually. 2) Small, detachable certificate that is removed from main certificate and presented for payment of bond interest. Bonds no longer have coupons attached as they are issued to an owner and interest is paid directly to the current bond owner.

COVER — To close out a short futures or option position.

CRACKING —Refining processes involving the thermal decomposition (pyrolysis) of heavier petroleum fractions to produce hydrocarbons of gasoline boiling point range. Catalytic cracking employs a catalyst to assist the process.

CRACK SPREAD —A futures market technique whereby a petroleum refiner buys crude oil futures, and at the same time sells gasoline and heating oil futures, thus insuring his profit margin for the specified future delivery month. The most common form is the 3-2-1 Crack Spread which seeks to simulate the commercial position of a petroleum refiner as a purchaser of crude oil and a seller of refined products. The transaction involves the price differential between futures in three volumes of crude oil, and two volumes of gasoline plus one volume of heating oil.

CRITICAL PATH METHOD (CPM) — The development for purposes of cost control and scheduling of the sequence of activities through a network that determines the minimum time required to complete the entire project.

CUMULATIVE PREFERRED STOCK —A feature of certain preferred shares, established by the issuing corporation, assuring the holder that if any preferred dividends are not paid by the corporation, such passed dividends plus any current dividend will be paid to the preferred stockholder before any dividend is paid to common shareholders. Almost all currently issued preferred shares are cumulative.

CUMULATIVE VOTING —Privilege occasionally granted to common stockholders by the corporation's charter. The privilege permits shareholders to allocate their votes in any manner they please. Thus, a holder of 100 shares—if five persons are to be elected to the board—may assign his 500 votes (100 shares times five vacancies for example) to a single candidate of his choice. Through cumulative voting, minority shareholders can, if they act in concert, make sure that at least one or more persons on the board will represent their interests.

CURB —Nickname for American Stock Exchange. Prior the 1940's, the ASE was known as the New York Curb Exchange because it operated outdoors on Broad Street—literally on the curb. Old-timers still refer to the ASE as the Curb.

CURRENCY IN CIRCULATION —Apopular misnomer for money in circulation. Currency consists only of coins and paper money. Money is a broader term and includes demand deposits, which are principally balances in checking accounts.

CURRENT ASSET — Item of value owned by a corporation that either is cash, or can be converted to cash within one year. The most common current asset entries on a corporation's balance sheet are cash, marketable securities, accounts receivable, and inventory.

CURRENT LIABILITY—Term used in balance sheet bookkeeping and financial statement analysis representing the sum of all debts, currently owed, that will become due within one year.

CURRENT RATIO —Common measurement of a corporation's liquidity. Ratio is computed by dividing its current assets by the current liabilities of the firm. For example, a corporation with current assets of $2.5 million and current liabilities of $1 million has a Current Ratio of 2.5. The generally accepted norm is between 2 and 1, although utilities are an exception and usually have a lower ratio.

CURTAILMENT —When a well, particularly a gas well, is forced to produce less per day than is specified in purchase contract.

CUTTINGS —Fragments of rock displaced from the bottom of the well at depth and brought to the surface while drilling. Cuttings are used to analyze the formation being penetrated although they are normally contaminated with cuttings from other previously penetrated formations up the hole.

CYCLE OIL —A non-specification product from a refining unit which is recycled through the unit to obtain additional conversion to the desired product.

D

DAISY CHAIN —The term used to describe a long sequence of buying and selling transactions involving a single cargo of crude oil, or product under which the final party in the chain finally takes wet delivery. (See Chapter IV.)

DAY RATE —Drilling contract in which the contractor charges a fixed rate per day, no matter how much (if any) footage is drilled. Day rates prevail in the higher-risk drilling areas such as the Overthrust, or the deep Anadarko Basin, and during a boom, when rigs and crews are difficult to obtain.

DD&A —(Depreciation, Depletion and Amortization)—This is a common abbreviation for non-cash charges including capital recovery and allowances for exhaustion of known reserves. (See Chapter VIII).

DEAD WEIGHT TONNAGE (DWT) —The maximum weight in long tons (2,240 lbs.) of a vessels cargo, including the weight of its stores, water and fuel (bunkers). Deadweight is the actual weight in tons of cargo, fuel, stores and ballast that a ship can carry before submerging to her load line, or Plimsoll mark. Fuel, stores and ballast represent only a fractional difference in how a vessel sits in the water. Deadweight thus becomes a fair indication of how much the ship can carry.

DEBENTURE — Longer-term debt instrument issued by a corporation and is unsecured by other collateral. Hence, only the good faith and credit standing of the issuer backs the security. Sometimes called a note if its maturity is less that 10 years.

DEBT SERVICE — Issuer's required payment of principal and interest on a debt. Usually computed on an annual basis until debt is repaid.

DEBT-TO-EQUITY RATIO —A measure of leverage in a corporation's financial structure. In other words, it is a ratio which measures the extent to which the firm has been financed by debt. Computation of the ratio varies. Less common: total long-term debt divided by stockholder's equity. More common: long-term debt plus par value of preferred stock divided by common stockholders' equity. Use of the term must be reviewed carefully to make sure which set of figures is being employed. The second method of calculation is more properly a measure of securities with fixed charges (bonds and preferred stock) against securities without fixed charges (common shares, or equity).

DECISION TREE —A graphical device for setting forth the relationships between decisions and chance events. (See Chapter VI).

DEDICATION OF RESERVES —The means of assuring an adequate supply of natural gas to a purchaser usually accomplished by means of a Natural Gas Purchase Contract between the producer and the consumer. A typical contract will provide that the "seller hereby dedicates to the performance of this contract all commercially recoverable gas located in, under, or hereinafter produced from the units, leases, and lands described in Exhibit 'A' (hereafter referred to as dedicated reserves)." Such contracts normally run for a period of at least twenty years. DECOMMISSIONING — The act of removing an offshore platform from drilling and production service either by physical removal, scuttling in place, or using it for some other non-oil purpose.

DEFERRED BONUS — Dollars paid to the landowner in return for his granting a lease, the payment for which is made in installments spread over a number of years, as distinguished from the usual mode of payment, which would be a lump sum on execution and delivery of the lease.

DEFERRED PRODUCTION PAYMENT — A production payment which does not commence until after a specific sum has been realized from production from the lease, or after a primary production payment.

DEFERRED TAX —A result of timing differences between recognition of a tax liability and its actual payment. This can occur when taxable income differs from financial book income, often because of different depreciation and amortization methods employed for tax and financial books. It can also occur when a company operates in more than its home country of incorporation, and the tax laws may differ somewhat between the various jurisdictions.

DEFLATORS —Factors used to convert all spending, or income to a base year , thereby netting out the effects of inflation.

DELAY RENTALS —The sums of money paid by the lessee to the lessor (subsequent to any bonus) drilling or production is not commenced on the property on or before the contractual date.

DELINEATION WELLS — Wells drilled in a stepout pattern from a discovery well to determine the extent of the oil or gas find and to establish the productive limits of the new reservoir.

DELIVERY—The satisfaction of a futures contract position by the tendering and receipt of the actual physical commodity (wet barrels).

DELIVERY MONTH — The month specified in a given futures contract for delivery of the actual physical spot or cash commodity (wet barrels).

DELIVERABILITY — A well's tested ability to deliver, or produce against the normal backpressure of a pipeline. This normally is not the same as the well's allowable production for proration purposes.

DEPLETION —Accounting term for a theoretical pool of funds set aside from a corporation's annual earnings for the replacement of a natural resource asset that is being used up. For example, oil from a well or reservoir, ore from a mine, etc. Because it is usually impossible to replace the asset currently as it is produced, the depletion allowance is usually given in the form of a deduction from taxable income of the corporation. Thus, it is a non-cash deduction from current revenue which provides for the recovery of a portion of the capitalized costs of an oil or gas, or other mineral property. Under the IRS Code in the U.S. a taxpayer who does not own a refinery or service station marketing outlets, nor produce over 1,000 BOPD, is allowed cost depletion or percentage depletion, whichever is greater. Depletion is computed on a property by property basis.

COST DEPLETION — This method computes depletion in relation to investment An accepted procedure is the units of production method in which capitalized items including acquisition costs are recovered in direct proportion to the amount of production during a given period.

PERCENTAGE DEPLETION — Statutory depletion allowance in the U.S. is a percentage of gross production income. Currently the allowance is 15% of gross revenue but limited to 100% of the net taxable income after deducting operating expenses and depreciation but before deducting depletion. Gross revenue is before state severance or production taxes. There is no limit to the aggregate percentage depletion which may be taken as to a property over its productive life.

DEPRECIATION — The theoretical sum of money that represents the loss of value of a tangible asset over time because of use, or obsolescence. The depreciation sum is deducted from the original purchase price of a fixed asset to provide its residual value for balance sheet accounting. It also comprises a non-cash deduction from current revenues which reduces taxable income. Several depreciation methods are used in the petroleum industry in determining corporate income and taxable income.

These methods are generally known as:
1. Straight line

2. Accelerated
    — Declining balance
    — Sum-of-the-Years-Digits (SYD)
    — ACRS (Accelerated Cost Recovery System)

3. Units of Production

DEMURRAGE — See Lay Time.

DERIVATIVE, FINANCIAL — See Financial Derivative

DERRICK—The load bearing structure, usually 100 feet, or more, in height used in drilling to support the drilling equipment and drillpipe. It is named for a famous 17th century hangman (Derrick) who practiced his trade in Tyburn, England
.

DERRICKMAN—The member of the drilling crew who works close to the top of the derrick as the drillpipe is being brought up out of the well or lowered into it.

DEVELOPMENT COSTS — Expenses incurred in drilling and developing a property for production.

DEVELOPMENT WELL — A well drilled in an area that already has producing oil or gas wells and has sufficient geological data to indicate a continuation of the production trend. Development wells are drilled for the purpose of completing the desired pattern of production. The likelihood of a development well being a producer is normally fairly high. FASB 19 defines them as those "wells drilled within the proved area of an oil or gas reservoir to the depth of a stratigraphic horizon known to be productive." A 'step-out well' drilled beyond the boundaries of a proven area under these rules are designated as 'exploratory wells.'

DILUTION—The effect on book value per common share, or earnings per share (EPS) if it is assumed that all convertible securities (bonds, debentures and preferred stock) are converted and/or all warrants and stock options are exercised.

DIRECTIONAL DRILLING—Drilling a wellbore at an angle to a predetermined subsurface target rather than vertically. Controlled directional drilling makes it possible to reach subsurface points laterally remote from the location at which the drill bit penetrates the earth.

DISCOUNT – l) As a noun: the dollar (or point) difference between the price of a security and its redemption value. For example, a bond with a face value of $1,000 is selling at a $50 discount if its current market value is $950. 2) As an adjective: used to describe a security selling at a discount or offered for sale at a discount. For example, Treasury bills are discount securities. 3) As a verb: the act of factoring the long-term effects of news into one's estimate of the present value of a security. For example, the stock has gone down in value because the market is discounting the news of a threatened labor strike. Also used as a past participle. For example, the news is discounted.

DISCOUNT RATE—A percentage rate at which future money is discounted. It is the interest rate used in the discounting process of economic analysis. It is the inverse of compounding and is used in computing the Present Value (or Worth) of a cash flow.

DISCOUNTED CASH FLOW — Cash flow that has been discounted time period by time period to determine the total cash flow's equivalent value at an earlier point in time, usually time zero. The resulting total discounted cash flow is then termed the Present Value (or Worth) of the cash flow.

DISCOUNTED CASH FLOW RATE OF RETURN (DCFR)—This is essentially internal rate of return (IRR). It is the compound interest rate whose discount factors will equate the present worth of a project's net cash flow to zero. It may also be thought of as the rate which causes the net sum of the discounted outflows and inflows to equal the net cash outlay in year zero. It is an important investment yardstick.

DISCOUNTED RETURN ON INVESTMENT (DROI) — This is the ratio of the project's Net Present Value to the Present Worth, or Value, of the project's Major Cash Investments (after-tax and after overhead and including investment and depreciation tax credits), discounted at some specified rate.

DISCOVERY WELL— An exploratory well that encounters a new and previously untested accumulation of petroleum; e.g. a successful wildcat well. A discovery well may be one that opened a new productive horizon in an established oil or gas field.

DISSOLVED GAS — Natural gases, mostly light molecular weight hydrocarbons which are dissolved in the liquid crude oil at the high pressure of the subsurface formation. Synonym: Solution gas.

DISSOLVED GAS DRIVE—A form of primary recovery of crude oil from the subsurface in which the evolution of dissolved gas responds to the lower pressure in the immediate vicinity of the producing well to drive crude oil out of the formation into the wellbore and thence to the surface.

DISTILLATE—Light liquid hydrocarbons, usually water-white or pale straw color and high API gravity, (about 60°) recovered from wet gas in an oilfield separator. The more modern term for this material is condensate, or natural gasoline.

DISTILLATION—A fundamental process of petroleum refining by which heat is applied to liquid crude oil causing it to vaporize and be separated into various components as they condense at different temperatures.

DISTRIBUTOR — A wholesaler who markets a product. In the petroleum industry, distributors are called salaried or commissioned agents if they work for a refiner, or jobbers if they are independents.

DIVIDEND—A distribution of cash from net profits (cash dividend), or shares of stock (stock dividend), made at the discretion of the Board of Directors to the equity share-holders of a corporation. Cash dividends are currently taxable to the recipient. Generally, stock dividends are not taxable in the U.S. at the time of receipt.

DIVIDEND PAYOUT RATIO — The amount of dividends relative to the company's net income or earnings per share. Companies with high dividend payout ratios are generally not considered as "growth" companies.

DIVISION ORDER — The instrument, signed by all parties in interest, which instructs the purchaser of oil and gas to be produced from a property as to how the proceeds of such sales are to be distributed to the various owners, whose fractional or decimal interests are set out therein; it includes all owners of the production proceeds, e.g. working interest, royalty, overrides and production payments.

DOODLEBUGGER—Oil explorationist who relies on unconventional aids, from tea leaves to chemical or electrical paraphernalia, and "black boxes." Sometimes doodlebuggers do strike oil. The term is also frequently applied to seismic crews within the language of the oilfield.

DOUBLE-DECLINING-BALANCE METHOD —Amethod of computing depreciation that accelerates depreciation in earlier years and reduces it in subsequent years. It starts with percentage of straight-line depreciation, doubles it, and applies the same percentage in subsequent years. For example, if straight-line depreciation is 10%, DDB is 20%.

DOWNSTREAM — Term applied to the refining and marketing phases of the petroleum industry. Petroleum is said to flow downstream from the wellhead to its final use. Upstream activities are concerned with finding and producing hydrocarbons in commercial quantities .

DRAFT — A negotiable instrument that will, if properly endorsed, transfer money from the account of the payer to the account of the payee. The significant difference between a check and a draft is that a check is an immediate debit on the books of the issuer at the time it is issued.Adraft is not technically considered a debit until it is presented to the issuer for payment.

DRAINAGE — Underground migration of crude oil or natural gas in the reservoir due to pressure sinks, or regional pressure drawdowns, caused by production from wells producing from the particular reservoir.

DRILLING CONTRACT — An agreement between the operator and drilling contractor setting forth the major items of concern to both parties for the drilling of a designed well at a specific location and a specified time of commencement. Contract terms may be set out primarily in feet of hole drilled, or in time on location, known as "day rate."

DRILLING MUD (Fluid)—Aqueous, or oil-based slurries or emulsions of varying densities, viscosities and thixotropies employed in rotary drilling, so called because the original drilling fluid was, indeed, mud. It circulates from its mixing pit down the drill-stem, flows into the hole through drill bit openings, and flows back up the annulus between the drill string and the sides of the hole into the mud pits. Its purpose is to carry cuttings to the surface, lubricate and cool the drill bit, stabilize the hole (thus preventing cave-ins), and, perhaps the most important, control sub-surface pressure through its weight. Subsurface formation pressures of 20,000 pounds per square inch have been encountered.

DRILLING PLATFORM — A permanent structure standing with its deck well above the surface of the water, with legs anchored to the ocean bottom, used to support offshore drilling equipment, from which wells are drilled and later produced.

DRILLING PROGRAM — A joint venture, partnership or other vehicle organized for the principal purpose of joint investment by multiple parties in exploration and/or drilling and development for oil, gas or other minerals.Acharacteristic of a "program" is that the prospects to be explored and drilled are not specifically defined at the start of the program. Programs may also be organized for, or include property purchases. Drilling programs may be offered to the public, or as a private placement. However, public offerings must be registered with the SEC unless otherwise exempted.

DRILLING RIG — All of the surface equipment used to drill a well.

DRILL VERSUS DROP—An incremental economic decision which compares the drilling of a well with the alternative of not doing so, and losing one's interest in the acreage.

DRILLER — The head man in charge of the drilling crew each "tour," or shift. The driller manipulates the controls and commands the operation.

DRILLPIPE — The tubular steel used to rotate the drill bit and circulate the drilling mud.

DRILL STRING — The complete column of individual drillpipe lengths, drill collars, stabilizers and other downhole equipment.

DRY GAS — Natural gas from which all liquid hydrocarbons have been removed, usually by processing through a "Gasoline Plant."

DRY HOLE — An unsuccessful well; i.e., one drilled to the objective depth without encountering commercial quantities of oil or gas. Synonym: Duster

DRY HOLE CONTRIBUTION — Money or interest in a property given to an operator as compensation for drilling a well on near by or offsetting acreage in which the contributor has no direct financial interest, but payable only in the event that the designated well reaches the agreed depth and is a dry hole. The guarantor of such payments receives well information whether or not the well is a dry hole.

DRY HOLE COSTS — A well's drilling expenses (excluding lease costs) up to the point of deciding whether or not to complete it for production. Up to that point in time the costs are essentially the same whether the well is abandoned as a dry hole or work continues in an effort to make it a future producer.

DUAL, OR MULTIPLE COMPLETION — A well which is completed for simultaneous production from two or more producing formations at different depths. This generally avoids the necessity and cost of drilling two wells. A dual completion, however, is more complex to complete, and also to produce.

DUBAI —A grade of crude oil (Approx. 36 API, 2% S) from the U.A.E. which is a major marker crude in the Persian Gulf.

DUBAI MARKET — The widely traded international forward market which emerged in the mid-1980's. Futures trading with in Dubai crude now exist in the IPE and Simex exchanges.

DYNAMIC POSITIONING—The technique of maintaining floating offshore equipment on location over the well by means of computer-controlled thruster motors, thus avoiding the need for anchors and allowing production in water depths too great for anchoring. The truster motors respond constantly to changes in wind, current, waves etc. in maintaining the unit in a constant position.

E

E&P (Exploration and Production) — The functions in the petroleum industry which are involved in the exploration and drilling for, and the production of crude oil and natural gas. This grouping of industry functions is also know as the "Upstream" (of the wellhead) portion.

EARNINGS—In financial terminology earnings is a synonym for net profit after tax, and often expressed on a per share basis.

EARNINGS PER SHARE (EPS) — Net income of a corporation after taxes and payments required to preferred shareholders. The term, primary earnings per share is used if the corporation also has convertible securities, or other common stock equivalents out standing. Fully diluted EPS also will reflect effect of conversion or exercise of stock options. For example, a company might have primary earnings of $4 per share and fully diluted earnings of $3.50. This means that right now all common shareholders have an EPS of $4, but if all the calls on the stock were exercised, the earnings would be $3.50. Calls do not include listed stock options which are contracts between a writer and a holder and are freely traded on the exchanges. Such transactions do not involve the issuing corporation.

EBITDA— Earnings Before Interest, Taxes, Depreciation and Amortization based upon annual report income statement data.

ECONOMIC INTEREST — Any interest or ownership of oil, gas or minerals in place, in which the owner must look solely to the proceeds derived from the extraction or production of such reserves, if, as, and when produced for a return on and of his capital. Types of economic interests common to the petroleum industry are:

CARRIED WORKING INTEREST — (See Carried Interest)

FEE INTEREST — An economic interest in fee acreage.

MINERAL INTEREST — An economic interest in the reserves in place without corresponding ownership of the surface of the land.

NET PROFITS INTEREST — An interest in a property which entitles the owner to receive a stated fraction of the net profits from that property as defined in the document creating the interest.

NET WORKING INTEREST — The full 8/8ths working interest of a property less the Royalty Interest and any Overriding Royalty and Production Payment interests, if applicable.

OVERRIDING ROYALTY INTEREST (ORR) — An economic interest created in addition to the royalty stated in the basic lease. ORR's can be made payable out of all, or a specified fraction of the revenue or production attributable to the net working interest. An override is limited by the duration of the lease under which it was created.

ROYALTY INTEREST — The share of minerals reserved in money, or in kind, free of expense, by the owner of the mineral interest or fee when leasing the property to another party for exploration and development.

WORKING INTEREST—The lessee's interest in a lease before deducting basic royalty and any overriding royalty and production payments. The full working interest thus includes the revenues due to the royalty owner which are normally spelled out in the 'Division Order' for the property. The working interest represents the lessee's participation in exploration, development and producing costs and any pooled, or joint interest activity involving the property.

ECONOMIC LIMIT—That point, both in time and producing rate, at which the net revenue to the working interest owner, of a well or producing property, before income tax, equates to its operating costs.

ENHANCED OIL RECOVERY (EOR)—Advanced recovery methods for crude oil which go beyond the more conventional secondary recovery techniques of gas injection and waterflooding. EOR methods currently being employed include steam injection, subsurface combustion, micellar, polymer or other surfactant flooding, miscible hydrocarbon displacement and carbon dioxide injection. Some low grade fields require one of the techniques even for initial production.

EFP (EXCHANGE FOR PHYSICALS) — A futures contract provision involving the delivery of physical product (wet barrels), which do not necessarily conform to contract specifications in all terms, from one market participant to another and a concomitant assumption of equal and opposite futures positions by the same participants.

EQUITY —Basically, the term refers to the titled ownership, right, or claim to an asset or group of assets. Equity can be defined as the residual value of a business after deducting any mortgages, and all of its liabilities. Since it represents the amount of the stockholder's ownership in the company, it comprises its capital stock, paid-in surplus and retained earnings. The term is also used for any security certificate which represents residual ownership in a corporation. If there is no preference in payment of dividends, the security is called common stock. Preferred stock is the designation employed for those securities to which dividend payment preference is granted.

EQUIVALENT BARREL—A unit representing one barrel of oil when the physical volume is natural gas. On a fuel value equivalency this is approximately 6 MCF per equivalent barrel.

EXAMINER —The title designating the person named by the Commissioners to preside over, and conduct hearings for the RRC of Texas. The title is often used in other states and jurisdictions, as well.

EMV (EXPECTED MONETARY VALUE, also EXPECTED VALUE) — The expected value of an operational outcome is the product of multiplying the conditional outcome by its probability of occurrence.

ESCROW —Assets placed with an independent third party to insure that all parties to a contract fulfill its terms. For example, a client purchases a contractually mutual fund at a sales charge of $500. The sales charge is wholly or partially refundable under certain circumstances. The $500 will be placed in escrow until the client either fulfills the conditions of the contract or rescinds it.

EURODOLLAR — Common term for dollars, either U.S. or Canadian, held by banks in European countries. Such dollars, originally received by European merchants for goods purchased by North American companies, are used to pay for inter-country trades or for petroleum purchases in which case, they become known as "petrodollars." When Eurodollars are used to pay for trade with the issuing country they revert to regular dollars on deposit with banks in North America. Many European debt securities are issued with payment promised in Eurodollars.

EXCHANGE AGREEMENT — An agreement by one producing or purchasing company to deliver oil, gas or NGL in terms of a specified daily volume, and ultimate total quantity, to another company at a specified location in exchange for oil, gas or NGL to be delivered by the latter to the former company at a different location. In the case of crude oil, the agreement may be set up because each has crude production closer to the other's refinery than to its own refinery, thus saving transportation. With natural gas, exchange agreements are more likely to be made because of proximity of gas production of one company to another firm's gas gathering system.

459

EXCISE TAX — A tax based upon any parameter other than profit. This would include ad valorem tax, property tax, use tax, production tax, severance tax, import tax, export tax, etc.

EXPECTED VALUE, or RETURN — The arithmetic mean or average of all possible outcomes where these outcomes are weighted by the probability that each will occur. (See EMV).

EXPENSE — Expenditure which provides a short term benefit (less than one year) and is deducted from revenue in the current period. Includes operating costs, noncash charges, exploratory dry hole expense, overhead and G&G, where they are not associated with a successful hydrocarbon discovery.

EXPENSE TRACKING AND CONTROL — Administrative and accounting procedures designed to ensure that expenditures are made wisely, and in accordance to plan.

EXPLORATION COSTS — Expenses incurred in performing geological and geophysical studies and in the positioning and drilling of test wells.

EXPLORATORY WELL — Generally speaking, wells drilled to find new production or to locate the limits of a hydrocarbon bearing reservoir. FASB 19 defines an EXPLORATORY WELL as a "well that is neither a development well, service well, or a stratigraphic test well as those terms are defined" by them. Synonym: Wildcat Well.

EXPONENTIAL (Constant Percentage) DECLINE—Production rate vs. time relationship wherein the change in production per unit time is a constant percentage of the producing rate.

EXPOSURE — Measures the amount of money outstanding at any point in time. Some companies seek to limit the amount of book assets in any one country.

EXTERNAL COMMON EQUITY — A new issue of common stock

F

FAILURE—An unsuccessful outcome due either to economic or physical causes; i.e., a dry hole or explosion. Probability of failure is one minus the probability of success.

FAIR MARKET VALUE —The value that an informed buyer would pay a willing seller for an item in an unrestricted market.

FARM-IN — If Company A wishes to earn an ownership interest in Company B's lands or leases, or to acquire assignment of the lease, it farms-in by drilling one or more wells, or performing some other pre-determined activity. CompanyBhas farmed out the land or lease, and may retain any of a variety of types and sizes of interests. Terms of such agreements vary considerably. Companies with substantial land positions may wish to farm out in order to prove up the acreage, i.e., have it explored and/or evaluated.

FARM-OUT AGREEMENT — A specific form of assignment in which the lessee grants a conditional interest to a third party usually in consideration for the drilling of a well within a specified time period on given acreage. This type of deal is frequently under taken where the lessee has a large block of acreage and does not wish to assume the whole cost or risk of exploring and developing it. In many instances this form of agreement will be between a major oil company who has assembled the large block and an independent operator who is happy to take on the acreage block already assembled. The terms: "Farmor" and "Farmee" are often used to denote the party granting the farmout, and the party taking the farmout, respectively.

FEE ROYALTY — The landowner's, or lessor's share of oil and gas production.

FEEDSTOCK — A refining term pertaining to the petroleum fraction, or raw material, introduced into a process unit.

FIELD — A geographical area with several producing oil or gas producing wells. In a single field there may be one continuous reservoir (petroleum accumulation) or several separate reservoirs at varying depths.

FINANCIAL ACCOUNTING STANDARDS BOARD (FASB)—An independent body of certified public accountants in the U.S. that studies bookkeeping methods and practices and publishes opinions about these practices. Publicly held corporations generally abide strictly by these standards in preparing their financial reports.

FINANCIAL BOOKS — The financial (accounting) records of a firm including its balance sheet and income statement. The financial books frequently differ in several important respects from its "tax" books.

FINANCIAL COMMUNITY — Popular term for the loosely related, but highly interdependent group of financial institutions including the very large commercial banks, investment banking houses, stock exchanges, exchange floor specialists, brokerage houses, security analysts, arbitrageurs, underwriters, and the regulators.

FINANCIAL CONSOLIDATION — Combining all financial holdings, both total and investment interest, and newly merged or acquired ownerships, at the corporate level.

FINANCIAL DERIVATIVE—A financial instrument (contract) that has no value of its own, but derive their value from the value of some other asset, i.e., stocks, bonds, currencies, and commodies. Generally, derivatives fall into two broad categories—contracts for future delivery at a specified price and options to buy or sell at a future date at a specified price. Derivatives may be listed on exchanges or traded privately.

FINDER'S FEE — Remuneration to a firm, or person who locates and refers business to another. For example, "The finder's fee for the referral of Company A to Company B in their merger was $500,000."

FINDING COSTS—The exploratory expenditures per barrel of oil equivalent spent in locating new reserves. The aggregate cost of all geological, geophysical, lease acquisition, exploratory drilling and other expenses required to find and prepare to produce a new discovery of oil or gas. Finding costs are generally expressed in dollars per barrel, per MCF, or per barrel equivalent of reserves discovered.

FISCAL YEAR — Any consecutive 12-month period of financial accountability for a corporation or other government agency. Often abbreviated FY with a date. For example, FY May 31. The firm's fiscal year goes from June 1 through May 31 of the following year.

FIXED ASSETS—Accounting term for assets of a corporation that are not readily saleable and which represent the depreciated value of property, equipment, and other tangible assets.

FLARING, OR VENTING — Flaring is burning of gas vented through a pipe or stack. A method of disposing of gas while the well is being completed, or afterwards, if there is no market or other use for it. Flaring is controlled by state and provincial regulatory bodies as a conservation measure and varies between jurisdictions. Venting, i.e., letting unwanted gas escape to the atmosphere unburned, is dangerous and generally prohibited.

FLASH REPORT—Periodic financial reports of a company, usually omitting a great deal of detail in order to expedite their completion and release.

FLOAT—Float is the time lag in the check-clearing process. Float may be advantageous to the check writer, or disadvantageous to the check depositor, depending on the number of days it takes for a check written to appear as a debit, or a check deposited to appear as a credit on the books of deposit of the banking institution. (See draft.)

FLOTATION COSTS — The underwriter's spread and issuing costs associated with the issuance and marketing of new securities.

FLOOR—A swap transaction in which the seller seeks to establish a minimum fixed price for his oil sales. This may also be accomplished through the use of futures and options.

FLOOR BROKER — An exchange member who executes orders for futures trades in the trading ring on the floor of a commodities exchange.

FOB (FREE ON BOARD)—A transaction in which the seller provides crude oil or product at an agreed unit price at a specified loading port within a specific time period. It is the responsibility of the buyer to arrange for the transportation and insurance and to lift the material within the allotted laytime.

FOOTAGE RATE — Drilling contract where the contractor charges strictly on the basis of footage drilled, no matter how little or how much progress is made on a given day. Depending on the terms this type of contract is usually advantageous to the operator or investor.

FORM 10K — Report that must be filed annually with the Securities and Exchange Commission in the U.S. by the reporting corporation. This includes all listed corporations and corporations with 500 or more share holders and with assets of $1 million or more. The SEC outlines information that must be included in report. Summary of 10k report must be included in the annual report that is to be sent to registered shareholders.

FORMATION FRACTURING (Fracing, Frac Job)—A method of stimulating production from an oil or gas well. A special fluid is pumped into the reservoir at very high pressure forcing open existing, or creating new fissures which facilitate flow to the wellbore. This is followed by pumping in propping materials(proppants) to keep the fissures open.

FORWARD MARKET—The process of buying or selling a specific crude oil for delivery at a future date. Not all forward market transactions result in physical delivery of the crude oil represented by the contract as the agreement for actual delivery may be canceled by subsequent selling of the contract. There is no clearing house, or marginal payment arrangement, in the forward market as distinct from futures trading in an organized commodities exchange.

FOSSIL FUELS — Those hydrocarbon fuels whose origin is believed to be plant and animal fossils, including petroleum, natural gas, coal, oil shale, tar sands, lignite, and peat.

FRACTION — One of several components obtained from petroleum through distillation or other refinery processes.

FREE WELL—A division of interest agreement in which one party drills one or more wells completely free of cost to the second party in return for an agreed economic interest in the property.

FULL CYCLE ECONOMICS—Performance of a project encompassing its entire life span regardless of the portion of the project already underway or completed.

FULLY DILUTED EARNINGS PER SHARE — Computation of corporate earnings applicable to each share of common stock that would be outstanding if all convertible and exchangeable securities were tendered for common shares at the beginning of the accounting period. If such conversion had not been made, and earnings were divided among the previously outstanding shares, the term "primary-earnings per share" is used.

FUNGIBLE—Any bearer instrument that is freely transferable and equal in all respects to any other instrument of the same class of the same issuer. For example, a dollar bill that you lend is paid back by another dollar bill. As a general rule, common stock of the same issuer is fungible. Other securities, with different maturities, call dates, or par value are not fungible. Fungible in the terminology of petroleum trading and transportation denotes interchangeability of crude oils or products that can be commingled for purposes of shipment or storage.

FUTURES CONTRACT — An agreement, reached in an organized futures trading "pit," to take or make delivery, at a future date, of a specified quantity and quality of goods, (e.g. crude oil, gasoline or heating oil) at a predetermined delivery point and at a price determined by public outcry.

FUTURES MARKET—A formal commodities exchange in which contracts for delivery of a specific type of crude oil and/or products in future months are traded. Only a very small proportion of the contracts result in actual delivery, and in some markets, there is only cash settlement. All of the markets include a clearinghouse and require daily margin payments on all positions in order to ensure the financial integrity of the operation. The market is open to all participants.

G

GAAP— Generally accepted accounting principles.

GANTT CHART — A chart showing the specific time sequences in which the activities that comprise a project are budgeted to be undertaken and completed.

GAS CAP — A gas producing layer above an oil producing layer in the same reservoir.

GASOHOL — A mixture of approximately 90 percent gasoline and 10 percent ethyl alcohol used as a motor fuel.

GASOIL — European designation for No.2 heating oil and diesel fuel.

GASOLINE—A light petroleum distillate with a boiling range of 30° to 100°C. It is produced for use in spark ignition internal combustion engines. It is also known in various parts of the world simply as gas, motor fuel, motor spirit, and petrol.

GASOLINE, STRAIGHT RUN — The untreated gasoline cut, or fraction, that is distilled directly from crude oil.

GAS OIL RATIO — The volume of natural gas at atmospheric temperature and pressure produced concurrently in association with a unit volume of crude oil.

GATHERING SYSTEM—A network of pipeline and other surface equipment employed to transport crude oil and/or natural gas to a central location.

GAUGER — The representative of the purchaser or pipeline who samples and measures the volumes of crude oil in lease tanks before and again after manually operating the valve in the storage tank which lets the oil flow into a truck or the pipeline company's gathering line.

GEOLOGICAL AND GEOPHYSICAL (G&G) — Costs of topographical, geographical, geological, and geophysical studies, rights of access to properties to conduct those studies and salaries and other expenses of geologists, geophysical crews and others conducting those studies.

GILT-EDGED—Term applied to the security of a corporation that has proved, over time, its ability to pay continuous dividends or interest. It refers more often to high-quality bonds than to equities.

GROUP OF EIGHT (G-8)—An organization of eight developed countries (Canada, France, Germany, Italy, Japan, Russia, United Kingdom, United States) Gulf Cooperation Council (GCC) — An organization of six Persian Gulf countries (Bahrain, Kuwait, Oman, Qatar, Saudi Arabia, and U.A.E.)

H

HEAVY CRUDE — Crude oil with a low API gravity (high specific gravity) due to the presence of a high proportion of heavy (high molecular weight) hydrocarbons.

HEDGING — The simultaneous initiation of equal and opposite positions in the cash and futures markets. Hedging is employed as a form of protection against adverse price movements in the cash market.

HUNDREDYEARSTORM —A combination of storm conditions (wave height and sustained wave conditions) that statistically, on average, should occur only once every hundred years in a specific area. Offshore structures are generally designed to withstand such a storm.

HURDLE RATE — The discount rate used in determining NPV. It must at least exceed the firm's cost of capital.

HYDROTREATING—A generic term for refinery processes designed to remove sulfur from crude oil by substituting hydrogen for sulfur in the molecular structure.

I

IDC — See Intangible Drilling Costs

IBOR—Inter-bank offering rate which is a key rate for international borrowing and lending; i.e., LIBOR (London) and KLIBOR (Kuala Lumpur, Malaysia).

IFRS — International financial-reporting standards

INCREMENTAL COST—The added cost attributable to a particular item or project versus some alternative.

INDEPENDENT — In North America the term generally applies to a non-integrated oil company, or individual operator, active in one or two sectors of the petroleum industry. These sectors, or functions, generally are considered as exploration and production, transportation, refining, and marketing. An independent producer might be active only in exploration and production. An independent marketer may buy product from a major or independent refineries and resell it under its own brand name. Internationally the term, Independent, is interpreted as meaning any company other than one of the multinational majors. Amerada Hess and Hunt would be considered independents.

INDEPENDENT PRODUCER — A production ownership status defined by the U.S. Tax authority (IRS) indicating a working interest owner who has less than 1.25 million dollars per calendar quarter in sales and less than 50,000 barrels of oil per day of refinery runs. Independent producers who have less than 1,000 BOPD of production can use percent depletion. (see DEPLETION)

INFILL DRILLING — The drilling of additional wells to increase the development density from say, 40 acres to 20 acres per well.

INFLATION—The rise in price charged for a good or service which is not attributable to any real change in the product. Inflation is generally measured and applied in economic forecasts as a percentage change over time. (See Chapter IV).

INJECTION WELL—Wells used in secondary recovery and enhanced oil recovery to inject or pump various fluids into the reservoir to increase recovery of oil or condensate.

INSIDER — Jargon for a person who (1) controls a corporation, (2) owns 10% or more of a company's stock, or (3) has inside information regarding the company, its financial condition and business prospects which is not entirely made public. The term is used most often of the directors and the elected officers of a corporation.

INSOLVENCY — The inability to meet interest payments or to repay debt at maturity.

INSTITUTIONAL INVESTOR — Industry term for an investor who, because of the size or frequency of transactions, is eligible for preferential commissions and other special transaction services. The term is often applied to banks, mutual funds, insurance companies, and large corporate investment accounts, although use of the term will vary from one brokerage firm to another.

INTANGIBLE DRILLING COSTS — Expenditures incurred by an operator for labor, fuel, repairs, hauling, and supplies used in drilling and completing a well for production. IDC's basically include any necessary cost associated with drilling which has no salvage value. IDC's can also be incurred in deepening or plugging back a previously drilled oil or gas well. This distinction is made primarily for income tax purposes.

INTEGRATION — An oil industry term denoting the degree in which one company participates in all phases of the petroleum industry, i.e., exploration and production, transportation, refining, and marketing.

INTEREST —
1. Financial— The cost of using money, expressed as a percentage rate per period of time, usually one year. Normally it is tax deductible as a normal cost of business.
2. Ownership—Share, right, or title in property, e.g. Mineral Interest, Working Interest, etc.

INTERNAL RATE OF RETURN (IRR)— Measures the effective rate of return earned by an investment as though the money had been loaned at that rate.

INVENTORIES — In the oil and gas producing industry inventories essentially consist of materials, such as casing and tubular goods for wells and lease flow lines, well heads and similar equipment and pipeline fill. Since production of oil and gas are moved off the property by pipeline there is little, if any buildup of production. This is in distinct contrast to most other types of business which produce a product.

INVESTMENT TAX CREDIT — A credit against taxes due in any given period on eligible investments. An incentive offered by governments to encourage capital investments.

INVESTMENT BANKER — A financial specialist who underwrites and distributes new securities and advises corporate clients about raising new funds. The investment banker, also known as a merchant banker in the U.K., is the middleman between the corporation issuing new securities and the investing public. Normally, one or more investment bankers agree to buy outright the new issue of stocks or bonds, and then resell it to individuals and institutions. IPE (International Petroleum Exchange) — The London oil futures market which trades in gas oil and Brent and Dubai crude oils and options.

ISOMERIZATION — A catalytic refining process whereby light hydrocarbons, primarily butanes, are merged into hydrocarbon molecules of the gasoline range.

J

JET FUEL— Kerosene type fuel; a quality kerosene product used primarily as fuel for turbojet aircraft engines.

JOINT VENTURE—An arrangement by which two or more parties, one of which may be a government, agree to cooperate on a project. This grouping of organizations differs from a standard partnership. In a joint venture, costs and profits are shared proportionately. Each co-venturer retains control over its own interest, including the right to sell it. Each venturer has unlimited liability only as to its interest. One party serves as the operator who is required to keep the other owners currently informed, and depending upon the terms of the agreement may also be required to obtain prior authorization from the co-venturers for certain types or magnitudes of expenditures on the joint property. When one of the joint venture parties is a government, their share of exploration and development expenditures are usually "carried" by the other parties, to be recovered out of the government's share of production.

K

KEROSENE— A refined petroleum distillate with a boiling range between gasoline and gas oil. It was the principal product of crude oil in the early days of the industry used as fuel for illuminating lamps. It is still used extensively in Developing Countries as stove and cooking fuel. Most of the industry's kerosene fraction is now sold as jet (aviation) fuel.

KICK—Awell is said to "kick" when the pressure encountered in a formation penetrated by the bit while drilling exceeds the hydrostatic pressure exerted by the column of drilling mud circulating through the hole. If uncontrolled, a kick can very quickly lead to a blowout.

KILL A WELL—To overcome a high-pressure surge by loading the hole, that is, by adding weighting materials to the drilling mud.

KNOCKOUT — A tank or fluid trap used to separate some, but not necessarily all, produced water from crude oil.

L

LANDMAN—A person whose primary duties are to secure oil and gas leases, and to maintain company relations with mineral and royalty owners. In North America he might also have been known as a "lease hound." In the international oil business he is known as a "negotiator."

LANDED COST — The actual cost of crude oil to a refiner including all costs from the wellhead, or other purchase point, to the refinery.

LAY TIME—Tanker charters provide for a certain number of days at the owner's expense in port for loading or unloading known as lay time. If the lay time, for example, is 12 days and the time counted in port is 15 days three days are charged to the charterer of the vessel as demurrage.

LEASE—The legal instrument by which a leasehold is created in minerals. Thus it is a contract that, for a stipulated sum, conveys to the operator the right to drill for and produce oil and gas. An oil lease should not be confused with a real estate lease of land or a building. In Alberta, where essentially all of the minerals are owned by the Crown, and somewhat different terms are employed with respect to obtaining and holding oil and gas rights, the drilling people frequently use the term, 'lease' to denote the land-cleared drillsite.

TOP LEASE—The term used to denote a lease taken on a property by a third party prior to the expiration of the primary term of the existing lease.

LEASE ACQUISITION COSTS—Bonus payments to obtain the lease. In Canada such costs resulting from competitive bidding have, in practice, been treated as nondeductible capital expenditures.

LEVERAGE, or GEARING—The ratio of total undiscounted future revenue, divided by the capital investment. It is sometimes considered to be a kind of reward/risk parameter. It has a high multiplier for the winner, but like playing the options market, has an equally high multiplier for the loser. The degree of leverage can have significant effect on per-share earnings of the common stock of a company if large amounts of the cash flow must be paid to bondholders as interest or to the owners of preferred shares as preferential dividends.

LEVERAGED COMPANY—A business enterprise whose capital structure comprises both equity and debt. Industrial corporations with more than one third of their capital in the form of long-term debt may be said to be highly leveraged.

LIBOR (London interbank offered rate)—The rate at which the most credit worthy banks can borrow money from eachother in several different currencies for a period of time from 1 day to 1 year. This is based upon a daily survey of 16 London banks, some of which are branches of banks from other countries. This is a key rate for international borrowing and lending.

LIFTING — Refers to the loading of petroleum into tankers or barges at a terminal or transshipping point.

LIFTING COSTS — Synonym for operating expenses, that is, those incurred in operating a lease and the equipment on it to produce oil and gas after exploration and development has been completed.

LIGHT CRUDE — Crude oil with a high API gravity due to a high proportion of lighter molecular weight more volatile hydrocarbons.

LIGHT ENDS — The more volatile petroleum fractions including propane, butanes and gasoline.

LIMIT—The maximum amount a futures price may advance or decline in any one day trading session.

LIMITED PARTNER—Amember of a partnership who has no vote in the management of the partnership. Limited partners are also called silent partners. Their potential loss is limited to their capital contribution, and usually they receive a fixed dollar return that is payable in full before the general partner shares in any profits.

LINE OF CREDIT—A banking arrangement that enables a prospective borrower to obtain a maximum sum of money at any time within a predetermined time period. For example, "The ABC Bank granted its client a $1 million line of credit for 6 months."

LINEAR DECLINE RATE—An exponential, constant percentage straight-line decline in oil production (log.) vs. time.

LIQUIDATION — 1) The process of converting securities or other assets into cash. 2) The dissolution of a company, with any cash remaining after sale of its assets and payment of all of its debts and obligations being distributed to its shareholders. 3) Closing out of futures and options positions.

LIQUIDITY — A measure of the ease with which assets may be converted into cash. A commodities market is said to be "liquid" when it has a high level of trading activity.

LIQUID ASSET—Cash and marketable securities are generally considered liquid assets. The expression is commonly used in the analysis of corporate financial statements, but it is also used of brokerage clients to determine the suitability of certain relatively illiquid investments for the portfolios. For example, "This investment is not suitable unless the client has $50,000 in other liquid assets."

LIQUIDITY RATIO — An infrequently used ratio that compares cash and marketable securities owned by a company to its current liabilities. It is probably the most strict of the measurements of corporate liquidity.

LIQUIFIED NATURAL GAS (LNG) — Natural gas (methane) which has been liquified by reducing its temperature to minus 258°F at atmospheric pressure.

LOAD OIL — Liquid hydrocarbons used in remedial operations on an oil or gas well, particularly oil injected into a well during fracturing operations.

LOG—
(1) A graphical presentation of the lithologic and/or stratigraphic zones traversed by a borehole.
(2) The similar presentation of the variation of some physical property in a well with depth, such as resistivity, self-potential, gamma ray intensity, sonic velocity, or hydrocarbon content of the mud returns or cuttings. (3) The record of formations penetrated, drilling progress, record of depth of encountering oil, water, gas, or other minerals, the record of size, specification and depth of casing set, and other recorded facts having to do with the drilling of a well.

LONG POSITION—Signifies ownership of securities. In the futures market, the position of a futures contract buyer whose purchase obligates him to accept delivery unless he liquidates his contract with an offsetting sale.

LONG-TERM DEBT — 1) A debt instrument with more than one year to maturity. 2) In financial statement analysis, long-term debt is variously used. It always means more than 1 year to maturity in balance sheet bookkeeping, but analysts may change their usage of long term to cover bonds of 5 or 7 years, or longer terms.

LPG (LIQUIFIED PETROLEUM GAS) — Principally propane and butane sometimes referred to as 'bottled gas.'

M

MAJOR OIL COMPANY— An arbitrary term usually defining a large integrated oil company with extensive operations in all sectors of the petroleum industry, both upstream and downstream, and whose marketing brand name is broadly recognized.

MAJOR CASH EXPENDITURES—This is a fairly common budgeting term which includes capital expenditures plus exploration dry hole and G&G costs.

MARGIN—1)Funds or good faith deposits posted during the trading life of a futures contract to guarantee fulfillment of contract obligations; 2)The operating profit representing the difference between refined product prices and the crude oil feedstock costs.

MARGINAL REINVESTMENT RATE—The IRR of the least acceptable project of average risk which the firm will accept for new investment.

MARGINAL WELL — A low capacity well, or one having a small productivity and approaching its economic limit of operation. Generally synonymous with the term, 'stripper well.' Marginal wells are specifically defined in some governmental statutes and may receive special allowables and prices.

MARKER CRUDE — A crude oil of widely known quality, which is widely traded, and available in good supply (generally considered to exceed 300,000 BOPD) against which other crude oils of lower or higher quality can be related in price.

MAXIMUM CASH OUTLAY (MAXIMUM CASH OUT-OF-POCKET)—The maximum after-tax cumulative net cash deficit over the life of a project.

MEDIAN — The value of a variable that divides the group into two equal parts—one comprising all values greater, and the other all values less than the median.

MER (MAXIMUM EFFICIENT RATE)—The highest rate at which a well or reservoir may be produced without causing physical waste in the reservoir. In practice, economics are usually considered and an MER below the minimum rate at which a well can be economically produced is not considered.

MERGER — Popular term for the combining of two or more businesses into a single operational entity. Merger generally implies a friendly union and states nothing about the name of the resulting company. Acquisitions may be friendly or unfriendly and generally result in the loss of corporate identity for the acquired company.

MIGRATION—The movement of oil or gas through the sub-surface, especially with reference to movement across lease boundaries or from one portion of the reservoir to another. The rate of movement varies with the permeability of the rock, the viscosity of the oil, existing pressure gradients and other factors.

MINERAL RIGHTS—Ownership of minerals under a tract (including crude oil and natural gas), which encompasses the right to explore, drill and produce such minerals, or assign such right in the form of a lease to another party. Such ownership may or may not be severed from land surface ownership. Title in fee simple means all rights are held by one owner; the fee in surface owner does not hold mineral rights. Minerals is loosely used to refer to mineral owner-ship and even, incorrectly, to royalty ownership. A mineral acre is the full mineral interest under one acre of land.

MINIMUM ROYALTY—A payment to be made regardless of production, or producing rate. Such payments are normally chargeable against future production, if and when, which accrue to the royalty interest.

MODE — The size of the variable that occurs most frequently within a distribution.

MONEY MARKET—1) An abstraction for the dealers and their communication network who trade short-term, relatively low risk, securities. 2) The securities traded in the money market consisting principally of securities such as U.S. Treasury bills, bankers' acceptances, commercial paper, and negotiable certificates of deposit.

MON (Motor Octane Number) — (See Octane Number)

MONTE CARLO METHOD—A technique employed to simulate random processes which provides a method for generating outcomes from the probability distribution.

MUD ENGINEER — An engineer, or technician who studies and supervises the preparation and modification of various fluids and emulsions, collectively called "mud," as used in rotary drilling.

N

NATIONAL OIL COMPANY (NOC) — An enterprise entirely owned and operated by the state in an industrial and commercial endeavor established for the exploitation of petroleum.

NATURAL GAS—Gaseous forms of hydrocarbons, principally methane, with minor amounts of ethane, propane, butanes, pentanes and hexanes-plus along with non-hydrocarbon impurities such as nitrogen, carbon dioxide and hydrogen sulfide.

NATURAL GAS LIQUIDS (NGL) — Hydrocarbons found in natural gas which may be extracted and sold separately as ethane, propane, normal and iso-butane or simply as natural gasoline.

NET ASSETS — An accounting term relating the dollar difference between a corporation's assets and its liabilities. When used in this sense, net assets also are called net worth, or stock holders' equity.

NET CURRENT ASSETS —The dollar difference, if a positive number, between the current assets of a corporation and its current liabilities. The quotient of dividing current assets by current liabilities is called the current ratio. Thus, if a corporation has current assets of $5 and current liabilities of $2, its net current assets are $3 and its current ratio is 2.5. Also called NET WORKING CAPITAL.

NET LIQUID ASSETS — The dollar difference, if any between a corporation's cash and readily marketable securities and its current liabilities. Thus, a corporation with $5 in cash and $5 in marketable securities and current liabilities of $7 has $3 in net liquid assets. Net liquid assets is the strictest measure of a corporation's ability to meet its current debt obligations.

NETBACK —A petroleum industry term referring to the price a refiner is willing to pay for a barrel of crude oil recognizing refinery configuration and efficiency, anticipated product prices and profit objectives.

NET INCOME — The remainder after deducting royalty and all expenses from gross revenue, exclusive of DD&A and capital.

NET PRESENT WORTH — Measures the capital created over and above the company's hurdle rate (or cost of capital).

471

NET PROFIT —The dollars remaining from revenue after deducting all related cash expenses and non-cash charges for the period, including income taxes.

NET PROFITS INTEREST —An interest in a property which entitles the owner to receive a stated percentage of the net profit as defined in the instrument creating the interest. He receives a share of the funds remaining after royalties and operating expenses have been paid. Like the overriding royalty (see below), it is "carved out" of the working interests and continues for the life of the lease.

NET REVENUE INTEREST (NRI) —The interest owner's share of production revenue. This may be represented by the fraction of the income the working interest owner receives as compensation for his drilling, development and production expenditures. For example, the net revenue interest to a lessee whose lease carries a normal 12.5% royalty and is burdened with a 2.5%ORRis 85 percent. In other words, the lessee owns a 100% working interest and an 85% net revenue interest.

NETWORTH —The dollar value by which assets exceed liabilities. Also called stockholders' equity when used on corporate balance sheets.

NON-ASSOCIATED GAS — Gas produced from reservoirs which contain hydrocarbons in only the gaseous phase at reservoir temperature and pressure. Some of these hydrocarbons may be produced as a liquid (condensate) due to a phase change caused by temperature and pressure reductions as they are brought to the surface

NONCASH CHARGES — Book charges against revenue, usually for the purpose of determining income taxes and/or financial book profit, which do not involve actual expenditure of cash during the current period.

NON-OPERATING INTEREST — Such owners are without operating rights; they do not bear any part of the drilling or production costs and have no control over these activities.

NON-RECOURSE DEBT — A form of borrowing in which the general assets of the borrower's firm are not exposed to claims of the lender. His claims are limited to the revenue generated by the particular project for which he furnished funding.

NOTE — Adebt security with a relatively limited maturity, as opposed to a long-term bond. The expression "note" is commonly employed for government securities with maturities of from 2 to 10 years. Privately placed corporate debt securities and non-callable shorter-term debt securities are commonly called notes. Municipal notes may be very short term or they may have maturities up to 2, 3, or even 4 years. In practice, a note is a debt security so-called by the issuer. No specific time limitations for maturity can be given as a universal rule.

NYMEX (New York Mercantile Exchange) —which includes among its commodity futures trading contracts in WTI (West Texas Intermediate) crude oil, home heating oil, propane, natural gas, and motor gasoline.

O

OCTANE NUMBER — A measure of anti-knock characteristic of a motor fuel measured in a standard internal combustion engine, with variable compression ratio, by varying the behavior of the fuel being tested (appearance of knock) with that under identical conditions of blends of two reference fuels. By definition the unit of knock intensity (octane number) is the volume percentage of iso-octane which must be mixed with n-heptane in order to match the knock intensity of the motor fuel undergoing test in a standard test engine. Higher octane ratings are more desirable in high performance engines due to the higher antiknock qualities. RON and MON (Research and Motor Octane Number) are measures of octane number determinations under simulated mild and severe service conditions, respectively.

OFFSET — The elimination of a current long or short position by the opposite transaction in a commodity exchange. A sale offsets a long position; a purchase offsets a short position.

OFFSET WELL — (1) A well drilled on an adjacent location to the original well. The distance from the first well to the offset well depends upon the state or provincial spacing regulations, and whether the original well was classified as an oil or gas well. (2)Awell drilled on one tract of land to prevent the migration of oil or gas onto an adjacent tract where a well has been drilled and is producing.

OIL-INDEXED LOAN —A borrowing in which repayment is tied to the level of designated crude oil prices.

OIL-INDEXED SECURITY — An equity interest, or debt instrument, sold to the investing public in which the return to the investor is linked, at least in part, to the published price of oil over a period of time.

OIL, OR PRODUCTION PAYMENT — A fractional share of production, free of costs of production, terminating when a specified dollar amount or volume of production is reached.

OPEC — ORGANIZATION OF PETROLEUM EXPORTING COUNTRIES — (See Chapter XI).

OPEN OUTCRY — The method of public auction for making verbal bids and offers for contracts in the trading pits or rings of a commodities exchange.

OPERATOR — The party holding all or a fraction of the operating, or working, rights in a property or lease and designated as manager of operations, such as exploration, drilling, production. If the operator is a contractor, the term used is contract operator.

OPERATING AGREEMENT — An agreement between concurrent co-owners for the operation of a property or group of properties. The joint operating agreement does not result in the creation of a separate tax entity. It must provide that each co-owner has the right to take his share of production in kind. These agreements often are undertaken among the owners of adjacent properties providing for their joint operation under the terms specified. In such cases each owner has the power to convey, and to make operating decisions as to his currently owned interest, except those powers which are limited by the stipulations of the agreement.

OPERATING COSTS — All costs related to production activities including fuel, materials, operating labor, maintenance and repairs. Workover costs incurred to maintain or increase production from an existing completion are also charged to expense.

OPTIONS —The right to buy or sell a commodity at a specified price in the future. There are two types of option: call and put. The former gives the buyer the right to purchase a contract at a certain price, before a certain date. The latter provides the buyer the right to sell a contract for a specified price before a certain date. After purchasing a specific option the buyer is thus able to choose whether or not to exercise the option depending upon future market conditions as they develop.

ORGANIZATION FOR ECONOMIC COOPERATION AND DEVELOPMENT (OECD) — A 24 nation organization based in Paris. (Austria, Australia, Belgium, Canada, Denmark, Finland, France, Germany, Greece, Iceland, Republic of Ireland, Italy, Japan, Mexico, Netherlands, New Zealand, Norway, Portugal, Spain, Sweden, Switzerland, Turkey, United Kingdom, United States).

OVERHEAD —All costs principally for services that are not directly identifiable with specific operations, but which are necessary costs of doing business, e.g. administrative, accounting, personnel and other staff department expense.

OVERRIDING ROYALTY (ORRI) — This type of interest differs significantly from the standard landowner's royalty in that it is payable to someone other than the mineral owner, often as landman's or geologist's incentive and/or reward. Carved out of operating interests, it pays its owner a share of gross production revenues free of exploration, drilling and production costs. Its term is coextensive with the operating interest from which it is derived.

P

PAPER BARRELS — A generic term for crude oil or products traded in forward or futures markets which are, or may be closed out by subsequent sales or settlements without physical, or "wet" delivery. (See Wet Barrels).

PARAFFINS —Aclass of hydrocarbons in which the carbon atoms in each molecule are linked in a straight chain as distinct from olefinic, or naphthenic ring configurations.

PAR VALUE — l) Used in connection with fixed-income securities to designate the value of bonds at maturity, or the price at which preferred shares may be redeemed, unless a premium price is established. Also called the Face Value. 2) Sometimes used with common shares as an identifier. For example, the $10 par common shares. This value may also be used on the balance sheet to designate the stated value of outstanding common shares. Par value is not a measure of market value for common shares. 3) Par Value may have tax implications in some jurisdictions.

PARTICIPANT — (1) In a Unit, an owner of an oil and gas interest who has contributed his interest to a drilling, exploratory, or fieldwide secondary recovery Unit. (2) A purchaser of an interest in a drilling program, fund, prospect, or property. In cases where the program is a limited partnership, participants are limited partners of the program; where a joint venture, participants are co-owners or venturers.

PARTICIPATION —1) The fractional interest owned in a Fieldwide Unit, or in a drilling or exploratory unit. 2) The fractional ownership, or number of units purchased by a participant in a Drilling Program, or Fund. 3) In a swap deal, "participation" refers to a type of arrangement in which the provider, i.e., the financial institution, grants the customer a share in some of the gains from market movements which favor the customer.

PARTICIPATING AREA —The part of a unitized area to which production is allocated as set out in the unit agreement.

PARTNERSHIPS

    GENERAL PARTNERSHIP —A business form in which the partners as joint co-owners share the management, profits and losses, assets and liabilities, which pass through to the individual co-owners. Each co-owner is subject to unlimited liability in person for the indebtedness and other liabilities for the entire partnership.

    LIMITED PARTNERSHIP — A business form having a General Partner and Limited Partners. The General Partner is responsible for the operation of the partnership, makes all decisions, manages the business without limitation, including making contracts and other commitments on behalf of the partnership. The General Partner has unlimited liability as to the partnership. The partnership agreement sets forth the obligations of the General Partner and the allocation of income and expenses as between the partners, general and limited. Limited Partners have no voice in management or operations and are liable only to the extent of their capital contributions. Profit and loss are allocated to partners on some sharing arrangement and flow through to the individual partners for income tax purposes.

PAYOUT (PAYBACK PERIOD) —Aneconomic term referring to the period of time required for the cumulative cash position of a particular ownership interest to reach zero.

PERFORATIONS — Series of holes made in casing and cement through which oil and gas flows from formation into wellbore and up to the surface. Holes are usually made with a shaped explosive charge placed in the wellbore and detonated from a surface location near the wellhead.

PERMEABILITY—A physical measure of how easily fluids may flow through porous rock; expressed in millidarcies.Atight rock, sand or formation will have low permeability in the order of tenths, or even hundredths of a millidarcy, and thus have very low capacity to produce. Wells in zones of this nature usually require fracturing or other artificial stimulation. An extremely permeable formation may measure several darcys of permeability.

PERT (Program Evaluation and Review Technique) —A planning and control program for defining and integrating activities and the time needed for each to meet a complex objective. It is usually presented in a network type of flow diagram. (See CPM).

PLUG AND ABANDON — The operation of sealing a well bore hole with cement and/or drilling mud as required by the government regulatory body in order to discontinue all further drilling and producing operations at the particular location and permanently abandoning a well.

PLUGBACK — To plug the bottom of a well with cement in order to produce from another more shallow formation.

POOL — A common but now less frequently used term for an oil field. (See Reservoir.)

POOLING —The consolidation or unitization of small tracts of land to form a drilling unit as required by state or provincial well spacing regulations.

POROSITY — Pore space in rock which enables it to hold or contain fluids; measured in percentage of void space, which can vary from near zero to 35% or more.

POSTED PRICE —The publicly posted price a major purchaser of crude oil is willing to pay per barrel for oil produced from a particular area or region.

POUR POINT —The temperature below which a petroleum fraction or product ceases to flow.

PRE-EXPORT FINANCING — A procedure for raising capital, most commonly used by exporting countries, in which loans are received as advance payment on future oil sales.

PREFERENTIAL RIGHT TO PURCHASE — A prior right of purchase specifically reserved to buy an oil and gas interest by meeting the terms and conditions of a proposed sale of that interest to any other party.

PREFERRED STOCK — A special class of stock (equity) frequently created as the result of a merger or acquisition. The dividend is fixed much like bond interest and may have priority over common stock dividends. (See Cumulative Preferred Stock). It may be senior to common stock in dissolution. It may be convertible into common stock at a set price. It may not have voting rights.

PRESENT WORTH (PRESENT VALUE) — The current value of future cash flows discounted at a specific interest rate.

PRESSURING AGENT — A hydrocarbon, usually normal butane, used to bring gasoline motor fuel blends up to an acceptable vapor pressure.

PRICE EARNINGS RATIO — The simple arithmetic ratio of the current market value of a share of capital stock to the per-share net earnings.

PRIMARY RECOVERY—Production from a reservoir, through flowing or pumping wells, because of the existence of natural energy within the reservoir. Recoveries of 10% to 35% of the original oil and gas in place are normally expected from primary production. The term may also refer to an engineering estimate of what the reservoir would have produced had secondaryrecovery measures not been implemented.

PRIME RATE — A preferential rate of interest on short-term loans offered by commercial banks to their most credit-worthy customers. Theoretically, it is the lowest rate of interest for bank loans that are not backed by items of value pledged as collateral.

PRIMARY TERM —The period specified in the lease during which it can be kept in effect by the lessee through the payment of rentals, or by drilling even though no oil or gas is being produced. The primary term of most leases is typically three to ten years.

PROBABILITY — Numerical estimate of the likelihood of an event occurring. (See Chapter VI).

PRINCIPAL AMOUNT — 1) The face amount or par value of any debt security. 2) The face amount of a loan.

PRODUCTION —The term used to denote the part of the petroleum industry that deals with the extraction of oil and gas from the ground; also, the volume of oil and/or gas produced, e.g. daily production, monthly production, flush production, settled production, primary production, etc.

> **GROSS PRODUCTION** — The total production from a property irrespective of how it may be divided up among the various economic interests.

> **WORKING INTEREST PRODUCTION** — Gross production multiplied by the fractional working interest which is owned. When the working interest is 100 percent, working interest and gross interest are the same value.

> **LEASEHOLD PRODUCTION** —Working interest production less royalty. This caption is equivalent to a working interest owner's net production.

> **NET PRODUCTION** — Production derived from ownership in a well equivalent to:
> (a) Working interests less royalty and any other outstanding interests, and/or
> (b) Other economic interests such as overriding royalties, net profit interests, etc.

> **SETTLEDPRODUCTION** —The production of an oil well or reservoir during the period when its decline in productivity is rather slow as distinguished from its initial period of FLUSH PRODUCTION.

> **ULTIMATE PRODUCTION** —The total accumulated production of oil and gas that is recovered from a well or reservoir during its productive life.

> **INDUSTRY BASIS PRODUCTION** —The total production in an area regardless of how it is divided among its various economic interests at the individual property level. This category is sometimes used to total a company's gross interest production in a particular state, province or producing basin.

PRODUCTION PLATFORM — An immobile, off-shore structure from which wells are drilled and/or produced. In general, wildcat wells are drilled from floating vessels (semisubmersible rigs and drill-ships) or from jackups, whose legs rest on the ocean floor; development wells are usually drilled from platforms.

PRODUCTION SHARING CONTRACT — An arrangement which divides the petroleum production between the host government and the contracting company after allowing a portion for cost recovery. It may also include a royalty provision.

PRODUCTION TAX (SEVERANCE TAX) —Atax levied by some U.S. states on oil and gas production, either as a percentage of the gross value, or as a specified amount per barrel of MCF. Another local tax is the 'ad valorem' tax which is not a production tax per se but a tax on the value of an oil property and/or its equipment. In some states and provinces the ad valorem tax is measured on the value of current production, in others it is on the basis of estimated reserves in the ground, and in some instances only on the value of the surface facilities.

PROFIT OIL —That portion of the production revenue under a production sharing agreement which remains for distribution to the parties after deduction of the cost oil. (See Cost Oil)

PROJECT FINANCING —Atype of non, or limited recourse financing in which the funds are loaned to the project and the lender does not have direct claim on any other assets of the owners of the project. This has become a popular method of financing the major offshore production installations in the North Sea.

PRORATION — As applied in the petroleum industry, the term pertains to the allocation of permissible oil and gas production among the properties or wells producing from a common reservoir, or among the fields in a particular state or province on some agreed and enforced basis. The factors most frequently employed as the bases for proration are acreage, number of wells, well potential, or a combination of these and other parameters.

PRORATION UNIT — An area attributed to an individual well by the regulatory agency in order to control the development drilling in an oil or gas field. It also becomes the acreage factor if one is employed in determining the well's allowable for proration purposes.

PROSPECT — A property or group of properties on which an operator intends to drill an exploratory well or a geologic anomaly of interest.

PROSPECTUS —The formal printed legal summary of a registration statement filed with, and approved by the SEC in conjunction with the public offering of nonexempt securities for a proposed new commercial venture. The prospectus, which contains the material information about the offering of securities, must be given to the original purchasers of the security no later than the confirmation date of their purchase.Acompany must be careful about what information is given out while financing is in process, otherwise it may be alleged to be trying to sell its securities without using a prospectus approved by the ruling regulatory agency.

> **PRELIMINARY PROSPECTUS** — A disclosure filing of a public offering under an S-1 registration with the SEC. It is called a "Red Herring" because of the required red print on the margin of the first page stating that the registration has not yet become effective and the securities being registered may not be sold or offers to buy accepted prior to the time the approved registration statement becomes effective.

> **(Effective) PROSPECTUS** — The disclosure statement of a public offering whose registration has become effective. Securities may then be offered through the prospectus as authorized and bought, sold and traded.

PUT —A type of option which gives the user the right to sell a contract on a futures market a commodity at a given price for a specified period of time in the future. (See "options" and "call.")

Q

QUALIFIED OPINION —A statement by the company's outside auditors making reference to an item or situation that may have material negative effect on the company's financial status.

R

RACK PRICE —Price charged by a supplier to a customer who buys transport truck lots at a terminal, on an FOB basis.

RANDOM NUMBERS —A set of numbers in which the probability of each number of the set occurring is equal to that of any other. Pseudo random numbers generated by a computer, or obtained from published tables are employed in the Monte Carlo Method of risk analysis.

RATE ACCELERATION PROJECT —Additional field development work incremental to a field's existing depletion plan undertaken for the predominant purpose of shifting the existing plan's expected production profile forward in time, rather than increasing the field's ultimate production.

RATABLE TAKE — This term refers to the orders sometimes issued by state or provincial regulatory agencies requiring common carriers, or purchasers (usually pipelines) to take production from all operators in proportion to the number of wells connected, or the capacity of those wells. Ratable take orders are normally instituted where total well capacity exceeds pipeline capacity to insure that all sellers are able to market their fair share of the oil or gas.

RATE OF RETURN — A measure of profitability of a business, or project recognizing financial returns in relation to the amount of money invested. There are a number of detailed ways of computing this economic decision tool. (See Chapter V)

RECOMPLETION —To rework a well so that it produces from a different formation, either shallower or deeper than the original completion. It is a capital expense in contrast to a well repair, or workover which leaves the well completed in the same zone or formation, and is treated as an expense.

RECOVERY FACTOR —That fraction of the total oil and/or gas in place in a reservoir which is indicated to be recoverable by known production techniques.

RED HERRING — Industry jargon for a preliminary prospectus. The name arises from the caveat, printed in red along the left border of the cover of the preliminary prospectus, warning the reader that the document may not contain all of the information about the issue and that some of the information may be changed before the final prospectus is issued following final review by the SEC. Also called red herring prospectus.

REFINERY TURNAROUND — The periodic shutdown of a complete refinery for maintenance and possible modification. This is often an annual undertaking.

REFINING MARGIN — The difference between the revenue from sale of products derived from crude oil and all the costs involved in its refining; including the price of the crude, freight, insurance, financing expense and refinery operating costs.

REGULATORY AGENCIES —Most states and provinces have an oil and gas commission to regulate the industry. Their names and responsibilities vary; one of the busiest is the Texas Railroad Commission, which guides operations in that heavily drilled state. Operators typically are required to apply for permits and file reports for all phases of drilling through transporting and refining. Regulations can cover the spacing of wells, amount of oil and gas that can be produced, as well as procedures for plugging and abandoning. Reports are required for spudding, depth, testing, completion or recompletion. In some states operators may request their reports to be kept confidential (see tight hole) for a certain period of time. In Alberta all data from wells on crown lands are made public after one year. Production data are generally public information in all jurisdictions.

RELIEF WELL — A well drilled to a high-pressure formation to control a blowout.

RESERVES —The unproduced but recoverable oil and/or gas in place in a geologic formation whose existence has been proven by production. Reserves are estimated by engineering studies as of a specific date, and may be classified as follows:

> **GROSS RESERVES** — Total reserves disregarding the economic interest breakdown.

> **NET RESERVES** — Reserves derived from ownership of
>   a) working interest less royalty or other outstanding interests and/or
>   b) other economic interests such as overriding royalties, production payments, etc.

> **INDICATED ADDITIONAL RESERVES** — Reserves which cannot yet be considered as "proved," but which are expected to be recoverable from properties already on production under the following situations:

>> (1) Known reservoirs expected to respond favorably to fluid injection and other improved recovery techniques where:

>>> (a) An improved recovery technique has been installed on the property but its effect cannot yet be fully evaluated; or

>>> (b) An improved technique has been installed but knowledge of reservoir characteristics and the results of a known technique installed in a similar situation are available for use in the estimating procedure.

>> (2) Natural gas liquids to be recovered from plant for plant improvements which are planned but not yet installed.

>> (3) Other situations where a memorandum reserve item is justified but where proved status has not yet been achieved.

> **POTENTIAL RESERVES** — Expected reserves at outset of drilling.

> **PROVED DEVELOPED RESERVES** — FASB19 defines Proved Developed Reserves as those which can be expected to be recovered through existing wells with existing equipment and operating methods. The term includes:

480

(a) proved developed producing reserves (those that are expected to be produced from existing completion intervals now open for production in existing wells), and

(b) proved developed nonproducing reserves (those that exist behind casing of existing wells or at minor depths below the bottoms of such wells, which are expected to be produced through these wells in the predictable future, where the cost of making such oil and gas available for production should be relatively small compared to the cost of a new well).

PROVED NOT DEVELOPED RESERVES — FASB 19 refers to reserves which are expected to be recovered from new wells on undrilled acreage, or from existing wells where a relatively major expenditure is required for recompletion. Reserves on undrilled acreage shall be limited to those drilling units offsetting productive units, which are reasonably certain of production when drilled. Proved reserves for other undrilled units can be claimed only where it can be demonstrated with certainty that there is continuity of production from the existing productive formation.

WORKING INTEREST RESERVES — Gross reserves times the fractional working interest owned. (When the working interest is 100 percent, working interest reserves equal gross reserves).

RESERVE/PRODUCTION RATIO (R/P) — The ratio, expressed in years, of booked reserves at year-end to production during the year.

RESERVOIR — A porous underground rock formation in which hydrocarbons have accumulated. Synonym: Trap

RESERVOIR VOLUME FACTOR (also Formation Volume Factor) — The ratio of the volume of a barrel of crude oil at reservoir pressure and temperature with its dissolved gas to the same barrel at atmospheric conditions.

RETAINED EARNINGS — Book earnings which are not paid out as dividends, but are retained by the company as additional investment in the business on behalf of the shareholders. Retained earnings comprise a part of the firm's equity capital.

RETURN ON EQUITY — The net income available to the common shareholders. It is most frequently stated as a ratio relating earned income to the common stockholder's investment.

The return on common equity is more specific. It uses net income minus preferred dividends divided by stockholders' equity after excluding the par value of outstanding preferred shares. Common abbreviation is ROE.

RETURN ON INVESTED CAPITAL —Measurement used in financial analysis to evaluate the rate of return on all sources of longterm capital. It is most commonly calculated by adding interest paid on bonds to net income after taxes and dividing by the par value of long-term bonds plus total stockholders' equity, that is, by the total capitalization of the company. The exact application of the formula may vary somewhat from industry to industry. Common abbreviation is ROI.

RETURN ON INVESTMENT (ROI) — An economic yardstick measure of project performance.
ROI — Cumulative net income over the Project Life divided by the total project investment.

REVERSIONARY INTEREST — This has similarity to a carried interest but differs in the sense that the type of interest held by each party changes after a specified set of conditions have been fulfilled. A typical example might be found in a farmout agreement. Lessee A might retain a 1/16th of 7/8ths ORR until B recovers $195,000 from drilling a successful farmin well. After this recovery, the override reverts to perhaps a one-half working interest in the well and lease, and the override is terminated. Under such an arrangement, A is said to have a reversionary one-half interest in the property.

REVOLVING LINE OF CREDIT — Term used in banking if a bank establishes a line of credit for a customer to be used when, and if needed. There often is a commitment fee paid by the potential borrower.

RING FENCE — A term denoting a provision in a host country's tax and accounting regulations which prevents charging the costs of exploration and other expense from a project which was unsuccessful against another similar project which was successful in finding oil or gas in commercial quantities.

RISK — The possibility of incurring economic loss or a reduced economic value. High risk is the chance of incurring a large loss, even if the probability of doing so is very small.

RISK ADJUSTMENT — The adjustment of the cash flow stream of a proposed project to reflect the consequences of varying probabilities that the project may be only partially as rewarding as the forecast.

ROACE — Return on average capital employed which is the accounting net income (profit) for the year divided by the average accounting assets of the company for the year. Accounting assets are original cost minus cumulative depreciation.

ROCE — This is the same ratio defined as ROACE, except that the asset value is the year end balance sheet value.

ROLL UP — A tax-free exchange of a partnership interest for a proportionate share interest in a corporation. Because partnership interest was taxed annually, most of the cash flow from this exchange is considered a return of the owner's capital and is not subject to tax.

RON (Research Octane Number) — See Octane Number

ROUGHNECK — A laborer who is a member of a drilling crew.

ROUSTABOUT — An oilfield laborer.

ROYALTY — Payments made to the landowner or mineral interest owner for the right to explore for, and produce petroleum after a discovery is made. Royalty payments are usually a percentage of the gross production. Royalty is free of any cost of development or the ongoing operation.

RUN TICKET — A record of the volume of crude oil transferred from a lease storage tank to a pipeline.

RVP (Reid Vapor Pressure) — See Vapor Pressure.

S

SALVAGE VALUE — The fair market value of an asset at the end of a specified period of usage. This may be scrap iron when an asset is no longer usable.

SBM (Single Buoy Mooring) —Used for loading, or unloading, tankers in the open sea. Also sometimes called single point mooring (SPM). The technique permits the loading or un-loading of tankers at locations where water depths prevent their approaching the shore. TheSBMpermits the tanker to moor and load or unload oil whatever the direction of wind or current and to swing at its mooring to present the least resistance to the prevailing conditions.

SCOUT —An individual who observes and reports on competitors' leasing and drilling and collects other desired information.

SECONDARY RECOVERY — Methods employed to increase the recovery of oil from the reservoir in a second effort after all that is economically recoverable by pumping the wells under primary depletion has been accomplished. The most common of these is waterflooding.

SECURITY —An instrument, usually freely transferable, that evidences ownership (stock) or creditorship (bond) in a corporation, a federal or state government, or its agency, or a legal trust. The courts also have included evidences of indirect ownership in the definition. For example, rights, warrants, options, and partnership participations. The term also is used of property, whether a security or not, pledged as collateral for a loan. For example, an issuer of bonds might pledge his plant and equipment as security for an issue of mortgage bonds.

SECURITY AND EXCHANGE COMMISSION (SEC) — The regulatory bodies in the federal governments of the U.S. and Canada, and its ten provinces, which oversee the issuance and trading of securities.

SEISMIC SURVEY —Petroleum exploration with the seismograph which detects the return of induced vibrations in the earth in order to locate potential petroleum-bearing formations.

SENIOR SECURITIES — Debt instruments that have first claim on a firm's assets (secured debt) or cash flow (unsecured debt) in the event of liquidation.

SENSITIVITY—Additional evaluation of a project with different values of the most sensitive assumptions.

SERVICE CONTRACT AGREEMENT —An agreement between a host country and an oil company under which the company engages in exploration and development under a contract, for an agreed fee, or other compensation.

SERVICE WELL — FASB 19 defines a service well as a "well drilled or completed for the purpose of supporting production from an existing field." Wells in this class are drilled for the specific purpose of: the injection of water, gas, steam, NGL, or air; for salt water disposal, water supply, subsurface observation, or for use in an in-situ combustion project.

SETTLEMENT PRICE — The most representative price prevailing in a given commodity futures contract near the close of daily trading. Settlement prices are used to determine margin calls and invoice prices for deliveries (of wet barrels).

SHORT POSITION — The position of a futures contract seller whose sale obligates him to deliver the commodity unless he liquidates his contract by an offsetting purchase.

SHRINKAGE —The term used to denote the reduction in volume of the natural gas production stream due to the extraction of natural gas liquids, NGL.

SHUT-IN — To close the valves at the wellhead so the well stops flowing or producing. A shut-in well may be waiting for a pipeline connection, for a gas sweetening plant to be built, or for the market to improve. It also simply refers to a well which has produced its month's allowable and is shut-in for the remainder of the period.

SHUT-IN ROYALTY —Payments made when a gas well, capable of commercial production is shut in for lack of market for the gas. This type of royalty provision, or some form of rental is usually required to prevent termination of the lease.

SIMEX (Singapore International Monetary Exchange) —A futures market which trades oil contracts in Dubai crude and high sulfur residual fuel oils.

SLIDING SCALE ROYALTY — A royalty which varies with the level of production, for example, a one-eighth royalty if the production is 100 BOPD or less, increasing to a 3/16ths royalty if the production is greater than 100 barrels per day.

SOUR CRUDE —Crude oil containing sulfur or sulfur compounds in sufficient concentration to require their removal. The following guidelines represent general classifications:

CRUDE	SULFUR PERCENT BY WEIGHT
Low Sulfur	Up to 0.5%
Medium	Sulfur 0.5% to 2.0%
High Sulfur	Greater than 2.0%

SOUR GAS —Natural gas which contains hydrogen sulfide. A sweetening plant must remove the H2S before the gas can be sold. Gas is generally considered sour if it contains 10 or more grains of H2S or 200 or more grains of total sulfur per MCF.

SPOT —A one-time open market transaction in which a commodity is purchased "on the spot" at current market rates. Spot purchase transactions generally involve individual cargoes or other reasonably small lots of crude oil or products. Spot sales are distinct from contract, or term, sales which specify a steady stream of the production or product over a period of time.

SPREAD —The monetary relationship between two prices, either for different crude oils, or for the same grade of oil at different time periods. These price differentials lie at the heart of most current oil trading on the exchanges since they tend to be less volatile than the absolute prices themselves.

SPUD — To start the drilling of a well.

"STANDARDIZED MEASURE "OF DISCOUNTED FUTURE NET CASH FLOWS —An Annual Report table required by the SEC of all U.S. oil companies and those non-U.S. oil companies whose stock is traded on a U.S. Stock Exchange. The preparation and format of the table is dictated by FASB 69. (See Chapter VIII).

STEP-OUT WELL —A well drilled adjacent to a productive tract but located in an area still unproven.

STOCKHOLDER'S EQUITY — The difference on the firm's Balance Sheet between the company's total assets and total liabilities. It represents the shareholder's ownership in the company and is sometimes referred to as the net worth or book value of the company.

STOCK —The shares of a company (equity) representing ownership, comprised of COMMON STOCK and PREFERRED STOCK. The number of "authorized" shares of a company stock is approved by the shareholders.

> **TREASURY STOCK** — Common stock which has been repurchased by the company from shareholders and carries no voting rights and pays no dividends. It may be used for acquisitions, stock options, or other corporate purposes.

> **AUTHORIZED UNISSUED STOCK** — Common stock shares which have been authorized by the shareholders but not yet sold or used for other corporate purpose. These shares carry no voting rights nor pay dividends until they are disbursed.

STRAIGHT RUN PRODUCTS — Any product of a refinery produced by simple fractional distillation.

STRAIGHT-LINE DEPRECIATION — A conservative accounting procedure that apportions a corporation's cost of a qualified asset over its useful lifetime in equal annual amounts. The amount to be depreciated is the actual cost of the asset when purchased. The annual amount of depreciation is subtracted from fixed assets on the balance sheet and from operating income on the income statement. Because depreciation is part of the normal cost of doing business, it reduces the company's tax liability. (See also Cash Flow.)

STRATIGRAPHIC TEST WELL — An FASB 19 term for a drilling effort, geologically directed, to obtain information pertaining to a specific geologic situation. Such wells are drilled usually without any intention of completing them for hydrocarbon production (e.g., C.O.S.T. wells).

STREET NAME —Popular term for a security registered in the name of a broker-dealer. For example, "The security is in street name." Securities owned by the broker-dealer are in street name, and securities owned by customers that have been deposited with the broker-dealer normally are register-ed in the name of the broker-dealer to facilitate transfer when the security is sold. Customer-owned securities in street name are said to have the broker-dealer as the nominee and the customer as the beneficial owner, i.e., the owner of the security and all of the rights pertaining thereto.

STRIPPER WELL — An oil well which according to I.R.S. definition yields 10 or fewer barrels of oil per day during any prior 12-month period. Stripper wells are generally pumped only intermittently, allowing time for the crude oil to accumulate in the wellbore during rest periods. A gas stripper is a designated gas well that produces no more than 60,000 cubic feet of gas a day during a 90-day production period and no more than one barrel of oil along with the gas.

SUNK COSTS — Those costs which have already been incurred, and which therefore do not affect yet-to-spend economics for a given project except for their effect on subsequent years.

SWAP — These are financial risk-management techniques in which two parties exchange differing market risk exposures in order to assure a known fixed price, usually for a period of several years. Short term swaps arrangements are more likely to be held and managed directly by a financial house for a fee.

SWEET CRUDE — A crude oil with a relatively low sulfur content, in contrast with a high sulfur "sour crude."

SYNTHETIC OILFIELD —An instrument devised by the Wall Street financial community to mimic the payout of an oilfield with regularly scheduled payments for a given volume of oil at current prices over a period of years. The purpose this type of arrangement is to provide an instrument for hedging in order to provide a firm supply of crude for many years into the future with little price volatility.

T

TAIL GATE — The delivery point for residue gas sales after plant processing.

TAKE-OR-PAY — A provision in gas sales contracts requiring the purchaser to pay for a minimum annual contract volume whether taken or not; the purchaser may, or may not have the right to take gas in succeeding years without additional payment, gas previously paid for but not taken, termed "make-up gas."

TANGIBLE, OR CAPITAL COSTS —Partnership costs required to be capitalized for federal income tax purposes in the U.S. (and recovered through depreciation and depletion), such as expenses for lease acquisition and related geological and/or geophysical work; all equipment, purchased for drilling, completion and production. Expenses related to dry holes or leases abandoned without production are generally considered non-capital costs. (See definition of CAPITAL ASSET)

TANK BATTERY — A group of at least two crude oil production storage tanks, usually of 1,000 barrel capacity each, located near the producing well(s) in an oilfield. A "Tank Farm" generally denotes a group of much larger storage tanks located at a junction or gathering point on an oil pipeline system.

TANKERS —Seagoing vessels for the transport of petroleum and its products. VLCCs (very large crude carriers) are tankers of 200,000 DWT, or larger. Vessels of over 300,000 DWT are generally referred to as ULCCs (ultra large crude carriers).

TARIFF — A schedule of rates charged by a pipeline for transporting crude oil or products.

TAX BOOKS — This is a broad term which refers to tax liability records of the firm. Differences frequently develop between financial and tax books, primarily due to timing differences in recognizing, or booking revenues and expenses including tax obligations, which may not be due until some time in the future depending upon the regulations which apply to the specific situation.

TEMPORARILY ABANDON — A well shut in and not likely to be returned to production without workover or recompletion, although a decision has not been made as to its possible Permanent abandonment.

TOP LEASE —A lease granted by a landowner during the existence of a recorded lease which will become effective if, and when the existing lease expires or is terminated.

THERMAL CRACKING —A petroleum refining process employing high temperatures and pressures to convert large hydrocarbon molecules into smaller ones in the gasoline motor fuel range.

THERMAL RECOVERY—Methods of increasing well productivity and recoverable oil by heat. Present techniques include:

1) intermittent steam injection, called "huff and puff" or "steam soak,"

2) continuous steam injection, or "steam flood," and

3) in-situ combustion, in which air is injected and a portion of the reservoir oil is burned to heat the formation. Thermal methods are more adaptable to low gravity oils (below 20 degrees API).

THIEF —Adevice lowered into a storage tank through the opening on the roof of the tank called a "thief hatch" in order to obtain a sample of the oil in storage. The term, "thiefing a tank" pertains to the actual sampling operation.

TIGHT HOLE —Drilling and completion information, and, in particular geologic data from a well is not publicly released by the operator. Frequently secrecy is maintained so that the operator may acquire surrounding acreage before news of a new discovery drives up the price. Tight holes are rumor mills.

TIGHT FORMATION —Ageologic zone of low permeability and thus low well productivity. Wells in such zones usually require formation fracturing or other stimulation. Typically, the productive capacity of a new well completed in a tight zone declines rapidly for several months or longer after the initial completion. Tight Formation Gas receives special proration and pricing treatment in some regulatory jurisdictions.

TOOLPUSHER — A foreman or supervisor in charge of one or more drilling rigs.

TOUR — (pronounced "tower") An eight-hour work shift at a drilling site.

TRANSFER PRICES — Prices set for the sale of crude oil and products between affiliated companies within a vertically integrated enterprise. These transfer prices do not necessarily relate directly to costs or external market forces and may ameliorate the tax burden in certain jurisdictions.

TREASURY STOCK — (See Stock).

TRIPPING — Pulling the drillpipe from the hole to change the bit and then running the drill string and new bit back in the hole. On deep wells, round trips or a trip may take 24 hours, that is, three eight-hour tours, or shifts.

TUBING—Anarrow diameter (often 2") pipe hung inside the casing of a well which facilitates the flow of the produced fluids, oil, gas, water, etc. to the surface.

487

TURNAROUND —The planned, periodic shutdown of a refinery for inspection and overhaul usually of a week or more duration.

TURNKEY—A style of drilling contract in which the contractor, for a fixed amount, furnishes all labor, materials and equipment, and does all the work necessary to drill a well, to casing point or to completion, so all the owner has to do is "turn the key." These agreements are generally used in areas where drilling problems are minimal and predictable, and costs fairly uniform. This type of contract differs from the normal drilling contract which bases the price on an agreed price per foot of hole drilled and/or a daily standby fee, with the operator assuming responsibility for the well completion and equipment.

U

ULCC (ULTRA LARGE CRUDE CARRIER) — See Tankers

UNDERWRITER — General industry term for a person who facilitates the public sale of securities by an issuer. The name is used as a synonym for an investment banker. It is also used to denote the sponsor, also called the wholesaler, of investment company securities. Specifically, an underwriter is any person who purchases securities from an issuer for purposes of resale.

UNIT —

1) The area incorporated in a field or reservoir unitization agreement.

2) DRILLING UNIT combining two or more small tracts of land so that there will besufficient acreage to qualify under applicable well spacing requirements.

3) PROGRAM UNIT The basic fractional or dollar amount offered for purchase to an investor in a drilling program.

UNITS OF PRODUCTION BASIS — Method of allocating depletion, depreciation, or amortization based on the number of barrels of oil or MCF of gas produced in a year as a fraction of the oil and gas reserves at the beginning of that year.

UNITIZATION —The process whereby the owners of adjacent properties pool their reserves and form a single unit for the operation of the properties. The operating costs, applicable overhead charges and revenues from the operation of this unit are divided on the basis established in the documentation of the Unit Agreement. The purpose of such an agreement is to produce the reserves more efficiently and thereby increase the profit to all of the Unit Participants. Unitization can be particularly important where secondary or enhanced recovery programs are anticipated.

UNSECURED LOAN — A loan made on the overall credit of the borrower, who does not pledge any assets as collateral to the lender.

UPSTREAM —Refers to the production of crude oil and natural gas, as distinguished from the downstream operations of refining and marketing.

V

VLCC (VERY LARGE CRUDE CARRIER) — See Tanker.

VAPOR PRESSURE —The measure of the surface pressure necessary to keep a liquid from vaporizing. RVP stands for Reid Vapor Pressure which is the numerical measure of vapor pressure by his method.

VAPORLOCK —The phenomenon of insufficient flow of motor fuel through the fuel pump of a gasoline engine due to its inability to pump liquid-gas mixtures.

VENTURE CAPITAL — A financial industry term for investments in new, untried business ventures with all of the financial risks inherent in such enterprise. Individuals or companies specializing in such investments are called venture capitalists and, as part of their compensation for such investments, usually demand a large portion of the equity ownership. Thus, if the risky enterprise prospers, they will be richly rewarded.

VISCOSITY — A measurement of a liquid's resistance to flow, usually decreasing with increasing temperatures.

W

WTI (WEST TEXAS INTERMEDIATE) CRUDE OIL — The principal 'marker' crude in North America. It is the contract grade for futures trading on the NYMEX. (See Chapter IV).

WATER DRIVE —A natural reservoir depletion mechanism which occurs because oil floats over water and artesian pressure pushes the oil to the pressure sink of the producing wells.

WATERFLOOD —Asecondary recovery method in which water is pumped through injection wells located in a pattern around the producing wells so as to sweep the moveable oil to them.

WELL COMPLETION — The activities involved in preparing a newly drilled well for production. This involves setting and cementing the production string of casing, perforating the casing, running the production string of tubing, hanging the well control valves (miffling up the Christmas tree), connecting the flow lines, and setting the lease storage tank battery.

WELLBORE — The hole in the earth drilled by a bit.

WELLHEAD —The installed equipment on the top of the tubing to maintain surface control of the well.

WET BARREL — An actual barrel of crude oil or product already physically in storage or afloat at the time of a given transaction, as opposed to a paper barrel, which appears only as a credit on an accountant's ledger. The term is used to distinguish a trading transaction which involves actual volumes of crude oil or product from forward and future trading in "paper barrels." (See paper barrels).

WET GAS — Natural gas that contains a relatively large amount of easily condensible hydrocarbons such as propane, butanes, pentanes and heavier.

WILDCAT — A popular term for an exploratory well.

WINDFALL PROFITS TAX — A taxation scheme in which the government takes an increasing share of the revenue from oil and gas production depending on the price obtained above a base level.

WORKING CAPITAL — Refers to that amount of capital in current use in the day-to-day operation of the business. It is probably best described as the firm's investment in short term assets, i.e., cash, short-term securities, accounts receivable, and inventories. Gross working capital is generally defined as the firm's total current assets. Net working capital comprises current assets minus current liabilities. If the term 'working capital' is employed without further qualification, it is understood to mean gross working capital.

WORKING INTEREST — See Interest

WORKOVER — Remedial operations on a completed well with the hope of restoring or increasing production from the same zone. This includes such work as deepening, plugging back, squeeze cementing, reperforating, fracturing, cleaning out, acidizing, oil squeezing, casing repair, etc. A workover to recomplete to another producing formation is called a Recompletion, and receives different accounting treatment from the well repair workovers, all of the later being expensed. Synonym: Well repair.

WORLDSCALE — (Worldwide Tanker Nominal Freight Scale) A schedule of shipping rates published by an independent body, covering costs of tanker transportation between two ports. The basic rate established for any given voyage expressed in U.S. dollars per ton is referred to as WS 100. Actual contract shipping rates are subject to negotiation with the agreed figure expressed as a fraction, or multiple of WS 100.

WRITE-DOWN (OR WRITE-OFF) —The company's accounting recognition of the loss of value of an asset beyond ordinary DD&A. It is a non-cash charge against income for that accounting period.

Y

YIELD — A general term for the percentage return on a security investment. Although the context often will qualify the term, it is important that the precise meaning is clear. For example, nominal (coupon) yield, current yield, yield to maturity, or yield to average life, and so on. (See also DIVIDEND, and RATE OF RETURN.)

Also, the volume of condensate recovered from a natural gas stream as a ratio to the initial volume of gas.

Z

Z-FACTOR (Gas Deviation Factor) —The measure of the amount that a "real" gas deviates, in terms of temperature and pressure, from perfect behavior of "ideal" gas which conforms to the Perfect Gas Law.

ZONE — A somewhat indefinite term generally denoting a stratigraphic interval in the subsurface having common rock characteristics.

ABBREVIATIONS EMPLOYED FREQUENTLY:

AFE	Authorization for Expenditure
AFIT	After Federal Income Tax
AFRA	Avg. Freight Rate Assessment
API	American Petroleum Institute
APR	Annual Percentage Rate (interest)
bbl	Barrel
BFIT	Before Federal Income Tax
BO	Barrels of Oil
boe	Barrels of Oil Equivalent
BOPD	Barrels of Oil per Day
BOPWD	Barrels of Oil per Well per Day
BPSD	Barrels per Scheduled Day
BD	Barrels per Day
BWD	Barrels per Well per Day
BROR	Book Rate of Return
BS&W	Basic Sediment & Water
BTU	British Thermal Unit
CAPEX	Capital Expenditures (Cash)
CIF	Cost, Insurance & Freight
DD&A	Depreciation, depletion and amortization
DROI	Discounted Return on Investment
DWT	Dead Weight Tonnage
e	Base of natural logarithms (2.7182818)
E&P	Exploration and Production
EDROI	Expected discounted return on investment
EIRR	Expected internal rate of return
EPS	Earnings per Share
EMV	Expected Monetary Value
EV	Expected Value
FASB	Financial Accounting Standards Board (U.S.A.)
FOB	Free on Board
FVIF	Future Value Interest Factor
FVIFA	Future Value Interest Factor of an Annuity
G-8	Group of eight developed countries
GCC	Gulf Cooperation Council
GOR	Gas Oil Ratio
IBOR	Inter-Bank Offering Rate
IDC	Intangible Drilling Costs
IPE	International Petroleum Exchange
IRR	Internal Rate of Return
LIBOR	London Inter-Bank Offering Rate
Ln	Natural logarithms (to the base e)
LPG	Liquified Petroleum Gas
MCF or MSCF	Thousand Standard Cubic Feet Gas
MER	Maximum Efficient Rate

MIPS	Monthly Income Preferred Securities
MMCF or MMSCF	Million Standard Cubic Feet Gas
NCF AT	Net cash flow after income tax
NCF BT	Net cash flow before income tax
NGL	Natural Gas Liquids
NIO	Net Income from Operations
NOC	National Oil Company
NPV	Net Present Value
NYMEX	New York Mercantile Exchange
OCIBT	Operating cash income before income tax
OECD	Organization for Economic Cooperation and Development
OPEC	Organization of Petroleum Exporting Countries
OPEX	Operating Expense
ORRI	Overriding Royalty Interest
P&A	Plug and Abandon
psi	Pounds per Square Inch
psia	Pounds per Square Inch, Absolute
PVIF	Present Value Interest Factor
PVIFA	Present Value Interest Factor of an Annuity
PVNCF	Present Value of Net Cash Flow
RI	Royalty Interest
ROI	Return on Investment
ROACE	Return On Average Capital Employed
ROCE	Return on Capital Employed
SBM	Single Buoy Mooring
TD	Total Depth (of a well)
ULCC	Ultra-Large Crude Carrier
VLCC	Very Large Crude Carrier
WI	Working Interest
WS	World Scale
WTI	West Texas Intermediate

REFERENCES

Langenkamp, R.D., *Illustrated Petroleum Reference Dictionary,* 2nd Ed., 1982, PennWell Publishing, Tulsa

Glossary of Words and Phrases Used in the Oil, Gas and Petrochemical Industries, 1980, Phillips Petroleum Company, Bartlesville, OK

—plus many other sources over the years, all of which are gratefully acknowledged.

NOMENCLATURE

Errors that last the shortest time are always the best.

—Molière

CHAPTER II

a = nominal decline rate, decimal per year
a_i = initial decline rate, decimal per year
Boi = initial reservoir volume factor for oil, RB/STB
e = base of natural logarithms (2.71828)
d = effective annual decline rate, fraction per year
h = exponent of hyperbolic decline, decimal between 0 and 1.0
Ln = natural logarithm
N_p = cumulative oil production, Bbl
q_a = producing rate at abandonment, BOPD
q_i = original production rate at time = 0, BOPD
q_t = production rate at end of time = t, BOPD
qec.lim. = production rate at the economic limit, BOPD
q_1 = original production rate at time = 0, BOPD
q_t = production rate at time = t, BOPD
R/P = beginning year's reserves divided by year's production
t = time, years
t_a = producing time to economic limit, years

Subscripts: i, t, and a designating original conditions, conditions at time "t", and final or abandonment conditions respectively

CHAPTER III

I_I = general inflation rate (decimal/year)
I_E = escalation rate above inflation (decimal/year)
p_{io} = value of crude oil at the base year ($/bbl)

p_{to} = value of crude oil at end of year, "t" (\$/bbl)

t = time, years

CHAPTER IV

C_C = capital investment in "constant dollars"

C_{cf} = capital investment in "Constant dollars" at some future date, \$

C_D = capital investment in "dollars of the day"

E_C = operating expenses in "constant dollars"

E_{cf} = operating expenses in "Constant dollars" at some future date, \$

E_D = operating expenses in "dollars of the day"

I_I = general rate of inflation, fraction per year

I_{Ex}, I_{Ey}, I_{Ez} = effective annual escalation rate of items x, y, and z above the general rate of inflation, fraction per year

I_{Ia}, I_{Ib}, I_{Ic} = general rates of inflation for time periods a, b, and c respectively

PI = annual productivity improvement (cost reduction), decimal

p_i = average price, or annual cost, for initial year

p_t = average price, or average cost, for year, "t"

p_{tx}, p_{ty}, p_{tz} = average price or cost of items "x, y, or z" in year "t"

p_{ix}, p_{iy}, p_{iz} = average price or cost of item "x, y, or z" in initial year

t = time, years

CHAPTER V

A = annual payment

$DROI$ = discounted return on investment

d = effective annual decline rate, fraction per year

e = base of natural logarithms

i = nominal rate of interest per period, expressed as a decimal

$i_{const.}$ = interest at zero inflation rate, decimal/year

I_I = national inflation rate, decimal per year

IRR = internal rate of return

j = nominal annual interest or discount rate, decimal

Ln = natural logarithm

M = monthly payment

m = number of interest compoundings per year

NPV = net present value

n = number of compounding periods

PIR = profit on investment (undiscounted), decimal

PV = present value of what is in brackets, dollars

$PVIF_{i,n}$ = present value interest factor for i and n

$FVIF_{i,n}$ = future value interest factor for i and n

$PVIFA_{i,n}$ = present value annuity factor for i and n

$\text{FVIFA}_{i,n}$ = future value annuity factor for i and n

R_1 = annual revenue or cash flow for the first year, dollars

ROI = return on investment, undiscounted

t = time, years

v_f = future value

v_{fn} = principal + accrued interest at time, "n"

v_i = the initial principal

v_p = principal (PV) amount

X = $(1 - d)(1 + I_I)/(1 + i)$, dimensionless

CHAPTER VI

DHC = all costs which would be incurred before the project is determined to be a failure (Dry Hole Cost), $

$E(C|F)$ = expected cost of failure, $

$E(I|S)$ = expected investment for successful outcomes, $

$E(P|S)$ = expected profit conditioned on success, $

NPV = expected net present value of all successful cases (Net Present Value),$

P_{cum} = cumulative probability at a value, decimal

P_f = probability of failure = $(1 - Ps)$, decimal

$P(x)$ = probability of event "x" occurring, decimal between 0 and 1.0

$P(y)$ = probability of event "y" occurring, decimal between 0 and 1.0

P_R = risk capacity, probability

P_s = probability of success = $(1 - Pf)$, decimal

$P_1, P_2, P_3...$ = probability of outcome 1, 2, 3... occurring

$R_1, R_2, R_3...$ = conditional value of outcome 1, 2, 3...

v = parameter value

v_{min} = minimum value

v_{ml} = most likely value

v_{max} = maximum value

CHAPTER VII

C_D = market value of debt

C_E = market value of common stock equity

C_P = market value of preferred stock

C_{RD} = cost of debt, %

C_{RE} = cost of common stock equity, %

C_{RP} = cost of preferred stock, %

C_{RW} = weighted average cost of capital, %

i_c = coupon interest rate

I_s = semi-annual interest payment, $ (coupon value)

IRR = internal rate of return, %

i = annual interest rate, decimal (yield)

n = number of semi-annual interest payments remaining to maturity

ORRI = overriding royalty interest, decimal

$PVIF_{i,n}$ = present value interest factor for i and n

$PVIFA_{i,n}$ = present value annuity factor for i and n

RI = royalty interest, decimal

T_R = tax rate, decimal

V_{Bond} = current value of bond (debenture)

v_f = face, or par value of the bond, $, usually $1,000

WI = working interest, decimal

CHAPTER VIII

C = cost of tangible capital item or leasehold cost

DD&A = depreciation, depletion, and amortization, dollars

D_{SL} = straight line depreciation

DB_n = declining balance depreciation for year "n"

M = multiplier (for double declining balance = 2)

N = year to switch from declining balance to straight line

n = depreciation year

Q_n = annual production (BOE) of year "n"

R_n = reserves at beginning of year "n"

t_L = depreciation life, years

U = ultimate recovery

UOP_n = unit-of-production depreciation or cost depletion

CHAPTER IX

DROI = discounted return on investment, decimal

IRR = internal rate of return, decimal

NPV = net present value, dollars

PIR = profit on investment (undiscounted), decimal

ROI = return on investment (undiscounted), decimal

CHAPTER X

DROI = discounted return on investment

$FVIFA_{i,n}$ = future value annuity factor for i and n

IRR = internal rate of return, %

i = nominal rate of interest per period, expressed as a decimal

n = number of interest periods

$PVIFA_{i,n}$ = present value annuity factor for i and n
$PVNCF_n$ = present value net cash flow for period n

Note: Symbols used throughout this text conform to those established by the Society of Petroleum Engineers as published in 1986 and the update of economic symbols printed in August, 1992. Where a standard symbol has not been established for a term used in this text, one was selected which is compatible with the standard symbols.

INDEX

God helps them that help themselves.

—PROVERB

NOTES